Vibrational Spectroscopy - Modern Trends

Vibrational Spectroscopy - Modern Trends

Edited by

A. J. Barnes and W. J. Orville - Thomas

Department of Chemistry and Applied Chemistry
University of Salford
Salford M5 4WT, Great Britain

ELSEVIER SCIENTIFIC PUBLISHING COMPANY
AMSTERDAM – OXFORD – NEW YORK 1977

ELSEVIER SCIENTIFIC PUBLISHING COMPANY
335 Jan van Galenstraat
P.O. Box 211, Amsterdam, The Netherlands

Distributors for the United States and Canada:

ELSEVIER NORTH-HOLLAND INC.
52, Vanderbilt Avenue
New York, N.Y. 10017

ISBN 0-444-41632-3

Printed in The Netherlands

PREFACE

Contrary to some ill-informed reports vibrational spectroscopy is alive and thriving, and progress in this area is still spectacular. This being so, a volume containing up-to-date accounts of the most fruitful areas was thought to be timely.

It is now nearly fifteen years since Mansel Davies produced his edited volume on "Infrared Spectroscopy and Molecular Structure". This book was warmly welcomed and has been constantly cited in the scientific literature since its publication. It was not written to provide a survey of recent advances, but to outline the basic principles, and to indicate the potential scope and the then current limitations in infrared spectroscopy. The present volume has been written with the intention of surveying recent advances in both infrared and Raman spectroscopy, and to illustrate where these techniques have made the greatest contributions to studies of molecular structure and molecular behaviour. It should be emphasised that in order to produce a volume at a reasonable cost, it was decided to use a camera ready copy format, and to restrict the authors to a maximum number of pages. In some cases this has entailed the authors in a great deal of difficult compression so far as their topic is concerned. The editors wish to record their thanks to the authors for the ready cooperation received. Any defects in the present text are therefore the responsibility of the editors.

It is instructive to look at the nationalities of the authors involved in this publication. As expected it is clear that new ideas and new techniques have their origins in many countries throughout the world. It is to be hoped that the readers of this volume will find that the proud claim made by scientists that their activity promotes international understanding has been achieved in this volume. In 1945 infrared spectroscopy was being vigorously promoted only in the U.S.A. and the U.K. It is gratifying to find therefore that this and its allied Raman spectroscopic technique are now so widely spread.

The contents of this book are not meant to be exhaustive. Rather it is hoped that the various chapters will whet the appetite of new entrants to the field of vibrational spectrosсopy and enable them to seek an area which will become important to them and so dominate their activity for a period of their scientific lives.

Authoritative accounts can only be provided by active leaders, and this explains the choice of authors. The editors wish to place on record their thanks to these dedicated people who have spared a considerable amount of time to cooperate in this venture.

We are grateful to authors and copyright owners for their permission to reproduce various figures.

Department of Chemistry
and Applied Chemistry,
University of Salford,
Salford M5 4WT.

A.J. Barnes
W.J. Orville-Thomas

June 1977

CONTENTS

SECTION C. THEORETICAL METHODS.

CONTRIBUTORS

S. Abbate	Istituto di Chimica delle Macromolecole del CNR, Via Alfonso Corti 12, Milan, Italy.
A.J. Barnes	Department of Chemistry and Applied Chemistry, University of Salford, Salford M5 4WT, G.B.
I.R. Beattie	Department of Chemistry, The University, Southampton SO9 5NH, G.B.
H.J. Bernstein	Division of Chemistry, National Research Council of Canada, Ottawa, Canada.
J.D. Black	Department of Chemistry, The University, Southampton SO9 5NH, G.B.
S. Califano	Istituto di Chimica Fisica, University of Florence, Florence, Italy.
R.J.H. Clark	Christopher Ingold Laboratories, University College London, 20 Gordon Street, London WC1H OAJ, G.B.
J.P. Coates	Perkin-Elmer Ltd., Beaconsfield, Bucks., G.B.
J.R. Durig	Department of Chemistry, University of South Carolina, Columbia, S.C. 29208, U.S.A.
W.H. Fletcher	Department of Chemistry, The University of Tennessee, Knoxville, Tennessee 37916, U.S.A.
W.O. George	The Polytechnic of Wales, Pontypridd, Glamorgan, G.B.
M. Gussoni	Istituto di Chimica delle Macromolecole, Via Alfonso Corti 12, Milan, Italy.
H.E. Hallam	Department of Chemistry, University College of Swansea, Swansea SA2 8PP, G.B.

N. Mohan	Institute of Chemistry, University of Dortmund, 46 Dortmund, West Germany.
A. Müller	Institute of Chemistry, University of Dortmund, 46 Dortmund, West Germany.
W.J. Orville-Thomas	Department of Chemistry and Applied Chemistry, University of Salford, Salford M5 4WT, G.B.
G.C. Pimentel	Department of Chemistry, University of California, Berkeley, California 94720, U.S.A.
G. Riley	Department of Chemistry and Applied Chemistry, University of Salford, Salford M5 4WT, G.B.
S. Sunder	Division of Chemistry, National Research Council of Canada, Ottawa, Canada.
S. Suzuki	Department of Chemistry and Applied Chemistry, University of Salford, Salford M5 4WT, G.B.
M. Tasumi	Department of Biophysics and Biochemistry, Faculty of Science, University of Tokyo, Bunkyo-ku, Tokyo, Japan.
M. Tsuboi	Faculty of Pharmaceutical Sciences, University of Tokyo, Bunkyo-ku, Tokyo, Japan.
J.J. Turner	Department of Inorganic Chemistry, The University, Newcastle-upon-Tyne NE1 7RU, G.B.
G. Zerbi	Istituto di Chimica, Università, Trieste, Italy.

CHAPTER 1

INTRODUCTION

H.E. Hallam

1. THE DEVELOPMENT OF INFRARED AND RAMAN SPECTROSCOPY

By 1950 almost all of the physical techniques now so widely used by chemists had been discovered. The classic paper by Coblentz [1] in 1906, based on even earlier work of Abney and Festing [2], included the infrared vibrational spectra of over one hundred organic compounds. Raman [3], based on Smekal's predictions [4], discovered his effect in 1924, using extremely primitive apparatus and in the 1930's the Raman spectra of many hundreds of compounds, organic and inorganic, were reported, mainly by Kohlrausch [5] (published in his 3 classic volumes in 1931, 1938 and 1943) and by Raman. Infrared however, due to the then experimental difficulties, lagged far behind and few spectra of polyatomic molecules were reported. The situation was reversed in the 1940's when war-time developments in electronics led to the mid-infrared region being opened up. The first commercial infrared spectrometers were marketed in the very late 1940's and Herzberg's influential book, 'Infrared and Raman Spectra of Polyatomic Molecules' appeared in 1945.

These early spectrometers, first single-beam but very soon double-beam, were prism instruments and used rock-salt prisms covering the 2 to 15 μm (5000-650 cm^{-1}) near and mid-infrared range. KBr prisms pushed this further to 400 cm^{-1} and other alkali halides to longer wavelengths still. Perfection of a reliable replication process for producing good copies of diffraction gratings (by Sayce in the National Physical Laboratories, Teddington) allowed the development of commercial high-resolution (0.2 cm^{-1}) infrared spectrometers in the late 1950's. This soon enabled the steady advance of the long-wavelength limit and grating monochromators made the far infrared routinely available. Many modern instruments scan automatically to 250 or 200 cm^{-1} and some to 30 cm^{-1}.

The past decade has seen the increasing use of interferometry rather than monochromators to separate frequencies, the initial impetus in the development of the principle of the Michelson interferometer stemming from the work of Gebbie and his group at the N.P.L. In conventional spectrometers, spectral scanning is sequential and high resolution is obtained by narrowing the slits though with a decrease in luminosity and a concomitant increase in signal-to-noise ratio. Interferometric spectroscopy, on the other hand, is multiplex, i.e. all frequencies are studied simultaneously; it makes better use of the available energy - and is therefore more sensitive and, in principle, quicker than conventional spectroscopy. The records presented are not spectra but interferograms, which have to be measured as a set of sampled ordinates of a function of (t) where t can be either time elapsed or path difference. The function f(t) is the Fourier transform of the spectral function $F(\nu)$ where ν is frequency. Thus we have the necessity of mathematical computation, preferably with a built-in computer, and hence the term Fourier transform spectroscopy.

On the theme of computerisation, recent developments of microprocessors are likely to add a new dimension to conventional spectroscopy. The microprocessor consists of a single integrated circuit on a small silicon chip, and costs ca. £20. A microcomputer can be assembled using two or three integrated circuit boards and is thus physically small and inexpensive. It can be programmed to perform control operations and arithmetical processing. By putting the program in a permanent store the computer becomes dedicated to performing a single programmed task, and can be connected to be part of the equipment it is to control. In the case of spectrometers the complete recording and optical systems can be placed under the direct control of the microcomputer. Both the monochromator and chart positions are tracked precisely, ensuring high accuracy and repeatability. This together with the storage capability allows spectral accumulation and computer averaging. The computer also monitors abscissa function to provide, for example, a variety of scale expansions, giving superior spectral detail. Such a facility allows acceptable S/N levels even for signals weaker than 1% transmission, thus greatly extending the capability of techniques involving adsorbed and matrix-isolated species. This microcomputer growth is ushering in a new era in instrumentation, offering the convenience, performance, ease of operation and versatility required for either routine survey work or high-resolution studies.

However, the major impact on spectroscopy in recent years has been the advent of the laser light source. The characteristics of laser radiation are the extreme directionality, the very high output powers, the polarisation, the extremely narrow line-widths and the coherence. Such a source had long been the Raman spectroscopist's dream and not surprisingly such continuously operating fixed-frequency laser sources as Ar^+ and other noble gas ions have revolutionised the technique. Direct-recording instruments have become available and the method is now firmly placed in its rightful place alongside infrared so that we are all now vibrational rather than infrared and/or Raman spectroscopists.

Additionally, the power of lasers has allowed many optical experiments to be performed which would not otherwise have been possible; in particular the non-linear phenomena such as the stimulated Raman effect and the inverse Raman effect from the use of the very great power of giant pulse lasers. Laser sources have also made possible the study of fluorescence in the infrared spectral region giving us more detailed knowledge of energy transfer processes among upper vibrational states.

However, the most recent development is the one which is likely to revolutionise infrared spectroscopy, viz., that of tunable lasers, allowing one to scan over a spectral range with all the advantageous properties of laser radiation. We are still some way from having a laser fully tunable over the whole infrared region but progress is currently so rapid in laser technology that, as with the Heisenberg Uncertainty Principle, the position has changed greatly in the time which has elapsed since commencing this sentence! Dye lasers which are readily available and which produce tunable radiation over the whole of the visible region are gradually being pushed further into the near and mid-infrared regions. Other devices such as spin-flip Raman lasers which operate within the mid-infrared are becoming tunable over ever-widening ranges.

It is in the area of high-resolution infrared spectroscopy that the impact will first be felt. The resolution of a conventional infrared spectrometer, as discussed above, is limited by the decreasing energy reaching the detector as the slit width is reduced; thus it inevitably has a resolution which is significantly worse than the Doppler-limited width. A comparison of this with the line-widths of typical tunable lasers (taken from the admirable review by Burdett and Poliakoff [6], in turn based on the data of Hinkley) is shown in Figure 1.

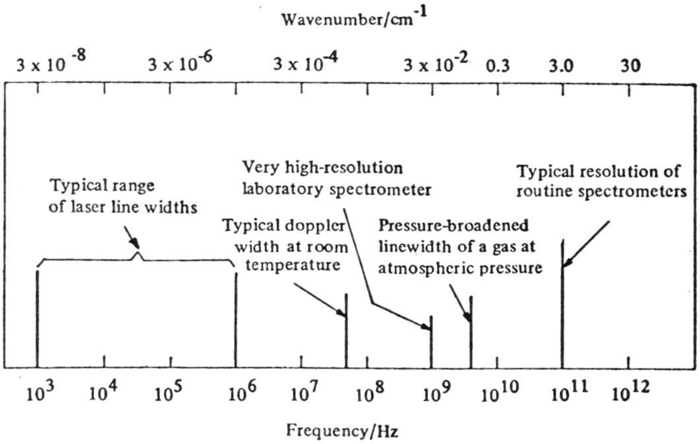

Figure 1. Resolution requirements for spectroscopy compared with conventional and tunable laser linewidths.

This clearly shows that lasers have a resolution capability several orders of magnitude better than conventional infrared spectrometers and also better than typical Doppler widths. It is not the role here to review the very high resolution studies which have stemmed from these experimental advances but such resolution is clearly leading to details of energy levels, offering a tremendous challenge to the theoreticians to provide adequate interpretations.

2. APPLICATIONS AND TECHNIQUES

Although today vibrational spectroscopy is a relatively poor technique for the full determination of a molecular structure it remains of considerable importance as one of the chemist's front-line tools. There are several reasons for this but principally because it is relatively cheap, and thus widely available, and because of its wide applicability. Both infrared and Raman spectroscopy may be applied routinely to molecules, radicals and ions, as gases, liquids/solutions and solids; and less routinely, but without unduly sophisticated apparatus, to surface films, to adsorbed species, and to species isolated in a matrix.

The elegant matrix isolation technique introduced by Pimentel [7] in 1954 has now become widespread in its application as a result of the considerable developments in cryogenic technology in the mid-1960's in the form of the availability firstly of open-cycle, and more recently of closed-cycle, mini-cryocoolers. The recording of an infrared and a Raman spectrum of a sample either as a pure solid or isolated in an argon or other matrix at a temperature of 10 K or lower now presents few problems. The immobilisation of a species in an inert matrix has been especially valuable for the investigation of highly reactive species. A vast amount of work on the identification and characterisation of free radicals and of high-temperature vaporising species has been carried out. From the outset, inorganic and physical chemists were quick to exploit the technique but it is only recently that organic chemists, and as yet only a few, have appreciated the potential of the technique.

This is best illustrated by the recent synthesis under cryogenic matrix conditions of the elusive molecule, cyclobutadiene. This molecule, the prototype anti-aromatic molecule, and the counterpart of benzene in the aromatic series, had defied the extensive efforts of organic chemists for a century, until 1972 when it was simultaneously reported synthesised in a matrix by Lin and Krantz [8] and by Chapman et al. [9]. This was achieved by the photolysis with a medium pressure mercury lamp of α-pyrone or of pyridine matrix-isolated in argon or nitrogen at 8 K. In addition to the well-known bands due to CO_2 or HCN, three new major features appeared which were attributed to cyclobutadiene: 1240 (ring C-C stretch), 650 (in-plane ring deformation), 570 (out-of-plane ring deformation) cm^{-1}.

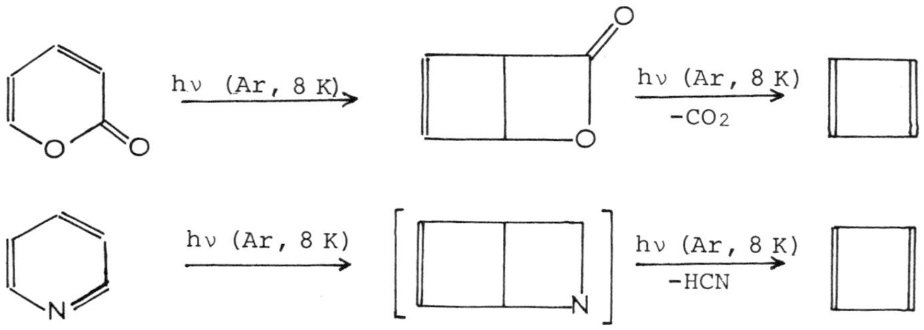

At the time of publication, the author was on sabbatical leave at the Physikalische Chemisches Institüt of the University of Marburg where, after a research seminar he gave on matrix isolation spectroscopy, a discussion was held with Professor G. Maier, of the Organische Chemisches Institüt, and Dr. B. Mann, on the possibility of similarly preparing cyclobutadiene from one of Professor Maier's precursors. Deposition and photolysis of the following anhydride in argon was performed by Dr. Mann in two experiments within two days.

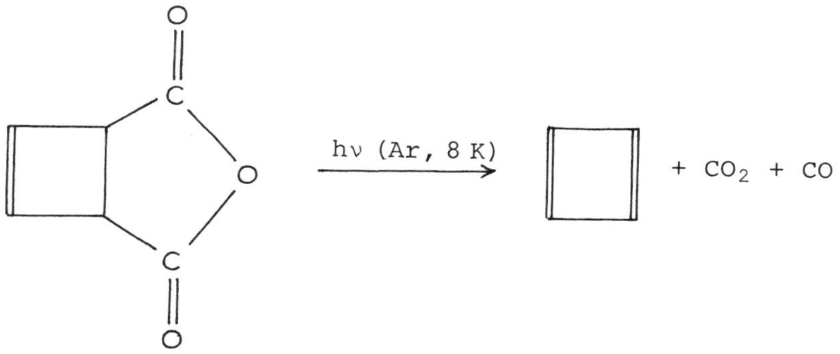

The spectral features of the photo-fragments CO and CO_2 were readily identified together with the same three bands as reported by Lin and Krantz and Chapman, et al. Such is the ease with which such syntheses and characterisations can be carried out. The work is reported as a footnote in an excellent review of the cyclobutadiene problem by Maier [10]. Benzyne is another example [11] of a recent organic synthesis. From these studies it is clear that a large area here awaits exploitation of the technique.

The availability of a wide selection of laser frequencies has opened up the technique of resonance Raman spectroscopy. Resonance Raman arises from the greatly increased transition probability as the exciting radiation freqeuncy approaches a strong electronic absorption frequency. For relatively simple species this will normally consist of a totally symmetric fundamental, greatly enhanced in intensity, together with a long series of its overtones. Intensity enhancements of several thousands can occur which are highly selective, and have led to many analytical applications. The technique has proved of considerable value for the investigation of chromophoric centres of large inorganic lattice structures and of many large biological molecules such as haemoglobin and vitamin B12.

Finally, applications continue to exploit the standard methods of microsampling but to increasingly lower detection limits. In gas analysis the use of multiple-path cells continues to be extended. The use of a pair of 40 metre cells (40 x 1 m) is the standard method for the determination by infrared of CO, saturated and unsaturated hydrocarbons, etc., in cylinder oxygen supplied to hospitals and for high-altitude aircraft use. This presents the ideal situation to the infrared analyst. Commerical portable instruments, such as the Miran (Wilks Scientific Corporation), are available which utilise pathlengths of several metres and can directly monitor pollutant gases with a sensitivity down to 0.1 to 1 p.p.m. An interesting example [12] of the use of this was the monitoring of the levels of anaesthetic gases (halothane and nitrous oxide) at different heights in a dental surgery during dental operations. Despite an authorised air extraction system, levels at the 5-foot contour were found to be unacceptably high with the consequential effect on the surgical staff!

To summarise, far from being over the hill as a technique, vibrational spectroscopy is currently in a very exciting phase which will be borne out by the developments, experimental and theoretical, described in the following chapters.

3. REFERENCES

1. W.W. Coblentz, Investigation of Infrared Spectra, Carnegie Inst. Pub., Washington, Vol. 35 (1905).
2. W. de W. Abney and Festing, Phil. Trans., 416 (1881).
3. C.V. Raman, Nature, 121 (1928) 619.
4. A. Smekal, Naturwiss., 11 (1923) 873.
5. K.W.F. Kohlrausch, Der Smekal-Raman Effekt, Berlin, (1931); ibid., Ergänzungband (1938); Ramanspecktren, Hand - u. Jahrb. der Chem. Physik., Vol. 9, VI, Leipzig (1943).
6. J.K. Burdett and M. Poliakoff, Chem. Soc. Rev., 3 (1974) 293.
7. D.A. Dows, G.C. Pimentel and E. Whittle, J. Chem. Phys., 22 (1954) 1943.
8. C.Y. Lin and A. Krantz, J. Chem. Soc. Chem. Comm., (1972) 1111.
9. O.L. Chapman, C.L. McIntosh and J. Pacansky, J. Amer. Chem. Soc., 95 (1973) 614.
10. G. Maier, Angew. Chemie Int. Ed., 13 (1974) 425.
11. O.L. Chapman, K. Mattes, C.L. McIntosh, J. Pacansky, G.V. Calder and G. Orr, J. Amer. Chem. Soc., 95 (1973) 6134.
12. G.A. Lane, Br. J. Anaesthesia, 48 (1976) 264.

CHAPTER 2

PRINCIPLES OF LASERS

J.J. Turner

1. INTRODUCTION

Much of the basic physics of lasers can be understood by considering a simple two-level non-degenerate energy level system

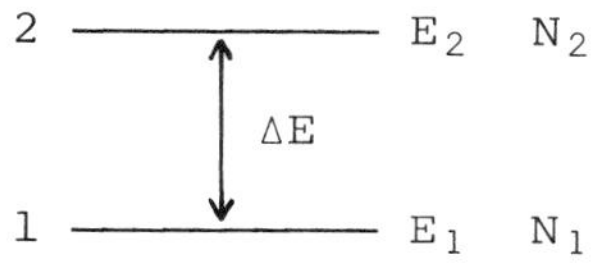

The diagram represents two levels of some atomic or molecular system X. In principle the energy separation (ΔE) can correspond to a frequency range from the microwave to the vacuum ultra-violet. From Boltzmann's Law, at equilibrium the ratio of the number of species in level 2 (N_2) to the number in level 1 (N_1) is given by

$$\frac{N_2}{N_1} = e^{-\Delta E/kT} \tag{1}$$

If light of frequency corresponding to the energy $\Delta E(h\nu_{12})$ falls on the system, energy is absorbed and hence species promoted to level 2 at a rate which depends on the number of species in level 1 (N_1), the intensity of the radiation (ρ) and the Einstein coefficient of induced absorption (B).

$$\text{Rate of absorption} = BN_1\rho \tag{2}$$

A species in the excited state 2 can relax to the state 1 by either radiative or non-radiative processes. If the latter is the overwhelming mechanism then laser action is impossible. In what follows we shall assume that the dominant processes are radiative, which can be spontaneous or stimulated. The rate of spontaneous decay from $2 \rightarrow 1$ and hence spontaneous emission of light corresponding to ΔE is simply:

$$\text{Rate of spontaneous decay} = AN_2 \tag{3}$$

where A is the Einstein coefficient for spontaneous emission. In stimulated emission, a photon of energy ΔE (i.e. $h\nu_{12}$) can stimulate the species in level 2 to emit radiation of ΔE according to:

$$h\nu_{12} + X(E_2) \rightarrow 2h\nu_{12} + X(E_1) \qquad (4)$$

$$\text{Rate of stimulated emission} = B'N_2\rho \qquad (5)$$

where B' is the Einstein coefficient for stimulated emission. Before considering the relative values of the three rates we note from equation (4) that we can amplify the light by stimulated emission since one photon has gone in and two have come out. Figure 1 is a classical attempt to contrast stimulated

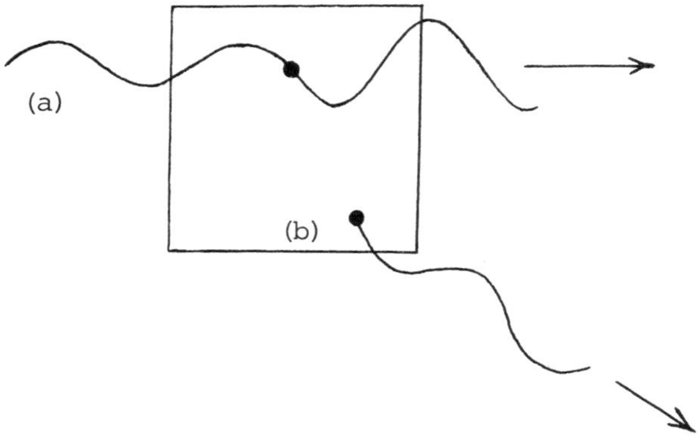

Figure 1. (a) Amplification by stimulated emission. (b) Spontaneous emission.

and spontaneous emission. The analogue of equation (4) is represented at the top; a wave of frequency ν and amplitude a interacts with an excited molecule of X, its amplitude is doubled but the wave retains the same phase and direction. In contrast the spontaneous emission in the lower half occurs in any direction at a random time and hence with any phase. Suppose however that a photon obtained by spontaneous emission is directed along a narrow tube containing a large number of X species in the excited state. There will be a continuous build up of amplitude due to stimulated emission as the wave travels along the tube. A more efficient way of producing stimulated emission is clearly to direct the light back and forward along a tube with partially reflecting mirrors at the ends. However this tube introduces some complications which we return to shortly.

We now consider how the rates of the three processes affect the possibility of laser action.

The rate of increase of the population of level 2 is

$$\text{Rate}_{1 \rightarrow 2} = BN_1\rho \qquad (6)$$

The rate of depopulation $2 \rightarrow 1$ is

$$\text{Rate}_{2 \rightarrow 1} = B'N_2\rho + AN_2 \qquad (7)$$

and at equilibrium therefore $BN_1\rho = B'N_2\rho + AN_2$. It can be shown that

$$A = \frac{8\pi h\nu^3}{c^3} B \tag{8}$$

in addition when levels 1 and 2 are non-degenerate $B' = B$. (9)
Laser action depends on light amplification by stimulated emission of radiation; the minimum requirement for this to occur is

$$BN_2\rho > BN_1\rho \tag{10}$$

otherwise absorption will dominate emission. That is there must be a population inversion,

$$N_2 > N_1 \tag{11}$$

and hence we are no longer in a thermodynamic equilibrium situation. How is such a population inversion maintained particularly when spontaneous emission is an important depopulating process? In the microwave region (say $\lambda = 1.25$ cm, i.e. $\nu/c = 0.80$ cm^{-1}) A $\sim 10^{-25}$B and so in equation (7) A can be ignored. Thus since spontaneous emission is so slow, it is possible to physically separate in a molecular beam those molecules in level 2 from those in level 1.

The components of the inversion doublet of the ground vibrational state of ammonia are separated by 0.8 cm^{-1} (Figure 2). At room temperature in the equilibrated gas $N_2 \simeq N_1$. In

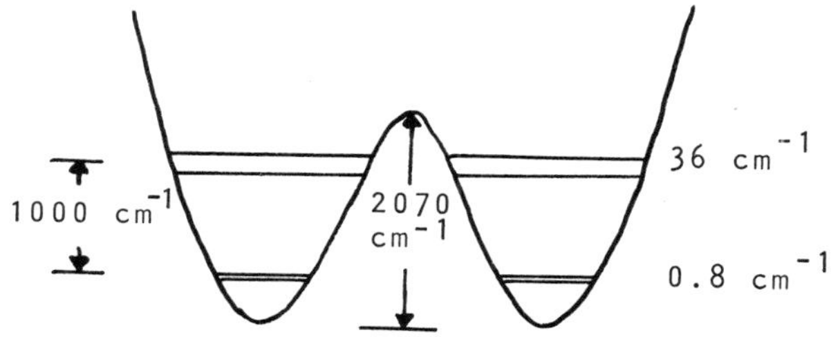

Figure 2. Energy diagram for the inversion coordinate of ammonia.

an inhomogeneous electric field molecules in level 2 pass down the axis of the field electrodes whereas those in level 1 are defocussed. The beam of molecules in level 2 can then be passed into a high-Q (Q = quality factor) microwave cavity and in view of the large (theoretically ∞) population inversion maser action can occur. (Maser ≡ microwave amplification by stimulated emission of radiation.)

In the visible region of the spectrum, A cannot be ignored, physical separation of states is impossible, and we must take a closer look at absorption and emission. In a spectroscopic experiment, schematically represented in Figure 3, the absorption spectrum of a Doppler-broadened gas might be obtained as shown in Figure 4(a). At any value of ν the ratio of the intensity of transmitted (I) to incident (I_0) light is given by

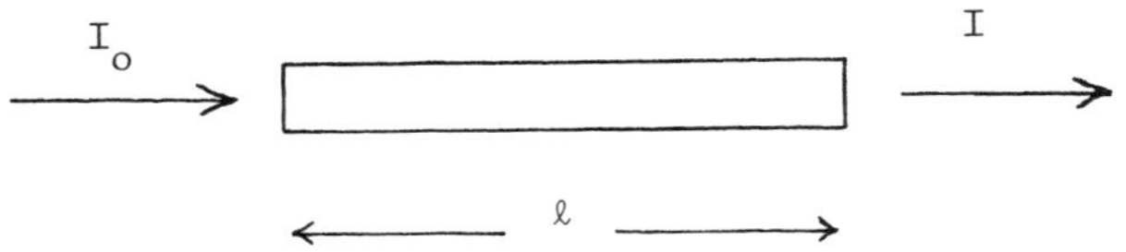

Figure 3. Schematic representation of spectroscopic experiment.

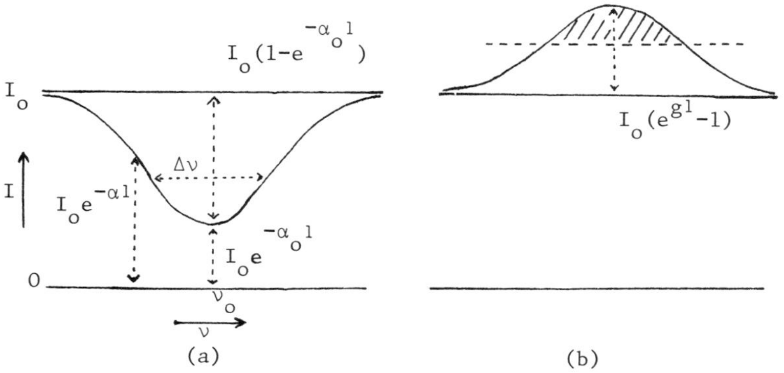

Figure 4. Spectra of Doppler-broadened gas system.

$$\frac{I}{I_o} = \exp(-\alpha \ell) \tag{12}$$

where α is the absorption coefficient and ℓ is the length of the absorbing medium. As light passes through the sample some is absorbed; also light is emitted by both stimulated and spontaneous emission. As noted previously the spontaneous emission will be in all directions and hence will make a negligible contribution in the direction of the beam. However the stimulated emission is only in this direction and must be included in the appropriate expressions. The area under the absorption curve is related to the Einstein coefficients by

$$\int \alpha d\nu = \frac{h\nu_o}{c} B(N_1 - N_2) \quad \left[= \frac{c^2}{8\pi\nu_o^2} A(N_1 - N_2)\right] \tag{13}$$

For a Doppler-broadened line profile, $\alpha = \alpha_o \exp(-k(\nu - \nu_o)^2)$, it can be shown that

$$\alpha_o = 2\sqrt{\frac{\ln 2}{\pi}} \frac{h\nu_o}{c} B \frac{(N_1 - N_2)}{\Delta\nu} \quad \left[= 2\sqrt{\frac{\ln 2}{\pi}} \frac{c^2}{8\pi\nu_o^2} A \frac{(N_1 - N_2)}{\Delta\nu}\right] \tag{14}$$

where $\Delta\nu$ is the line width at half-peak height. If $N_2 > N_1$ then α becomes negative because the stimulated emission is contributing more than the absorption is removing. That is, in

$$\frac{I}{I_0} = \exp(-\alpha\ell) \tag{15}$$

$-\alpha$ is positive and $I/I_0 > 1$. This is more usually written in this case as

$$\frac{I}{I_0} = \exp(g\ell) \tag{16}$$

where g (= $-\alpha$) is the gain coefficient. The "absorption" spectrum is now more like Figure 4(b). Thus the absolute minimum requirement for amplification is that $g > 0$ (i.e. $N_2 > N_1$). However, this ignores loss effects due to reflections, diffraction, scattering etc. For laser action the gain must exceed the loss i.e.

$$e^{g\ell} - 1 > \text{loss} \tag{17}$$

For small gain and loss this is approximately

$$g\ell > \text{loss} \qquad (\equiv -\alpha\ell > \text{loss}) \tag{18}$$

So at the centre of the band this means,

$$\left(2\sqrt{\frac{\ln 2}{\pi}}\,\frac{h\nu_0}{c}\,B\,\frac{(N_2 - N_1)}{\Delta\nu}\right)\ell > \text{loss}$$

$$\text{i.e. } (N_2 - N_1) > \frac{\text{loss}}{\ell}\,\frac{c}{2h\nu_0}\sqrt{\frac{\pi}{\ln 2}}\,\frac{\Delta\nu}{B} \tag{19}$$

The requirement is thus more stringent than simply $(N_2 - N_1) > 0$. This can be diagrammatically represented in Figure 4(b). The horizontal dotted line represents the minimum value of the gain required; laser action will only occur for gain greater than this, i.e. in the shaded area.

It is well known that laser emission is frequently much narrower than the half-width of the gain profile. To see how this arises we consider the optical cavity. So far we have considered the amplification taking place in a very long tube with a single pass of light. Of course, normally an optical cavity with partially reflecting mirrors is used; in this case the wave is reflected back and forth in the cavity being amplified at each traverse. However, several passes of the cavity will result in destructive interference unless the mirror separation is an exact multiple of $\lambda/2$ (λ = wavelength)*. Suppose the cavity is 1 m long ($\equiv 10^9$ nm) and λ = 500 nm. Then there are 4 million $\lambda/2$'s in the cavity length. This has important consequences. By how much would λ have to change to alter this

---

* This is essentially the same argument as is used in the derivation of the form of radiation from a black-body.

number by unity?

$$3{,}999{,}999 \times \frac{\lambda + \delta\lambda}{2} = 10^9$$

$$\therefore \quad \delta\lambda = .000125 \text{ nm}$$

That is adjacent axial modes are separated by .000125 nm, or .01 cm^{-1} or 150 MHz. These modes* can be superimposed on the gain profile as shown in Figure 5.

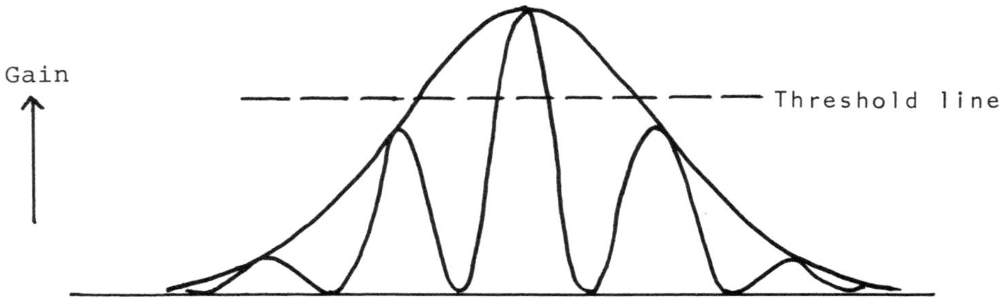

Figure 5. Cavity modes - schematic.

Since a typical gain profile half-width is say 1000 MHz the laser emission will consist of a large number of different modes, provided the gain peak is higher than the minimum required (dotted line). In fact it is possible that only one mode will operate and thus the laser output will be very narrow.

Before considering specific systems enough points have been covered to illustrate some of the important properties of lasers and to hint at the potential of such properties.

(a) Monochromaticity; narrow line width.
Enormous potential in spectroscopy particularly if it is possible to tune the laser frequency (see Chapter 5).
(b) Intensity.
Also important in spectroscopy in signal/noise consideration. Central to exploitation of non-linear effects, e.g. frequency doubling, non-linear photochemistry.
(c) Collimation.
Spectroscopy at a distance.
(d) Coherence.

2. LASERS WITH MORE THAN 2 LEVELS

The discussion so far has been restricted to a 2-level non-degenerate system. If level 1 is g_1 fold degenerate and level 2 g_2 fold degenerate then some of the equations must be modified. These are listed in the Appendix; the most

---

* If the beam is not exactly perpendicular to the two mirrors other modes are possible, but are outside the scope of this discussion.

significant point is that the condition for inversion of population is no longer simply $N_2 > N_1$ but becomes

$$\frac{g_1}{g_2} N_2 > N_1 \qquad (11a)$$

Secondly a laser operating on simply two levels is almost impossible to achieve. Level 1 is invariably the ground state in this case and it is extremely difficult to obtain population inversion. The simplest alternative is a 3 level system shown diagramatically below

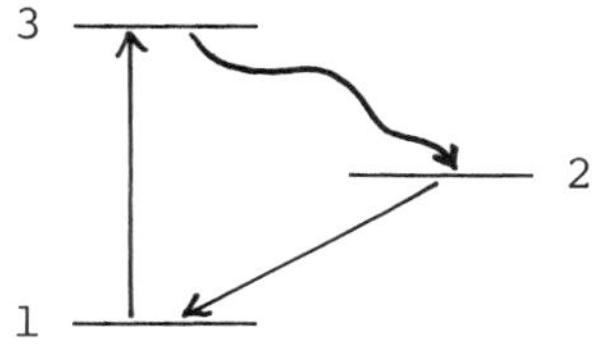

Level 3 is pumped directly from 1; there is crossover to 2 and laser action 2 → 1. However we still require more than half the species to be in level 2, but if 2 is a metastable state the transition 2 → 1 is forbidden and this is possible. In the examples which follow the lasers operate on 3 or more levels.

3. CONTINUOUS WAVE He/Ne LASER

This, the first continuous wave (cw) gas laser to operate, has become quite ubiquitous. A simplified energy level scheme is shown in Figure 6.

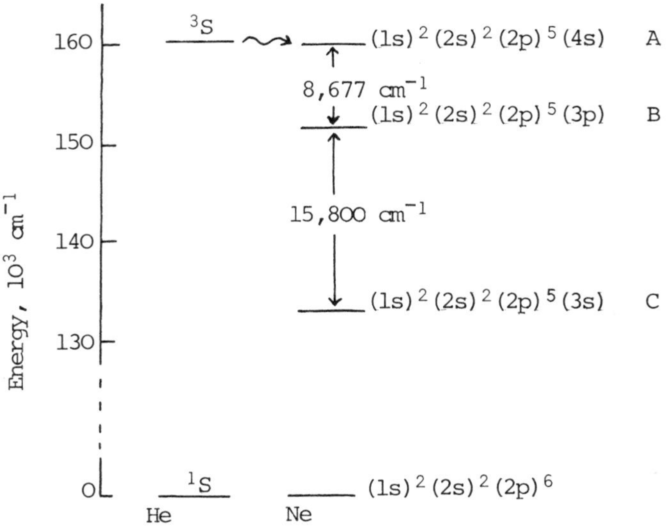

Figure 6. Energy levels for 8677 cm^{-1} He/Ne laser.

In a He/Ne gas mixture (0.1 mm Ne, 1 mm He), an electric discharge causes efficient excitation of He to the ^3S level, whereas Ne atoms are not efficiently excited. Since the $^3S \rightarrow {}^1S$ transition is forbidden, the state is long lived and by collision the energy can be transferred to the almost identical energy state of Ne $(1s)^2(2s)^2(2p)^5(4s)^1$ as shown. This creates a population inversion between A and B and laser emission at 8677 cm^{-1} is observed. Other laser lines can also be observed by varying the conditions; in particular B $\rightarrow$ C at 15,800 cm^{-1} (632.8 nm). The apparatus used for this is shown schematically in Figure 7.

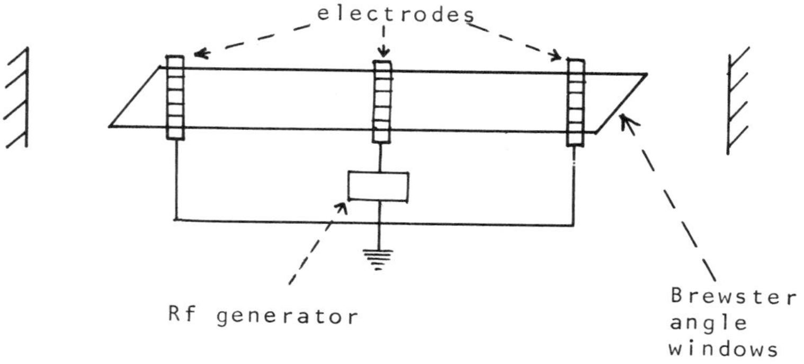

Figure 7. Schematic illustration of discharge tube for He/Ne laser.

The Brewster angle windows ensure that one plane of polarisation is transmitted. The power of the conventional laser emitting at 632.8 nm is 0.5 - 50 mW with very high stability and a line-width of 10^{-6}nm.

4. Q-SWITCHED CO_2 LASER

The CO_2 laser operates in the 10 μm region of the infrared and large models can produce enormous power. CO_2 lasers can be pulsed or continuous; in the latter case a $He/N_2/CO_2$ gas mixture is used and the appropriate vibrational energy levels are shown in Figure 8.

Laser action depends on the population inversion involving 00°1 as the upper level. This is achieved in an electric discharge in a cavity not unlike the He/Ne set-up, but is very much easier to make since gain is high, and the long wavelength makes optical tolerances much less stringent. Population inversion occurs by direct excitation by electron impact from the ground state, by recombination and cascade processes from higher energy levels and, when N_2 is present, by resonant transfer by collision with vibrationally excited N_2 molecules. In the cw laser system there is an equilibrium between the rate of filling the 00°1 level and the rate of depopulating. If one of the end mirrors (cf. Figure 7) is turned away from the axis, laser action will cease and for a short time the population of 00°1 will increase because the rate of depopulation has

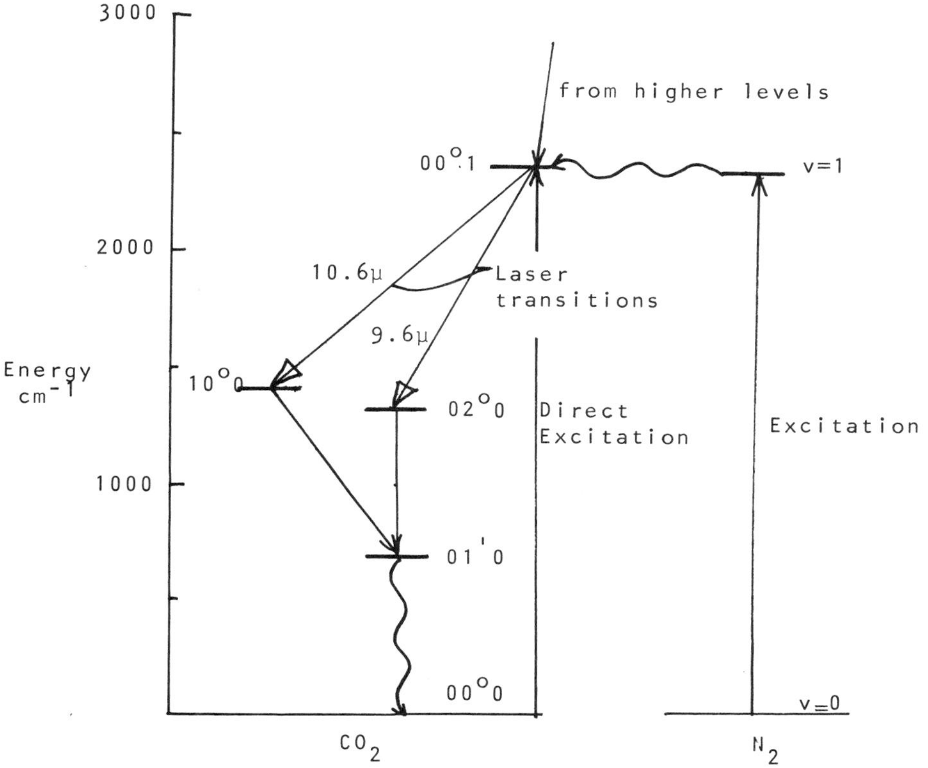

Figure 8. Energy levels for CO_2 laser.

suddenly dropped. If the mirror is then readjusted laser action will occur with very much increased intensity since the population inversion has increased. Usually a rotating mirror replaces one of the fixed mirrors. On a crude approximation all the normal cw output is concentrated in bursts which emit over the short time the laser mirrors are aligned. In this way power as high as kilowatts during the emitted flash can be obtained. The technique is know as Q-switching or Q-spoiling since the cavity quality factor is destroyed except when the mirrors are aligned.

5. RUBY LASER

We have so far considered gas lasers; in fact the ruby laser was the first optical laser developed. (Ruby is crystalline aluminium oxide with 0.05% of Cr^{3+} in place of Al^{3+}.) It is a solid state device and operates on the Cr^{3+} levels shown in Figure 9. When light from a Xe flash lamp falls on the rod of ruby the transitions to states $^4T_1$ and $^4T_2$ readily occur. There is inter-system crossing to both components of the ^2E

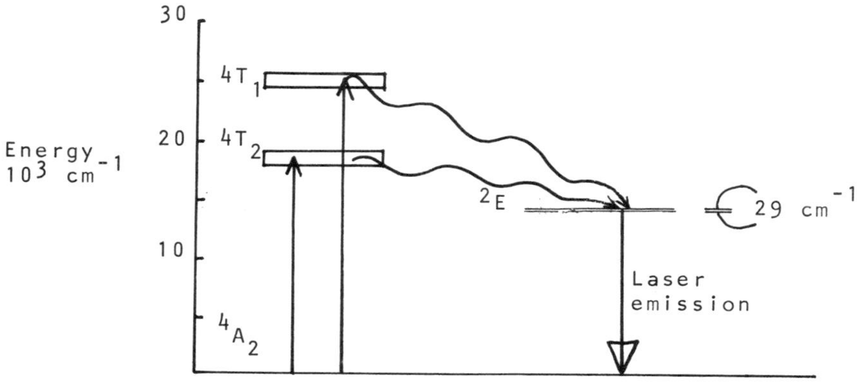

Figure 9. Energy levels for ruby laser.

state. Spontaneous transitions from ^2E to the ground state are spin forbidden so there is a rapid build up of population inversion between ^2E and $^4A_2$. A spontaneously decaying photon stimulates emission which is amplified in the optical cavity in the usual way. With a pulse duration of $\sim$1 ms the output power is approximately 1 kW. However the output consists of a series of spikes; this is because during the lifetime of the flash the ^2E level is populated and then depopulated by laser action several times. Not surprisingly it is possible to get all the energy into one "spike" by Q-switching to give powers of 10^4 kW during the $\sim 10^{-8}$ sec.

6. MODE-LOCKED Nd/YAG LASER

The problem with ruby is that it operates on a transition to the ground state. Thus it is necessary to take at least half the molecules (ignoring degeneracies) into the upper state to obtain population inversion. This is a disadvantage of a 3-level system ($^4T_1$ and $^4T_2$ are essentially one state for this purpose). However Nd^{3+} ions (usually in yttrium aluminium garnet (YAG) or glass) can operate in a 4-level manner - shown schematically in Figure 10. On flashing the Nd/YAG, many higher levels are populated. There is rapid non-radiative decay to the $^4F_{3/2}$ state and laser action occurs on several lines of which the most important is the transition to the lowest $^4I_{11/2}$ level giving laser emission at 1.06 μm (1060 nm). Laser action is very efficient since the $^4I_{11/2}$ is not the ground state and an enormous population inversion $^4F_{3/2}$: $^4I_{11/2}$ can be achieved. Such high powers can be obtained that Nd/glass lasers are the favourite candidates for laser fusion experiments.

Conventional Q-switching produced pulses of length $\sim 10^{-8}$ s; it is possible by the technique of mode-locking to obtain a train of pulses each of length as short as a few picoseconds and each separated in time by the traverse time of the cavity. The pulse width is approximately determined by the inverse of the half-width of the gain curve i.e. $\Delta t \sim 1/\Delta\nu$. So for a gas

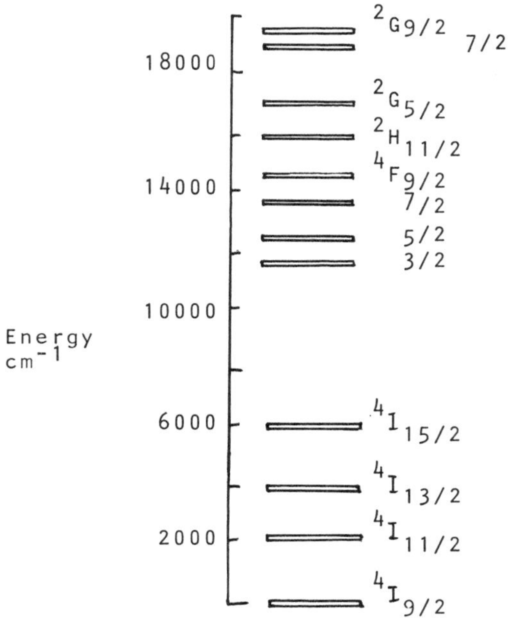

Figure 10. Energy levels of Nd^{3+}; the levels are actually split by the crystal field.

laser with $\Delta\nu \sim 10^9$ Hz, $\Delta t \sim 1$ nsec which is not much better than Q-switching. However with Nd/glass, $\Delta\nu \sim 10^{12}$ Hz so $\Delta t \sim 1$ ps.

The importance of mode-locked lasers in chemistry is the application to kinetic spectroscopy of very short lived species such as excited singlet states of organic molecules.

7. DYE LASERS

These are of immense importance and becoming increasingly so, because of the tunable properties of the output. They are based on a whole series of organic dyes in solution of which Rhodamine 6G is the classic. Figure 11 shows part of the energy level system of a typical organic dye. The absorption spectrum of such a dye shows very broad bands due to transitions from the lowest vibrational level(s) of the singlet ground state (S_0) to the higher singlet states S_1, S_2 .... There is very rapid radiationless decay to the lowest vibrational level of S_1. From here there is either inter-system crossing to the lowest triplet state (T_1); or non-radiative decay to S_0; or fluorescence to several of the higher vibrational levels of S_0. There is in effect a 4-level system $S_0(o) \rightarrow S_1/S_2/.. \rightarrow S_1(o) \rightarrow S_0(n)$ and laser emission can occur. The triplet state introduces a complication but this is readily overcome by using a triplet quencher.

Dye lasers can be either pulsed or, more recently, cw and

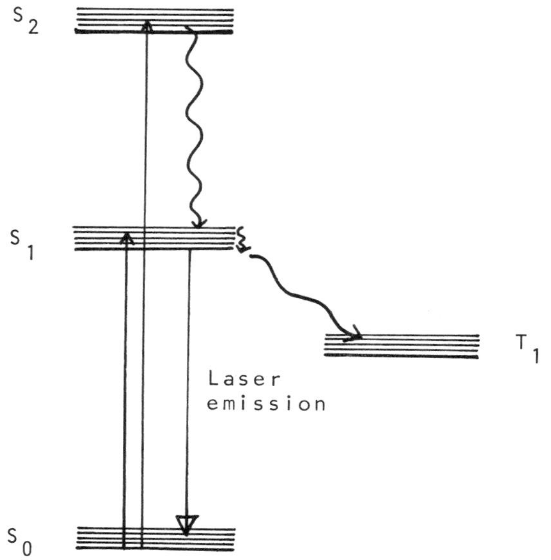

Figure 11. Dye laser energy level scheme.

are invariably pumped by another laser such as pulsed nitrogen or argon ion cw lasers.

It is interesting to see how the tunability arises. The vibrationally excited states of S_0 are broadened by solvent perturbation and the fluorescence is extremely broad and featureless. Over this area the loss factors in the laser cavity are not constant and indeed can be deliberately modified by changing for instance the mirror reflectivity. Thus the particular wavelength at which the laser threshold is first exceeded is a variable. Replacing one of the end mirrors with a grating produces narrow line-width laser output. By changing the dye, or the solvent, or the concentration or by varying the Q of the cavity (as above) it is possible to have tunable laser emission covering a large part of the UV/visible region.

8. CHEMICAL LASERS

The details of this type of laser fall outside the scope of this chapter. Suffice it to remark that here the population inversion is derived from chemical reactions. These lasers can be made extremely powerful.

Acknowledgment

I am particularly grateful to Dr. A. MacNeish for discussions on the material in this chapter.

9. REFERENCES

No specific references are given in this text but the following articles and books are helpful.

W.J. Jones, Lasers, Quart. Rev. Chem. Soc., 1 (1969) 73.
S.R. Leone and C.B. Moore, Laser Sources, Chap. 1 in C.B. Moore (Editor), Chemical and Biological Applications of Lasers, Vol. 1, Academic Press, 1974.
B.A. Lengyel, Lasers, Wiley, 1971.
A.E. Siegman, An Introduction to Lasers and Masers, McGraw-Hill, 1971.
O.S. Heavens, Lasers, Duckworth, 1971.

10. APPENDIX

Equations modified if degeneracies of levels 1 and 2 are g_1 and g_2.

$$\frac{N_2}{N_1} = \frac{g_2}{g_1} e^{-\Delta E/kT} \tag{1a}$$

$$g_1 B = g_2 B' \tag{9a}$$

$$\frac{g_1}{g_2} N_2 > N_1 \tag{11a}$$

$$\int \alpha d\nu = \frac{h\nu_0}{c} B(N_1 - \frac{g_1}{g_2} N_2) \tag{13a}$$

$$\alpha_0 = 2 \sqrt{\frac{\ln 2}{\pi}} \frac{h\nu_0}{c} \frac{B(N_1 - \frac{g_1}{g_2} N_2)}{\Delta\nu} \tag{14a}$$

$$\frac{g_1}{g_2}(N_2 - N_1) > \frac{loss}{\ell} \frac{c}{2h\nu_0} \sqrt{\frac{\pi}{\ln 2}} \frac{\Delta\nu}{B} \tag{19a}$$

CHAPTER 3

NON-LINEAR RAMAN EFFECTS

I.R. Beattie and J.D. Black

The term non-linear implies the property measured is dependent on some power of the field. For Rayleigh scattering we usually write the induced polarisation as

$$P = \alpha E \tag{1}$$

The induced polarisation P is proportional to the field E and α is a constant (the polarisability) relating the two. In Rayleigh and Raman scattering the intensity of the scattered light is proportional to the intensity of the incident light. (The intensity of light is proportional to the square of the maximum amplitude E_0 of the wave $E = E_0 \cos \omega t$.)

As the light intensity is increased so non-linear effects appear in the induced polarisation and a power series

$$P = \alpha E + \tfrac{1}{2}\beta E^2 + \tfrac{1}{6}\gamma E^3 \text{ ......} \tag{2}$$

may be used to represent the relationship with the field. For a centrosymmetric crystal changing the direction of E must also change the direction of P and hence β must be identically zero. It will be useful to return to this approach to non-linear phenomena at a later stage. For the present, we shall use the terminology more familiar to chemists involving molecules and photons [1].

1. STIMULATED RAMAN EFFECT

In a classical sense molecules may be considered not to rotate during the lifetime of a vibrational transition so that in vibrational Raman spectroscopy the molecules are considered as fixed in space. Consider a spherical top such as CCl_4 with incident light plane polarised along Z (laboratory fixed) and propagating along Y. For ν_1 the resultant Raman scattering will be plane polarised along Z and will propagate in the form of a toroid ($\cos^2\theta$ law) so that the intensity is a maximum in a disc in the XY plane and is zero in the Z direction. In a normal Raman experiment, the Raman radiation may be thought of as incoherent photons or noise. If the intensity of the beam of light incident upon a finite sample is gradually increased it is clear that the flux of Raman photons in the direction of the laser beam (+Y and -Y) will correspondingly increase. Under these conditions, the initially incoherent photons (noise) may have sufficient flux density to start to cause appreciable stimulated emission of Raman photons. It is immediately apparent for CCl_4 that the most intense flux of Raman photons in the Y direction will occur for ν_1 (rather than for ν_2, ν_3 or ν_4). We may thus obtain stimulated Raman spectra at the frequency ν_1. This causes pump depletion and hence other modes are not normally stimulated.

In the stimulated Raman experiment the Raman photons cause stimulated emission (see Chapter 2) from the intermediate level pumped by the incident laser beam ν_0. Clearly the polarisation characteristics of the incident laser beam and the stimulated Raman beam need not be identical, although they will be in the case of a completely polarised Raman band such as ν_1 of CCl_4. In most cases, one transition - usually the most intense, narrow, polarised band - is selectively pumped. In addition to stimulated emission at the Stokes frequency for ν_1 of CCl_4, other phenomena will be observed giving shifted frequencies at $2\nu_1$, $3\nu_1$ .... and also anti-Stokes frequencies (in cones around the central beam). The reasons for these frequencies and the directional properties will become clear later [2].

2. INVERSE RAMAN EFFECT

The inverse Raman effect is closely related to the stimulated Raman effect. The significant difference is that stimulated emission of Raman photons is caused by an intense laser beam, not by build up of Raman photons to threshold [3]. In an inverse Raman experiment, the sample is irradiated by two colinear laser beams, one at ν_0 (tunable) and one at ν_s (corresponding to the frequency of the Stokes photon for the particular Raman transition given by $\nu_o - \nu_s = \nu_R$). Both lasers must be operating below the thresholds at which stimulated Raman effects would be observed. Under these conditions, <u>absorption</u> occurs in the laser beam at ν_o, given by an analogous expression to Beer's Law

$$I = I_o e^{-gc\ell} \tag{3}$$

where I refers to the intensity of the laser beam at ν_o (initial intensity I_o) after traversing a distance ℓ through the medium. The "extinction coefficient" g is given by

$$g = (P_s/n_s{}^2)(10^7\nu_o/4hc^2\Gamma\nu_s{}^4) \qquad |d\sigma/d\Omega| \tag{4}$$

where P_s = power per unit area (W cm^{-2}) for the Stokes photons,
n_s = refractive index at ν_s (cm^{-1}),
$\frac{d\sigma}{d\Omega}$ = Raman scattering cross section,
Γ = half width at half maximum height for the Raman transition.

Thus as the laser at ν_o is tuned through successive Raman transitions ($\nu_o - \nu_s = \nu_R$ where ν_R refers to a Raman active vibration) absorptions are seen in ν_o. Note the directional nature of the effect and the lack of threshold. The chief disadvantage of this technique is that one is looking for absorption, which is inherently less sensitive than detection of photons. However the exponential dependence on length (leading to the possibility of long path length of multi-reflection cells) and possibility of discrimination against fluorescence and background thermal radiation makes inverse Raman spectroscopy of some interest to the chemist. It is also of importance in "picosecond spectroscopy" [4].

3. THE HYPER-RAMAN EFFECT

In equation (2) the term $\frac{1}{2}\beta E^2$ clearly results in $\cos^2\omega$ terms and hence in $\cos 2\omega$ terms. Thus β, the hyperpolarisability, is the factor that enables second harmonic generation. A chemist's way of visualising this process on a molecular scale is shown in Figure 1. Note that another method of terminology

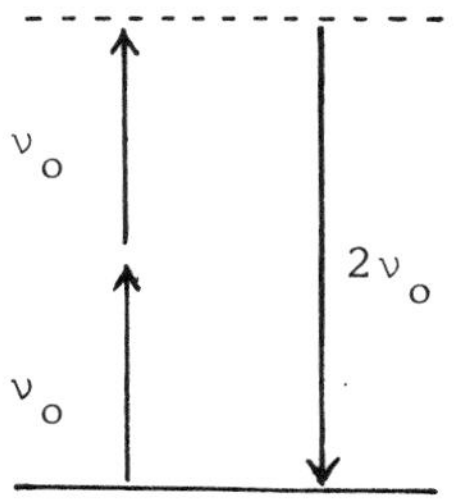

Figure 1. Second harmonic generation schematically.

would be that three-wave mixing had occurred: two ν_0 photons annihilated with the creation of a photon at $2\nu_0$. This is termed a parametric process - the molecule ends the process in the same state as that in which it started. This has important results, one of these is that momentum must be conserved. In a crystal such as potassium deuterium phosphate (KD*P), which is used for frequency doubling, the refractive index at ν_0 will, in general, not be the same as that at $2\nu_0$ (due to dispersion). The momentum in a photon may be expressed as $\hbar K$ where K is the wave vector given by

$$K = 2\pi\nu \frac{n}{c} \quad \text{so that } \hbar K = h\nu \frac{n}{c} \tag{5}$$

where n is the refractive index. The requirement for conservation of wave-vector or phase matching for second harmonic generation may be written

$$\vec{K}_{\nu_0} + \vec{K}_{\nu_0} = \vec{K}_{2\nu_0} \tag{6}$$

The problem thus reduces to one of refractive index matching at the fundamental and the second harmonic to provide sustained growth of the harmonic in a crystal. It will be seen that under phase-matching conditions, the wave fronts ν_0 and $2\nu_0$ proceed through the medium at the same rate. Normally the fundamental and the second harmonic are polarised in planes that lie perpendicular to one another (type 1 phase-matching).

The conventional vibrational Raman effect depends on the change of polarisability of a molecule during a vibration. Put more formally $(\partial\alpha/\partial Q)_0 \neq 0$ where Q refers to the normal coordinate. In an exactly similar way the hyper-Raman effect depends on $(\partial\beta/\partial Q)_o \neq 0$. The Raman tensor ($\alpha$) is a second rank tensor having nine components of which at most six are independent in the normal vibrational Raman effect. The tensor for the hyper Raman effect (β) is of third rank and has twenty-seven components of which a maximum of ten are independent. The activity is governed by the transformation properties of ijk terms (cf. the ij terms for the Raman effect) where ij and k refer to the Cartesian axes defining the polarisation directions of the two incident photons and one scattered photon. For bands forbidden in the infrared effect depolarisation ratios of 2/3 are found using polarised incident light. By contrast bands also permitted in the infrared effect have depolarisation ratios of less than 2/3 and may be termed polarised. The effect is extremely weak and has proved experimentally very difficult. It is a technique for the connoisseur, not for the average chemist [5]. As with the Raman effect resonance enhancement may occur in the region of an absorption band.

4. FOUR WAVE MIXING (CARS)

In the same way that the hyperpolarisability β may be considered to allow three wave mixing (including second harmonic generation), the term γ allows four wave mixing including third harmonic generation. The simultaneous irradiation of a medium with two lasers ν_o and ν_t can cause four-wave mixing so that light of frequency $(2\nu_o - \nu_t)$ is emitted. Because this is a parametric process phase matching is necessary, the resultant light is coherent and may form a diffraction limited beam, spatially related to ν_o and ν_t. Before proceeding with this more formally it is interesting to consider the phenomenon from a molecular viewpoint (in a way which the physicist would find relatively unacceptable).

Consider an assembly of molecules with one narrow Raman transition well removed from any electronic absorptions or other Raman bands. The conventional Raman process is illustrated in Figure 2(a). This is a two-photon process. The corresponding CARS process is illustrated in Figure 2(b) where it has been assumed that $(\nu_o - \nu_t)$ has been tuned precisely to the Raman transition frequency. Under these conditions of resonance there is an enormous increase in the probability of another ν_o photon being annihilated, the molecule then returning to the ground state with the emission of light at frequency $(2\nu_o - \nu_t = \nu_o + \nu_R = \nu_a)$ (where ν_R is a vibrational Raman frequency and ν_a is the frequency of the anti-Stokes radiation). Because the molecule returns to the ground state, ideally the phase matching requirement must be met (Figure 3) and the resultant anti-Stokes radiation is emitted into a coherent diffraction limited beam, providing the input laser beams are also diffraction limited. Hence the name Coherent Anti-Stokes Raman Spectroscopy (CARS). The experiment thus involves the simultaneous irradiation of the sample by two lasers ν_o and ν_t, one or both of which must be tunable. If the phase-matching condition is met in a dispersing medium ν_o, ν_t and ν_a will all be spatially separated from one another. An alternative to phase matching

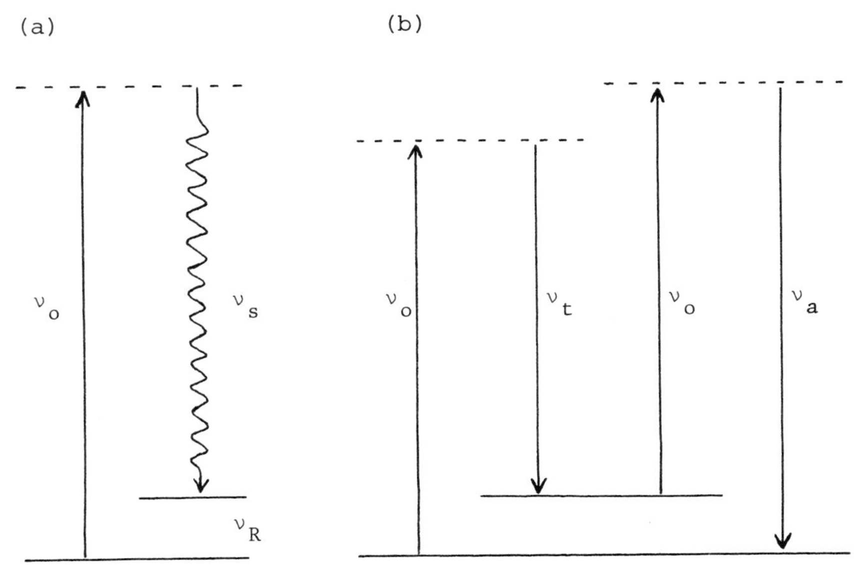

Figure 2. (a) conventional Raman process (b) CARS process at resonance. Note that ν_t represents stimulated emission of a photon by the tunable laser.

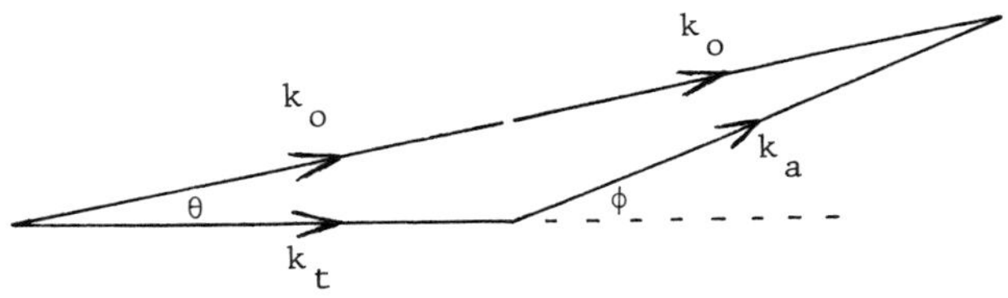

Figure 3. Phase matching for CARS. θ = angle between incident beams, ϕ = angle between ν_a and ν_t.

is to use co-linear beams which are tightly focussed. In the experiments so far carried out at Southampton by J. Black, T.R. Gilson and L. Laycock, ν_o has been a frequency doubled Nd/YAG laser operating TEM_{00} at approximately 0.2 MW peak power and with a pulse duration of about 15 nanoseconds; ν_t has been a Chromatix CMX4 flash-lamp pumped tunable dye laser normally working on Rhodamine 6G with peak powers up to 10 kW and a pulse duration of around two microseconds. ν_o lies at 532 nm or 18,800 cm^{-1} (green), ν_t peaks at 600 nm and is tunable from roughly 580 to 640 nm or 17,240 to 15,620 cm^{-1} (i.e. from yellow through orange to red). Thus $(2\nu_o - \nu_t)$ covers the range 20,360 to 21,980 cm^{-1} (491 to 455 nm) which is in the blue region. The range† of vibrational frequencies around ν_o

† This range is somewhat optimistic and refers to an ideal system. Probably 3000 to 2000 cm^{-1} is a more realistic generalisation.

(ν_R) is given by (ν_o - ν_t) i.e. 3180 to 1560 cm^{-1}.

Visually the observer sees two beams entering the sample (one yellow and one green) at a small angle to one another (to allow for phase matching), crossing in the sample and re-emerging as two distinct beams. In addition, spatially separated from these two is a new beam (blue) corresponding to anti-Stokes radiation. Figure 4 shows the first spectrum obtained at

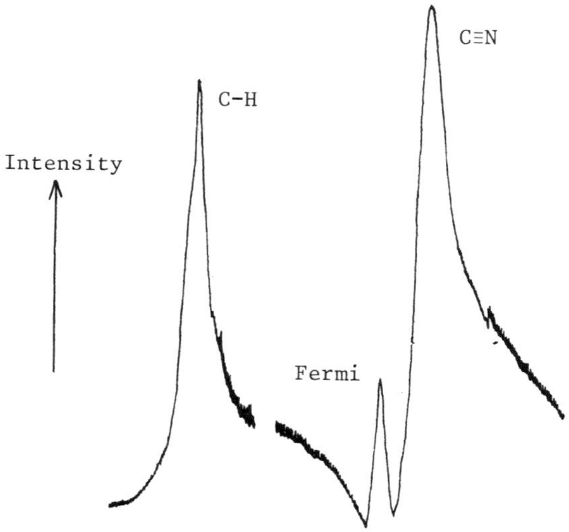

Figure 4. First CARS spectrum obtained at Southampton, showing acetonitrile (courtesy Dr. J. Black, Dr. T.R. Gilson and Mr. L. Laycock).

Southampton, using acetonitrile as the sample and with the experimental arrangement shown in Figure 5. The period from having the two lasers in the laboratory to obtaining a spectrum was less than a month.

On the elementary picture outlined above the intensity of a CARS transition is dependent on the square of the Raman scattering cross-section. Thus all Raman active bands are also CARS active. This is confirmed by more precise theoretical considerations [6]. The selection rules for CARS are thus identical to those for Raman scattering, using linearly polarised light.* Because the final state of the atom in the CARS process is the same as the initial state, the atoms undergoing the process are not distinguished from the others. As a result of this, the matrix elements must be summed for the different atoms <u>before</u> squaring to obtain the total intensity. Thus the intensity is proportional to the number density squared, not to the number density.

Many of the papers in this area are written by physicists

---

* Assuming the transition is isolated and the frequencies are well removed from any electronic transition.

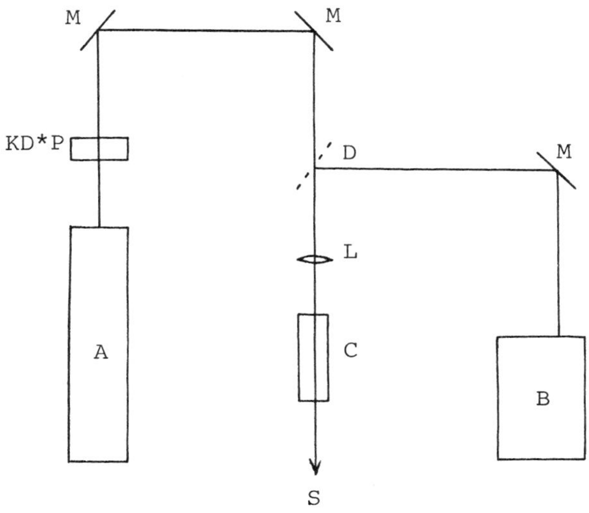

Figure 5. Diagramatic representation of elementary CARS experiment at Southampton using co-linear beams. A = Nd/YAG pump ν_o, B = CMX4 laser ν_t, M = mirrors, D = dielectric mirror, L = lens, KD*P = frequency doubler, S = spectrometer, C = cell.

who make use of the terminology χ_3 (the third order susceptibility) in preference to $\frac{1}{6}\gamma$.† The efficiency of the CARS process may be defined in terms of the intensity of the anti-Stokes radiation:

$$I_a = \frac{4\pi^2\nu_a^2}{\varepsilon_o^2 c^2} \cdot \frac{1}{n_a n_o^2 n_t} \left|\tfrac{3}{4}\chi_{CARS}\right|^2 I_o^2 I_t \ell^2 \qquad (7)$$

where n_a, n_o, n_t refer to the refractive indices at ν_a, ν_o, ν_t (cm^{-1})
ε_o is the permittivity, 8.854×10^{-14} Farad cm^{-1}
c is the velocity of light, 3×10^{10} cm sec^{-1}
I_a, I_o, I_t are in Watt cm^{-2}
ℓ is the length of sample in cm,
where it is assumed that the beams are not focussed and that phase matching is perfect.

$|\chi_{CARS}|$ refers to the susceptibility at resonance:

---

† The relationship between χ_3 and $\frac{1}{6}\gamma$ is complicated and definitions vary from author to author [7]. As used above in equation (2), γ is the static second hyperpolarisability.

$$|\chi_{CARS}| = \frac{\varepsilon_o}{3\pi^2 hc} \cdot \frac{N}{\nu_o \nu_t^{\,3}\Gamma} \left(\frac{d\sigma}{d\Omega}\right) \tag{8}$$

where $\left(\frac{d\sigma}{d\Omega}\right)$ is the Raman scattering cross-section cm^2 steradian^{-1}
Γ is the half-width at half maximum height (cm^{-1})
N is the number of scattering molecules cm^{-3}

For tight focussing the phase-matching condition becomes much less critical. The confocal parameter is given by

$$b = 2\pi w_o^{\,2}/\lambda \tag{9}$$

where w_o is the minimum beam waist and λ is the wavelength. Thus $\pi w_o^{\,2} b$ may be considered to define a cylindrical volume of length b and with a radius given by the minimum beam waist. Noting that intensity is power per unit area and putting $b = \ell$, the power (in Watts) at the antiStokes frequency then becomes

$$P_a = \frac{16\pi^2 \nu_a^{\,2}}{\varepsilon_o^{\,2} c^2} \frac{\nu_o^{\,2}}{n_a n_o^{\,2} n_t} \left|\tfrac{3}{4}\chi_{CARS}\right|^2 P_o^{\,2} P_t \tag{10}$$

In familiar units the intensity of the anti-Stokes radiation (in Watts cm^{-2}) for a non-focussed beam is:

$$I_a = \frac{1}{(\pi c)^4 (2h)^2} \cdot \frac{1}{n_a n_o^{\,2} n_t} \left(\frac{\nu_a}{\nu_o^{\,3}\nu_t}\right)^2 \left(\frac{N}{\Gamma}\right)^2 \left(\frac{d\sigma}{d\Omega}\right)^2 I_o^{\,2} I_t \ell^2 \frac{\sin^2(\frac{\Delta k \ell}{2})}{(\frac{\Delta k \ell}{2})^2} \tag{11}$$

where Δk is the phase mis-match.

The above outline is based on a discussion with M.A. Yuratich [8]. The non-resonant part of the susceptibility, which is present in all materials, has been neglected. Because of the dependence of P_a on $P_o^{\,2}.P_t$, it is customary to work with high peak power, pulsed lasers.

We may summarise the above discussion and extend it by noting [6,8]: for non-absorbing materials the selection rules for CARS and for Raman spectroscopy are identical. The intensity of an (isolated) Raman active transition is proportional to (a) the square of the Raman scattering cross-section, (b) the square of the inverse of the normal Raman half-width and (c) the square of the concentration of scattering centers. The intensity is also proportional to the product of the laser powers $P_o^{\,2}.P_t$.

The CARS radiation ideally forms a coherent, diffraction limited beam, whereas normal Raman radiation is incoherent and approximately spread out over 4π steradians. For discrimination against background radiation (fluorescent or thermal) CARS shows considerable promise. Even where at first sight the photon efficiency might appear to be rather low relative to the Raman effect, the extremely high efficiency of collection may still make CARS attractive.

Recently a very exciting experiment has been described by Barrett and Begley [9], using two cw lasers - one a tunable dye laser (ν_t) of 36 mW and one an argon ion laser (ν_0) of 460 mW. They obtained a beautiful spectrum of ν_1 of methane gas at atmospheric pressure. This immediately leads to the possibility of doing very high resolution spectroscopy as line-widths of the order of 0.001 cm^{-1} are relatively easily obtained in cw dye lasers. Extending the logic of this discussion we may note that pure rotational Raman lines are frequently intense. They are also depolarised. Because the incident photons ν_0 and ν_0 are in the same direction as the emitted photons ν_t and ν_a, the Doppler velocity will cancel (as noted by Stoicheff [10] for pure rotational Raman scattering in the forward direction). For CARS there is thus the possibility of carrying out high resolution rotational spectroscopy in the absense of Doppler broadening at pressures of less than one torr. Theoretically all molecules other than spherical tops exhibit a rotational Raman spectrum.

Finally we may note that where a molecular transition cannot be considered to be isolated, or where there is an appreciable non-resonant background susceptibility, then the concentration dependence may deviate from the normal squared relationship. This occurs because the total susceptibility must be squared - leading to cross terms. Thus in dilute solutions a linear dependence on concentration may become apparent. Although at first sight this looks attractive, it must be remembered that the limit to detectability will be dictated by the ability to see the resonant signal above the non-resonant background of the solvent [11].

In this connection the very recently discovered Raman induced Kerr effect is of interest [12]. Again the sample may be irradiated by two lasers ν_0 and ν_t. Both beams must be polarised and overlapped in the medium in time and space. The tunable beam ν_t is initially stopped by a polariser oriented so that it passes only light with a component perpendicular to the polarisation of ν_t. When ν_0 illuminates the sample simultaneously with ν_t then it is found that ν_t is transmitted if $\nu_0 \pm \nu_t = \nu_R$ (where again ν_R is a Raman active vibrational frequency). It is possible that the non-resonant background susceptibility may be eliminated in this technique - but it is early days!

5. CONCLUSIONS

Of all the techniques outlined above, CARS at the moment looks the most exciting. It will not replace - or displace - Raman spectroscopy, but it will be invaluable in certain applications. Certainly it looks very attractive to a high temperature chemist.

Acknowledgments

I am glad to acknowledge helpful discussions with Mr. Michael Yuratich and Dr. Al Harvey on CARS.

6. REFERENCES

1. For an excellent review on non-linear optics see Y.R. Shen, Rev. Mod. Phys., 48 (1976) 1; see also P.D. Maker and R.W. Terhune, Phys. Rev., 137A (1965) 801.
2. For papers on stimulated Raman see: G.E. Eckhardt, R.W. Hellwarth, F.J. McClung, S.E. Schwarz, D. Weiner and E.J. Woodbury, Phys. Rev. Lett., 9 (1962) 455; Y.R. Shen and N. Bloembergen, Phys. Rev., 137A (1965) 1787.
3. For an excellent account of Inverse Raman Spectroscopy see E.S. Yeung, J. Mol. Spectry., 53 (1974) 379; see also W. Jones and B.P. Stoicheff, Phys. Rev. Letters, 13 (1964) 657.
4. R.R. Alfano and S.L. Shapiro, Chem. Phys. Letters, 8 (1971) 631.
5. For papers on the hyper-Raman effect see: R.W. Terhune, P.D. Maker and C.M. Savage, Phys. Rev. Letters, 14 (1965) 681; S.J. Cyvin, J.E. Rauch and J.C. Decius, J. Chem. Phys., 43 (1965) 4083; D.A. Long and L. Stanton, Proc. Roy. Soc., A318 (1970) 441.
6. M.A. Yuratich and D.C. Hanna, IX International Quantum Electronics Conference, Amsterdam, June 1976.
7. For a clear discussion see M.P. Bogaard and B.J. Orr, in A.D. Buckingham (Editor), Int. Rev. Sci., Phys. Chem. Series 2, Volume 2, Butterworth, London, 1975.
8. M.A. Yuratich, personal communication.
9. J.J. Barrett and R.F. Begley, Appl. Phys. Letters, 27 (1975) 129.
10. B.P. Stoicheff, J. Mol. Spectry., 33 (1970) 183.
11. The following is a selection of references on CARS:
    P.R. Régnier and J.P. Taran, Appl. Phys. Lett., 23 (1973) 240.
    R.F. Begley, A.B. Harvey, R.L. Byer and B.S. Hudson, J. Chem. Phys., 61 (1974) 2466.
    I. Itzkar and D.A. Leonard, Appl. Phys. Lett., 26 (1975) 106.
    F. Moya, S.A.J. Fruet and J.P. Taran, Optics Comm., 13 (1975) 169.
    R.F. Begley, A.B. Harvey, R.L. Byer and B.S. Hudson, International Laboratory, Jan./Feb. 1975, p. 11.
    S.A. Akhmanov and N.I. Koroteev, Sov. Phys. JETP, 40 (1975) 650.
    I. Chabay, G.K. Klauminzer and B.S. Hudson, Appl. Phys. Lett., 28 (1976) 27.
    J.P. Taran, Tunable Lasers and Applications, Conference, Loen, 1976.
12. D. Heiman, R.V. Hellwarth, D.M. Levenson and G. Martin, Phys. Rev. Lett., 36 (1976) 189; M.D. Levenson and J.V. Song, J. Opt. Soc. Am., 66 (1976) 641; J.J. Song and M.D. Levenson, 5th International Conference on Raman Spectroscopy, Freiburg, 1976.

7. ADDENDUM

In later experiments carried out at Southampton by J.D. Black, T.R. Gilson and L. Laycock, ν_o has been obtained from the frequency doubled output of a single mode Nd/YAG laser (532 nm) or stimulated Raman (SRS) generated by this laser in liquid oxygen (580 nm) or liquid nitrogen (607 nm). Peak powers are around 0.2 MW at 532 nm with up to 50% conversion to the SRS shifted frequencies. The pulse length is approximately 15 ns corresponding to a pulse energy of 3 mJ at 532 nm. ν_t has been obtained from a Chromatix CMX4 flashlamp pumped dye laser using Rhodamine 6G or Rhodamine B with peak powers 1-10 kW and a pulse duration of about 1 μs. In principle (using fresh dye) vibrational frequencies (ν_R) anywhere in the region 0-3760 cm^{-1} can be studied (see Table 1). To date

Table 1. ν_R ranges for different combinations of dye and SRS medium.

SRS medium	ν_o (cm^{-1})	dye	ν_t range (cm^{-1})	ν_R range (cm^{-1})
None	18,800	Rhodamine 6G	17,240 - 15,620	1560 - 3180
None	18,800	Rhodamine B	16,130 - 15,040	2670 - 3760
Liquid O_2	17,248	Rhodamine 6G	17,240 - 15,620	8 - 1628
Liquid O_2	17,248	Rhodamine B	16,130 - 15,040	1118 - 2208
Liquid N_2	16,469	Rhodamine 6G	17,240 - 15,620	0 - 849
Liquid N_2	16,469	Rhodamine B	16,130 - 15,040	359 - 1029

spectra have been obtained in the region 2000-3000 cm^{-1} for both liquids and gases. In the 60-400 cm^{-1} region using either liquid nitrogen or liquid oxygen as the stimulated Raman scattering medium, we have recently obtained the spectrum of liquid $SnCl_4$, showing all four fundamentals (Figure 6).

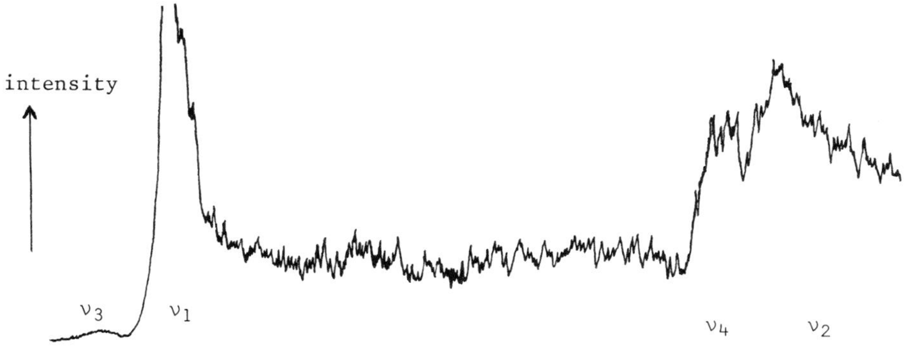

Figure 6. CARS spectrum of liquid tin tetrachloride (Courtesy Dr. J. Black, Dr. T.R. Gilson and Mr. L. Laycock).

CHAPTER 4

INFRARED FLUORESCENCE

George C. Pimentel

The advent of powerful laser sources has made possible the excitation of molecules into selected vibrational states and in high concentrations. One of the notable advances that has resulted has been the reawakening of the study of energy transfer out of and among vibrational degrees of freedom. Thus we are developing a better understanding of the role and importance of vibrational degrees of freedom as a non-equilibrium system re-distributes its energy to approach a Boltzmann equilibrium state. An obvious application of such information lies in the optimisation of laser systems, including chemically pumped lasers. More fundamental, however, is the understanding of energy movement in the course of a chemical reaction. By laser excitation and fluorescence monitoring of specific state occupancies, it is possible to learn how vibrational excitation of a reactant influences reaction rate, to investigate the importance of the product's vibrational degrees of freedom as immediate energy recipients in exothermic reactions, and to sense the participation of spectator molecules in the diversion of such product vibrational excitation into rotational and translational degrees of freedom.

Such fluorescence studies, both electronic and vibrational, draw heavily upon our understanding of vibrational motions in molecules; the concepts of normal modes, energy levels, anharmonicities, Fermi resonances, and transition probabilities are all of relevance. Furthermore, the techniques and instrumentation familiar to vibrational spectroscopists are generally applicable. Hence, it behoves vibrational spectroscopists to be aware of recent progress in laser-induced fluorescence and to consider the possible devotion of their own expertise to the interesting problems to be investigated.

1. USE OF LASERS AS EXCITATION SOURCES

Listed in Table 1 are four ways in which lasers can be used to excite vibrational states of molecules. Prototype examples of each will be discussed, as indicated in the last two columns. In Type I studies, a visible or ultraviolet laser is used to excite a vibrational state through a stimulated Raman process. This implies that there is no need for the laser frequency to coincide with a transition of the molecule to be excited - hence the term "non-coincidence excitation". The stimulated Raman process requires high pump powers and, for gases, high pressures. In Type II, a molecule is excited directly through infrared absorption, exploiting an accidental coincidence between a convenient laser frequency and a molecule that absorbs

Table 1. Prototype examples of the ways in which lasers can be used as excitation sources

Laser pumping method	Prototype studies	Laser used	Sample excited
I. Non-coincidence	De Martini and Ducuing [1]	Q-switched ruby	H_2 (gas, 10 atm)
II. Accidental coincidence	Yardley and Moore [2]	He/Ne, 3.39 μm	CH_4 (gas)
	Abouaf-Marguin et al [3]	CO_2	NH_3 (in solid Ar)
	Dubost et al [4]	CO_2 (doubled)	CO (in solid Ar)
III. Resonant pump	Javan and co-workers [5,6]	CO_2	CO_2 (gas)
	Chen and Moore [7]	HCl	HCl (gas)
IV. Tunable pump	Allamandola and Nibler [8]	Ar^+, dye	C_2^- (in solid Ar)

at that frequency. For gases, applicability is limited by the narrow line widths, which make close coincidences rare, and, in general, the experimenter must work with molecules selected because they absorb in the same spectral region as emitted by one of the known lasers. In the methods classified as Type III, the molecule excited is identical to the emitting molecule in the laser source. Frequency coincidence is, then, perfect, but again the experimenter must tailor his choice of molecules for study to those suggested by the existing list of lasers.

Studies of Type IV, using a tunable laser, bypass many of the limitations of the earlier techniques. As high power, tunable lasers become available, the experimenter can attempt to excite any energy level transition suggested by his interest. Since this technology is now becoming available, fluorescence studies can be expected to flourish.

2. GAS PHASE FLUORESCENCE STUDIES OF VIBRATIONAL RELAXATION

H_2 : De Martini and Ducuing [1] focussed a high power, Q-switched ruby laser into hydrogen gas at 10 atmospheres pressure. Through the stimulated Raman effect, they were able to excite about 10^{16} H_2 molecules to v = 1 in 20 ns. The decay of anti-Stokes radiation was measured using spontaneous Raman scattering to determine the vibrational relaxation time.

CH_4 : Yardley and Moore [2] mechanically chopped at a variable frequency a He/Ne laser operating at 3.39 μm to excite periodically ν_3 of gaseous methane (the asymmetric C-H stretching mode at 3020 cm^{-1}). They observed infrared fluorescence both from ν_3 and from ν_4 (the triply degenerate bending mode at 1305 cm^{-1}). The phase shifts between excitation and fluorescence were measured as a function of chopping frequency to determine the rate constants for the rapid disappearance of ν_3 excitation (the relaxation time at one atm pressure is 7 ns) and the rate constant for the slower disappearance of ν_4 excitation (the relaxation time at one atm pressure is 1.9 μs).

The rapid loss of ν_3 fluorescence includes all loss processes, of course, but it is dominated by vibration-to-vibration (V → V) energy transfer processes. The most important are undoubtedly those with the smallest energy discrepancies:

$$CH_4(\nu_3) + CH_4 \rightarrow CH_4(2\nu_2) + CH_4 + \Delta E = -52\ cm^{-1}$$

$$\rightarrow CH_4(\nu_2 + \nu_4) + CH_4 + \Delta E = +196\ cm^{-1}$$

$$\rightarrow CH_4(2\nu_4) + CH_4 + \Delta E = +420\ cm^{-1}$$

$$\rightarrow CH_4(\nu_2) + CH_4(\nu_2) + \Delta E = -47\ cm^{-1}$$

The slower rate of loss of ν_4 excitation is attributed to the fact that further degradation of the vibrational energy requires vibration-to-rotation and translation (V → R,T) energy transfer processes since ν_4 is the lowest vibrational frequency of methane. Thus, the V → R,T relaxation of ν_4 is more than two orders of magnitude slower than the V → V relaxation of ν_3, a typical contrast between these two types of energy transfer.

CO_2 : A CO_2 laser can operate on a number of lines in either of the two vibrational transitions (00°1)→(10°0) or (00°1)→(02°0). Since the fractional Boltzmann occupancy of the (10°0) level at 300 K is about 10^{-3}, CO_2 laser light can be used to excite the (00°1) level in a CO_2 sample in a separate cell. Javan and his co-workers [5,6] have carried out such experiments, observing fluorescence in the (00°1)→(00°0) transition. They also monitored occupancy of the (02°0) level through its absorption of the (00°1)→(02°0) emission from a second, probe CO_2 laser.

A number of interesting findings resulted. The CO_2 excitation was found to diffuse from the site of excitation faster than CO_2 itself diffuses, indicative of (V → V) resonant energy transfer. Wall deactivation was found to be measureable: the probability of loss of (00°1) excitation is 0.22 per collision. (There is no way to tell if some of the energy is left in lower energy vibrational modes of the CO_2 molecule.) Again, near-resonant V → V energy transfer processes were found to be very fast: the (10°0)→(02°0) + 100 cm^{-1} exchange requires less than 10 collisions, on the average, while the (02°0)→(01°0) - 49 cm^{-1} exchange requires an average of about 50 collisions.

HCl : Chen and Moore [7] used the Cl + HI → $HCl^{\dagger}$ + I chemical laser to produce 10-20 μs laser pulses of about 0.01 to 0.03 Joules energy. About 40% of the laser emission occurred in v = 1 → 0 transitions, primarily $P_1(9)$, $P_1(10)$ and $P_1(11)$.

The pulse laser emission was absorbed in an HCl gas sample from which vibrational fluorescence was measured. The fluorescence was readily detected with a 3 x 10 mm Au-Ge detector at 77 K (NEP = 10^{-7} Watts) protected from the laser light by a narrow-band filter transmitting only in the v = 1 → 0 R branch. Various collision partners were added to measure their effectiveness as vibrational deactivators. When DCl was added, $DCl^{\ddagger}$ (v = 1 → 0) fluorescence could be observed, as well. A number of rate constants for V → V and V → R,T processes were measured. Some of these rates are collected in Tables 2 and 3 (as characterised by the average number of kinetic collisions for energy exchange) and discussed below.

Table 2. Vibrational relaxation of $HCl^{\ddagger}$ (v = 1) by V → R energy transfer

Collision partners	Z^a
$HCl^{\ddagger}$-HCl	7,900
$DCl^{\ddagger}$-DCl	26,000
$DCl^{\ddagger}$-HCl	11,000
$HCl^{\ddagger}$-n-H_2	106,000
$HCl^{\ddagger}$-p-H_2	106,000
$HCl^{\ddagger}$-H_2O	~10
$HCl^{\ddagger}$-M^b	$>3.10^6$

a $Z = P^{-1}$ where $P = \sigma/\sigma_{kin}$, $\sigma_{kin} = \pi(d_{HCl} + d_{AB})^2/4$ and $\sigma = k/n\bar{v}$, k in s^{-1} $torr^{-1}$, n = molecules cm^{-3}, and $\bar{v}$ is the average velocity.

b M = He, Ne, Ar

Table 3. Vibrational relaxation of $HCl^{\ddagger}$ (v = 1) by V → V energy transfer

Collision partners	Z	ΔE (cm^{-1})
$HCl^{\ddagger}$-CH_4	106	-30 (ν_1), -133 (ν_3)
-D_2	143	108
-HBr	166	327
-DCl	2,500	795
-N_2	9,000	555

Examining, first, Table 2, we see that HCl deactivates $HCl^{\ddagger}$ about three times as rapidly as DCl deactivates $DCl^{\ddagger}$. This is in qualitative contradiction to the classical expectation that the $DCl^{\ddagger}$-DCl deactivation would be three orders of magnitude the more rapid. The observed rates can be readily understood only if the dominant deactivation process transfers the vibrational energy into rotation (rather than into translation). Chen and Moore [7] further interpret the $HCl^{\ddagger}$-DCl and $DCl^{\ddagger}$-HCl rates as an indication that the molecule that loses the vibrational energy retains most of the energy, as it is transferred into rotation.

The data for deactivation of $HCl^{\ddagger}$ by hydrogen contrast the effectiveness of normal and para-hydrogen. Within experimental accuracy, there is no difference between the two relaxation rates. Since p-H_2 can involve only even rotational states, it can be concluded that rotational excitation of the H_2 collision partner plays no important role in the deactivation of $HCl^{\ddagger}$, even though it is also considered to be predominantly a V $\rightarrow$ R process (though the evidence is less direct).

As a final example of vibrational relaxation, Table 2 shows that only about 10 H_2O collisions are needed to relax $HCl^{\ddagger}$. Chen and Moore observe that both this unusually high effectiveness and that of HCl can be attributed to the attractive interactions associated with hydrogen bonding.

Turning, now, to Table 3, the deactivation of the $HCl^{\ddagger}$ ($v = 1$) by CH_4, D_2, and HBr is seen to require less than 200 collisions. Plainly the explanation lies in the close coincidence between vibrational modes of the collisional partner, as indicated by the energy discrepancies (in the last column) which are comparable to collisional energies, as measured by kT (at 300 K, kT $\sim$200 cm^{-1}). Thus V $\rightarrow$ V transfer is involved.

The remaining two examples are also considered to be V $\rightarrow$ V transfer processes (because of their rapidity) and they show how the deactivation probability drops as the energy discrepancy becomes larger. They also show that specific, attractive interactions, such as are present between $HCl^{\ddagger}$ and DCl, can overcome a rather large energy discrepancy. This is considered to be the reason that DCl requires far fewer collisions to deactivate $HCl^{\ddagger}$ than does N_2, despite the much higher energy discrepancy.

3. MATRIX PHASE FLUORESCENCE STUDIES OF VIBRATIONAL RELAXATION

Evidence that vibrational relaxation might be slow in low temperature (4 to 20 K) matrix samples was signalled in the electronic fluorescence studies by Tinti and Robinson [9] in 1968. Their spectra of the $N_2 A^3\Sigma_u^+ \rightarrow X^1\Sigma_g^+$ transition of N_2 isolated in inert gas solids showed the fluorescence occurred from excited vibrational states of the upper electronic states with maximum intensity at $v' = 6$. From the electronic lifetime, they were able to place the lifetime for vibrational deactivation in the range 0.4 to 3.2 μs. In a similar way, Shirk and Bass [10], in 1970, estimated a lifetime of about 5 ns for the vibrational relaxation of matrix-isolated CuO. These lifetimes revealed the possibility of fluorescence study of vibrational deactivation in cryogenic samples.

CO in solid Ar : Dubost et al [4] have vibrationally excited carbon monoxide in solid argon (and in neon) at 6 to 10 K and observed infrared fluorescence. They attribute absorption at 2138.4 cm^{-1} to CO isolated in solid argon; the observed half-width is about 0.8 cm^{-1}. They used as a pumping source a Q-switched, frequency-doubled CO_2 laser. The Q-switching was obtained with a rotating cavity mirror at 40 to 200 Hz, giving 1-2 kW pulses of 150 to 500 ns duration. A grating in the optical cavity was used to select the R(8) transition in the $(00°1)\rightarrow(02°0)$ band; this transition is at 1070.43 cm^{-1}. A 9 mm thick tellurium crystal doubled this to 2140.86 cm^{-1} with about 1% efficienty. The doubled frequency was focussed with a CaF_2 lens (5 cm focal length) onto the matrix sample. Fluorescence light was observed at a 60° angle to the direction of excitation, as collected by a 4 cm focal length CaF_2 lens focussing the radiation onto a Au-Ge photo-conductive detector with a 20 μs response time.

The fluorescence displayed a 4 ms lifetime in solid argon at 6 K, thus approaching the radiative lifetime of gaseous CO (about 10 ms). The lifetime was reduced by an order of magnitude if the window temperature supporting the matrix sample was raised only 2 K.

Most remarkable in this study was the observation that CO emitted light in the fundamental spectral region from excited states as high as v = 6 or 7, even though the pumping necessarily initiated in the $v = 0 \rightarrow 1$ excitation. Dubost et al. concluded that anharmonic "up-pumping" was taking place through exothermic reactions such as

$$CO(v=1) + CO(v=1) \rightarrow CO(v=2) + CO(v=0) + \Delta E_{1,1;\ 2,0}$$

$$CO(v) + CO(v=1) \rightarrow CO(v+1) + CO(v=0) + \Delta E_{v,1;\ v+1,0}$$

Because of anharmonicity, these reactions tend to "pool" the energy, transferring the anharmonic energy discrepancies, $\Delta E_{v,1;\ v+1,0}$ into the Ar lattice phonon modes. Because the ambient matrix temperature is so low, $\Delta E_{v,1;\ v+1,0} >> kT$, so the reverse reactions cannot occur. Of course, these reactions cannot occur unless the vibrational relaxation times are long compared to the time for energy migration to permit the energy pooling reactions to occur.

Striking support for this interpretation is provided by the observation that a substantial fraction of fluorescence occurs from ^{13}CO, despite the low concentration of this isotope at natural abundance. Obviously, the isotopic frequency shift again provides an exothermic reaction path with ΔE much larger than kT at T = 6 K.

$$^{12}CO(v=1) + {}^{13}CO(v=0) \rightarrow {}^{12}CO(v=0) + {}^{13}CO(v=1) + \Delta E = 47\ cm^{-1}$$

As could be expected, the fluorescence from these laser-excited matrix samples decreased as the concentration was decreased, slowing down the migration process by which excited molecules come into proximity. Furthermore, the fluorescence was severely quenched by matrix impurities.

NH_3 in solid N_2 : Abouaf-Marguin et al. [3] also attempted similar studies using NH_3 suspended in solid nitrogen at 8 K. In an N_2 matrix, NH_3 absorbs at 969.5 cm^{-1} and it neither rotates nor does it invert [11]. With a band width of 1.7 cm^{-1} at 8 K, matrix-isolated ammonia can be excited with a CO_2 laser operating on the Q-switched R(10) and R(12) of the (00°1)→(10°0) transitions. In this case, despite a gas-phase radiative lifetime of 66 ms, NH_3 fluorescence radiation was not detected by a Au-Ge detector at 77 K.

In the absence of fluorescence, these workers conducted a double resonance experiment, using a second, cw CO_2 laser as a probe. The probe laser polarisation was oriented perpendicular to that of the Q-switched, pump laser and focussed onto a Cu-Ge detector at 20 K. Evidence of NH_3 excitation (at N_2/NH_3 = 500) was provided by an increase of the sample transmittance immediately after the Q-switched pulse, provided the pump frequency, ν_s, and the probe frequency, ν_p, were the same. The temporal behaviour of this transmittance revealed the relaxation behaviour.

With $\nu_s = \nu_p = 969.1$ cm^{-1}, the bleaching recovered with a rapid time constant τ_1 near 2 μs. Surprisingly, the transmittance decreased to a minimum somewhat <u>below</u> that of the original sample, and then recovered to its original transmittance in a few hundred microseconds. With $\nu_s = \nu_p = 970.5$ cm^{-1}, again two such time constants characterised the recovery, but at this frequency, the transmittance did not "overshoot". With $\nu_s = 970.5$ cm^{-1} and $\nu_p = 969.1$ cm^{-1}, only a weak decrease in transmittance occurred, and only the slow recovery process was observed.

As the N_2/NH_3 ratio was increased from 20 to 2000, the short time constant τ_1 increased perhaps two-fold. As temperature was raised, the increase of transparency diminished; at about 11 K, the rapid decay was no longer measureable.

Abouaf-Marguin et al. [3] attribute τ_1 to relaxation of the vibrationally excited NH_3 molecules and conclude, since ν_s and ν_p must be identical, that the probe is detecting the same molecules excited by the pump. During relaxation, vibrational energy is converted into heat, warming the sample slowly. The slower recovery process is attributed to the thermal recovery of the matrix sample.

C_2^- in solid Ar : The fluorescence decay of pulse-excited C_2^- in inert gas matrix samples has been under concerted study by Nibler and co-workers [8,12]. In the most recent work, acetylene was photolysed with vacuum ultraviolet light (H_2 resonance lamp) during deposition of a C_2H_2-Ar mixture at 3-8 mmol h^{-1}, at 14 K, and with Ar/C_2H_2 = 100 to 20,000. The matrix sample so produced displays absorptions at 238.2 nm attributed to C_2 and two at 521 and 473 nm attributed to the 0" → 0' and 0" → 1' transitions in the $X^2\Sigma_g^+ \rightarrow B^2\Sigma_u^+$ system of C_2^-.

The sample was exposed to pulsed laser light from an argon ion laser at frequency ν_p. This pump laser provides pulses of about 50 mW, a rise time of 100 ns, a width of 450 μs, and a 1000 Hz repetition rate. The pump pulse is long compared to the fluorescent lifetime (< 10 μs). A second laser is used as a probe with a variable delay. The probe laser, at frequency ν_{pr}, is a tunable dye laser pumped by a 4 W Ar^+ laser. The probe laser provides about 100 mW and a 14 μs half-width. Fluorescence is measured with a Cary 82 Raman spectrometer; photon counting is used.

If the pump laser is used alone at 472.7 nm (ν_p = 21,150 cm^{-1}), there is sufficient photon energy to raise C_2^- to the v' = 1 level of the upper electronic state, which is only 10 cm^{-1} lower in energy (21,140 cm^{-1}). The fluorescence spectrum shows a strong emission progression for v' = 1 → v" with v" = 0 to 3. (This shows that the v' = 1 state has a long lifetime compared to that of the electronic transition.) There is also much weaker 0' → v" fluorescence. When the probe laser is used as well, it is tuned to 574.3 nm (17,413 cm^{-1}), close to the v" = 1 → v' = 0 transition (in N_2 matrix, at 17,339 cm^{-1}; in Ar matrix, at 17,417 cm^{-1}). The simultaneous irradiation of the matrix C_2^- sample with ν_p and ν_{pr} <u>stimulates</u> fluorescence from the v' = 0 → v" = 0 transition. The intensity of this stimulated fluorescence decreases exponentially as the time delay increases between ν_{pr} and ν_p. The stimulated v' = 0 → v" = 0 emission is plainly due to absorption of the probe laser frequency due to occupancy of the v" = 1 level that persists for hundreds of microseconds after the pumping process (absorption of 472.7 nm up to the v' = 1 state, fluorescence back to the v" = 1 state) has been interrupted.

In this fashion, Allamandola and Nibler [8] measured a lifetime of 0.2 ns for the v" = 1 state of C_2^- in solid argon. Surprisingly, they measured an even longer lifetime, 1.2 ns, for the v" = 2 state. They investigated the effect of concentration from starting Ar/C_2H_2 ratios ranging from 50 to 20,000 and observed an order of magnitude decrease in the lifetime as concentration increased.

4. SUMMARY

The prototype fluorescence studies described here open new avenues for studying energy transfer into and out of molecular vibrational degrees of freedom. They are made possible by technological developments of the last decade, including a variety of powerful laser sources, rapid response and highly sensitive infrared detectors, and computer-aided averaging techniques. Lasers give us much higher intensities and virtually monochromatic light sources that operate on the sub-nanosecond time scale. As these powerful tools become generally available, and particularly as the lasers become tunable, the types of chemical problems that can be addressed through fluorescence measurements will surely increase. We are bound to learn more about the role of vibrational degrees of freedom in chemistry and vibrational spectroscopists will wish to contribute to this progress.

5. REFERENCES

1. F. De Martini and J. Ducuing, Phys. Rev. Letters, 17 (1966) 117.
2. J.T. Yardley and C.B. Moore, J. Chem. Phys., 45 (1966) 1066; 49 (1968) 1111.
3. L. Abouaf-Marguin, H. Dubost and F. Legay, Chem. Phys. Letters, 22 (1973) 603.
4. H. Dubost, L. Abouaf-Marguin and F. Legay, Phys. Rev. Letters, 29 (1972) 145.
5. L.O. Hocker, M.A. Kovacs, C.K. Rhodes, G.W. Flynn and A. Javan, Phys. Rev. Letters, 17 (1966) 233.
6. M. Kovacs, D.R. Rao and A. Javan, J. Chem. Phys., 48 (1968) 3339.
7. H-L. Chen and C.B. Moore, J. Chem. Phys., 54 (1971) 4072.
8. L.S. Allamandola and J.W. Nibler, Chem. Phys. Letters, 28 (1974) 355.
9. D.S. Tinti and G.W. Robinson, J. Chem. Phys., 44 (1967) 3229.
10. J.S. Shirk and A. Bass, J. Chem. Phys., 52 (1969) 1894.
11. M.O. Bulanin, M. Van Thiel and G.C. Pimentel, J. Chem. Phys., 36 (1962) 500.
12. V. Bondybey and J.W. Nibler, J. Chem. Phys., 56 (1972) 4719.

CHAPTER 5

TUNABLE INFRARED LASERS

J.J. Turner

1. INTRODUCTION

Conventional IR Spectroscopy

"Conventional" infrared spectrometers are of two kinds - dispersive and interferometric. A typical dispersive spectrometer has a polychromatic Nernst glower source, slits, dispersing agent (prism or grating), sample compartment and detector. Double beam spectroscopy employs a chopper which sends the beam alternately through sample and reference cells and the two signals are combined on the detector; comparison of sample and reference is achieved by either null or ratio-recording methods. The important properties of such a spectrometer are resolution, signal/noise and speed, which are of course all inter-related. The great disadvantage of such an instrument is that only a small fraction of the output energy of the source is actually used at each stage of the scan. For a typical instrument the energy in a 1 cm^{-1} band width might be as low as 10^{-6} Watts. If the slits are narrowed to increase the resolution the energy is lowered further and noise problems become increasingly important.

This low energy presents problems when spectra have to be recorded extremely quickly such as when IR spectra of transients are required. This point is taken up at the end of the chapter.

Interferometric infrared instruments, developed by Gebbie, Connes and others were originally produced for the far IR. More recently several mid-IR instruments have appeared. The advantage of these Fourier transform spectrometers over dispersive instruments is that most of the light from the polychromatic source, over the appropriate spectral range, is being used all the time the interferogram is being recorded. Thus as a rough comparison a Fourier transform instrument will produce good spectra with $\sim$1/50 of the light needed for a dispersive monochromator. The advantage for the detection of weak signals is clear; moreover high resolution spectra can be obtained quickly without destroying the S/N ratio.

With both grating and interferometric instruments digital handling, spectral summation, etc, greatly improve spectral performance but nonetheless the limitation of polychromatic sources limits the potential performance. Tunable IR lasers promise enormous advantages.

Infrared Photochemistry

There is great current interest in the use of IR lasers for photochemistry. Absorption of IR radiation populates excited vibrational energy levels; provided the lifetimes in the excited levels are sufficiently long it is possible to promote specific chemical reactions, including isotope separation. These effects are only observed at very high power densities (typically for isotope separation 10^9-10^{10} W cm^{-2}) and so high power pulsed lasers are required. The pulsed CO_2 system has been the most popular choice but of course the chemical system has to be chosen to match the laser rather than the reverse, even allowing for the possibility of selecting one of the several output lines. High power tunable IR lasers would have an enormous impact on the field.

2. TUNING

Although we are primarily concerned with "genuinely" tunable (i.e. >100 cm^{-1} tuning range) lasers there are other ways in which spectroscopy can be achieved.

Molecular Tuning

If the laser frequency is fixed then spectroscopic scanning requires continuous alteration of the molecular energy level separation. This can be achieved in appropriate cases by the application of electric or magnetic fields - Stark or Zeeman tuning. The tuning range is usually very small except for Zeeman tuning of paramagnetic free radicals. An example of the latter is recent work [1] on the HCO radical; a field of 0.1 T (1 kGauss) brings the transition into coincidence with the D_2O laser line at 92.8 cm^{-1}.

Perturbation laser tuning

By this we mean the alteration of the output of an essentially fixed frequency laser by applying some external perturbation. The distinction from genuinely tunable lasers is somewhat arbitrary but we might include for example the Zeeman tuning of the 3.39 μm (2950 cm^{-1}) line of a He-Ne laser; the tuning range is very limited ∿0.05 cm^{-1} per 0.1 T but very high resolution spectra in the C-H stretching region of several hydrocarbons have been studied.

"Genuinely" tunable IR lasers [2,3]

There are basically two types of laser under this heading; those which are primary lasers in that they can be pumped by incoherent stimulation, and secondary lasers, so called because operation depends on pumping by another laser. The dye laser, operating in the UV/visible spectral region is an example of the former; infrared diode lasers are a second example. Secondary tunable IR lasers include the Spin Flip Raman Laser and several systems dependent on non-linear devices.

A further point is that in Chapter 2 we noted the cavity mode structure of laser output. Because of this it is usually not possible to continuously tune lasers over the whole accessible spectral range, but "mode-hopping" occurs. However this complication is well understood and we shall ignore it here.

(a) The diode laser

Figure 1 shows a schematic plot of energy against density

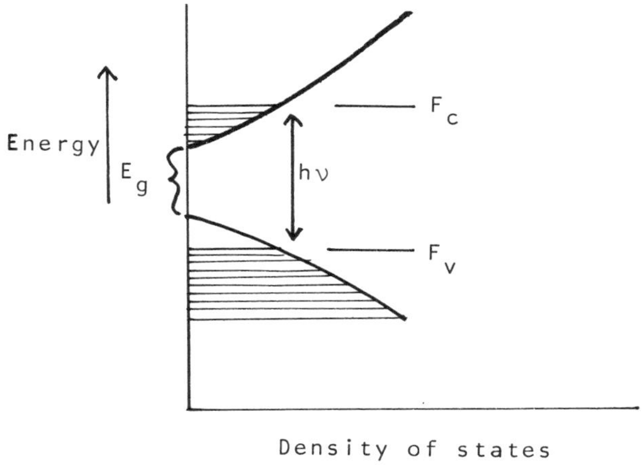

Figure 1. Intrinsic semiconductor; energy vs. density of states; population inversion at T = 0 K.

of states for an intrinsic semiconductor; the diagram has been drawn to illustrate an inversion population at T = 0 K (the picture is slightly more complex if T > 0 K). Stimulated emission will occur if the energy of the incident photon lies between the energy gap (E_g) and (F_c - F_v). This suggests the possibility of laser action if such a population inversion can be achieved and if the semiconductor can be incorporated in an optical cavity in the usual way. The most popular method of producing a population inversion uses electron injection at a p-n junction. This is illustrated in Figure 2.

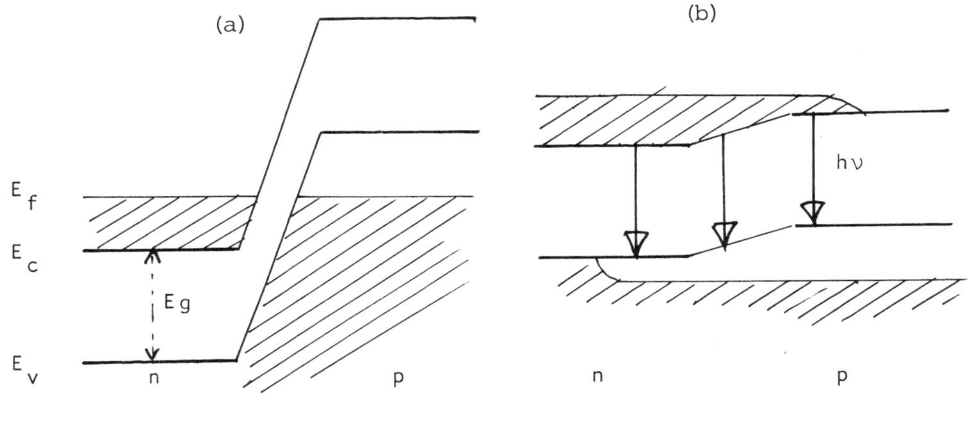

Figure 2. Production of population inversion at n/p junction (a) zero field (b) field on.

The n-type in Figure 2(a) has excess electrons and hence is filled up to the Fermi level in the conduction band. The p-type is deficient in electrons. At the p-n junction the relative energies are as illustrated since thermodynamics requires the Fermi level to be the same on both sides. On applying a voltage across the junction (Figure 2(b)) the barrier is reduced, a population inversion exists and electrons can flow n → p emitting photons with energy approximately equal to the band gap E_g. Because the maximum optical gain occurs in the plane of the junction the cavity is constructed as shown schematically in Figure 3. Semiconductor materials such as GaAs, InAs and

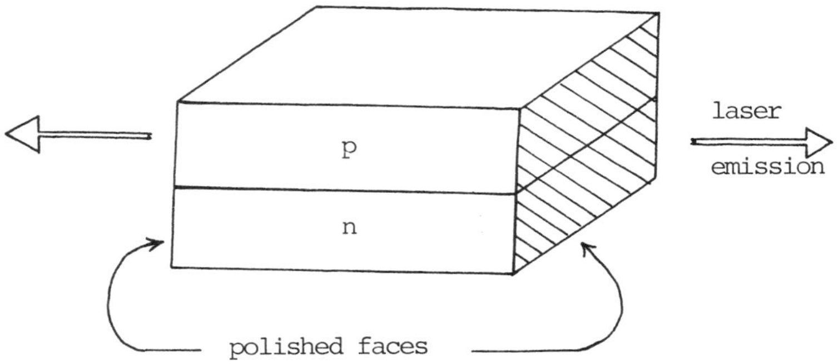

Figure 3. Schematic of p/n crystal laser.

ternary systems such as $Pb_{1-x}Sn_xTe$ have band gaps in the infrared region and doping of both n- and p- character is feasible.

The radiation can be tuned in several ways - by altering the temperature, pressure or current - and the centre of the tuning range can be shifted by altering the composition of the ternary systems. The power from such lasers is usually very small (a few μWatts) but the linewidth can be very narrow ($\sim 10^{-6}$ cm^{-1}) over the tuning range of an individual diode of about 40 cm^{-1}. One of the classic spectra obtained by such a device is shown in Figure 4 where it is compared with a conventional grating spectrum.

The great problem in this field is that the devices are extremely difficult to manufacture although spectrometers based on them are beginning to appear.

(b) The Spin Flip Raman Laser (SFRL)

The SFRL is also based on a semiconductor device but in this case a population inversion is not required; however the system must be driven by a laser pump. Figure 5 shows the effect on the energy levels of InSb of applying a magnetic field.

The Landau levels are split into "spin-up" and "spin-down" levels with energy separation given by

$$\Delta E = g\beta B \qquad (1)$$

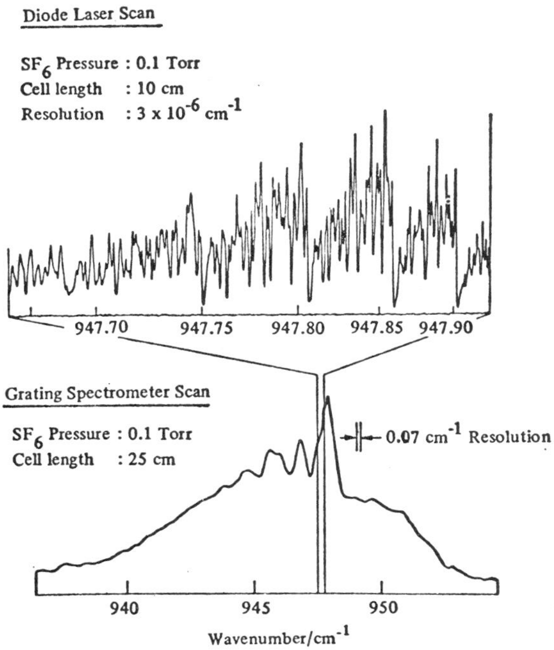

Figure 4. A section of the IR vibrational spectrum of ν_3 of SF_6 using a diode laser (above) compared with the best available spectrum (below), obtained using a very high-resolution conventional spectrometer. (Reproduced from ref. 3 with permission.)

where β is the Bohr magneton, B the value of the external magnetic field and g is the usual g value, which for InSb is -45 (cf. free electron $\sim$2). Thus the energy separation changes by about 20 cm^{-1} per Tesla. At low temperature the population distribution is as shown and it is possible by using a powerful laser to cause Stokes scattering by which molecules in 0,$\downarrow$ are promoted to 0,$\uparrow$ by a process very similar to the usual stimulated Raman effect. The pump lasers have usually been either cw CO operating around 5 μm, or pulsed CO_2 operating at either 10 μm or frequency doubled to 5 μm using a Te crystal. The frequency of the output radiation is given by

$$\nu_{output} = \nu_{input} \pm mg\beta B \qquad (2)$$

the $\pm$ because it is possible to obtain both Stokes and anti-Stokes scattering and the m because it has been possible to obtain higher order Stokes than the first. Typical tuning curves are shown in Figure 6.

The crystal of InSb sits in a cryostat in a powerful tunable magnetic field. The crystal has ends polished in the usual way and the output light is separated from the unconverted pump radiation by a monochromator. These systems are extremely

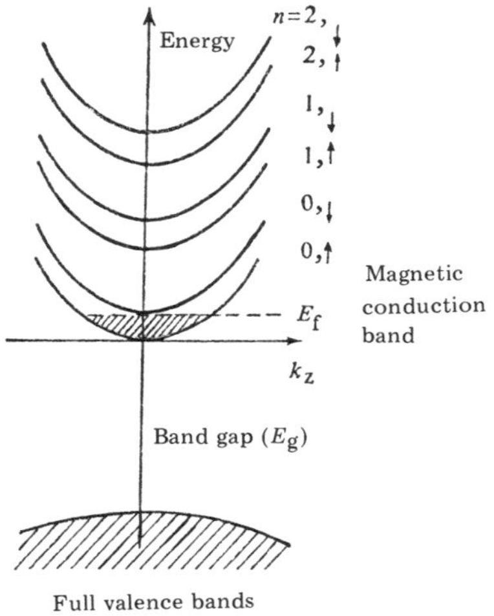

Figure 5. The structure of the conduction band of InSb in a magnetic field, illustrating the splitting of the Landau levels (designated by integral values n) into spin-up and spin-down levels. E_f if the Fermi level. (Reproduced from ref. 3 with permission.)

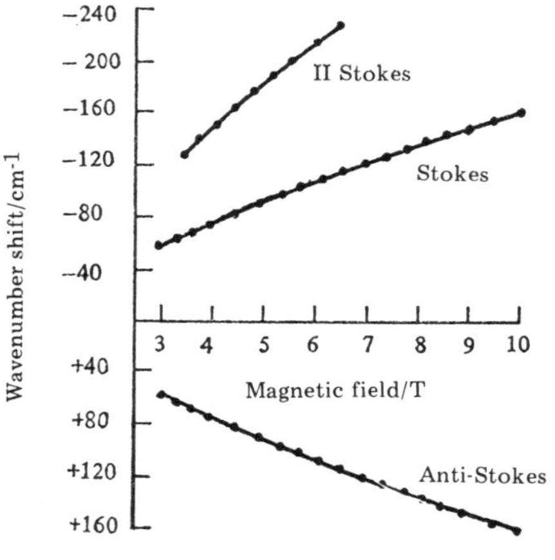

Figure 6. Typical tuning curve of a pulsed spin-flip laser. Note that the slope of the double-Stokes curve is approximately twice that of the Stokes. (Reproduced from ref. 3 with permission.)

difficult to construct and operate since so much has to be working properly at the same time. There are problems of multi-lining, optical coupling, pump laser instability and so on. Nevertheless the developing technology is extremely promising and one might hope for the following characteristics:

Pump	CO	CO_2
Tuning range (cm^{-1})	1950 - 1700	1950 - 1700 1050 - 850
Power	500 mW	kW during pulse
Linewidth	10^{-5} cm^{-1}	0.01 cm^{-1}

(c) Non-linear devices

The polarisation induced in a material by the oscillating electrical field due to light is given by

$$\vec{P} = \chi_1\vec{E} + \chi_2\vec{E}.\vec{E} + \chi_3\vec{E}.\vec{E}.\vec{E}. + \ldots\ldots. \qquad (3)$$

where the χ's are coefficients which decrease rapidly in magnitude with order ($\chi_{n+1}/\chi_n \sim 10^{-9}$). The first, linear, term gives rise to the dielectric constant and hence the refractive index. In high fields the non-linear second term becomes significant and results in harmonic generation and frequency mixing. For example if two beams are incident on the sample ($E_1\cos\omega_1 t$ and $E_2\cos\omega_2 t$) then the polarisation due to χ_2 will be

$$P = \frac{E_1E_2\chi_2}{2}[\cos(\omega_1 + \omega_2)t + \cos(\omega_1 - \omega_2)t]$$

$$+ \frac{E_1{}^2\chi_2}{2}(1 + \cos2\omega_1 t] + \frac{E_2{}^2\chi_2}{2}[1 + \cos2\omega_2 t] \qquad (4)$$

Thus additional frequencies have appeared at $\omega_1 + \omega_2$, $\omega_1 - \omega_2$, $2\omega_1$ and $2\omega_2$. It is the advent of lasers with their enormous power that enables us to take advantage of this second order effect. (The third order effect is also important but is outside the scope of this discussion, cf. Chapter 3.) There is however a complication; for maximum interaction and generation of, say, the harmonic of ω_1 we require the phase of ω_1 to be the same as the phase of $2\omega_1$. This means, because of momentum conservation,

$$\vec{K}_\omega + \vec{K}_\omega = \vec{K}_{2\omega} \qquad (5)$$

where $\vec{K}$ is the wave vector ($\equiv 2\pi\vec{r}/\lambda$, where $\vec{r}$ is a unit vector in the direction of the wave). For the fundamental and harmonic to be co-linear

$$2K_\omega = K_{2\omega} \qquad (6)$$

which implies that the refractive index of the material is the same for ω and 2ω; this is unlikely, and therefore in general the wave vectors are not co-linear. There is an ingenious way round this problem. In a birefringent crystal, at a particular angle of propagation to the optic axis, the refractive index of the extraordinary ray for the harmonic equals the refractive index of the ordinary ray for the fundamental (see references for further discussion).

Mixing Methods

If a tunable dye and fixed frequency laser are incident on a non-linear crystal (i.e. one with a high value of χ_2) then tunable IR radiation can be generated by frequency difference generation. For instance with dye laser, argon ion laser and $LiNbO_3$ crystal the frequency difference can be tuned over the range 2-4.5 μm. The power is modest (∿μWatts) but the frequency stability is impressive (0.0006 $cm^{-1} \equiv$ 20 MHz). The three frequencies have to be phase-matched and this is achieved by rotating the crystal as the dye laser is tuned.

Optic Parametric Oscillator

The polarisability equation is strictly speaking incomplete since in principle a whole infinity of frequencies can be generated from one input frequency. The sample already has a spontaneous noise profile and the input frequency will mix with the noise according to

$$\omega_i = \omega_p - \omega_s \tag{7}$$

where by convention ω_p is the pump frequency, ω_s is the noise signal frequency and ω_i is the idler frequency. The phase matching condition is

$$\vec{K}_i = \vec{K}_p - \vec{K}_s \tag{8}$$

Thus for each set of crystal conditions specific values of $\vec{K}_i$ and $\vec{K}_s$ will apply for a particular $\vec{K}_p$. This forms the basis of tuning the device since the phase-matching conditions can be altered by rotation of the crystal, by heating or compression. The crystal sits in a resonant cavity which amplifies the ω_i or ω_s or both.

3. APPLICATIONS OF TUNABLE IR LASERS

(a) High resolution spectroscopy

One example, SF_6, has already been given in Figure 4. A further example, taken from our work at Newcastle [4], is shown in Figure 7. The advantage of such methods over conventional methods needs no emphasising. In fact, assuming that the pressure is sufficiently low for pressure broadening to be unimportant, the resolution is more likely to be limited by the Doppler width of the absorption band than by the spectrometer. The Doppler width varies as $\nu(T/M)^{\frac{1}{2}}$ where ν is the frequency, T the temperature and M the molar mass. At 1000 cm^{-1}, 300 K and with M = 40, the linewidth would be 30 MHz $\equiv 10^{-3}cm^{-1}$.

There is a clever technique for cancelling out the Doppler broadening known as Lamb Dip or Saturation spectroscopy. Laser light, powerful enough to saturate the absorption feature is directed through the sample cell and then is reflected back

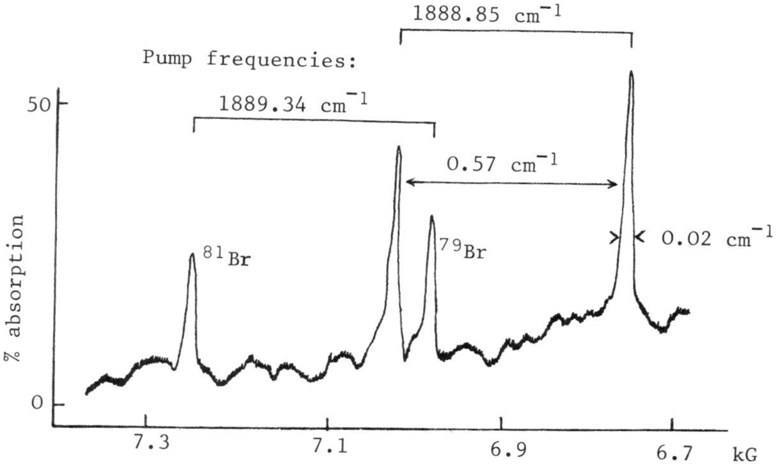

Figure 7. Double beam transmission spectrum (1 metre cells) of DBr (2 Torr) showing multi-lining of 1888 cm^{-1} pump line.

along the same path. If the absorption frequency of a stationary molecule is ν_0, then if the component of molecular velocity in the direction of the laser beam is ν, the actual absorption frequency is $(\nu_0 - v/c)$ in the forward direction and $(\nu_0 + v/c)$ in the return direction. Only those molecules which have zero component will be saturated by the laser in both directions. Thus the centre of the Doppler broadened absorption band shows a narrow dip. This technique has mostly been applied in atomic spectroscopy but there are examples in vibration spectroscopy.

(b) Sensitivity

In principle tunable IR lasers offer enormous gains in signal/noise and hence the possibility of detecting very weak signals. In practice the S/N calculations can be very misleading if no account is taken, for example, of pump laser jitter. The trouble with this method of detection is that one is looking for a small perturbation of the intense light falling on the detector.

The optoacoustic cell works on a different principle. The gas sample cell contains a microphone. The input laser beam is chopped at 100 Hz; if the sample absorbs any radiation this is almost immediately released as thermal energy so that a 100 Hz shock wave hits the microphone. Thus only radiation actually absorbed is effective in generating a signal. In this way extremely sensitive dilution is possible, conceivably as high as 1 part in 10^{10} for NO in dry air using an SFRL system, with obvious applications to pollution monitoring.

(c) Transient spectroscopy

Conventional flash photolysis combined with UV/visible photographic detection has provided an enormous amount of information concerning the structure of transients. Similar experiments in the IR region are restricted by the lack of photographic sensitivity in this region. Pimentel combined flash generation with a modified IR instrument which has a very rapidly rotating grating. After ten years of development work he and his colleagues [5] obtained the IR spectrum of CH_3 with resolution of 0.6 cm^{-1} over the range 450 to 740 cm^{-1} with a scanning speed of 1 cm^{-1} per μs (see Chapter 8).

Johnston [6] developed an alternative approach for a system which can be reproducibly perturbed from equilibrium by chopped photolysis. When combined with very sophisticated phase sensitive detection it was possible to obtain the IR spectra of such important intermediates as HO_2.

These methods are extremely difficult. In principle a pulsed tunable IR laser (e.g. SFRL) has many advantages and we end this chapter by predicting that such development will mark one of the most important applications of tunable IR lasers.

Acknowledgments

I should particularly like to thank Professor D.H. Whiffen and Drs. A. MacNeish, J.K. Burdett and M. Poliakoff for many discussions of relevance to this chapter.

4. REFERENCES

1. J.M. Cook, K.M. Evenson, C.J. Howard and R.F. Curl, J. Chem. Phys., 64 (1976) 1381.
2. M.J. Colles and C.R. Pidgeon, "Tunable Lasers", Reports on Progress in Physics, 38 (1975) 329.
3. J.K. Burdett and M. Poliakoff, "Tunable Lasers", Quart. Rev. Chem. Soc., 3 (1974) 293.
4. P.G. Buckley, A. MacNeish, J.K. Burdett, J.H. Carpenter, M. Poliakoff, J.J. Turner, D.H. Whiffen, unpublished data.
5. L.Y. Tan, A.M. Wincer and G.C. Pimentel, J. Chem. Phys., 57 (1972) 4078.
6. T.T. Paukert and H.S. Johnston, J. Chem. Phys., 56 (1972) 2824.

Other useful references:

7. D.H. Whiffen, Lasers in Infrared Spectroscopy, Proc. Inst. of Petroleum Molecular Spectroscopy Conference, Durham, 1976.
8. R.G. Brewer and A. Mooradian (Editors), Laser Spectroscopy, Plenum, 1974.

CHAPTER 6

FOURIER TRANSFORM SPECTROSCOPY

A.J. Barnes

1. INTRODUCTION

The basic purpose of a spectrometer is to analyse incoming radiation to give a plot of the power falling on the detector as a function of wavenumber, i.e. the spectrum. The radiation is an electromagnetic field fluctuating as a function of time; Fourier has shown that time-dependent fluctuations can be broken down into a set of cosine and sine waves of different frequencies. Thus the spectrum is the Fourier transform of the fluctuating electromagnetic field.

Spectroscopy is performed by dividing the incoming radiation into a number of beams, subjecting each beam to a different time delay, and recombining the beams so that interference occurs. A prism gives an infinite number of beams whose recombination leads to destructive interference in every direction except one, for a particular wavelength. Thus the spectrum is obtained directly from the spatial separation of the outgoing beams of different wavelengths. A diffraction grating works in a similar manner except that now the number of beams is finite (the number of grooves on the grating) and more than one output maximum is obtained for each wavelength. These are the different orders of the grating and must be separated if the spectrum of the radiation is to be obtained. Using a diffraction grating the resolution achieved by the spectrometer is determined primarily by the width of the slit used to define the band of wavelengths which falls on the detector.

2. THEORY OF INTERFEROMETRY

Interference between beams of radiation is possible only if there are at least two beams, thus the simplest possible way of performing spectroscopy is to use a two beam interferometer such as the Michelson interferometer (Figure 1). The intensity falling on the detector is a function of the optical path difference δ between the two beams [1].

$$I(\delta) = \int_0^\infty B(\bar{\nu})[\tfrac{1}{2} + \cos(2\pi\bar{\nu}\delta)]d\bar{\nu} \tag{1}$$

where $B(\bar{\nu})$ is the spectral power density at wavenumber ν. Subtracting the first term in the integral, which represents the total power, gives

$$I'(\delta) = I(\delta) - \tfrac{1}{2}\int_0^\infty B(\bar{\nu})d\bar{\nu} = \int_0^\infty B(\bar{\nu})\cos(2\pi\bar{\nu}\delta)d\bar{\nu} \tag{2}$$

A plot of $I(\delta)$, or $I'(\delta)$, against δ is known as an

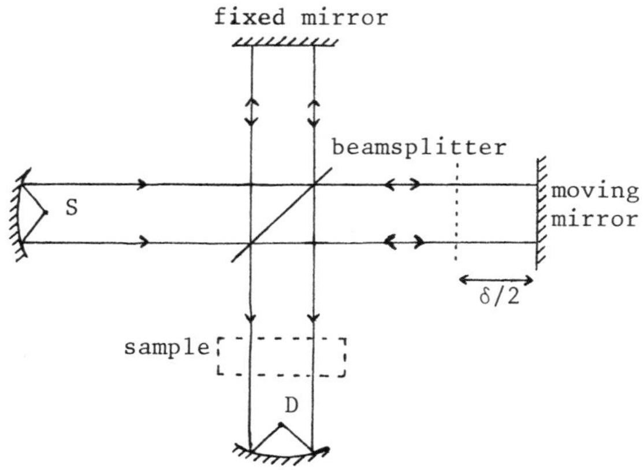

Figure 1. Michelson interferometer (S = source, D = detector).

interferogram. Interferograms for monochromatic radiation and for white light are shown in Figure 2. That for monochromatic radiation is a simple cosine wave whereas that for white light shows a central grand maximum because all wavelengths reinforce at zero path difference. The interferogram contains all the spectral information, but in a form which is not directly intelligible: to obtain the spectrum it is necessary to calculate the Fourier transform of Equation (2)

$$B(\bar{\nu}) \propto \int_{-\infty}^{+\infty} I'(\delta)\cos(2\pi\bar{\nu}\delta)\,d\delta \tag{3}$$

In practice, it is of course only possible to record the interferogram over a restricted range of path differences. If δ_{max} is the greatest path difference employed, the transform becomes

$$B(\bar{\nu}) \propto \int_{-\delta_{max}}^{+\delta_{max}} I'(\delta)\cos(2\pi\bar{\nu}\delta)\,d\delta \tag{4}$$

The effect of this truncation of the interferogram is shown in Figure 3 for monochromatic radiation. Suppression of the subsidiary maxima introduced by the finite integral is achieved by a mathematical procedure known as apodisation. The subsidiary maxima result from the sudden cut-off at δ_{max}, thus a function $f(\delta)$ is introduced, for example

$$f(\delta) = \cos(\pi\delta/2\delta_{max}) \tag{5}$$

which reduces the ordinate smoothly to zero after δ_{max}. Apodisation unfortunately has the secondary effect of broadening the band, giving lower resolution in the apodised than in the unapodised spectrum. The optimum resolution available with an unapodised spectrum is given by $\Delta\bar{\nu} = 0.7/D$, where D is the

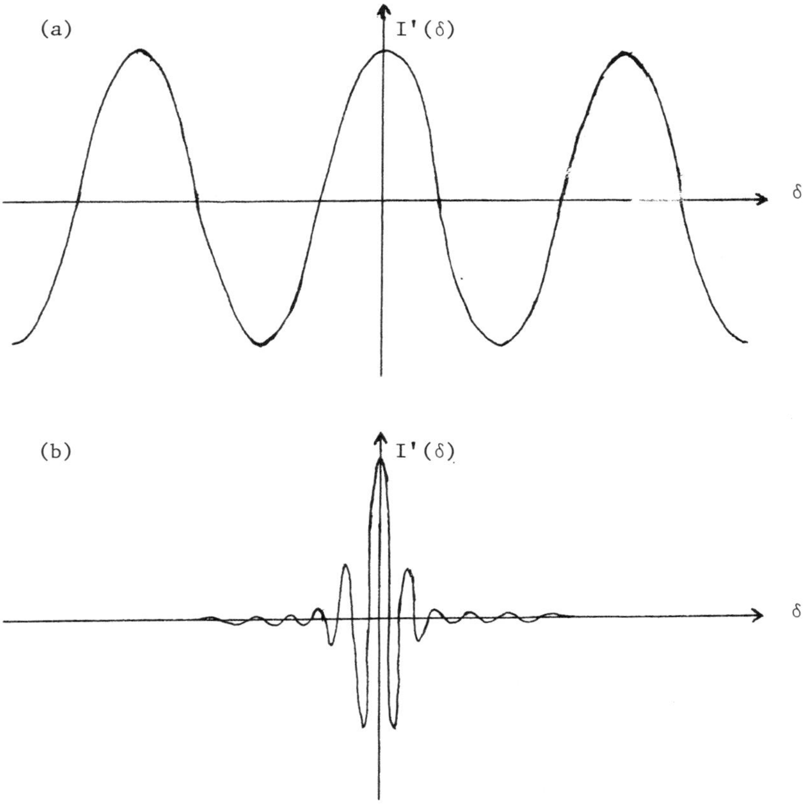

Figure 2. Interferograms for (a) monochromatic radiation and (b) white light.

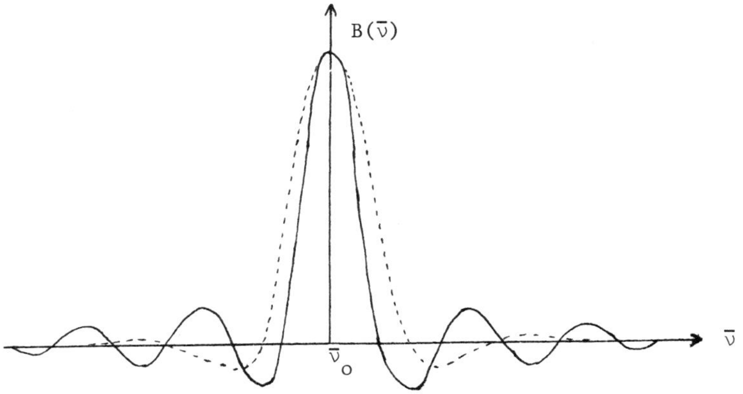

Figure 3. Apodised (...) and unapodised (—) Fourier transforms of a truncated interferogram for monochromatic radiation.

maximum optical path difference introduced, reducing to $\Delta\bar{\nu} \sim 1/D$ for an apodised spectrum [2].

A practical spectrometer has to record data at finite sampling intervals. This may be conveniently achieved by stepping the moving mirror through successive fixed distances, giving an optical path increment $\Delta\delta$. The integral in Equation (4) is then replaced by a summation

$$B(\bar{\nu}) \propto \Delta\delta \sum_{-n}^{+n} I'(i\Delta\delta)\cos(2\pi\nu i\Delta\delta) \tag{6}$$

or

$$B(\bar{\nu}) \propto 2\Delta\delta \sum_{0}^{n} I'(i\Delta\delta)\cos(2\pi\nu i\Delta\delta) \tag{7}$$

if data is collected on only one side of zero path difference (single-sided operation). The effect of the finite sampling interval is to reduce the range of wavenumbers for which meaningful spectral information is obtained to $\bar{\nu} = 0 \rightarrow (2\Delta\delta)^{-1}$ and spurious spectral information exists above the $(2\Delta\delta)^{-1}$ limit. This phenomenon is known as "aliasing" and determines the maximum stepping interval that can be used in a particular wavenumber range.

An alternative to stepping the moving mirror is to scan it rapidly at constant velocity v [3]. Equation (2) now becomes

$$I'(t) = \int_0^{\infty} B(\bar{\nu})\cos(4\pi\bar{\nu}vt)\,d\bar{\nu} \tag{8}$$

since the optical path difference is given by $\delta = 2vt$. The total time for a given scan is determined by the maximum optical path difference introduced (D). The Fourier transform for single-sided operation is then

$$B(\bar{\nu}) \propto 2 \int_0^{D/2v} I'(t)\cos(4\pi\bar{\nu}vt)\,dt \tag{9}$$

Weak spectral signals in rapid scan Fourier transform spectroscopy require that the interferogram be recorded repeatedly and the results (either the interferograms or the computed spectra) averaged to achieve the required signal-to-noise ratio.

3. INTERFEROMETERS VS. DISPERSIVE SPECTROMETERS

An interferometer, or Fourier transform spectrometer, gives certain inherent advantages over a conventional dispersive spectrometer:

<u>Multiplex advantage (Fellgett's advantage [4])</u>

An interferometer provides information about the entire spectral range during the entire period of the measurement, whereas a dispersive spectrometer provides information only about the narrow wavenumber region which falls within the exit slit of the monochromator at any given time. Assuming that the noise is independent of the signal, the improvement in signal-to-noise ratio arising from the multiplex advantage is given by $(N)^{\frac{1}{2}}$, where N is the number of spectral elements scanned. For a 4000 to 400 cm^{-1} scan at a resolution of 1 cm^{-1}, the theoretical improvement in signal-to-noise would be 60.

Throughput advantage (Jacquinot's advantage [5])

The interferometer can operate with a large circular aperture, and using large solid angles at the source and at the detector, whereas a dispersive spectrometer requires long, narrow slits to achieve adequate resolution. Quantitatively, up to 200 times more power can be put through an interferometer than through a good grating spectrometer.

Additional advantages

Several additional advantages follow from the multiplex and throughput advantages, for example:

(a) large resolving power. Since the resolving power depends principally on the maximum optical path difference introduced, high resolving power can be achieved by using large mirror movements (up to 2 m has been used in the interferometers constructed by the Connes's [6]). Also, unlike dispersive spectrometers, the wavenumber resolution is constant over the spectral range scanned.
(b) high wavenumber accuracy. The wavenumber accuracy is determined by the precision with which the position of the moving mirror can be measured.
(c) fast scan time and large wavenumber range are possible.
(d) reduced stray light problems.

The need to have a computer available to carry out the Fourier transform gives all the advantages of computer processing of spectra, discussed in Chapter 10 in relation to grating spectrometers. Thus multi-scanning can be used to improve signal-to-noise ratio and linear absorbance scale expansion and smoothing carried out. Storage of spectra enables difference spectra to be calculated. A disadvantage of Fourier transform spectroscopy, apart from the necessity of computation to obtain the spectrum, is that it is normally a single beam technique - thus comparison of sample and reference has always to be performed by computer subtraction.

4. THE DEVELOPMENT OF FOURIER TRANSFORM SPECTROMETERS

The Michelson interferometer dates back to the nineteenth century, but at that time the technolgoy to develop Fourier transform spectroscopy did not exist. It was not until the early 1950's that Fourier transform spectroscopy became an experimental reality with the work of J. Strong, H.A. Gebbie and G.A. Vanasse [7]. Later Gebbie's group at the National Physical Laboratory developed Fourier transform spectrometers, based on the Michelson interferometer [8], which started to become available commercially from about 1962. These covered the far infrared region of the spectrum, using polyethylene terephthalate (Melinex or Mylar) beam splitters. An alternative to the Michelson interferometer in the far infrared is the lamellar grating interferometer [7,9] (Figure 4). This type of

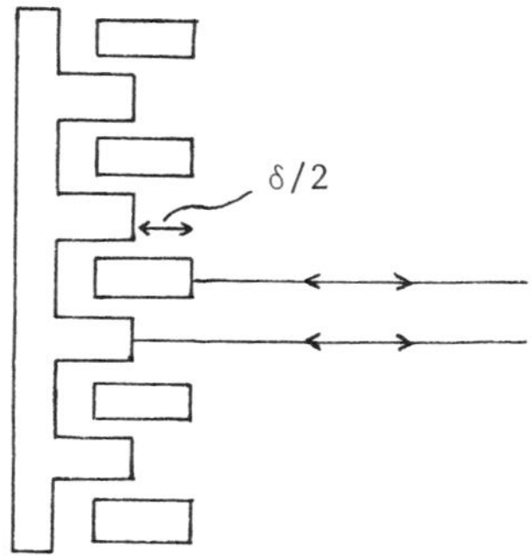

Figure 4. Lamellar grating interferometer.

interferometer has practical advantages over Michelson interferometers in the region below about 100 cm^{-1} since it is easy to achieve nearly perfectly reflecting mirror surfaces but almost impossible to make perfect beam splitters.

The requirements on accuracy of mirror movement are less stringent in the longer wavelength far infrared region than in the mid-infrared, also the performance of grating spectrometers in the source energy limited far infrared is relatively poor. It is consequently not surprising that Fourier transform spectrometers covering the far infrared were the first to appear even though neither the multiplex nor the throughput advantage is as great in this region as in the mid-infrared. Developments in technology and particularly in data processing, with the use of minicomputers both to carry out the transform of the interferogram (either using the fast Fourier transform, based on the Cooley-Tukey algorithm [10], or by real-time analysis) and to allow the interferogram to be recorded using the rapid scanning technique, led to the appearance of commercial mid- and near-infrared Fourier transform spectrometers. These use thin films of iron oxide, silicon or germanium deposited on a supporting substrate, such as calcium fluoride or potassium bromide, as beam splitters. The moving mirror is monitored by a reference interferometer, using the central grand maximum of the interferogram from a white light source as a marker combined with fringe counting of the interferogram from a monochromatic source such as a He/Ne laser (cf. Figure 2). This allows the position of the moving mirror to be located to better than a quarter of a wavelength of the monitoring signal, i.e. 0.16 μm for the 633 nm He/Ne laser.

5. APPLICATIONS

As pointed out by Jacquinot [11], although interferometers are always superior to grating spectrometers, the number of problems which can be investigated only by interferometric methods is relatively small. Most problems can be investigated by either method and the technique used will depend on the instrument available at the time. The applications discussed here are selected as examples where interferometric methods offer a significant advantage, and are not intended to be a comprehensive survey. More detail on instrumentation and applications can be found in the recent book by Griffiths [12].

GC-IR

Perhaps the most spectacular application of rapid scanning mid-infrared Fourier transform spectrometers is to "on the fly" identification of fractions separated by a gas chromatograph. Two basic experimental arrangements are used, one in which the sample is trapped in a light-pipe for the length of time required for the measurement of the spectrum, the other in which the carrier gas is constantly flowing through the light-pipe. The optimum dimensions of the light-pipe are strongly dependent on which arrangement is used. If the trapping technique is used, it is desirable that the whole fraction should be present in the light-pipe and this can be achieved by sample concentration techniques, whereas for flow through measurement the volume of the cell needs to be matched to the volume of the GC fraction.

In a typical system, the effluent gases from the chromatograph are passed into a heated light-pipe. Non-destructive GC detectors, such as the thermal conductivity type, are preferable otherwise the gas flow has to be split so that not all the sample passes through the light-pipe. The signal from the GC detector is monitored. When it rises above a certain threshold level, interferograms are recorded and signal-averaged until it falls below the threshold (a time delay is introduced to take account of the time taken for the gas to flow from the GC detector to the light-pipe). The signal-averaged interferogram is stored in the computer's memory and the spectrometer waits for the next peak in the chromatogram. A background spectrum of the empty light-pipe is recorded and then the interferogram of each GC fraction successively recalled, transformed and ratioed against the background. Using this system, spectra from as little as 1 μg of sample have been measured without trapping [12].

Emission spectroscopy

Emission spectra can only be measured when there is a difference in temperature between the source and the detector. Normally the source is held at a higher temperature than the detector, but equally measurements could be made with the source cooler than the detector. Fourier transform spectroscopy is sufficiently sensitive that qualitative data can easily be obtained even from sources close to ambient temepratute. The data have to be corrected for any background emission and also for the relative spectral response of the spectrometer at each wavenumber. The latter can be obtained by comparison of the experimental spectrum of a black body at a known temperature with the calculated black body emission. One problem with measuring emission spectra of heated samples is that of temperature gradients in the sample giving anomalous features in the spectra of bulk samples. If the sample is heated from below, the lower surface is at a higher temperature than the upper surface and radiation emitted from below can be absorbed before it reaches the upper surface. Thus a band which appears in emission for a thin film can appear as an absorption on a general black body background for a bulk sample [13].

An obvious application of infrared emission spectroscopy is in the remote identification and quantitative determination of gases, for example pollutants in stack gases or species present in the upper atmosphere. Infrared emission spectroscopy has also been applied in the laboratory to, for example, surface studies and microsampling, and also to the study of infrared chemiluminescence.

Transient species

The use of rapid scanning interferometers has enabled absorption spectra of species that are present in the infrared beam for only a few seconds, e.g. in "on the fly" analysis of GC fractions, to be easily measured. Provided that the scan time of the interferometer is short compared with the lifetime of the species, spectra of transient intermediates can be obtained similarly in a flow-through cell. Mantz [14] has recently described a technique by which detailed spectral and kinetic information about very short-lived species could be obtained using simultaneous spectral and temporal multiplexing. Electronic timing circuitry and specialised software enabled the interferogram representing a specific time in the evolution of the system kinetics to be decoded. Spectra of 2 μs "windows", 400 μs apart, were reported starting at an arbitrarily chosen time (40 μs) after the peak ultraviolet intensity of a lamp operating at 250 flashes per second. Acetone was used as the starting meterial. Ketene could be identified in the spectra from a band with an origin at 3138 cm^{-1} and other features between 2800 and 2400 cm^{-1} appeared in emission at 40 μs, in absorption by 1.24 ms and again in emission after 3.64 ms. Conventional Fourier transform spectra showed only carbon monoxide in the spectra of the decomposition products.

Low transmittance samples

An area where Fourier transform spectroscopy gives considerable advantages over conventional infrared spectroscopy is that of samples which absorb or scatter the radiation from the source to such an extent that only a small proportion reaches the detector. Examples of such applications are solutions in strongly absorbing solvents such as water, surface studies and microsampling.

Determination of optical constants

The position of the sample for the determination of its absorption spectrum with a Michelson interferometer is shown in Figure 1. If the sample is instead positioned in one arm of the interferometer, as shown in Figure 5, the symmetry of the arrangement is lost. The grand maximum of the interferogram is displaced from zero path difference because of the dispersion of the sample, the displacement being given by

$$\Delta\delta = 2(\bar{n} - 1)t \qquad (10)$$

where $\bar{n}$ is the mean refractive index of the sample and t is its

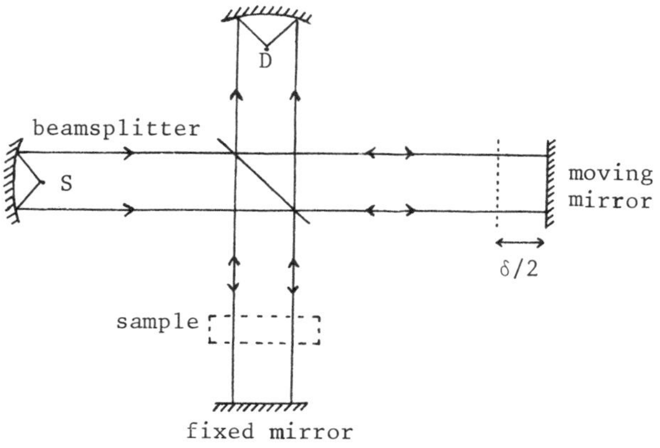

Figure 5. Position of the sample in a Michelson interferometer for the determination of refractive index spectra.

thickness. Additionally, if the sample exhibits anomalous dispersion, characteristic features appear in the interferogram only on the positive path difference side of zero displacement (Figure 6). The complex Fourier transform of the interferogram gives [15] the wavenumber variation of the complex refractive index $\hat{n}(\bar{\nu})$, defined as

$$\hat{n}(\bar{\nu}) = n(\bar{\nu}) - i\alpha(\bar{\nu})/4\pi\bar{\nu} \qquad (11)$$

where $n(\bar{\nu})$ is the real part of the refractive index and $\alpha(\bar{\nu})$ is the absorption coefficient. The evaluation of the optical constants of a material requires the absolute refractive index to be determined, but only the variation in the real part of the refractive index (i.e. $n(\bar{\nu}) - \bar{n}$) is required to determine integrated absorption intensities of spectral bands (see Chapter 13).

6. REFERENCES

1. For a fuller account see, for example, R.J. Bell, Introductory Fourier Transform Spectroscopy, Academic Press, New York, 1972; G.W. Chantry, Submillimetre Spectroscopy, Academic Press, London, 1971.
2. G.W. Chantry and J.W. Fleming, Infrared Physics, 16 (1976) 655.
3. L. Mertz, J. Phys. Coll. C2, Suppl. 3-4, 28 (1967) 88.
4. P. Fellgett, J. Phys. Radium, 19 (1958) 187.
5. P. Jacquinot and J.C. Dufour, J. Rech. C.N.R.S., 6 (1948) 91.
6. J. Connes, H. Delouis, P. Connes, G. Guelachvili, J.-P. Maillard and G. Michel, Nouv. Rev. Opt. Appl., 1 (1970) 3.

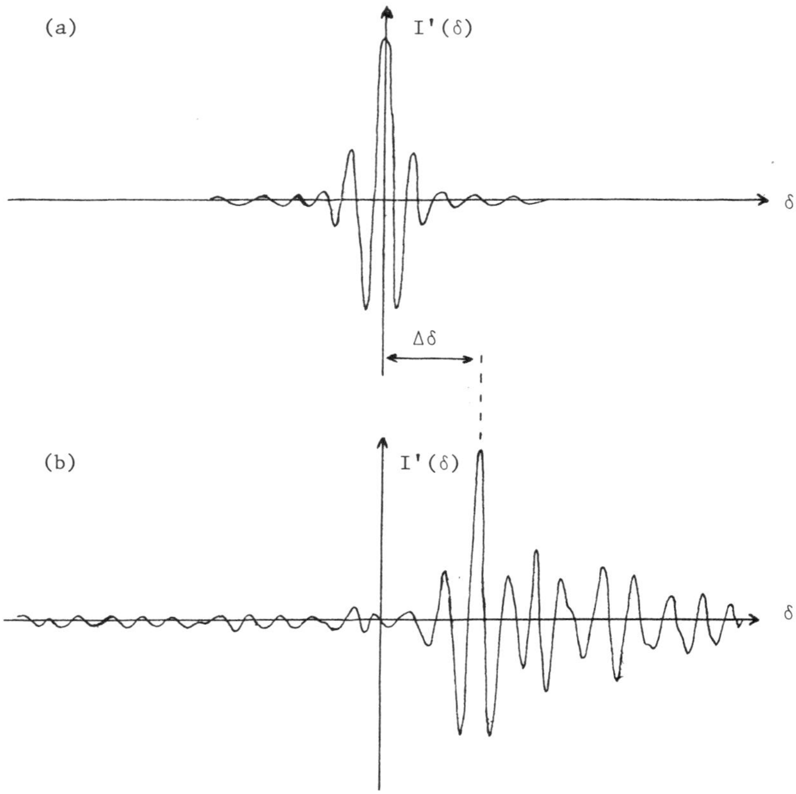

Figure 6. (a) Symmetrical background interferogram; (b) unsymmetrical interferogram due to a dispersive sample in one beam of the interferometer.

7. H.A. Gebbie and G.A. Vanasse, Nature, 178 (1956) 432; J. Strong, J. Opt. Soc. Amer., 47 (1957) 354; J. Strong and G.A. Vanasse, J. Opt. Soc. Amer., 49 (1959) 844.
8. H.A. Gebbie, in J.R. Singer (Editor), Advances in Quantum Electronics, Columbia University Press, New York, 1961, p. 155.
9. L. Genzel and R. Weber, Z. angew. Phys., 10 (1958) 127, 195.
10. J.W. Cooley and J.W. Tukey, Math. Comput., 19 (1965) 297.
11. P. Jacquinot, Appl. Optics, 8 (1969) 497.
12. P.R. Griffiths, Chemical Infrared Fourier Transform Spectroscopy, Wiley-Interscience, New York, 1975.
13. P.R. Griffiths, Appl. Spectr., 26 (1972) 73.
14. A.W. Mantz, Appl. Spectr., 30 (1976) 459.
15. J. Chamberlain, J.E. Gibbs and H.A. Gebbie, Infrared Phys., 9 (1969) 185.

CHAPTER 7

MATRIX ISOLATION

A.J. Barnes and H.E. Hallam

1. INTRODUCTION

Matrix isolation is a technique for trapping isolated molecules of the species to be studied in a large excess of an inert material by rapid condensation at a low temperature, so that the diluent forms a rigid cage or matrix. At a sufficiently low temperature, diffusion of the solute species is prevented and thus reactive species may be stabilised for leisurely spectroscopic examination. The noble gases (primarily argon) and nitrogen are the most widely employed matrix materials.

Infrared spectroscopy of matrix-isolated substances was first reported by Whittle, Dows and Pimentel [1] in 1954, and the subsequent development of the technique owes much to Pimentel and his co-workers. Despite an early study of methane in a krypton matrix [2] (using a mercury arc source), matrix isolation Raman spectroscopy had to await the advent of readily available gas laser sources. Rapid development has followed the pioneering paper by Shirk and Claassen [3]. The matrix isolation technique is not of course confined to vibrational spectroscopy, and has been successfully applied in conjunction with visible/ultraviolet, electron spin resonance, Mössbauer and magnetic circular dichroism spectroscopies.

This chapter will give a broad survey of the techniques and applications of infrared and Raman matrix isolation spectroscopy; the subject is covered in greater depth by a recent book [4].

2. EXPERIMENTAL TECHNIQUES

Cryostat

Matrix isolation spectroscopy generally requires the sample to be cooled to temperatures in the range 4-20 K. The simplest method of attaining such temperatures is by the use of liquid hydrogen (20 K) or liquid helium (4 K) as the coolants in a double Dewar cryostat, the outer Dewar being filled with liquid nitrogen to minimise heat leakage. This type of cryostat was used for all early matrix isolation studies, and is still widely used where there is a readily available source of liquid refrigerant.

In the 1960's miniature open-cycle cryostats, operating on the Joule-Thomson principle, became commercially available. Hydrogen gas is cooled to liquid nitrogen temperature and then expanded through a nozzle. The Joule-Thomson cooling effect leads to eventual liquefaction of the hydrogen giving a base temperature of 20 K. Addition of a further cooling stage using helium gas gives a base temperature of 4 K. (In either case the base temperature may be lowered slightly by pumping on the liquid.) Although the cooling power of these cryostats is much lower than the double Dewar type, the temperature may be controlled more easily allowing variation over the range 4 to 300 K.

An alternative method of refrigeration is the closed-cycle system, based on some variant of the Stirling Refrigeration Cycle. A two stage refrigerator, using helium gas as the working fluid, can attain a base temperature of ca. 10 K (although the cooling power decreases with decreasing temperature). Again temperature control is readily achieved by balancing the refrigeration with a heater. Despite a somewhat higher capital cost, this type of cryostat has much lower running costs. The commercial availability of miniature closed-cycle refrigerators has led to an "explosion" in the use of matrix isolation spectroscopy over the last few years.

Vacuum shroud

An efficient pumping system is vital for matrix isolation work, not only to minimise heat leakage but also to prevent impurities from influencing the spectra. An oil diffusion pump (mercury is inadvisable for Raman matrix isolation systems since it may give rise to fluorescence problems), backed by a rotary pump and a liquid nitrogen vapour trap are normally used. A typical vacuum shroud is illustrated in Figure 1.

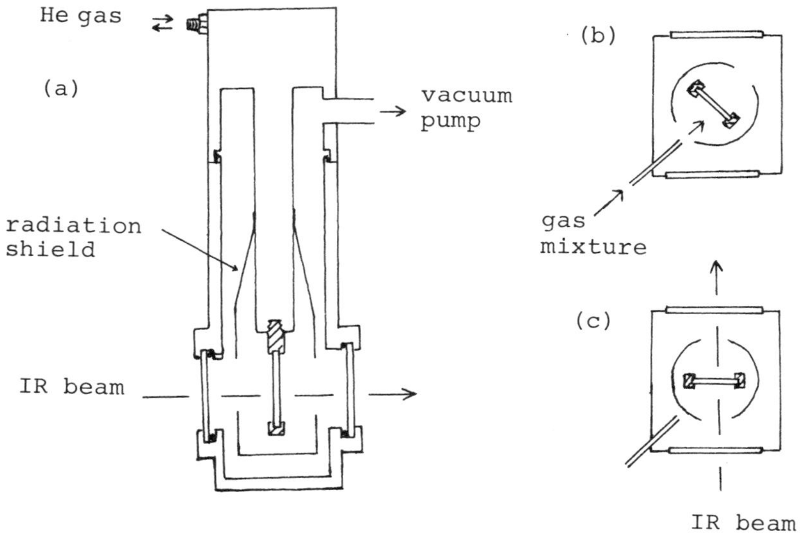

Figure 1. (a) Cryostat and vacuum shroud; (b) arrangement for deposition; (c) arrangement for spectral scanning.

Alkali halide windows are used in the infrared, while in the far infrared region high density polyethylene outer windows and a silicon deposition window (because of the poor thermal conductivity of polyethylene) are required. Raman spectroscopic studies have generally been carried out on a matrix deposited on a highly polished metal block, but it has recently been reported [5] that it is possible to obtain good quality infrared and Raman spectra off the same matrix deposited on an angled alkali halide plate.

Sample deposition

The simplest form of the matrix isolation experiment is to mix the solute and matrix gases in the desired proportions in a vacuum line using standard manometric techniques. The concentration is usually expressed as the matrix:solute ratio (M/S). The gas mixture is then sprayed, through a needle valve to control the flow rate, onto the deposition window (or block) which is turned at right angles to the gas inlet (Figure 1(b)). This deposition technique is known as "slow spray on", flow rates of between 1 and 15 mmol h^{-1} of gas mixture being used. After a sufficient quantity of the gas mixture has been deposited, the needle valve is closed and the cryostat rotated in the vacuum shroud to bring the deposition window in line with the outer windows (Figure 1(c)) for spectral scanning.

An alternative technique is pulsed deposition [6], in which a series of small quantities of the gas mixture is deposited in rapid succession. The heat released by each pulse anneals the matrix, giving samples which scatter light less than those deposited by the slow spray on technique. A comparative study [7] of the two methods showed that the pulsed technique does not lead to poorer isolation of solute molecules. Pulsed deposition on a metal block tends to lead to cracking of the matrix, and a modified technique has been suggested [8] to obtain clear matrices for Raman studies.

Volatile liquids and solids (e.g. iodine) may be handled in the same way as gases, but involatile materials require a rather different technique. The solid solute is heated to an appropriate temperature in some type of furnace to obtain a beam of vapour directed at the deposition window, and simultaneously the matrix gas is sprayed on the window. Organic materials may be evaporated in a simple glass tube wrapped with heating wire, while metals and other inorganic species require Knudsen cell effusion techniques. Moskovits and Ozin [9] have described a method of determining the concentration of solute in such a matrix by weighing the material effusing in the opposite direction to the deposition window (through a similar geometry of radiation shield collimators) on a quartz microbalance.

Unstable molecules and free radicals may be generated by a wide variety of techniques, either in the gas phase immediately prior to deposition or in situ after deposition. Examples of a number of methods of generation are given in Table 1.

Table 1. Methods of generating unstable species.

1. In situ photolysis of a single precursor, e.g. CH_3 from the vacuum u.v. photolysis of CH_4 in argon.
2. Atom-molecule reactions, with the atom produced by in situ photolysis, e.g. HCO from the photolysis of HI + CO in argon or from the photolysis of HI in carbon monoxide (reactive matrix technique).
3. Metal atom-molecule reactions, e.g. CCl_3 from $CHCl_3$ + Li in argon.
4. Gas phase pyrolysis prior to deposition, e.g. CH_3 from CH_3I.
5. Gas phase microwave discharge prior to deposition, e.g. O_3 from O_2 in argon.

3. MATRIX EFFECTS

An understanding of the various effects that the matrix can have on the vibrational spectrum of the solute molecule is vital, since the spectra of many of the species studied by the matrix isolation technique are previously unknown or markedly different from those obtainable in other phases (i.e. monomer instead of polymer). There are many instances in the matrix isolation literature of misinterpretation of spectra due to neglect of the possibilities of matrix effects.

The most obvious matrix effect is that the vibrational levels of the solute molecule will be perturbed by the matrix and thus the vibrational frequencies will be shifted from their gas phase values. Additionally, multiplet band structure may result from rotation of the solute molecule in its trapping site, alternative trapping sites in the matrix, aggregation of the solute, or from lifting of the degeneracy of vibrational transitions. Perturbation by the matrix environment may induce inactive modes (e.g. the hydrogen fundamental vibration is observed in the i.r. spectrum in argon matrices) and similarly inactive matrix lattice modes (e.g. in argon) may be induced by the presence of the solute.

Matrix shift

Just as solutes exhibit solvent shifts in room temperature solution spectra, so vibrational bands of molecules trapped in matrices exhibit matrix shifts. The same factors - electrostatic, inductive, dispersive and specific interactions - contribute to both solvent and matrix shifts, but the rigidity of the matrix cage means that repulsion forces also contribute to matrix shifts. Thus high frequency stretching modes usually shift to lower frequency, while low frequency bending modes often shift to higher frequency particularly in matrices with unsymmetrical trapping sites. The shifts are normally quite small in inert matrices such as argon or nitrogen, but strongly interacting matrices (e.g. HCl in a C_2H_4 matrix) or highly polar solute molecules (e.g. LiF) give rise to large frequency shifts. Figure 2 illustrates the progressive shift of the HCl stretching frequency as the polarity of the matrix increases.

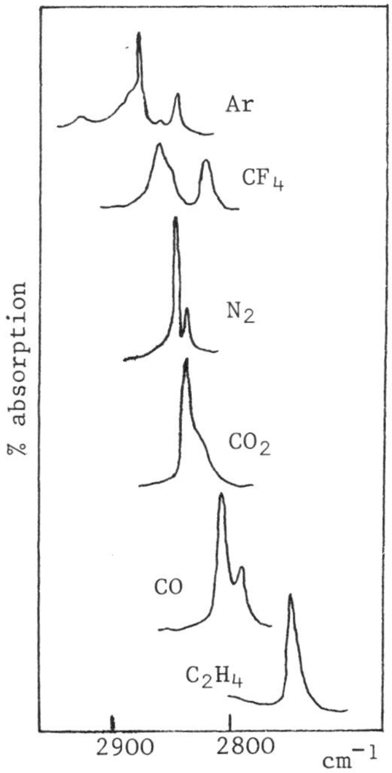

Figure 2. Monomer i.r. absorption region of HCl in various matrices at 20 K (reproduced with permission from A.J. Barnes, H.E. Hallam and G.F. Scrimshaw, Trans. Faraday Soc., 65 (1969) 3159).

Rotation

Molecules trapped in low temperature matrices may show doublet or multiplet band structure for a variety of reasons. The characteristic feature of rotational structure on a vibrational band is that it exhibits reversible temperature dependence, due to changes in the population of rotational energy levels as the temperature is increased. This is illustrated for DCl in an argon matrix in Figure 3.

A number of small molecules have been shown to rotate in noble gas matrices - hydrogen (i.r., Raman) hydrogen halides (i.r., far i.r.), water (i.r., far i.r.), ammonia (i.r., far i.r.) and methane (i.r., Raman). Additionally several small free radical species, e.g. NH_2, also exhibit rotational structure in argon matrices. Restricted rotation has been postulated in a number of other cases, e.g. sulphur dioxide in krypton, ammonia in nitrogen, but these are less well established. In general, rotation seems not to occur in matrices such as nitrogen with unsymmetrical trapping sites.

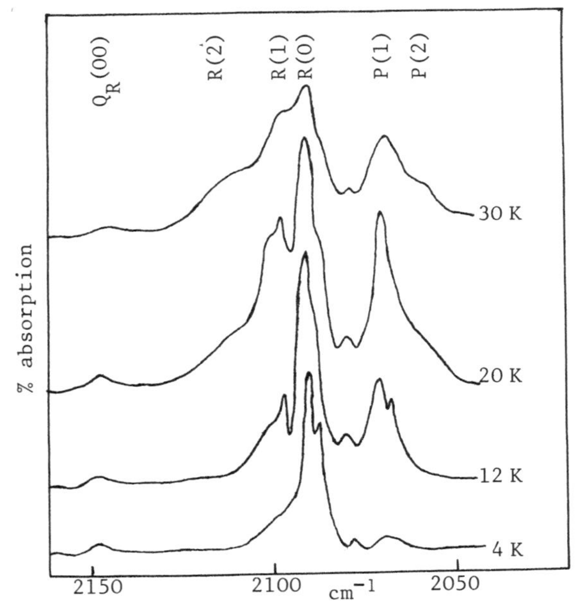

Figure 3. Temperature dependence of rotational-vibrational lines of monomeric DCl in an argon matrix (reproduced with permission from A.J. Barnes, Rev. Anal. Chem., 1 (1972) 193).

Multiple trapping sites

It is generally assumed that solute molecules occupy a substitutional site formed by removal of one or more matrix atoms or molecules. A very small solute such as hydrogen fluoride could conceivably fit into an octahedral interstitial site (which has a diameter <2 Å in the commonly used matrix materials). Since a solid grown by rapid condensation from the vapour is unlikely to be ideally crystalline, dislocation sites and grain boundaries offer possible trapping sites. Nevertheless, the trapping site occupied by most molecules must be well-defined since vibrational bands of matrix-isolated species are normally very sharp.

Where a solute molecule can be trapped in more than one type of site (e.g. in a single or a double substitutional site), each vibrational mode will generally exhibit more than one band since the interaction with the matrix environment will be different for each of the trapping sites. However, mutiple bands can arise from other causes, e.g. rotation or aggregation. The reversible temperature dependence of rotational lines allows these to be readily distinguished, but careful concentration studies are necessary to distinguish multiple trapping sites from aggregation effects. Other possible causes of multiple bands include interaction with impurities, particularly nitrogen or water.

The distinctive feature of doublets due to alternative trapping sites is that the relative intensity of the bands is usually not affected by changing the solute concentration, but may often be changed by annealing the matrix at higher temperatures for a few minutes. This process allows some diffusion of the solute molecules to occur, leading to aggregate formation, and it is found that monomer is lost selectively from one of the trapping sites. This is illustrated for hydrogen chloride in a nitrogen matrix in Figure 4.

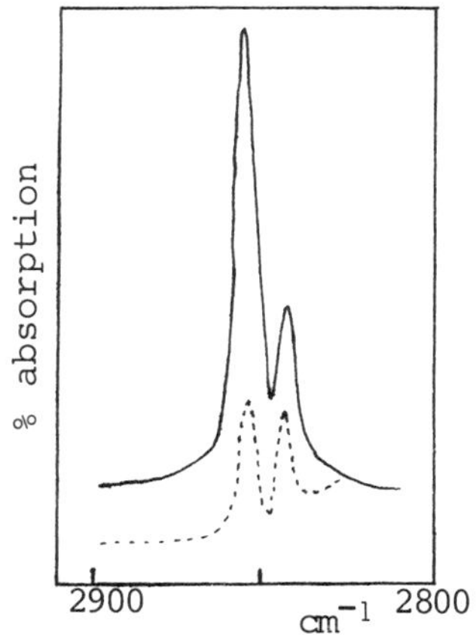

Figure 4. Monomer i.r. absorption region of HCl in a nitrogen matrix at 20 K; the dotted spectrum is after annealing at 35 K for a few minutes.

Aggregation

Even where the interaction between solute molecules is small, a group of two or more solute molecules will give rise to a frequency slightly different from that of the monomer. Solutes which exhibit strong intermolecular interactions (e.g. hydrogen bonding substances) may give very large shifts on aggregation. Unlike multiple trapping site effects, aggregation may be eliminated by reducing the concentration of the solute in the matrix - unless of course appreciable dimerisation occurs in the gas phase, as for carboxylic acids.

The intensification of XH stretching modes in the infrared spectrum on hydrogen bonding means that high dilutions (matrix to solute ratios, M/S, of 1000 or greater) must be used to minimise absorption due to dimeric species. In the Raman spectrum no such intensification occurs and thus the monomer band may be predominant at concentrations as high as M/S = 100. This is fortunate since it is not normally possible to obtain Raman spectra at high M/S because of the intrinsic weakness of the Raman effect. Figure 5 compares the infrared and Raman spectra of the ND_2 stretching region of methylamine-d_2 in argon matrices.

Statistically, a molecule occupying a single substitutional site should be 99% isolated at M/S = 100 if only nearest neighbour interactions are considered. However the M/S ratio required to isolate larger molecules increases rapidly as the

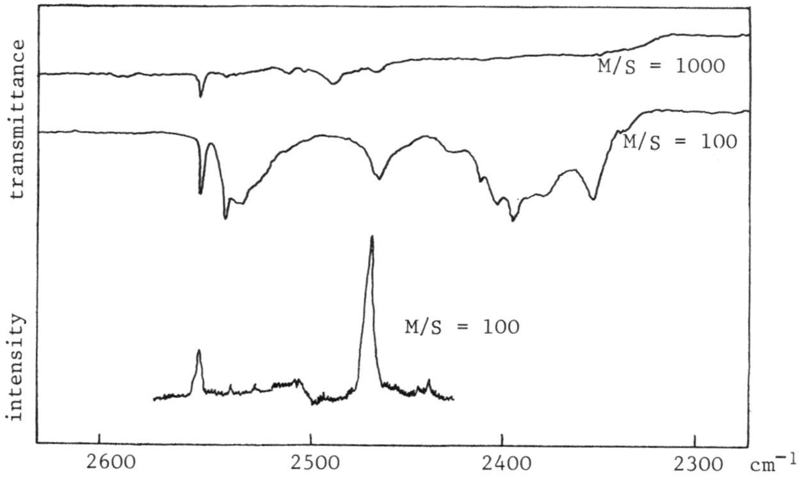

Figure 5. Infrared and Raman spectra of the ND_2 stretching region of CH_3ND_2 in nitrogen matrices (reproduced with permission from ref. 11).

number of matrix atoms or molecules displaced to accommodate the solute increases.

There are innumerable examples in the matrix isolation literature of aggregate bands being incorrectly attributed to the monomeric solute, thus it is important always to check for possible concentration dependence in matrix isolation spectra. A related problem is the possibility that impurities may have perturbed the spectrum - both nitrogen and water interact strongly with many solutes and great care should be taken to check for the possibility of leaks in the vacuum system.

4. APPLICATIONS

The matrix isolation technique offers a number of advantages over more conventional spectroscopic techniques. The most obvious is the ability to stabilise free radicals and unstable molecules, and the applications of matrix isolation in this field are described in Chapter 8. The use of the technique to study vaporising molecules, another type of reactive species, is discussed in Chapter 9. In this chapter we shall confine our attention to stable solute molecules, where the principal advantage lies in the small bandwidths obtained - due partly to the use of low temperatures and partly to the reduction in intermolecular interactions compared with other condensed phases (rotational structure, which broadens vapour phase bands, is absent from matrix spectra except for a few small molecules). Thus near-degenerate bands, which overlap completely even in the vapour phase or in dilute solution at room temperature, may be resolved. This leads to important applications in vibrational analysis and conformational isomerisn. A further application, in which increasing interest has been shown recently, is in the study of intermolecular interactions.

By deliberately increasing the concentration of solute, different aggregated species may be trapped for spectroscopic study, while mixed solute studies enable molecular complexes to be examined.

Vibrational analysis

The potential of the matrix isolation technique in vibrational assignment problems is best illustrated by a specific example. The location of the NH_2 twisting mode of methylamine has been the subject of controversy, frequencies ranging from 1455 cm^{-1} to 995 cm^{-1} having been suggested. Wolff and Ludwig [10] observed a band in the gas phase spectrum of a mixture of CH_3NH_2 and CH_3ND_2 which they assigned to the NHD twisting mode of CH_3NHD. A complete assignment of the CH_3NHD spectrum in the gas phase would not be possible because of overlapping bands of CH_3NH_2 and CH_3ND_2, but it could be obtained from matrix spectra [11]. Figure 6 shows the amino wagging modes of the three isotopic species of methylamine and the NHD twist of CH_3NHD in a nitrogen matrix. A normal coordinate analysis carried out using the argon matrix frequencies for the three isotopic species enabled the NH_2 twisting mode of CH_3NH_2 to be positioned at ca. 1030 cm^{-1} (although no trace of any absorption could be observed in this region).

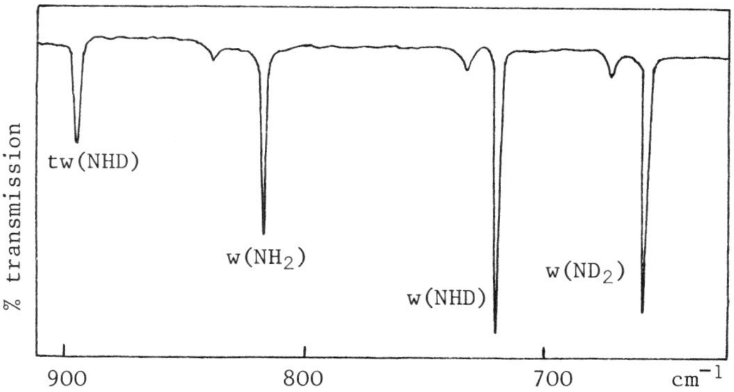

Figure 6. NH_2 wagging, NHD twisting and wagging, and ND_2 wagging modes from the i.r. spectrum of a mixture of CH_3NH_2 and CH_3ND_2 in a nitrogen matrix (M/S 1000) (reproduced with permission from ref. 11).

The matrix isolation technique is of obvious value in obtaining vibrational assignments for monomer species of unstable species or hydrogen bonding substances. It can also be of great assistance in the assignment of stable substances which do not exhibit strong intermolecular interactions. An early study of this type was of carbonyl chloride, where the 575 cm^{-1} gas phase band was found to be split into three components (two fundamentals and an overtone) in an argon matrix [12].

Conformational isomerism

Conformational isomers can often be identified from small splittings of vibrational bands in matrix spectra, whereas in spectra from other phases the bandwidths may be greater than the separation of the bands due to the two conformers. A classical illustration is ethanol (Figure 7), where a complete

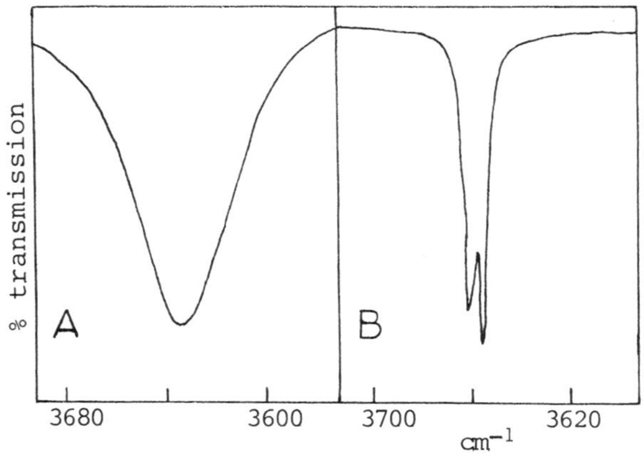

Figure 7. Monomer OH stretching absorption of C_2H_5OH in: (A) carbon tetrachloride solution; (B) argon matrix (reproduced with permission from A.J. Barnes, Rev. Anal. Chem., 1 (1972) 193).

vibrational assignment can be made for the trans (I) and gauche (II) conformers from the matrix spectra [13].

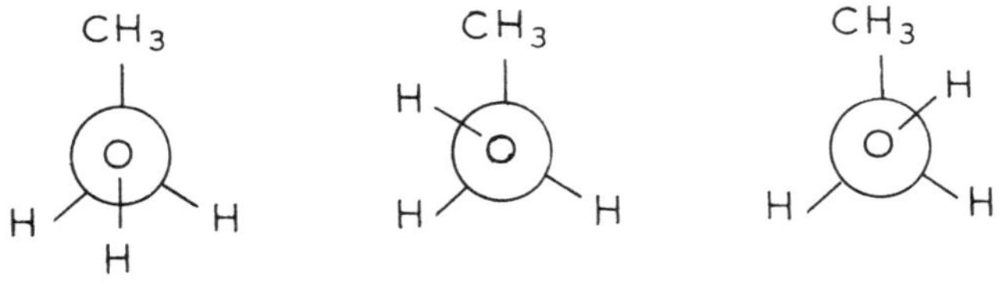

I II

It is usually assumed that the conformational equilibrium existing in the gas phase prior to deposition is trapped out. Huber-Walchli and Günthard [14] have obtained the energy difference between the trans and gauche conformers of 1,2-difluoroethane by trapping out the equilibria in the gas phase at two different temperatures. Similarly unstable conformers can be generated by photolysis (e.g. cis-tert-butyl nitrite from the more stable trans conformer [15]) in low-temperature matrices, a process which relies on a non-equilibrium distribution being trapped out. Other systems, however, are clearly in equilibrium at the temperature of the matrix. 1,1,1,3,3,3-Hexafluoropropan-2-ol [16] exhibits two conformers (C_1 and C_s) in the vapour phase, nitrogen or carbon monoxide matrices, but only the C_s

conformer in an argon matrix. Allyl alcohol and allylamine [17] behave similarly, and for these compounds annealing the nitrogen matrix at ca. 35 K causes the less stable conformer to disappear.

It has been suggested [18] that these variations may be at least partially accounted for in terms of the temperature variation of the rate of interconversion of isomers. Applying the Arrhenius equation, assuming an A factor of 10^{12}, it is found that at a temperature of 20 K (the temperature most commonly used for matrix isolation) a molecule with a barrier to rotation in the matrix of less than 6 kJ mol^{-1} will be in equilibrium at the temperature of the matrix. A slightly higher barrier would inhibit interconversion at 20 K, but allow it to occur at 30-35 K. Recent results show that, at least in one example, there is an appreciable increase in the barrier to internal rotation on trapping a molecule in an inert solid matrix (about 25% for the torsion of methylamine [11]) and the barrier may well vary between matrices.

From the above, it would be expected that molecules with a low barrier to rotation, such as the aliphatic alcohols (the barrier to rotation of the hydroxyl group is ca. 5 kJ mol^{-1}), should be in equilibrium at the temperature of the matrix. Results obtained [18] for several small aliphatic alcohols suggest that this is so, and give estimates of the energy difference between the trans and gauche conformers, e.g. 30 J mol^{-1} for propan-2-ol. Comparison gas phase data is available only for propan-2-ol, whose energy difference is given [19] as 100 J mol^{-1}.

Some years ago, nitrous acid was found to undergo infrared induced cis-trans isomerisation in the beam of the spectrometer [20] (see also Chapter 8). A tunable infrared laser could be used [21] in an attempt to perturb the distribution of conformers, thus making it possible to distinguish between equilibrium and non-equilibrium situations in the matrix as well as allowing the production of relatively high concentrations of less stable conformers (when the system is not in equilibrium at the temperature of the matrix). This appears to be yet another area where the tunable infrared laser could give exciting new results.

Self-association of hydrogen bonding substances

The matrix isolation technique is ideally suited to the study of hydrogen bonding. It is difficult to follow the self-association of hydrogen bonding substances in the vapour phase because of overlapping by the rotational structure of the monomer band and because of the low concentration of the dimer species. In condensed phases at room temperature, the large bandwidths of hydrogen-bonded modes make it virtually impossible to identify multimers of different size. In a low temperature matrix, on the other hand, the small bandwidths allow bands due to different multimer species to be resolved and identified from the dependence of their intensities on concentration. The structures of multimer species may be elucidated from (a) the presence or absence of bands due to the terminal X-H group of an open chain multimer and (b) the effects of isotopic substitution (particularly partial substitution).

A recent study [22] of the self-association of ammonia illustrates these points. The structure of the dimer (Figure 8)

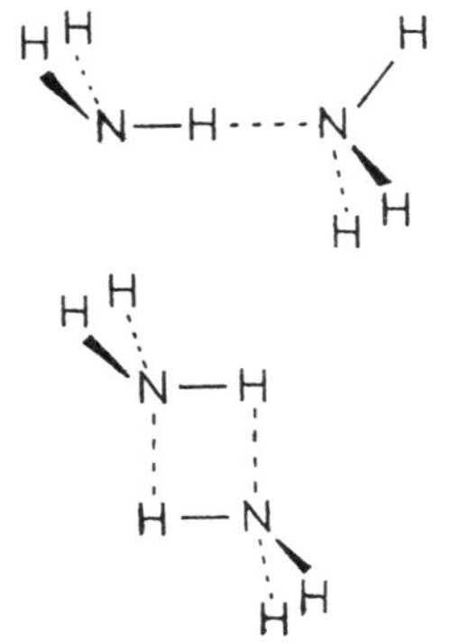

Figure 8. Structures of open chain and cyclic dimers of ammonia.

has been the subject of controversy, and previous matrix studies have not been entirely conclusive. Raman spectra in the region of the symmetric stretching mode in both argon and nitrogen matrices exhibit a side band which may be attributed to the free NH_3 group of an open chain dimer. The symmetric bending mode in a nitrogen matrix has a high frequency shoulder which may be assigned similarly. These indications that the structure of the dimer is open chain are supported by studies of a $^{14}NH_3/^{15}NH_3$ mixture, which gave doublets for all the dimer bands (a cyclic structure should give triplets).

As well as dimer bands, absorptions due to trimers and higher multimer species may be identified. With sufficient data to enable a force field to be calculated, changes in the force constants from dimer to trimer may be obtained and correlated with the effects on the bonding in the species of adding a second hydrogen bond. Such detailed studies of small multimers may lead to an understanding of the enormous increases in bandwidth and intensity on forming large hydrogen-bonded multimers.

Molecular complexes

It has long been appreciated that nitrogen can form molecular complexes with many solutes in low-temperature matrices, and care must be taken to check that spectral features are not due to such complexes with nitrogen (or water) from traces of air in the vacuum shroud. The HCl-N_2 complex is well known, and a detailed study [23] of mixtures of HCl and nitrogen in argon matrices showed that there were in fact three features in the spectra due to mixed multimers of HCl and nitrogen:
(1) the HCl-N_2 mixed dimer, (2) a mixed trimer of the form HCl-N_2-HCl and (3) a mixed trimer of the form HCl-HCl-N_2.

Similar complexes are formed between any of the hydrogen halides and nitrogen, carbon monoxide or carbon dioxide.

Matrix isolation is a convenient technique for studying molecular complexes free from interaction with other solute molecules. When the interaction is weak, the solutes may be pre-mixed in the gas phase prior to deposition but for strongly interacting solutes (e.g. ammonia and hydrogen chloride) the simultaneous deposition technique must be used. The proportion of AB pairs may be optimised relative to other species (such as AA or BB dimers) by using suitable concentrations of the two solutes, A and B. In this way the interaction of the hydrogen halides with a wide range of solute molecules [24,25] has been studied. A particularly interesting complex of hydrogen chloride is that with ammonia, first reported by Ault and Pimentel [26]. This strongly hydrogen-bonded complex exhibits [27] startling differences in the appearance of the spectra in argon and nitrogen matrices (Figure 9), which have yet to be satisfactorily explained.

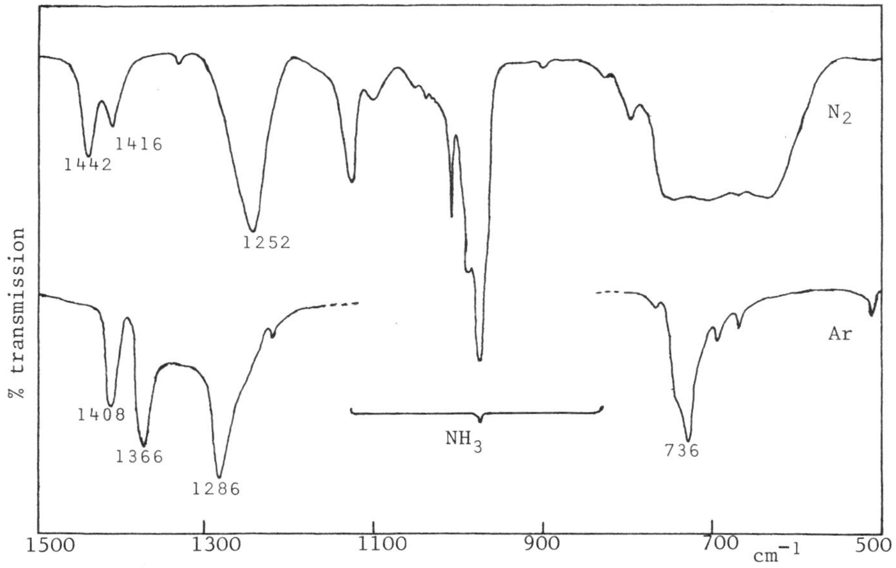

Figure 9. Comparison of i.r. spectra of ammonia-hydrogen chloride complexes in argon and nitrogen matrices (reproduced with permission from ref. 27).

Other molecular complexes studied by matrix isolation vibrational spectroscopy include the π-bonded alkene-halogen and benzene-halogen complexes. Although benzene-halogen complexes have been extensively studied in solution, liquid and solid phases, it is by no means certain that a 1:1 complex is observed under such conditions. Comparison with theoretical predictions, relating to the isolated complex, may consequently not be meaningful. Fredin and co-workers [28] have examined several such complexes by matrix isolation infrared (and visible/ultraviolet) spectroscopy and deduced their probable structures.

More detailed information may be obtained by combining infrared and Raman spectroscopy since many modes of, for example, an ethylene-halogen complex will be either inactive or very weak in the infrared spectrum. A preliminary study of the ethylene-chlorine complex by matrix isolation Raman spectroscopy has recently been reported [29].

5. SUMMARY

As a result of the ready availability of miniature refrigerators, matrix isolation is now a well-established and widely used technique. The range of applications is demonstrated by the fact that matrix isolation studies are discussed in several other parts of this book (principally Chapters 4, 8, 9, 12 and 22). Matrix isolation spectroscopy is capable of giving results not obtainable by any other technique, and there are still many areas of chemistry on which the technique may shed new light.

6. REFERENCES

1. E. Whittle, D.A. Dows and G.C. Pimentel, J. Chem. Phys., 22 (1954) 1943.
2. A. Cabana, A. Anderson and R. Savoie, J. Chem. Phys., 42 (1965) 1122.
3. J.S. Shirk and H.H. Claassen, J. Chem. Phys., 54 (1971) 3237.
4. H.E. Hallam (Editor), Vibrational Spectroscopy of Trapped Species, Wiley, London, 1973.
5. J.M. Grzybowski, B.R. Carr, B.M. Chadwick, D.G. Cobbold and D.A. Long, J. Raman Spectr., 4 (1976) 421.
6. M.M. Rochkind, Environ. Sci. Technol., 1 (1967) 434.
7. R.N. Perutz and J.J. Turner, J. Chem. Soc. Faraday II, 69 (1973) 452.
8. A.J. Barnes, J.C. Bignall and C.J. Purnell, J. Raman Spectr., 4 (1975) 159.
9. M. Moskovits and G.A. Ozin, Appl. Spectr., 26 (1972) 481.
10. H. Wolff and H. Ludwig, J. Chem. Phys., 56 (1972) 5278.
11. C.J. Purnell, A.J. Barnes, S. Suzuki, D.F. Ball and W.J. Orville-Thomas, Chem. Phys., 12 (1976) 77.
12. E. Catalano and K.S. Pitzer, J. Amer. Chem. Soc., 80 (1958) 1054.
13. A.J. Barnes and H.E. Hallam, Trans. Faraday Soc., 66 (1970) 1932.
14. P. Huber-Walchli and H.H. Günthard, Chem. Phys. Lett., 30 (1975) 347.
15. A.J. Barnes, H.E. Hallam, S. Waring and J.R. Armstrong, J. Chem. Soc., Faraday II, 72 (1976) 1.
16. A.J. Barnes and J. Murto, J. Chem. Soc. Faraday II, 68 (1972) 1642.
17. B. Silvi, F. Froment, J. Corset and J.P. Perchard, Chem. Phys. Lett., 18 (1973) 561; B. Silvi and J.P. Perchard, Spectrochim. Acta A, 32 (1976) 11, 23.
18. A.J. Barnes and G.C. Whittle, in Molecular Spectroscopy of Dense Phases - Proceedings of the 12th European Congress on Molecular Spectroscopy, Elsevier, Amsterdam, 1976.

19. F. Inagki, I. Harada and T. Shimanouchi, J. Mol. Spectr., 46 (1973) 381.
20. R.T. Hall and G.C. Pimentel, J. Chem. Phys., 38 (1963) 1889.
21. J.J. Turner, personal communication.
22. A.J. Barnes, C.J. Purnell and S. Suzuki, to be published.
23. A.J. Barnes, H.E. Hallam and G.F. Scrimshaw, Trans. Faraday Soc., 65 (1969) 3172.
24. A.J. Barnes, J.B. Davies, H.E. Hallam and J.D.R. Howells, J. Chem. Soc. Faraday II, 69 (1973) 246.
25. B.S. Ault, E. Steinbeck and G.C. Pimentel, J. Phys. Chem., 79 (1975) 615.
26. B.S. Ault and G.C. Pimentel, J. Phys. Chem., 77 (1973) 1649.
27. S. Suzuki, A.J. Barnes, D. Cowieson, Z. Mielke and C.J. Purnell, in Molecular Spectroscopy of Dense Phases - Proceedings of the 12th European Congress on Molecular Spectroscopy, Elsevier, Amsterdam, 1976.
28. e.g. L. Fredin and B. Nelander, Mol. Phys., 27 (1974) 885.
29. A.J. Barnes, D. Cowieson and S. Suzuki, Proc. 5th Int. Conf. Raman Spectroscopy, Verlag, Freiburg, 1976.

CHAPTER 8

TECHNIQUES FOR STUDYING HIGHLY REACTIVE AND UNSTABLE SPECIES

George C. Pimentel

The ubiquitous presence in chemistry laboratories of infrared and Raman spectrometers attests to their value as a chemist's diagnostic tool. These spectroscopic aids contribute to his ability to detect the presence and amount of substances, to gain some knowledge of their molecular bonding and structure, and to investigate quantitatively their thermodynamic and kinetic behaviour. Because of this general utility of vibrational spectroscopy, much effort has been devoted to the challenge of extending its applicability to the study of highly reactive and relatively unstable species. The special problems of transient existence and low concentrations under normal conditions have required some ingenuity. However, over the last two decades, significant progress has been made and, as tunable lasers become available, there will surely be new techniques added.

For the purposes of this discussion, a "highly reactive and unstable species" will be taken to mean either a molecular species that would have only transient existence or one that can have only extremely low equilibrium concentration under normal conditions of temperature and pressure. This transiency might be associated with extreme reactivity, as in the case of free radicals, or it might be the result of extremely weak bonding, as exampled by reaction intermediates like $O_2 \cdot NO$ in the oxidation of NO to NO_2. A literature survey will not be attempted, but rather, a few prototype examples will be described in sufficient detail to illustrate techniques in use - their advantages and their limitations.

1. MATRIX ISOLATION

In the matrix isolation method, a suspension of a reactive molecule is prepared in a cryogenic, inert matrix to permit leisurely spectroscopic study (see Chapter 7). The technique has been richly successful, as indicated by the existence of two recent books [1,2] and a number of reviews [3,4] to summarise the published work of this type. Meyer's book includes a convenient listing of the unusual molecules that have been studied in matrix suspension, comprehensive through July, 1969.

A novel application of the matrix method, the study of reactions as they take place in the matrix at the cryogenic temperature, remains to be exploited. With high power, tunable lasers, there is the potentiality for radiatively stimulating reactions between trapped reactants. The conditions would simulate high temperature vibrational occupancies without the accompanying translational excitation. It is possible that local functional groups in a complex molecule could be excited to reactivity while other, fragile parts of the same molecule are held quiet. Such speculations are yet to be realised, but the unique possibilities for control of reactions warrants an examination of a few of the pioneering works.

The slow cis → trans isomerisation of nitrous acid, HONO, in solid nitrogen at 20 K was first observed by Baldeschwieler and Pimentel [5] and then elucidated by Hall and Pimentel [6]. The slow conversion of the cis- form to the more stable trans-form was found to take place under infrared irradiation (ν = 3650-3200 cm^{-1}) and then the reaction can be reversed to give a random mixture through electronic excitation in the visible-UV region. Evidently, when the cis-HONO absorbs a quantum of excitation in the O-H stretching mode, there is sufficient energy, sufficient coupling to the torsional mode, and sufficient lifetime before matrix deactivation to permit the isomerisation to take place. The randomisation under electronic excitation suggests that the upper electronic state has no barrier to rotation (or is non-planar) and deactivation can leave the molecule in either configuration.

Another early detection [7] of radiation-induced isomerisation chemistry was connected with the suspension of NO_2 in solid N_2. Clarification was finally provided by Varetti and Pimentel [8], who showed that NO and NO_2 together in a matrix cage could be converted through near-infrared irradiation, 700-900 nm, from one structure, characterised by one set of infrared frequencies, to another structure characterised by an entirely different spectrum. Once again, the matrix reaction proved to be reversible under electronic excitation, in the ultra-violet range 370-480 nm.

A third study, again from the Berkeley laboratories, remains to be elucidated. Bondybey and Pimentel [9] passed H_2 and Ar or Kr through a glow discharge immediately in front of a cold window at 15 K. The resulting matrix sample absorbed at a single frequency which, by isotopic studies, was shown to be due to a species containing a single hydrogen atom. Prolonged infrared irradiation at the absorbing frequency caused the irreversible loss of the absorbing species, deuterium more slowly than hydrogen. There remains uncertainty about the absorber. Bondybey and Pimentel [9] considered the possibility that H^+ had been trapped, perhaps in the form of an ion, HAr^+ or HAr_2^+, but concluded that the H atom was more likely to be neutral, as did Ogilvie [10]. Milligan and Jacox [11] present arguments that an ion is involved, proposing HAr_4^+. There seems to be no doubt that the absorbing species contains only one H atom and that its disappearance is promoted through its own infrared absorption.

One more investigation of matrix reactions, still in progress, will be described. The gas phase chemiluminescent reaction between NO and O_3 has received much attention recently because of the possibility of perturbing the reaction rate by vibrational excitation of one or the other of the reactants [12,13] using CO_2 or CO lasers. The reaction proceeds, at room temperature, by two paths characterised by different activation energies [14]:

$$O_3(\tilde{X}\ ^1A_1) + NO(X\ ^2\pi_{3/2}) \rightarrow O_2(X\ ^3\Sigma_g^-) + NO_2^*(\tilde{A}\ ^2B_1)\ E_A^* = 4.18\ \text{kcal mol}^{-1}$$

$$O_3(\tilde{X}\ ^1A_1) + NO(X\ ^2\pi_{1/2}) \rightarrow O_2(X\ ^3\Sigma_g^-) + NO_2(\tilde{X}\ ^2A_1)\ E_A = 2.46\ \text{kcal mol}^{-1}$$

The X $^2\pi_{3/2}$ state of NO lies 121.1 cm^{-1} (346 cal) above the X $^2\pi_{1/2}$ state.

Lucas [15] has prepared solid N_2 matrix samples at 10 K containing both O_3 and NO at concentrations near $O_3/NO/N_2$ = 1/1/200, where perhaps 10% of the NO molecules will be found with an O_3 nearest neighbour. A dual jet, simultaneous deposition was carried out to prevent gas phase pre-reaction. The expectation was that the reaction would not proceed at 10 K and, hence, it would be possible to investigate the effectiveness of selective vibrational excitation in prompting the reaction.

In practice, the spectra of the matrix samples always display absorption due to NO_2 at 1617 cm^{-1}, attributable to pre-reaction. This absorption grows slowly but continuously without exposure to infrared (or any other) radiation. Over a period of 10 to 20 hours, the NO_2 absorption might be increased three-fold, while no reproducibly correlated changes occur anywhere else in the spectrum (e.g. there is no change in the intensity or half-width of the H_2O absorption inevitably detectable at 1600 cm^{-1}). The NO_2 absorption shows the proper ^{18}O isotopic shift and it is not obtained in the absence of either NO or O_3.

The thermal occupancy of the NO (X $^2\pi_{3/2}$) state at 10 K is 2.7 x 10^{-8} that of the ground state. Hence the chemiluminescent reaction should be quenched, irrespective of its activation energy. The second reaction, producing ground state products with a 2.46 kcal mol^{-1} activation energy, would be slower by a factor of 10^{52} than at 300 K insofar as the activation energy is concerned.

It can be argued that the A factor in an Arrhenius representation of the rate constant is very much higher in the lattice cage because the gas phase collision frequency is now replaced by the lattice cage vibrational frequency. At room temperature and 0.1 torr partial pressures, an NO molecule would experience about 10^6 collisions per second with an O_3 molecule. In the lattice cage, if the lattice cage frequency were in the range 1/3 to 30 cm^{-1}, the collisions with one of the cage walls (where an O_3 molecule might be lodged) would be in the range 10^{10} to 10^{12} sec^{-1}. This increase in collision frequency, by a factor of 10^4 to 10^6, while substantial, is overwhelmed by the expected 10^{-52} activation energy factor. There is no immediate explanation of the observed reaction rate at 10 K in terms of the gas-phase rate behaviour.

Careful measurements of the growth of NO_2 at various temperatures in the range 10 to 20 K revealed an apparent activation energy of 50 to 100 cal mol^{-1}. This magnitude seems to be too small to attribute to diffusion from cage-to cage, which might be nearer to the energy needed to break the lattice, as suggested by the heat of sublimation, 1500 cal mol^{-1}. A more reasonable interpretation might be that an orientational movement within the cage is providing the activation energy. Whatever the model used, it remains unclear how to rationalise the observed reaction rate with an Arrhenius behaviour connected to the gas phase parameters. We are presently considering an oxygen atom, tunnelling explanation.

In summary, there is now enough evidence to show that reactions with low activation energies can be induced under matrix conditions with vibrational excitation. The use of infrared lasers as pumping sources should expand greatly the number and types of reactions that can be so studied. The examples discussed suggest that the chemistry of new species will become accessible and that new phenomena will be encountered.

2. GAS PHASE STUDY OF WEAK COMPLEXES

Ewing and co-workers [16,17] have recently given impetus to the study of weakly bound complexes through the use of long path multiple reflection cells that can be operated at temperatures as low as 100 K. This opens an avenue toward the determination of thermodynamic quantities for these interesting species through the temperature dependence of the equilibrium constant. Unfortunately, special problems arise in the interpretation due to the rather small bond energies being measured. The potentialities and problems are illustrated in a recent determination of the bond energy for the Ar.HCl complex [18,19].

Infrared detection of the Ar.HCl complex was first reported by Rank and co-workers [20]. With a single low-temperature measurement, they estimated the bond energy to be 1.1 ±0.2 kcal mol^{-1}, but this value has not won general acceptance. Instead, a well-depth of about 0.4 kcal mol^{-1} is quoted, based on theoretical considerations [21,22] and on a molecular beam scattering study by Farrar and Lee [23]. We decided to repeat the infrared study with a more complete temperature study to resolve the discrepancy if possible.

A large multiple reflection cell that provides path lengths up to 2 kilometers at temperatures as low as 180 K was used [24]. The physical length of the cell is 10 m and it is 1 m in diameter. To sustain the large number of reflections needed, the mirrors are gold-deposited under specially high vacuum, $<10^{-9}$ torr, to give reflectances above 99%. Furthermore, signal-to-noise was improved by use of an electrically heated carbon rod as the light source. With currents in the range 100-150 amperes at 10-12 volts, black body radiative intensities of 2600-2700 K could be maintained without difficulty [19].

The absorptions of Ar.HCl reported by Rank et al [20] in the HCl null-gap region were observed and shown to follow Beer's Law within experimental accuracy. There was no contribution due to HCl dimer; the intensities of the two features seen under about 1 cm^{-1} resolution (at 2887 and 2879 cm^{-1}) depended upon each of the HCl and Ar partial pressures to the first power.

Over the temperature range 190 to 300 K, the change of intensity provides information about the temperature dependence of the quantity K', the product of the equilibrium constant K and the extinction coefficient ε.

$$K' = K.\varepsilon = \frac{O.D.(complex)}{\ell.[HCl][Ar]}$$

ℓ = path length

It is conventional to assume that the extinction coefficient is independent of temperature, so that the temperature dependence of K' can be taken to be that of K. Then, a plot of ln K' vs 1/T gives a quantity ΔH_{app} which is related to $\Delta H°$, the bond energy of the complex. This "apparent ΔH" would be equal to $\Delta H°$ if the heat capacities of reactants and products were identical (and, of course, if ε is temperature-independent). Unfortunately, for weak complexes (with low values of $\Delta H°$) it is not safe to assume that the ΔC_p contribution to ΔH_{app} will be negligible. Hence, $\Delta H°$ can be obtained from ΔH_{app} only if the heat capacity of the complex can be calculated. This requires some knowledge or model of the vibrational degrees of freedom of the complex. These may be difficult to measure and equally difficult to estimate.

Fortunately, the molecular beam electric resonance measurements by Novick et al [25,26] on Ar.HCl provide information about the moment of inertia, the heavy atom stretching mode, and, with less confidence, the librational mode. By no means does this aid answer all of the questions posed by the heat capacity calculation. Judgements must be made about the extent of interaction among the internal modes, the anharmonicities, and the extent to which the librational and rotational energy level manifolds should be truncated [17] to avoid internal energies that exceed the bond dissociation energy. Nevertheless, it is possible to rationalise a set of energy levels and to test parametrically the sensitivity of the deduced well depth to the energy level choice.

In practice, the assumed well depth limits the number of bound vibrational energy levels, certainly in the heavy atom stretching mode and, probably as well, in the librational and overall rotational motions because of coupling. For a given set of vibrational frequencies and energy level truncations, there is a range of well depths that permits reproduction of the experimental ΔH_{app}. However, even the most extreme assumptions (i.e. no interactions and, hence, no truncations) indicate that the well depth must exceed 700 cal mol^{-1}. The most appealing model, that no single mode can carry energy in excess of the bond dissociation energy, gives a well depth near 1.2 kcal mol^{-1}. Thus, the discrepancy with the low molecular beam value, 0.4 kcal mol^{-1}, is not resolved. The disparity cannot be put aside as probably spurious due to experimental errors nor to neglect of vibrational contributions to the heat capacity.

The interpretation of each type of data contains some vulnerable premises. The molecular beam rainbow scattering was explained [23] with an isotropic interaction between Ar and HCl.

This isotropic potential function is, of course, inconsistent with the Ar.HCl structure deduced by Novick et al [25,26]. The infrared data, on the other hand, are interpreted with the assumption that the extinction coefficient of the H-Cl stretching motion is independent of temperature. However, it seems likely, from hydrogen bonding behaviours, that this extinction coefficient will tend to decrease as the librational mode is excited, carrying the hydrogen atom further from a linear configuration. This means that the infrared experiment tends to emphasise the levels deeper in the potential well. Unfortunately, this difficulty, if real, does not work in the direction of explaining the discrepancy with the 400 calorie well depth. It seems, rather, that molecular beam scattering might accentuate the librationally excited complexes and hence tend to underestimate the well depth in weakly bound complexes. The infrared detection might accentuate the unexcited molecules and, hence, sample more deeply in the potential well. These problems of data interpretation are probably not unique to Ar.HCl but rather, will prove to be generally encountered for very weakly bound complexes. Insofar as spectroscopic studies are concerned, better understanding of the interactions among the internal degrees of freedom and their absorption characteristics will be needed to permit confident derivation of thermodynamic properties.

3. GAS PHASE INFRARED STUDY OF TRANSIENT SPECIES BY RAPID SCAN TECHNIQUES

The development of infrared spectrometers that brought scan times down to the 10 to 200 μs time range while preserving diagnostically useful spectral resolution has been summarised [27] to the year 1968. The optimistic expectations expressed at that time cannot be said to have been fully realised. The Berkeley laboratories, where such techniques were pioneered, have remained active, but productivity has been hard won. Table 1 lists the free radicals that have been studied - the

Table 1. Transient molecules studied by rapid scan infrared spectroscopy

Transient molecule	Year	Notes
CF_2	1964	Herr [28]: first gaseous IR detection
ClCOOH	1964	Herr [28]: first IR detection
CF_3	1966	Carlson [29]: first IR detection
ClCOOH	1967	Jensen [30]: kinetics of decomposition
CF_3	1970	Ogawa [31]: kinetics of recombination
CF_2	1971	LeFohn [32]: rotational structure resolved
CH_3	1972	Tan [33]: first gaseous IR detection
CCl_3	1975	J. Herr [34]: first gaseous IR detection

list is not long after a decade of intensive effort. Meanwhile, the instrument designed by Hexter and Hand [35] has been retired from use.

The most recent work conducted at Berkeley has been to extend and improve the spectra of the methyl radical in the region 400 to 700 cm^{-1}. This region contains the ν_2 vibrational frequency, the out-of-plane bending mode (for a planar structure) or the "umbrella mode" (for a pyramidal configuration). For gaseous CH_3, absorption was found [34] at 607 and 603 cm^{-1} while for CD_3, absorptions seemed to be identifiable at 461 and possibly 457 cm^{-1}. Unfortunately, a poor signal to noise ratio prevented positive conclusions both about the origin of the doublet splitting for CH_3 and about the reality of the possible splitting for CD_3. The question is one of great interest in view of the possible (though not favoured)[34] interpretation that two absorptions (at 607 and 603 cm^{-1}) are observed because of inversion doubling, as is the case in the spectrum of ammonia.

To gain more definitive spectra, a substantial improvement in signal-to-noise was needed. Such a gain has been achieved by Dr. Frank Prochaska [36] through the development of a flash photolysis flash-lamp as an infrared spectroscopic source. Preliminary experiments showed that a conventional flash lamp plasma is optically thin in the infrared spectral region 10 to 25 μm. However, by using a 50-60 cm long flash plasma viewed end on, it was possible to sustain for 50 to 100 μs black body temperatures considerably higher than those attainable with even a 180 ampere carbon arc. For example, by discharging 1000 Joules through 5 torr of xenon, plasma black body temperatures near 10,000 K were measured and sustained for 125 μs. With 2500 Joules and 15 torr of argon, temperatures exceeding 14,000 K were established. Helium gas provided the most intense infrared emissions, reaching 20,000 K, but this brightness was lost quickly after 60 or 70 μs. The conclusion was that use of a flash lamp plasma as an infrared light source could provide for 100 μs an intensity at 700 cm^{-1} three to four times greater than the brightest source known to us before, the 180 ampere, 60 volt carbon arc whose crater provides a 4500 K black body emissivity in this spectral region.

The improved performance resulting from this more intense light source was coupled with another gain associated with background interference by atmospheric water vapour in the long wavelength region. The elimination of the carbon arc collection optics shortened considerably the optical path that needed to be purged. This was of particular importance in the repetition of the CD_3 spectrum since H_2O rotational features are plentiful in the 400-500 cm^{-1} range.

The more definitive spectra now in hand for the gaseous CH_3 methyl radical confirm the previously reported doublet at 607.0 and 603.3 cm^{-1} and add a third component, at 599.7 cm^{-1}. The CD_3 radical spectrum, however, now appears rather different. With the better purge, the feature reported at 461 cm^{-1} is no longer observed; presumably it was due to H_2O rotational absorption. On the other hand, the second proposed but dubious band at 457 cm^{-1} has been confirmed (at 457.4 cm^{-1}) and a new second feature has been detected at 454.8 cm^{-1}. Thus we now have reliable spectra of ν_2 for both CH_3 and CD_3. The two new pieces of evidence, the third component in the CH_3 spectrum and the relatively large spacing between the two CD_3 features provide firm evidence against an inversion doubling interpretation. We must be observing hot band or upper stage transitions. Vibrationally hot methyl radicals are evidently formed in the photolysis process.

These successes with rapid scan infrared spectroscopy are satisfying and unique. However, we expect that a decade from now, they will be regarded as only the first stumbling steps in the infrared study of transient species. Laser sources will shortly render dispersion optical techniques obsolete in all dimensions: time scale, spectral resolution, and sensitivity. Tunable lasers are not yet easily used for diagnostic spectroscopy, but we are on the doorstep. When perfected, they will give access to the sub-microsecond time range and with resolutions better than 0.1 cm^{-1}. This considerable narrowing of bandwidth carries with it greatly improved sensitivity. It can be predicted that ten years from now, Table 1 will be much longer and it will summarise a new era in our understanding of the structure and properties of transient molecules.

4. SUMMARY

The techniques presently available to the spectroscopist for the study of highly reactive molecules have been displayed through case studies of recent completion. The matrix isolation method is now well developed and of growing application. A major obstacle to its wider use, the handling and expense of liquid helium and hydrogen, has been removed through the development of commercial closed cycle cryostats. Infrared spectroscopy at cryogenic temperatures can now be made routine.

By no means can a similar description be applied at this time to rapid scan infrared spectroscopy. Nevertheless, its future looks exciting. It is evident that further development of rapid scan spectroscopy will be based upon the unique capabilities of tunable lasers. Such equipment is presently quite expensive, of limited spectral range, and tedious to use. However, the prospects in view are breathtaking. There is every reason to hope that within the next few years it should be possible to shorten scan times by three orders of magnitude (from, say, 200 μs to 200 ns) and, simultaneously, to improve spectral resolution by more than one order of magnitude (from one or two cm^{-1} spectral slit width to band widths in the range 0.1 to 0.001 cm^{-1}). New types of information should become available to furnish new insights in the nature and behaviour of highly reactive molecules.

5. REFERENCES

1. B. Meyer, Low Temperature Spectroscopy, Amer. Elsevier Publishing Co., Inc., New York, 1971.
2. H.E. Hallam (Editor), Vibrational Spectroscopy of Trapped Species, J. Wiley, London, 1973.

3. L. Andrew, Infrared Spectra of Free Radicals and Chemical Intermediates in Inert Matrices, Ann. Rev. Phys. Chem., 22 (1971) 109.
4. D.E. Milligan and M.E. Jacox, Spectra of Radicals, in Physical Chemistry, Vol. 4, Academic Press, New York, 1970, Chapter 5.
5. J.D. Baldeschwieler and G.C. Pimentel, J. Chem. Phys., 33 (1960) 1008.
6. R.T. Hall and G.C. Pimentel, J. Chem. Phys., 38(1963) 1889.
7. J.D. Baldeschwieler, Ph.D. dissertation, University of California, Berkeley, 1959.
8. E.L. Varetti and G.C. Pimentel, J. Chem. Phys., 55 (1971) 3813.
9. V. Bondybey and G.C. Pimentel, J. Chem. Phys., 56 (1972) 3832.
10. J.F. Ogilvie, J. Chem. Phys., 59 (1973) 3871.
11. D.E. Milligan and M.E. Jacox, private communication.
12. R.J. Gordon and M.C. Lin, Chem. Phys. Lett., 22 (1973) 262.
13. W. Braun, M.J. Kuryle, A. Kaldor and R.P. Wayne, J. Chem. Phys., 61 (1974) 461.
14. M.A.A. Clyne, B.A. Thrush and R.P. Wayne, Trans. Faraday Soc., 60 (1964) 359.
15. D. Lucas, Ph.D. dissertation, University of California, Berkeley, 1976.
16. G. Henderson and G.E. Ewing, Mol. Phys., 27 (1974) 903; C.H. Bibart and G.E. Ewing, J. Chem. Phys., 61 (1974) 1284; C.A. Long and G.E. Ewing, J. Chem. Phys., 58 (1973) 4824.
17. G. Henderson and G.E. Ewing, J. Chem. Phys., 59 (1973) 2280.
18. A.W. Miziolek and G.C. Pimentel, J. Chem. Phys., in press.
19. A.W. Miziolek, Ph.D. dissertation, University of California, Berkeley, 1976.
20. D.H. Rank, B.S. Rao and T.A. Wiggins, J. Chem. Phys., 37 (1962) 2511; D.H. Rank, P. Sitaram, W.A. Glickman and T.A. Wiggins, J. Chem. Phys., 39 (1963) 2673.
21. W.B. Neilsen and R.G. Gordon, J. Chem. Phys., 58 (1973) 4131, 4149.
22. A.M. Dunker and R.G. Gordon, J. Chem. Phys., 64 (1976) 354.
23. J.M. Farrar and Y.T. Lee, Chem. Phys. Lett., 26 (1974) 428.
24. D. Horn and G.C. Pimentel, Applied Optics, 10 (1971) 1892.
25. S.E. Novick, P. Davies, S.J. Harris and W. Klemperer, J. Chem. Phys., 59 (1973) 2273.
26. S.E. Novick, S.J. Harris, K.C. Janda and W. Klemperer, Can. J. Phys., 53 (1975) 2007.
27. G.C. Pimentel, Appl. Optics, 7 (1968) 2155.
28. K.C. Herr and G.C. Pimentel, Appl. Opt., 4 (1965) 25.
29. G.A. Carlson and G.C. Pimentel, J. Chem. Phys., 44 (1966) 4053.
30. R.J. Jensen and G.C. Pimentel, J. Phys. Chem., 71 (1967) 1803.
31. T. Ogawa, G.A. Carlson and G.C. Pimentel, J. Phys. Chem., 74 (1970) 2090.
32. A.S. LeFohn and G.C. Pimentel, J. Chem. Phys., 55 (1971) 1213.
33. L.Y. Tan, A.M. Winer and G.C. Pimentel, J. Chem. Phys., 57 (1972) 4028.
34. J.J. Herr, Ph.D. dissertation, University of California, Berkeley, 1976.

35. R.M. Hexter and C.W. Hand, Appl. Opt., 7 (1968) 2161.
36. F. Prochaska, Ph.D. dissertation, University of California, Berkeley, 1976.

CHAPTER 9

TECHNIQUES FOR STUDYING HIGH TEMPERATURE SPECIES

I.R. Beattie

1. THE CHARACTERISATION OF HIGH TEMPERATURE VAPOURS

From a fundamental point of view, the study of high temperature systems is of particular interest to the inorganic chemist. It is a method of studying discrete molecules obtained from involatile solids; the shape and stability of such species; the nature of their polymerisation; their reactions with substrates. From a technological viewpoint, interest in high temperatures stems from a whole variety of areas: flames, plasmas, corrosion of turbine blades and rocket motors, vapour-phase transport in crystal growth, etc. In this chapter we shall deal with high temperature vapours, largely from a fundamental point of view, although the applications will be readily apparent.

The sources of the high temperature species may be broken into two main headings: static and dynamic systems. Within these frameworks there are sub-headings principally based on the Knudsen cell (dynamic) or closed systems in furnaces (static). We shall begin by briefly considering techniques for studying high temperature vapours.

The Knudsen cell

For an ideal orifice of unit area the number of particles passing through in unit time is given by [1]:

$$\left(\frac{dN}{dt}\right)_\theta = \left(\frac{n\bar{v}}{4\pi}\right)\cos\theta\, d\omega$$

where n = number of particles per unit volume (in the Knudsen cell).
$\bar{v}$ = average molecular velocity
θ = angle of measurement related to the normal to the orifice
$d\omega$ = solid angle of measurement

This leads to the cosine law relating the ratio of the number of particles which pass through at an angle θ to the number measured along the normal (for the same solid angle). For finite orifices corrections for non-ideality are necessary [2], usually by the introduction of so-called "Clausing transmission factors".

Mass spectrometry

This has undoubtedly been the most important technique for characterising high temperature species. The usual source is a Knudsen cell. The crude molecular beam from the inlet is ionised in the source chamber (normally by an electron beam) and passes into electrostatic and magnetic analyser systems. Because of the high sensitivity of the technique (detection levels of the order of 10^{-11} torr) a relatively low flux is required from the Knudsen cell. Analysis relies on determination of charge to mass ratios and hence isomers are not distinguished. However, the major difficulties lie in (i) fragmentation of molecules (including polymers to give "monomer" peaks for example) during the ionisation process and (ii) ion-molecule reactions to form higher molecular weight aggregates. The second difficulty can be largely eliminated by working at sufficiently low ionisation pressures, although it must be realised that the efficiency of the ionisation process is low (perhaps one molecule in one thousand forming an ion). Clearly if the purpose of the study is to differentiate monomer, dimer, trimer .... in the vapour, then ion-molecule reactions or fragmentation during ionisation can lead to erroneous conclusions being drawn from the observations.

Three main approaches have been made to the problem of fragmentation during ionisation: a study of the mass spectrum as a function of electron source energy - on the basis that less fragmentation will occur at lower accelerating potentials; the examination of the distribution of polymers/monomers as a function of angle of observation relative to the axis of the Knudsen cell orifice [3]; the use of a chopped molecular beam followed by time of flight of <u>neutrals</u> prior to ionisation, with phased detection of the ions[4]. In the last case it is assumed that ions of the same phase difference (relative to the source chopper) originate from the same initial molecular species. In this way it is possible in principle to distinguish "monomers" resulting from fragmentation reactions from those caused by ionisation of truely monomeric species. A basic first assumption is that all molecular species initially are at thermal equilibrium and hence the initial velocities will be dependent on $M^{-\frac{1}{2}}$, where M is the total mass of the molecule. (It will also be remembered from the kinetic theory that there is a range of molecular velocities at any given temperature for a group of molecules of the same mass. This may lead to overlap of different ions in the phase detection. The authors note that under the conditions of their experiment the velocity distribution departs markedly from a Maxwellian function [3]).

Electric deflection

Molecules showing a second order Stark effect are focussed in a quadrupolar field [5]. Ideally the molecule shows a restoring force proportional to the displacement of the molecule from the axis of the field. There is a close analogy with the normal lens in geometric optics. The molecular beam is studied using a conventional mass spectrometer or a modulated molecular beam source mass spectrometer [3]. A stop is then introduced to shield the detector (mass spectrometer) from the source. The signal should drop to close to zero - apart from molecules scattered to the detector. The quadrupole field is then applied. If an increased signal is observed the molecules possess a dipole. For molecules possessing a first order Stark effect the lens analogy is not accurate, although the quadrupole still retains spatial gathering power. (All molecules are polarisable and the effect of this polarisability will be to cause some defocussing.)

This is a non-quantitative technique, in terms of bond lengths and angles. In essence it detects electric dipoles. The theory is complex because of the possibility of vibrationally induced dipoles [6]. Note also that it suffers all the disadvantages associated with the detector - a mass spectrometer. However some very elegant work by Klemperer and co-workers [5] drew attention to the fact that all the monomeric barium dihalides are bent, for example.

Electron diffraction

Electrons are scattered by the atomic potential (essentially nuclei and electrons). It is thus possible by an analysis of the coherent molecular scattering to obtain a measure of the internuclear distances in a molecule. There are several major disadvantages to electron diffraction on high temperature molecules. Firstly the composition of the beam may not be known. For example, polymer/monomer formation, disproportionation, decomposition or reaction with the container/orifice may lead to erroneous interpretation of the observed diffraction pattern [7]. The conditions of an electron diffraction experiment normally require examination of the issuing jet close to the orifice of the "Knudsen cell" at pressures up to 1 torr. This is quite unlike the conditions in a mass spectrometer used to analyse the constituents of the effusing vapour from a thermally equilibrated Knudsen cell. Secondly, fogging of the detector (normally a photographic plate) by the high temperature gases, etc, will reduce the effective signal to noise ratio. Finally, the pattern observed refers to the molecule averaged over all vibrational motion. It is easier to deform an angle than to deform a bond. Thus bonded distances may be well determined (although similar distances in the molecule may not be resolved) while non-bonded distances will undergo so-called "bond shrinkage" due to the vibrational motion [8].

An example will clarify the problem. Molecular $HgCl_2$ is almost certainly a linear symmetrical molecule. Data from electron diffraction studies on the gas lead to [9]:

Hg —— Cl = 2.252 ±0.005 Å

Cl ... Cl = 4.48 ±0.04 Å

thus giving a shrinkage of 0.024 ±0.045 Å on the Cl ... Cl distance. The authors conclude the molecule is linear within 16°.

Spectroscopic techniques

Rotational analysis of a molecule in principle enables determination of the moments of inertia via m_n/r_n^2 terms where r_n refers to the distance of the nth mass, m_n, from the inertia axes. Thus by suitable isotopic substitution it is possible to define the molecular shape.

For the average high temperature chemist, pure rotational

spectroscopy cannot be regarded in any sense as routine. It is a specialist's technique. Microwave spectroscopy (applicable where the molecule has a dipole moment) requires the design of special high temperature cells. Commercially available spectrometers are relatively inflexible and usually rely on Stark modulation which is somewhat restrictive in waveguide design. More fundamentally the population of a whole range of rotational transitions, coupled with Doppler broadening, may make the analysis of such spectra very difficult. It must be realised that even for room temperature measurements the rotational analysis of an asymmetric rotor is by no means straightforward.

The selection rules for rotational Raman spectroscopy are much less restrictive than those for absorption spectroscopy. All molecules other than spherical tops show a pure rotational Raman spectrum. Although this is very attractive the experimental difficulties are considerable because of the high resolution required. For example, even for $^{11}BF_3$ (a symmetric rotor) the spacings are only 1.4 cm^{-1}. Remembering the m/r^2 dependence, the problem with a molecule such as $GaCl_3$ immediately becomes evident. The use of Coherent Anti-Stokes Raman Scattering (CARS) (see Chapter 3) may lead to a revolutionary new approach to pure rotational spectroscopy. Further, the Doppler velocity is cancelled out (as in rotational Raman scattering observed in the forward direction [10]). Inorganic chemists do not require molecular constants to the degree of sophistication of the chemical physicist. A low resolution rotational spectrum to define a symmetric top or to define a bond angle to a degree would be very exciting [11].

It is, of course, in principle possible to carry out a rotational analysis either of a rotation/vibration spectrum or of a vibrational band in the electronic spectrum. Such sophistication is usually only possible for rather small molecules, routinely only for diatomics in electronic spectroscopy. Again the use of narrow line tunable lasers together with observation of fluorescence is revolutionising this area of spectroscopy. The beautiful work on cooled molecular beams is of particular significance here [12]. By expansion through an orifice from a high pressure region to a low pressure region the rotational and vibrational energy of a molecule may be transformed largely into translational energy. In this way beam temperatures down to 0.4 K (rotational) and 50 K (vibrational) may be obtained. However it is still not a routine technique for the inorganic chemist. He is left with direct vibrational spectroscopy on high temperature gases or on species isolated in a matrix.

The attraction of direct spectroscopy on a high temperature gas is that no sampling probe is necessary [13]. Where scattering occurs from a spatially well-defined volume (Raman, fluorescence or CARS for example) the sample may readily be analysed at a variety of points even where it is a dynamic system such as a flame or a plasma. Using pulsed laser techniques (down to a few picoseconds for example) the fluctuations in composition as a function of time may also be probed. In this connection it must be remembered that a pulse length of a picosecond means an uncertainty in the frequency of roughly the inverse of the time (10^{12} Hz) - in practice about 10 cm^{-1}.†

2. HIGH TEMPERATURE SPECTROSCOPY

At temperatures up to 1,000°C in systems that do not react with silica Raman spectroscopy of gaseous species at pressures of the order of atmospheric is routine. There are of course several problems which may be encountered:

(1) fluorescence (Figure 1 shows the result of an attempt [15]

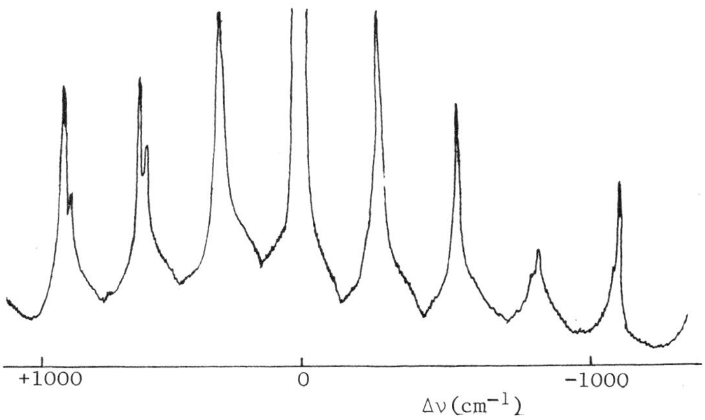

Figure 1. Resonance fluorescence from 488 nm irradiation of gaseous $PbCl_2$ at 1000°C [13,15] (reproduced with permission from ref. 15).

to take the Raman spectrum of gaseous $PbCl_2$).

(2) background thermal radiation, which can be minimised by careful cell and furnace design. (The maximum in the emission curve of a black body for the temperatures under discussion here lies in the infrared region. It is thus preferable to use short wavelength excitation for Raman spectroscopy. The 488 nm and 514.5 nm radiation of the argon laser is excellent for this approach.) It is worthwhile remembering that the intensity of the radiation at any given frequency is (in the short wavelength approximation) roughly exponentially dependent on absolute temperature.

(3) unexpected reactions such as one which caused the author to make an error [16]:

$$4AlCl_3(g) + 3SiO_2(s) \rightarrow 3SiCl_4(g) + 2Al_2O_3(s)$$

† Other techniques such as photoelectron spectroscopy have been used on high temperature molecules. However in at least one molecule of interest to inorganic chemists (Tl_2F_2) the results were more consistent with the model subsequently shown to be the <u>least</u> likely by matrix isolation spectroscopy [14].

In addition to this, it is necessary to make sure that all the material is in the gas-phase. It is very easy to accidentally take the Raman spectrum of solid or melt rather than gas. It is also important to note that a pressure of one atmosphere at room temperature means a concentration of roughly one twentieth molar. At 900°C it is nearer one hundredth molar for the same pressure. Figure 2 shows a furnace which can be turned out in

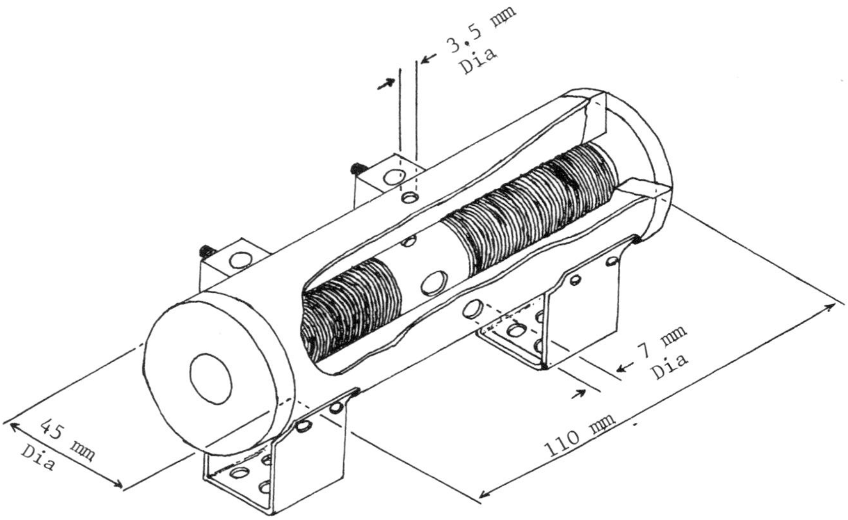

Figure 2. The furnace for Raman spectroscopy of gases (reproduced with permission from ref. 17).

a few hours by a modest workshop. Figures 3 and 4 show representative spectra which may be obtained routinely in glass or silica ampoules sealed off directly from a vacuum line and inserted into the furnace of Figure 2. In certain cases it may be desirable to study the molecular weight of the sample, using conventional vapour density techniques for example, under conditions identical to those used for the Raman experiment.

Where reaction with silica occurs or where temperatures much above 1100°C are required (when silica starts to devitrify fairly rapidly) sapphire represents an attractive alternative. Figure 5 shows progressive changes in our cell designs using sapphire as the basic material of manufacture. It must be admitted that above 1100°C background radiation is becoming very serious. The applications of CARS may revolutionise this area of spectroscopy.

The corresponding experiment in the infrared effect is much more difficult because of problems of window materials such as silicon, polythene, cesium iodide, silver chloride and diamond. Features to consider are reactivity of windows (cesium iodide), poor thermal behaviour (polythene) and difficulty of sealing to glass or metal (diamond). Many attempts have been made to utilise flowing gas streams to carry material to the cell and

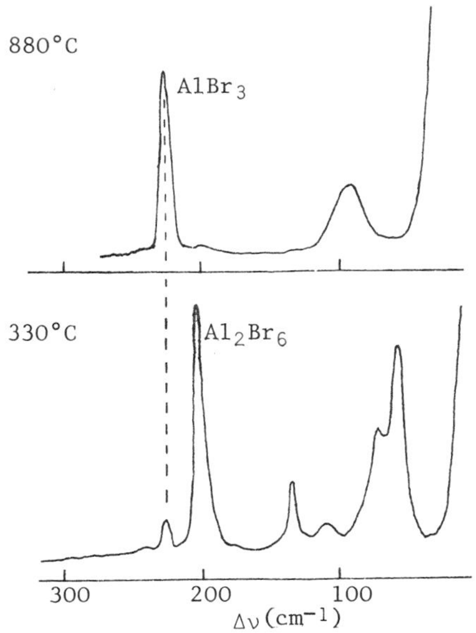

Figure 3. Gas-phase Raman spectra of aluminium tribromide as a function of temperature [16-18] (after ref. 16).

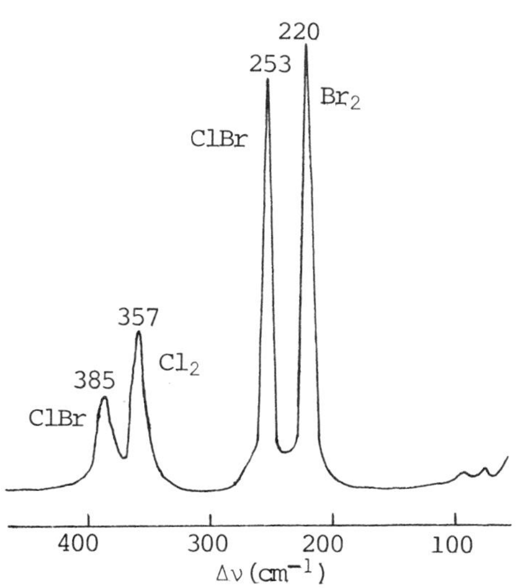

Figure 4. Gas-phase Raman spectrum of an equimolar mixture of $HgCl_2$ and $HgBr_2$ [13,19] (reproduced with permission from ref. 19).

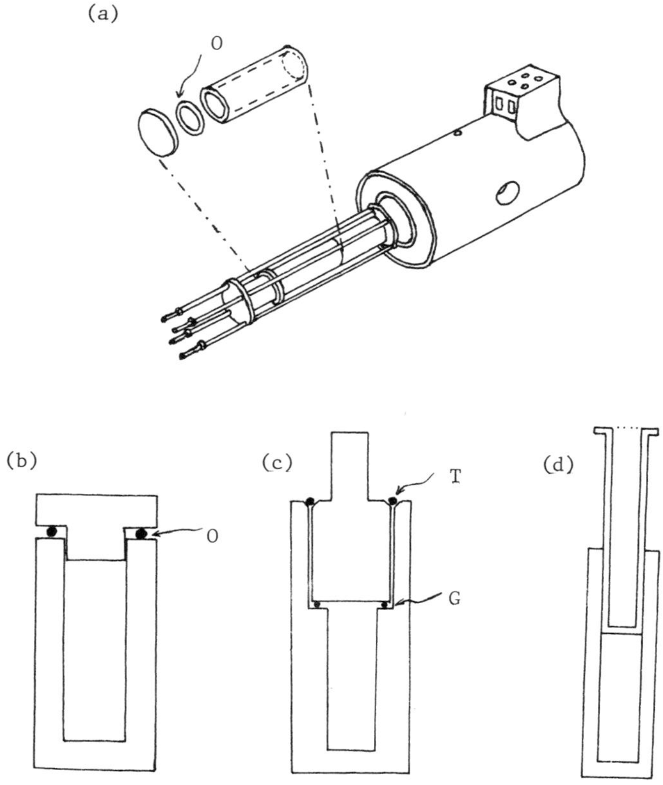

Figure 5. Designs of sapphire high temperature gas-phase Raman cells (a) O = thin teflon washer (reproduced with permission from ref. 20); (b) O = gas-filled monel (possibly gold plated) O-ring [21,22]; (c) T = water-cooled teflon O-ring, G = gold O-ring [23]; (d) in this cell there are no O-rings, the plunger is a tight (hypodermic-like) fit in the barrel.

away from the windows. The author has found these difficult to handle (for a survey of these cells see reference 23). Silicon will seal directly into borosilicate glass and a thermal gradient cell such as the one shown in Figure 6 enables sample temperatures to 1000°C to be attained. The material under study must have sufficient vapour pressure not to condense on the silicon windows (whose transmittance decreases rapidly with increasing temperature - the highest usable temperature routinely being about 275°C). There is, of course, a relatively small volume of cooler gas at either end of the hot central portion where the concentration of material will be appreciably higher than in the central portion. An alternative approach is to use cold windows insulated from the hot gas by a "barrier"

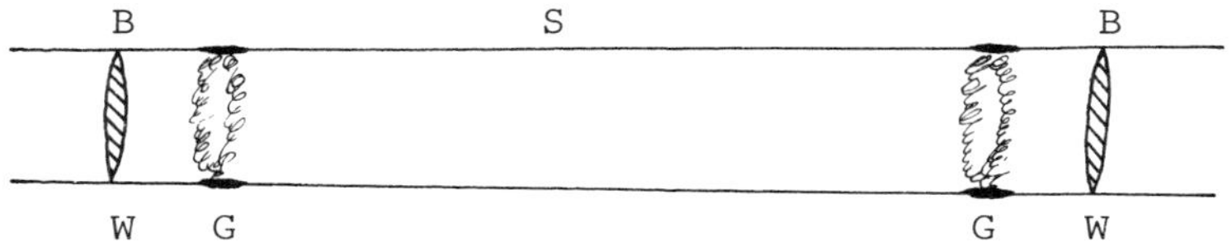

Figure 6. High temperature infrared cell. W = silicon window, G = graded seal, B = borosilicate glass, S = vitreous silica.

of inert gas. Such cells are liable to produce fogs, smokes or mists (although such problems become much less troublesome at lower pressures). Again the introduction of tunable infrared lasers will enable smaller tube diameters to be used and hence allow greater control of diffusion by the sample towards the windows.

In examining spectra from high temperature species several factors must be borne in mind. Firstly, the integrated intensity of an infrared band is independent of temperature. This is not true of a Raman band where the integrated intensity is proportional to

$$\left(\frac{I_{\nu_0}}{1 - e^{-h\nu/kT}}\right)$$

where I_{ν_0} is the intensity of the fundamental. In general there will also be problems associated with hot bands, broadening of rotational wings and Doppler and pressure broadening. Thus the average high temperature spectrum of a gas leads to relatively broad bands. However Q-branches are much less affected than the rotational wings. Since these are the bands normally seen strongly in the Raman effect (and occasionally in the infrared effect) high temperature Raman spectroscopy is an important technique because of the ease of the experiment and the relatively narrow lines.

3. MATRIX ISOLATION SPECTROSCOPY

The attractions of matrix isolation spectroscopy for the high temperature chemist are: the ability to trap species from a vapour and to study them at leisure; the lower temperatures necessary for volatilisation of refractory materials compared to those necessary to obtain pressures high enough to enable gas-phase spectra to be obtained (notably in the Raman effect); the narrowness of the lines (frequently less than 1 cm^{-1}) ideally enabling the resolution of isotopic species; the possibility of the direct synthesis of molecules of interest from metal atom reactions for example.

One of the major difficulties in matrix isolation spectroscopy is the identification of a species present in a matrix both in terms of empirical formula and charge. A combination of gas-phase and matrix isolation infrared and Raman spectroscopy may be particularly valuable from this point of view. Figure 7 summarises such data for the linear molecule $HgCl_2$.

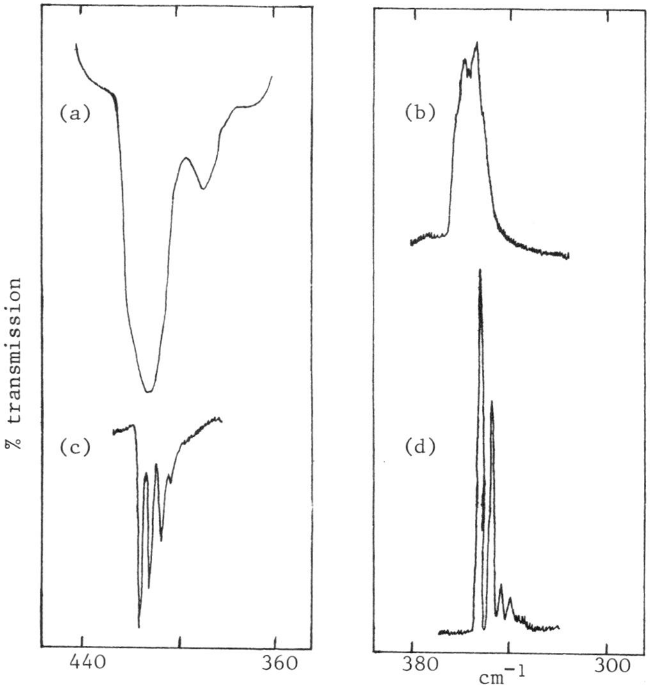

Figure 7. Spectra of mercury (II) chloride under various experimental conditions [21] (a) infrared gas [24] (b) Raman gas [25] (c) matrix infrared [26] (d) matrix Raman [27].

The pattern expected for the relative intensity of the isotopic molecules $^{35}ClHg^{35}Cl$:$^{35}ClHg^{37}Cl$:$^{37}ClHg^{37}Cl$ in the Raman effect is given by the relative abundances of ^{35}Cl and ^{37}Cl (roughly 3 to 1) raised to the power of the number of (equivalent) chlorine atoms in the molecule, giving a ratio of 9:6:1 for the isotopic molecules.

For a ligand atom attached to a hypothetical metal atom of infinite mass the span of the isotope splitting between ligand masses L_1 and L_2 is $\nu_1 |(L_1/L_2)^{\frac{1}{2}} - 1|$ where ν_1 refers to the frequency of the isotopic molecule containing only L_1. Thus as the mass of the ligand increases (and the frequency of the vibration under observation decreases) so the splitting becomes increasingly difficult to observe. Very unfortunately, fluorine has no naturally occurring isotopes other than ^{19}F. However the $^{35}Cl/^{37}Cl$ naturally occurring species and the availability of ^{18}O enable the high temperature chemist to study chlorides and oxides with relative ease.

The compound gallium trichloride is dimeric in the crystal, one Ga_2Cl_6 unit being formed by two tetrahedra ($GaCl_4$) sharing an edge [28]. In the gas-phase this compound exists at temperatures up to about 1000° as a mixture of dimer and monomer [16,29]. Figure 8(a) shows the symmetric stretching mode of

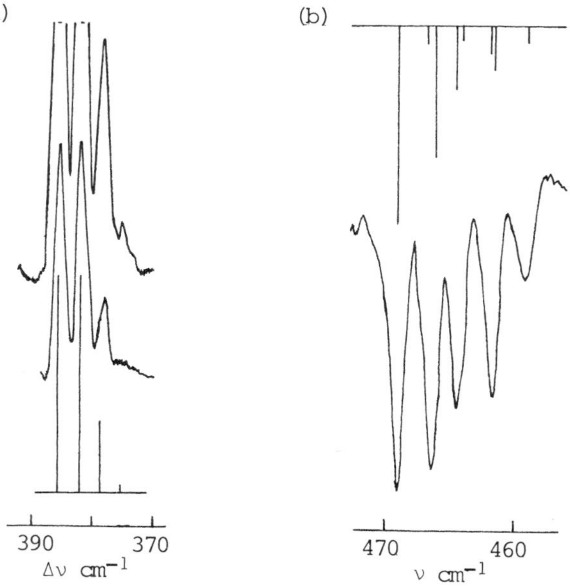

Figure 8. High resolution study of (a) ν_1 and (b) ν_3 of $GaCl_3$ isolated in an argon matrix (line diagrams are calculated)(reproduced with permission from ref. 30).

$GaCl_3$, ν_1 in the Raman spectrum for the matrix isolated material. Note the 27:27:9:1 splitting confirming the presence of three equivalent chlorine atoms in this molecule and hence of a C_3 axis [30].

At first sight it might be expected that there would be a similar distribution in the infrared effect. This is not so, the infrared active anti-symmetric stretching vibration ν_3 of a molecule MCl_3 with a C_3 axis (z) is degenerate (x ≡ y). Although $M^{35}Cl_3$ and $M^{37}Cl_3$ have degenerate ν_3 vibrations, the molecules $M^{35}Cl_2{}^{37}Cl$ and $M^{37}Cl_2{}^{35}Cl$ are of C_{2v} symmetry and the degeneracy is lifted. The relative abundances (and hence intensities) and frequencies of the molecular types may readily be calculated. Figure 8(b) shows theory and calculation [30] for ν_3 of $GaCl_3$. The agreement is excellent. (Note that gallium moves in ν_3 and that there are two isotopes ^{69}Ga and ^{71}Ga of relative abundances, 3:2, leading to a five line pattern.) The deduction of the planarity of $GaCl_3$ derives from selection rules and it is here that the weakness of the approach is evident. Neglecting anharmonicity, assuming that stretching modes may be factored from deformations of the same symmetry and further considering stretching to cause solely dipole changes along the bond directions, it is formally possible to calculate the relative intensities of ν_1 and ν_3 in the infrared effect for $GaCl_3$ for various ClGaCl bond angles. The procedure is open to criticism. For $GaCl_3$ ν_1 is not observed in the infrared effect and ν_3 is not observed in the Raman effect.

Some elegant work has been carried out on metal oxides. Figure 9 shows the high resolution infrared spectrum of SnO_2

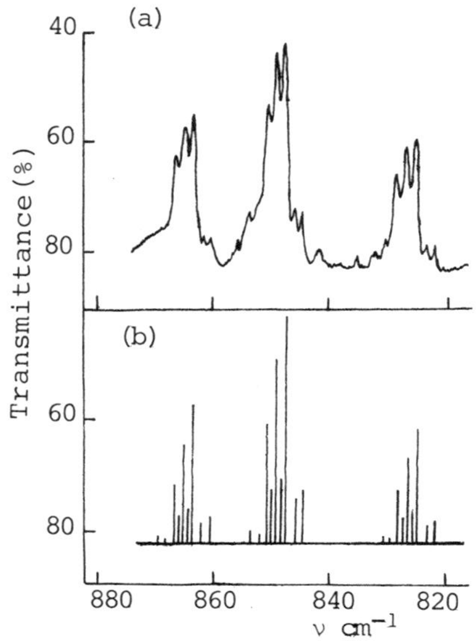

Figure 9. (a) Observed and (b) calculated infrared spectrum of SnO_2 with ^{18}O enrichment (reproduced with permission from ref. 31).

synthesised from tin atoms and molecular oxygen [31] (isolated in a krypton matrix using ^{18}O enrichment). The observed spectrum fits closely with that calculated on the basis of a linear symmetrical molecule. Related isotopic studies on $(SnO)_n$ species identified monomer, dimer, trimer and tetramer. Controlled warm up of the matrix (diffusion studies) is of value in following polymerisation phenomena [32] and confirming the identification of bands associated with the various species.

Recently Ogden [33] has been able to rationalise a range of data on M_3X_3 ring trimers including $(LiF)_3$ and $(PN)_3$. In all cases the calculations when linked to the spectra demonstrate the presence of a C_3 axis and suggest that a planar ring is adopted.

4. CONCLUSIONS

For the chemist interested in high temperature molecules up to 1000°C gas-phase Raman spectroscopy and matrix isolation infrared spectroscopy may be regarded as routine techniques. Matrix isolation Raman spectroscopy is rewarding but appreciably more difficult. The introduction of tunable infrared lasers may revive interest in gas-phase infrared spectroscopy of high temperature molecules under selected conditions. A resolution of 10^{-6} cm^{-1} with a frequency range of 4,000 to 280 cm^{-1} can only be described as exciting.

5. REFERENCES

1. M. Knudsen, Ann. Phys. (Leipzig), 28 (1909) 75, 999; R.D. Present, Kinetic Theory of Gases, McGraw Hill, New York, 1958.
2. P. Clausing, Z. Phys., 66 (1930) 471; Ann. Phys. (Leipzig), 12 (1932) 961.
3. See, for example, R.T. Grimley, L.C. Wagner and P.M. Castle, J. Phys. Chem., 79 (1975) 302, and references therein.
4. M.J. Vasile, F.A. Stevie and W.E. Falconer, International Journal of Mass Spectrometry and Ion Physics, 17 (1975) 195.
5. See, for example, L. Wharton, R.A. Berg and W. Klemperer, J. Chem. Phys., 39 (1963) 2023; R.A. Berg, L. Wharton and W. Klemperer, J. Chem. Phys., 43 (1965) 2416.
6. A.A. Muenter and T.R. Dyke, J. Chem. Phys., 63 (1975) 1224; I.M. Mills, J.K.G. Watson and W.L. Smith, Mol. Phys., 16 (1969) 329.
7. See, for example, P.A. Akishin, N.G. Rambidi and V.P. Spiridonov, in J.L. Margrave (Editor), The Characterisation of High Temperature Vapours, Wiley, New York, 1967.
8. For a discussion of electron diffraction see S.J. Cyvin (Editor), Molecular Structure and Vibrations, Elsevier, Amsterdam, 1972; Molecular Structure by Diffraction Methods, Chemical Society Specialist Periodical Reports, London, 1973, especially a paper by A.G. Robiette (p. 160).
9. K. Kashiwabara, S. Konaka and M. Kimura, Bull. Chem. Soc. Japan, 46 (1973) 410.
10. B.P. Stoicheff, J. Mol. Spec., 33 (1970) 183.
11. For a related discussion see, M.S. Farag and R.K. Bohn, J. Chem. Phys., 62 (1975) 3946.
12. See, for example, R.E. Smalley, L. Wharton and D.H. Levy, J. Chem. Phys., 63 (1975) 4977.
13. See, for example, I.R. Beattie, in A.J. Downs, D.A. Long and L.A.K. Staveley (Editors), Essays in Structural Chemistry, Macmillan, London, 1971.
14. M.L. Lesiecki and J.W. Nibler, J. Chem. Phys., 63 (1975) 3452.
15. I.R. Beattie and R.O. Perry, J. Chem. Soc. A, (1970) 2429.
16. I.R. Beattie and J.R. Horder, J. Chem. Soc. A, (1969) 2655.
17. I.R. Beattie, Molecular Spectroscopy, Institute of Petroleum, London, 1971.
18. J. Barnes and T.R. Gilson, personal communication.
19. I.R. Beattie and J.R. Horder, J. Chem. Soc. A, (1970) 2433.
20. L.E. Alexander and I.R. Beattie, J. Chem. Soc. Dalton, (1972) 1745.
21. C. Barraclough, I.R. Beattie and D. Everett, in J.R. Durig (Editor), Vibrational Spectra and Structure, Volume 5, Elsevier, Amsterdam, 1976.
22. P.J. Jones, personal communication.
23. I.R. Beattie, S.B. Brumbach, D. Everett, R. Moss and D. Nelson, Faraday Symposia of the Chemical Society, No. 8, (1973) 107.
24. A. Büchler, W. Klemperer and A.G. Emslie, J. Chem. Phys., 36 (1962) 2499.

25. R.J.H. Clark and D.M. Rippon, J. Chem. Soc., Faraday II, (1973) 1497.
26. A. Loewenschuss, A. Ron and O. Schnepp, J. Chem. Phys., 50 (1969) 2502.
27. H.E. Blayden, personal communication.
28. S.C. Wallwork and I.J. Worrall, J. Chem. Soc., (1965) 1816.
29. A.W. Laubengayer and F.B. Schirmer, J. Amer. Chem. Soc., 62 (1940) 1578.
30. I.R. Beattie, H.E. Blayden, S.M. Hall, S.N. Jenny and J.S. Ogden, J. Chem. Dalton, (1976) 666.
31. J.S. Anderson, A. Bos and J.S. Ogden, Chem. Comm., (1971) 1381.
32. J.S. Ogden and M.J. Ricks, J. Chem. Phys., 53 (1970) 896.
33. J.S. Ogden, personal communication.

CHAPTER 10

TRACE ANALYSIS BY INFRARED SPECTROSCOPY

W.O. George and J.P. Coates

1. INTRODUCTION

The purpose of the present chapter is to consider special methods of quantitative analysis by infrared of systems at trace levels and involving problems of a wide range of chemical, environmental and medicinal importance.

A spectrometer produces a record of variation in intensity of electromagnetic radiation with frequency after interaction with the sample being studied. This may be observed in emission or absorption. Traditionally emission spectroscopy has been used for trace analysis because of the high dynamic range of a detection system based on a positive measure of emitted energy. A limitation is that the sample must be excited thermally or electrically before an emission spectrum can be produced and many molecules decompose under these conditions. The main application of emission spectroscopy is for trace analysis of atomic systems from observation of their electronic spectrum in the visible or ultraviolet.

Absorption spectroscopy has traditionally been practised over a limited dynamic range because the measurement consists of the percentage of the energy of the source transmitted by the sample and the measurement has normally been made to best advantage in the 80-10% range. The advantage of absorption spectroscopy is that the vast majority of molecules are stable under conditions at which spectra are measured particularly in the infrared region. The present chapter considers methods by which the dynamic range of an absorption process in the infrared range may be increased using a ratio recording spectrometer rather than an optical null servo recording instrument.

2. MEASUREMENT OF ABSORPTION

The basis of quantitive measurements in infrared and other forms of absorption spectroscopy is the Beer-Lambert law which can be expressed in the form

$$\log_{10}(I_0/I) = \varepsilon C \ell = A \qquad (1)$$

where I_0 and I are the intensities of the incident and transmitted radiation respectively, C the molar concentration of the absorbing species, ℓ is the thickness of the absorbing layer, ε the extinction coefficient and A is the absorbance.

The Beer-Lambert law may be derived as follows:-
For a narrow beam of monochromatic radiation passing through a

sample consider a collision between a single photon and a molecule. The probability of transmission of the photon through the molecule may be defined as p on a scale 0 to 1.

Consider a series of collisions between a single photon and N molecules in the beam's path. The probability of transmission is now reduced to p^N.

Fraction of radiation transmitted by all molecules = Probability of transmission of a single photon

$$(I/I_o) = p^N \qquad (2)$$

N is proportional to the concentration, C, of the molecules in the beam and to the path-length, ℓ, of the sample

$$(I/I_o) = p^{kC\ell} \qquad (3)$$

$$\log_{10}(I_o/I) = (-k \log_{10} p)C\ell \qquad (4)$$

$$A = \varepsilon C\ell \qquad (5)$$

where $-k \log_{10} p = \varepsilon$ since $p < 1$.
The physical nature of the extinction coefficient, ε, emerges from its units. If C is in mol dm^{-3} and ℓ is in mm the units of $C\ell$ are mol m^{-2} and since A is dimensionless the units of ε are m^2 mol^{-1}. Hence ε is an effective cross-sectional area of a molecule.

Since the derivation is based on simple probability concepts and no assumptions are required there is no basis for true deviation from the Beer-Lambert law. If data is reported to show deviation it would be prudent to consider this as "apparent" rather than "real" and to consider whether the parameters related by the law are properly measured.

For example consider the measurement of absorbance and concentration of a series of solutions of benzyl alcohol in carbon tetrachloride [1]. Benzyl alcohol may be characterised by bands in the infrared near 3600 and 3300 cm^{-1}. According to the Beer-Lambert law a plot of absorbance of these bands against concentration should lead to a straight line of slope $\varepsilon\ell$. By contrast a plot of the absorbance of the higher wavenumber band against concentration leads to an apparent negative deviation whereas a plot of the absorbance of the lower wavenumber band leads to an apparent positive deviation (Figure 1). The reason for these departures is that benzyl alcohol participates in hydrogen bonding equilibria:

$$n\ C_6H_5CH_2OH \rightleftharpoons (C_6H_5CH_2OH)_n$$

The band near 3600 cm^{-1} in the infrared is assigned to the unbonded form of benzyl alcohol. At increasing concentrations proportionately less of this species is present and the true concentration and absorbance is lowered, leading to an apparent negative deviation. Conversely the band at 3300 cm^{-1} in the infrared is assigned to the bonded, associated form of benzyl alcohol. At increasing concentrations proportionately more of this species is present and the true concentration of associated

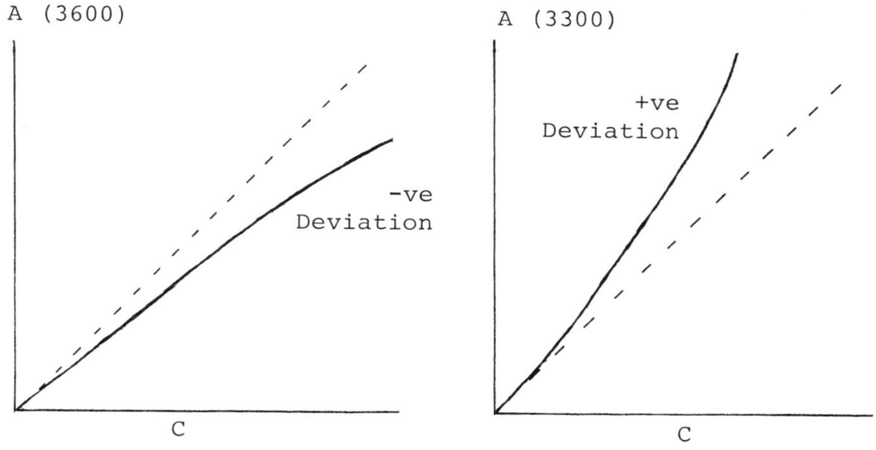

Figure 1. Beer-Lambert law plots for benzyl alcohol.

species giving rise to the form or forms leading to these absorptions is raised and an apparent positive deviation is observed.

In both cases the apparent deviations are caused by failure to measure the concentration of the actual species responsible for the absorptions. This may be regarded as a chemical reason for apparent departure from the Beer-Lambert law. Other examples occur wherever there is a change in the nature of the absorbing species with concentration. Alternatively, apparent deviations may be caused by failure to measure absorbance accurately. These may be regarded as physical reasons. Measurement of absorbance may be bedevilled by instrumental factors of the following type:-

Pen response

With optical null servo systems the difference in energy between two beams becomes the difference between two very small signals at low transmittance (high absorbance) values. Strong absorbances will then be measured with much lower precisions than moderate absorbances.

Attenuator errors

The accuracy of optical null servo systems is determined by the mechanical precision of a moving wedge attenuator. This is normally a comb-shaped device but rotating star-wheels have been used for improved precision. Since measured transmittance is in effect the distance moved by the attenuator, this movement must be linearly related to the percentage of energy transmitted over the entire range. Small changes in transmittance cannot be accurately measured because of mechanical inertia and measurements of weak absorbance are normally subject to error.

Stray light

Any spectrometer should ideally accurately measure radiation at each frequency. This is clearly an impossible task. Stray light is a measure of the departure from the ideal and is the percentage of measured energy arising from frequencies other than the one at which the instrument is set. It may only be measured approximately (e.g. using a sample opaque to certain frequencies) but it always exists and leads to errors in absorbance measurements.

Sample radiation

Samples, in particular solids and liquids, may emit significant amounts of infrared radiation which leads to a false value for the apparent absorption or transmittance. For example, the strong band near 700 cm^{-1} in a polystyrene film emits radiation which leads to an apparent transmittance of 1-2%. This shows up in the same way as stray light. It may be removed by using an instrument which chops the radiation before as well as after the sample.

Resolution

If a band is recorded at insufficient resolution the measured peak absorbance is lower than the true value.

3. SELECTION OF INSTRUMENT CONDITIONS

The accuracy and precision of measurement depends on the optimum choice of parameters such as electronic gain (determining pen response), slit width (determining resolution) and scan time. Important qualities are clearly high resolution, wavenumber accuracy, absorbance accuracy, facilities for expanding both scales.

Ratio recording rather than the null balancing servo system improves accuracy of absorbance measurements, particularly at extremes of transmittance regions because the pen is fully activated and live throughout. Pre and post-sample chopping also improves accuracy of absorbance by eliminating radiation arising from emission by the sample and other extraneous sources.

Most infrared spectra are recorded with a linear transmittance scale on the ordinate axis. In this case the 100% transmittance reading is set at I_0 and the measured percentage transmittance is T and is set at I. This is converted to an absorbance value $A = \log(100/T)$ for any quantitative measurement since A is linear with concentration and pathlength. Alternatively an infrared spectrometer may record spectra on a linear absorbance scale which may be more convenient for quantitative measurements. In this chapter spectra are recorded in either transmittance units (absorption bands downwards) or in absorbance units (absorption bands upwards) using a Perkin Elmer Model 580 infrared spectrometer in all cases.

4. TRACE ANALYSIS IN CARBON TETRACHLORIDE SOLUTIONS

Carbon tetrachloride is a valuable solvent because it is reasonably transparent to infrared radiation. Normal polar solvents absorb strongly at pathlengths greater than 0.1 to 1.0 mm. Because the highest wavenumber value fundamental occurs below 800 cm^{-1} it follows that the highest binary combination will be less than 1600 cm^{-1} and the highest ternary combination will occur at less than 2400 cm^{-1}. The implication of these factors are shown in Figure 2. Regions of very low

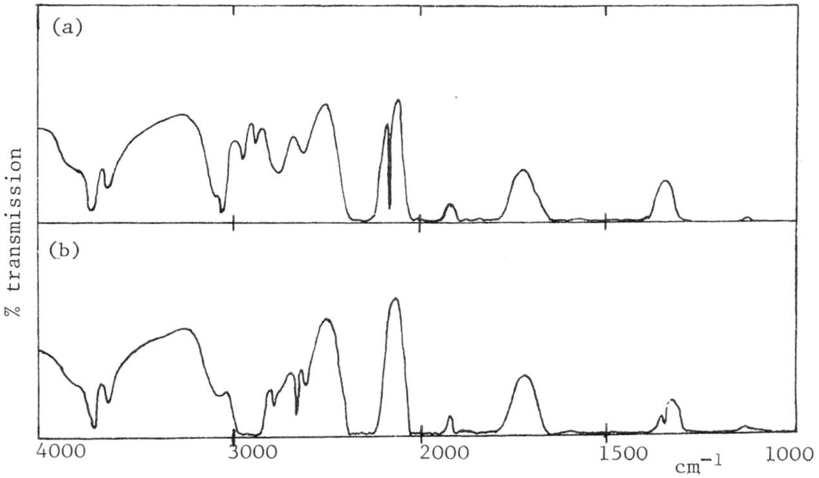

Figure 2. Infrared spectra of (a) "technical" and (b) "Spectrosol" carbon tetrachloride at 100 mm path length.

transparency occur up to 1600 cm^{-1}, moderately low transparency up to 2400 cm^{-1} and reasonable transparency above 2400 cm^{-1}.

Comparison of unpurified samples of CCl_4 labelled "Spectrosol" and "technical" reveal interesting differences. The former has additional bands due to saturated hydrocarbons at 2900, 2800 and 1360 cm^{-1}. Also bands at 2660 and 2800 cm^{-1}. The latter has additional bands at 3100 cm^{-1} due to aromatic hydrocarbons, also bands at 2160 and 2170 cm^{-1}. These hydrocarbon bands can be diminished by fractional distillation of the sample. Both spectra reveal bands at 3620 and 3715 cm^{-1} which are due to traces of water and which can be reduced by adding silica gel or other suitable dessicant.

Determination of trace quantities of hydrocarbon and carbonyl compounds

Hydrocarbon is a common pollutant from oil spills and other sources. When present as a contaminant or in suspension in water it may be readily extracted into CCl_4. From Figure 2 and by taking appropriate values of extinction coefficients of saturated and unsaturated hydrocarbons it is feasible to work out the limits of detection of hydrocarbons in solutions of the carbon tetrachloride extract and to estimate trace amounts of hydrocarbon.

The region of transparency between 1200 and 1800 cm^{-1} enables very low quantities of carbonyl compound to be estimated in carbon tetrachloride solution.

Spectra of acetoacetanilides in very dilute CCl_4 solution

Being polar, acetoacetanilides are only sparingly soluble in CCl_4. A saturated solution contains typically 75 mg/100 ml. Figures 3 and 4 show spectra of saturated solutions of

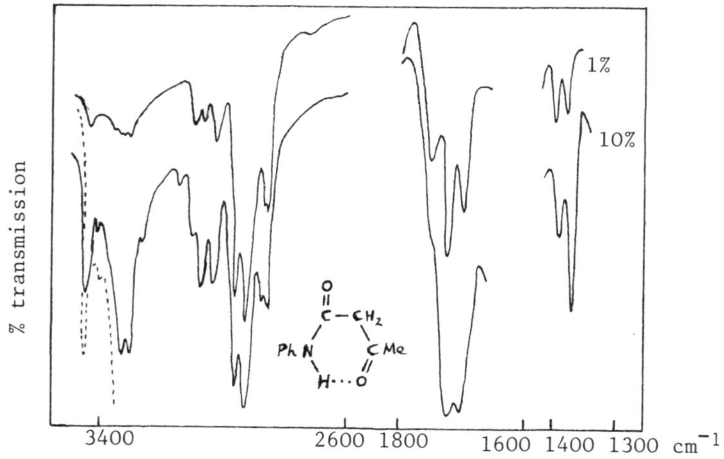

Figure 3. Infrared spectra of acetoacetanilide in CCl_4 solution.

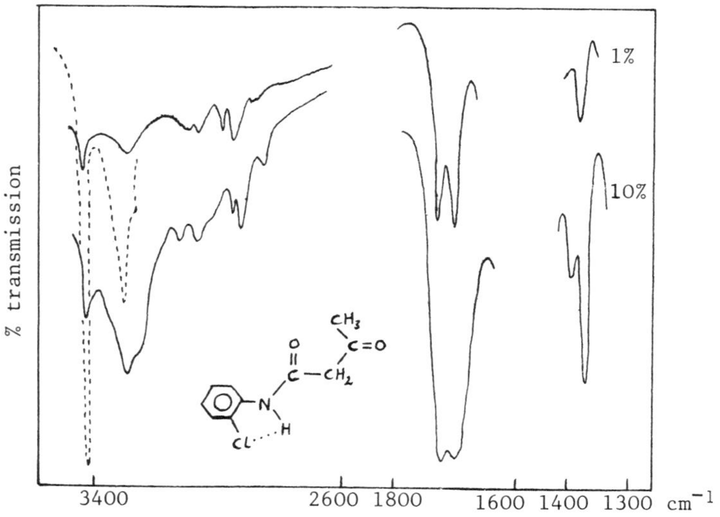

Figure 4. Infrared spectra of o-chloroacetoacetanilide in CCl_4 solution.

acetoacetanilide and o-chloroacetoacetanilide further diluted by 10% and 1% leading to concentrations of ca. 7.5 x 10^{-3} and ca. 7.5 x 10^{-4}% at 100 mm path length. Three regions of infrared transparency are shown: 2700-3700 cm^{-1}, 1600-1800 cm^{-1} and 1300-1450 cm^{-1}. These are shown at 10% and 1% of saturated solution. In addition the highest wavenumber region is repeated at the lower concentration but at 10-fold absorbance scale expansion.

In the infrared spectrum of acetoacetanilide bands at 3440 and 3320 cm^{-1} are assigned to free and bonded N-H stretching modes respectively. A ten-fold dilution reveals no change in the relative intensity of these bands. It follows that the bonding is probably intramolecular to the acetyl carbonyl groups, as shown in Figure 3.

Examination of the carbonyl region shows three bands which may be expected to belong to the unbonded acetyl (1743 cm^{-1}), bonded acetyl (1716 cm^{-1}) and the amide carbonyl (1692 cm^{-1}).

In the infrared spectrum of o-chloroacetoacetanilide bands at 3420 and 3292 cm^{-1} are assigned to N-H stretching modes as in the unsubstituted compound but it is seen that there is a marked concentration dependence of the relative intensities of these modes. This suggests there is intermolecular hydrogen bonding between N-H and C=O groupings as shown in Figure 4. In accord with this only a very weak shoulder at 1740 cm^{-1} is apparent in the carbonyl region.

A possible rationale is that the existence of an adjacent chloro group competes with the acetyl carbonyl for the N-H. The lone pair electrons on the chlorine may attract the proton away from those of the oxygen which are thereby more readily available for intermolecular association.

5. ESTIMATION OF TRANS FATTY ACIDS IN LIPIDS EXTRACTED FROM P.M. HEART TISSUE

Heart disease in the Western world is more significant as a killer disease than cancer, particularly in early to middle age. Many factors have been attributed to its incidence and it is widely agreed that excessive dietary fat has adverse effect. The quality of the fat may also be of importance and the view has been expressed that "animal" fats vis-a-vis "vegetable" fats are suspect on the grounds

(a) they contain lower % unsaturated components,
(b) they contain higher amounts of cholesterol.

On these grounds it has for some 20 years been suggested in particular that butter fat (low unsaturation) should be replaced by magarine (mostly ex vegetable sources).

A recent study by L.H. Thomas [2] has focussed attention in a different direction. Margarine and "shortenings" are of course manufactured by hydrogenation whereby unsaturated liquid oils are converted into more saturated solid or semi-solid fats. In this process, cis double-bond acids are easily isomerised to the trans isomer. Trans acids are very rare in natural fats but may thus be present in margarine to ca. 30-40%.

A more recent tendency is to use fish oils for margarine production. These contain high amounts of polyunsaturated C_{20} and C_{22} acids which are converted by the hydrogenation to the mono-unsaturated acids 20:1. These, too, are present in natural fats in low amount - except in rape-seed oil.

Thomas expressed the opinion that these "unnatural" components - trans acids and higher mono-enoic acids - may be harmful as such. By comparing mortality rates from arteriosclerotic disease, ischaemic heart disease and cerebro-vascular disease in the various standard regions and social classes in the U.K. with consumption of hydrogenated fat (HF) he showed that there is, in fact, a positive correlation.

In Figure 5(a) an infrared spectrum is shown of a lipid

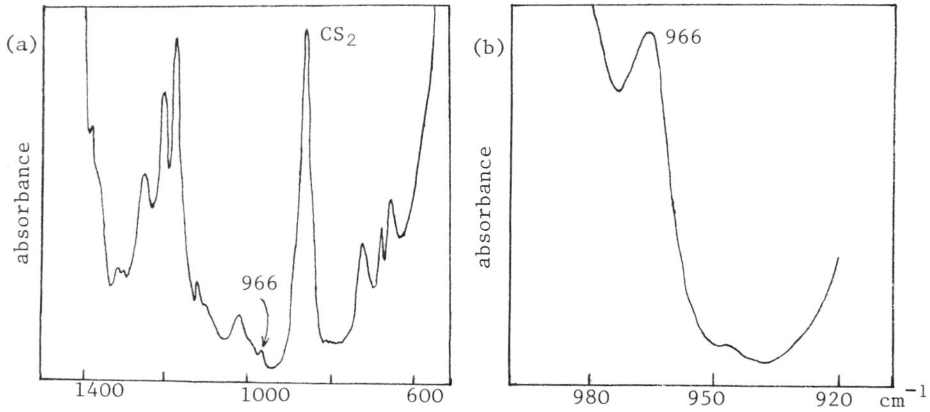

Figure 5. Infrared spectra of lipid extracted into CS_2 from P.M. heart tissue.

sample extracted from heart tissue and dissolved in CS_2. A very weak band at 966 cm^{-1} is assigned to the out-of-phase trans CH=CH deformation. Figure 5(b) shows the same band after absorbance scale expansion of 20 times. This leads to at least 10 times increase in precision. The repeatability is illustrated by superimposing 10 repeat scans.

There are two methods described for calibration of this analysis. Allen [3] has indicated how a rapid estimation may be made by relating the absorbance of the trans unsaturation band at 966 cm^{-1} to another band in the spectrum (1170 cm^{-1} for esters or 934 cm^{-1} for acids). To calculate the per cent isolated trans double bonds in the sample the ratio of these values is substituted into a linear equation developed from samples of known composition. The method is rapid since the sample need not be weighed nor made up to a known volume. Huang and Firestone [4] have compared the above rapid method with the tentative procedure of the American Oil Chemists Society based on calibration using weighed samples of known composition and found the latter significantly more accurate.

6. ESTIMATION OF QUARTZ IN COAL DUST AND ASBESTOS TRACES

In mining communities chest diesases caused by coal dust are commonplace. Quartz or silica entering the lung causes silicosis and pneumoconiosis. In the N.C.B. laboratories at Pontypridd samples are collected from surrounding coal mines by drawing measured volumes of air at coal-faces through PVC filters supported on metal grids. The infrared spectrum of the dust can be recorded directly by putting the filter and holder directly in the beam, using a specially made mount (Gravimatric Dust Sample Type 113A, C.F. Casella & Co. Ltd.).

Toma and Goldberg [5] have considered the measurement of alpha quartz at the 1-5 mg level after collection on membrane filters and consider the problems associated with (1) variation of the absorbance of various filters of the same type (2) inhomogeneity of the dust distributed on the filter and (3) the decrease in concentration of a given weight of alpha quartz measured directly on an exposed 33 mm diameter filter as compared to the concentration obtained on a 13 mm diameter pellet. Gillieson and Farrell [6] have discussed the interference of bands arising from other minerals. Freedman, Toma and Lang [7] have improved the sensitivity by transfer of sample from the collection filter to a small area of a second filter and compared the methods of infrared and X-ray diffraction methods at levels below 10 μg of quartz.

A related problem is associated with the measurement of airborne asbestos since certain forms of asbestos have been considered to have an association with cancer. Asbestos has bands similar to those in α-quartz since both are derived from Si-O vibrations. Beckett, Middleton and Dodgson [8] have shown that the three principal forms of asbestos, chrysotile, amosite and crocidolite have different infrared spectra and have developed a rapid technique for measuring quantities down to 10 μg of single varieties recovered from the lungs of rats.

The infrared spectra of the filter and its support are compared in Figure 6 with that of quartz. It can be seen that the infrared spectrum of quartz has distinctive bands at 780 and 800 cm^{-1}, which is a region of reasonable transparency in the filter medium. The pair of bands at 780 and 800 cm^{-1} are observed in the spectrum of coal dust and can be used to estimate the amount of quartz present. Figure 7 shows the spectrum of coal dust in the region 760 and 800 cm^{-1} expanded in both scales and repeated by re-running 10 scans. Using calibration data provided the concentrations of quartz in coal dust can be estimated.

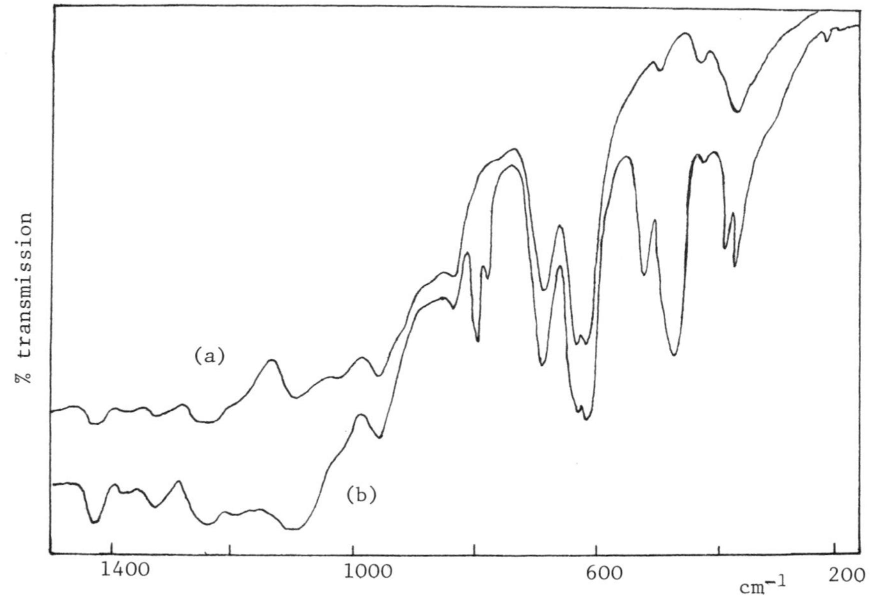

Figure 6. Infrared spectra of (a) PVC filter and (b) 3.02 mg quartz on filter.

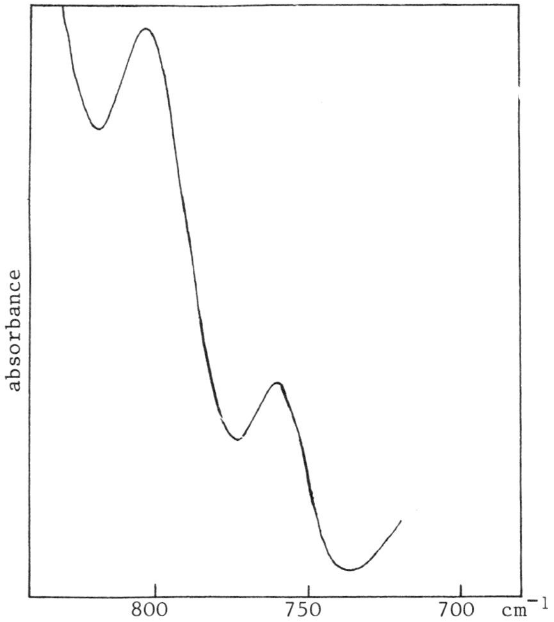

Figure 7. Infrared spectrum of coal dust on PVC filter, absorbance scale expanded x 20.

7. THE APPLICATION OF COMPUTER DATA HANDLING

All previous discussions in this chapter are based on results obtained directly from a high performance spectrophotometer. The following sections illustrate how the quality of spectroscopic data obtained from this instrument may be enhanced when it is coupled on-line to a dedicated computer. This approach extends the scope of conventional analysis to give lower detection limits, superior spectrum quality and the elimination of matrix interferences. The two main areas of computer application to trace analysis are:

(a) the improvement of spectra obtained from poorly transmitting samples (typically less than 5% T) and
(b) the enhancement of weak absorption bands that originate from a low sample/matrix ratio.

In case (b) data can be expanded from a maximum peak intensity of 1% T to give a full scale, undistorted spectrum. An example of the overall improvement is demonstrated in Figure 8, which

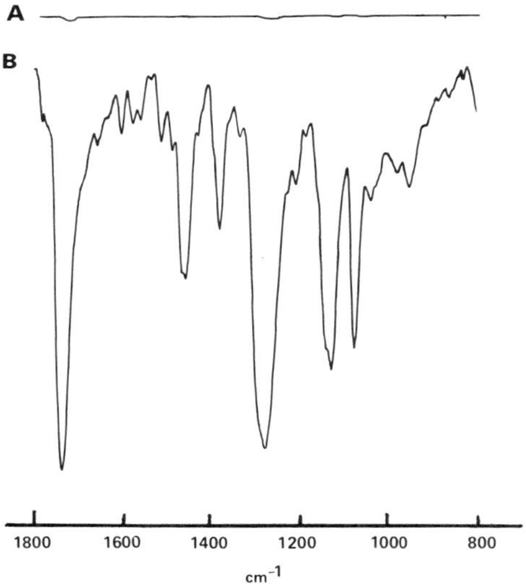

Figure 8. A, Evaporated residue (estimated weight 5 μg); B, computer enhanced spectrum.

is the spectrum produced by a trace quantity of an extracted contaminant from the surface of an electrical component. The original data (Figure 8A), recorded from approximately 5 μg of material deposited on a standard, 25 mm, potassium bromide window, gives a maximum peak intensity of 0.75% T. The spectrum in Figure 8B results from computer processing of this data and is produced by large ordinate expansion (200 x) applied to the absorbance form of the original spectrum - i.e. a simulation of 200 x increase in conçentration. It must be pointed out that this type of experiment can only be performed successfully on a ratio recording infrared instrument since the more conventional optical null instruments suffer from lack of sensitivity to small variations in spectrum intensity. This reduced sensitivity of conventional equipment is attributed to the "mechanical dead band" that exists in the optical wedge-servo system used to generate the spectral data.

The following computer facilities are utilised to achieve the objectives specified in (a) and (b):

Ordinate expansion (with absorbance/transmittance interconversion)

Certain physical limitations restrict the scope of normal instrumental ordinate expansion to a relatively narrow range - for example, a typical range covered by a high performance instrument is 0.25 x to 20 x (fixed expansion). Computer expansion, however, has greater flexibility - for example,

(a) it is continously variable (i.e. not necessarily integer),
(b) it may be operated over a wide range to give expansions as high as 200 x (or even larger),
(c) it can be applied by a specified amount - important for quantitative considerations.

An additional advantage is gained by expanding the data in absorbance and presenting the spectrum on chart in linear transmittance format. The importance of this technique is the possibility for direct comparison of the expanded spectrum with standard reference spectra. An illustration of the relative effects of normal instrumental (linear transmittance) versus computer (linear absorbance) expansion on a weak spectrum of phenacetin is given in Table 1. The relative intensities of the bands at 1250 cm^{-1} and 1270 cm^{-1} reflect the spectral distortion associated with instrumental (linear transmittance) expansion.

Table 1. Effect of different methods of ordinate expansion on the spectrum of ca. 50 μg phenacetin in KBr (100 mg).

Expansion	Band ratio* (1250/1270 cm^{-1})
5 x, Instrumental - linear transmittance	1 : 6.25
10 x, Computer - linear absorbance	1 : 2.56
10 x, Increase in sample concentration	1 : 2.50

* Calculated from the absorbance values of the two bands

Digital smoothing

High instrumental noise levels are often associated with large ordinate scale expansion and trace/micro sample analysis. Even a noise level as low as 0.1% T will produce excessively high noise after a large expansion - for example 50 x expansion would give rise to 5% T noise. It is also important to appreciate that after a linear absorbance expansion, a logarithmic increase in noise is experienced in the transmittance spectrum. This is observed as an increase in baseline noise relative to peak noise (i.e. has greatest disturbance at high transmittance /low absorbance values). Under normal circumstances the elimination of noise is necessary to improve the readability of spectral information and occasionally for the sake of a "cosmetic" appearance. This may be conveniently carried out on a recorded spectrum by the application of digital smoothing - a moving point, polynomial averaging function. The software package, used for the computer examples in this chapter, utilises a quadratic function which can be manipulated to give different degrees of noise reduction. The desired "smoothing level" is chosen from seven data "windows" which are defined by the number of data points used for the average i.e. 5, 13, 19, 25, 37, 49 and 169 point functions. For most routine applications, the 13, 19 and 25 point functions will reduce instrumental noise to an acceptable level without significant spectral distortion (N.B. this will depend on the data interval chosen for the recorded data). The relative effects of the 13 and 25 point functions on a noisy spectrum (4-5% T peak to peak noise) are demonstrated in Figure 9.

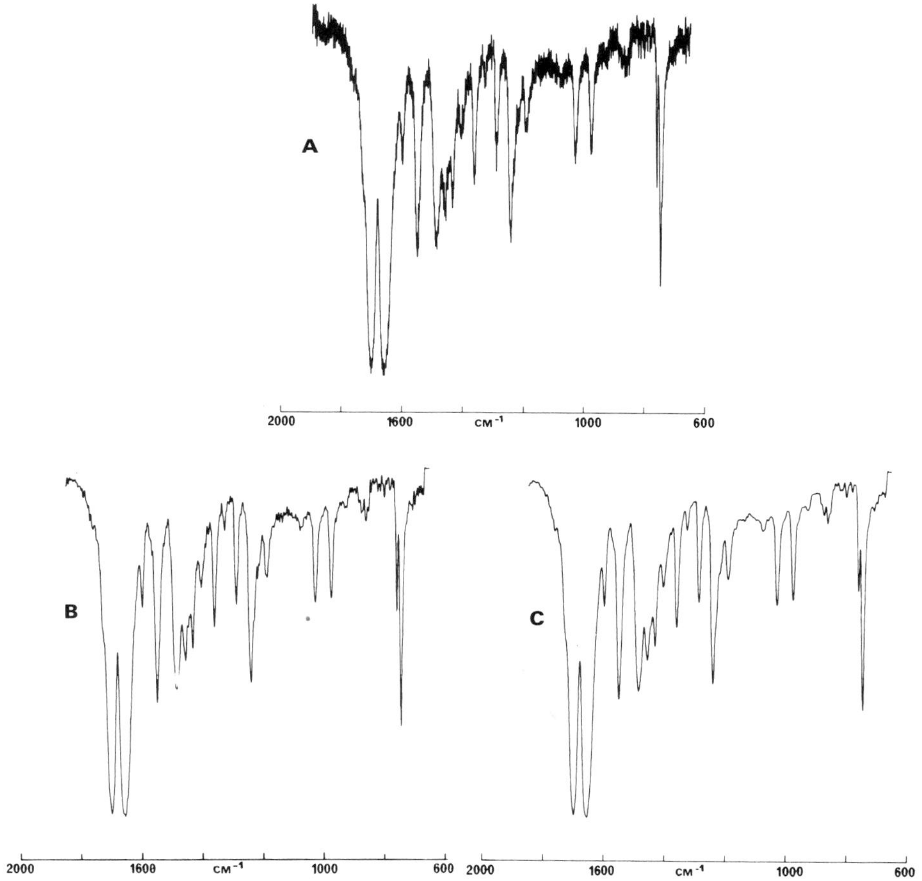

Figure 9. A, Spectrum of caffeine (in a KBr disc) with a high noise level; B, spectrum A after 13 point smooth; C, spectrum A after 25 point smooth.

If data is obscured or hidden within the noise it will probably remain hidden after digital smoothing since at this level data and noise have equal weighting in the averaging process. The significance of such data, relative to the noise, may be increased by spectral accumulation (CAT-ing) - see below.

Difference (absorbance subtraction)

A major hindrance to large ordinate expansions is the presence of a sloping/uneven baseline or a highly absorbing background. This may arise from one or more of the following factors:

(i) Scatter
(ii) Interference fringes
(iii) Solvents or solvent residues
(iv) Impurities
(v) Cell cut-off
(vi) Optical disturbances caused by an accessory
(vii) Absorptions from a bulk matrix

In cases (i) and (ii) the effects are variable and cannot be easily reproduced. All the remaining factors, however, are reproducible and may be eliminated by computer subtraction or difference. The subtraction is manipulated in absorbance (equivalent to a transmittance ratio) and has provision for automatic or manual scaling to compensate for intensity variations of the background.

One advantage of the computer method, compared to conventional differential infrared spectroscopy, is that it may be used in conjunction with any method of sample preparation. Straightforward application of the technique is analogous to solvent compensation with a variable pathlength cell. Recent experiments with compounds containing carbonyl groups, examined as dilute carbon tetrachloride solutions in a similar cell, have suggested that specific infrared bands, such as the carbonyl, can be detected down to 50 ppm (and possibly lower). With selected solvents, the method lends itself to the study of chromatographic fractions, e.g.:

components separated by liquid chromatography (HPLC)
trapped fractions from gas chromatography
compounds isolated by extraction from thin layer chromatography (TLC) plates.

If a fraction is contained in a poor infrared solvent - such as water/methanol mixtures - it is advisable to isolate the material by evaporation, freeze drying or solvent extraction. The analysis is continued in an infrared grade solvent.

Earlier in this chapter, reference is made to the application of infrared analysis to the estimation of quartz in coal dust from samples collected on PVC membrane filters. Computer difference is a convenient method for the elimination of the filter background since the autoscaling feature may be utilised to compensate for variations in filter thickness. A spectrum of α-quartz, free from any background interference, is easily obtained by this technique. Similar experiments with actual samples also indicate the presence of the α-quartz bands (800 cm^{-1}/780 cm^{-1}) after background subtraction. The resultant difference spectra are complicated by the presence of other silicate minerals - in particular kaolinite. This is illustrated by the spectra in Figure 10 obtained from a typical

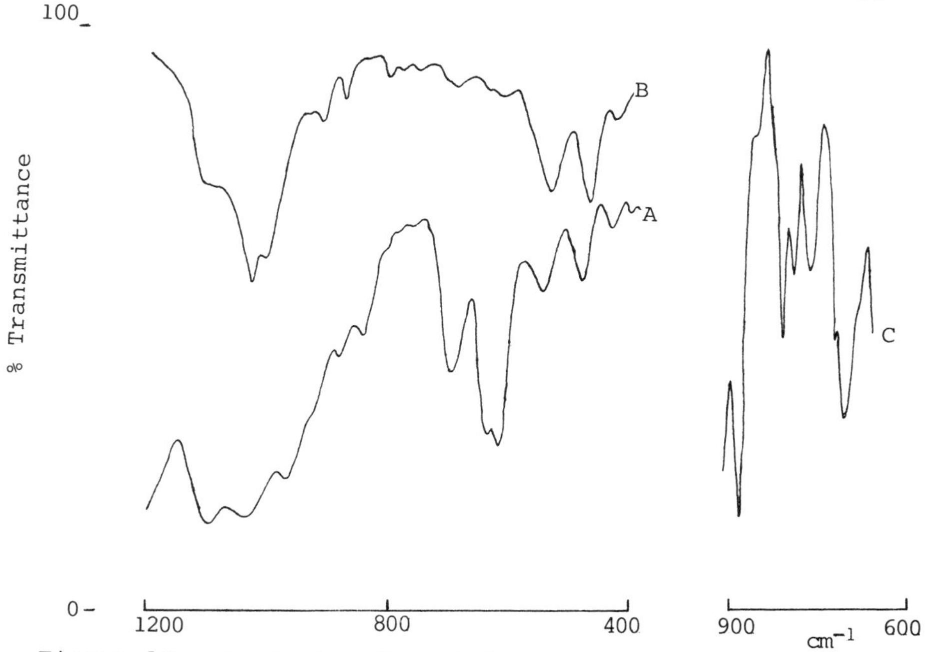

Figure 10. Analysis of coal dust on membrane filters: A, sample of coal dust on PVC filter; B, spectrum A after computer subtraction of filter spectrum; C, spectrum B expanded 20 x in absorbance.

sample of coal dust. Figures 10A/B are the original spectrum and the normal difference spectrum respectively. After 20 times ordinate expansion in absorbance, spectrum 10C (transmittance spectrum) is produced. Obviously, with additional standard samples, the kaolinite interference could also be removed by a further computer subtraction.

Spectral accumulation

On occasions, a spectrum may be too weak to distinguish the absorption bands from the background. The significance of such data is often improved by the application of spectral accumulation. The technique is well suited to the study of trace quantities of adsorbed species on solid substrates as illustrated in the following experiments.

In Figure 11, a trace quantity of a substituted aromatic hydrocarbon adsorbed on carbon black is examined in the form of a potassium bromide disc. Under normal conditions, the maximum transmission of this sample is approxmately 0.5% T over the range 3700-2700 cm^{-1} - spectrum 11A. After 64 accumulations, followed by some expansion and smoothing, clearly defined bands from the hydrocarbon (aromatic/aliphatic C-H) are observed. The broad band centred at 3600 cm^{-1} is assigned to -OH present

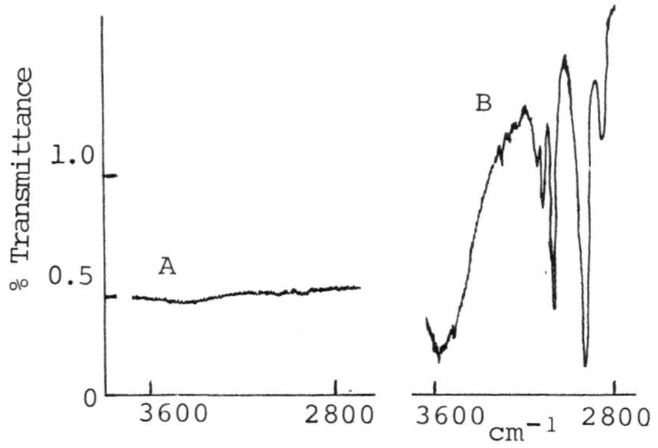

Figure 11. A, Normal transmission spectrum with instrumental expansion; B, computer replot from 64 accumulated scans with scale expansion and smoothing.

as water either adsorbed on the carbon surface or within the potassium bromide matrix.

In a second example, two metal specimens are studied by surface reflectance - Figure 12. Before accumulation there is

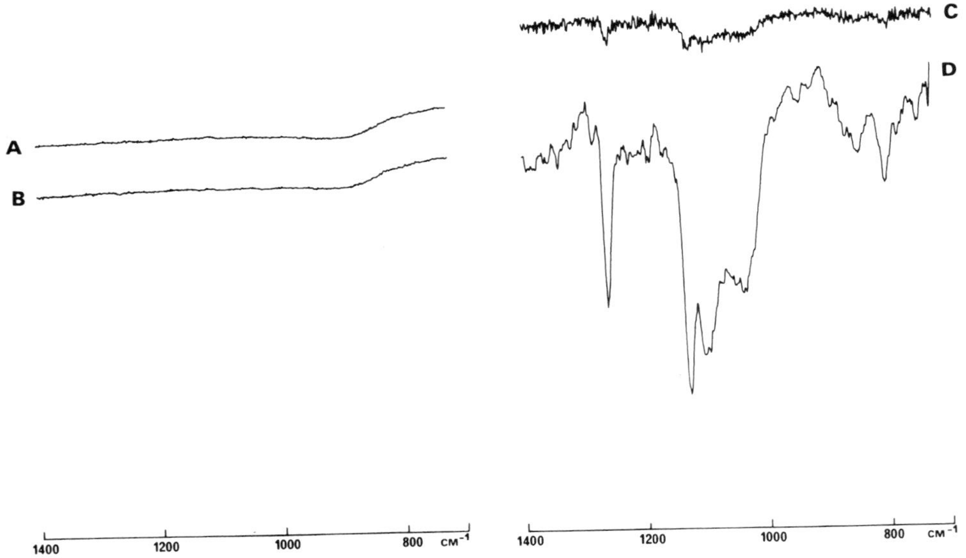

Figure 12. A, reflectance spectrum of metal surface; B, spectrum after surface treatment; C, difference spectrum from 15 accumulations of A and B; D, spectrum C scale expanded (10 x) and smoothed (25 point).

little evidence of any significant differences between the two surfaces (spectra 12A/B). A difference is apparent, however, after a spectrum accumulation of each surface, followed by computer subtraction, as seen in Figure 12C. Further processing of this data, i.e. 10 times ordinate expansion and a 25 point smooth, yields a spectrum which is characteristic of a silicone oil. The magnitude of this data indicates that the material has an approximate film thickness of 2 nm.

Baseline manipulation

In the section dealing with computer difference (see above), a reference was made to problems experienced with spectra that contain irregular baselines. Typically, these arise from poor transmission, grating and/or filter steps, or scatter often generated by micro samples or micro sampling accessories. When these artefacts directly result from the use of an accessory, they may be removed by computer subtraction of a background spectrum of the accessory. If, however, both samples and accessory contribute to the baseline deformation, the difference procedure will not give adequate compensation. In such cases, careful software manipulation, i.e. the use of selective baseline shifts (carried out in absorbance) and slope removal by a 2 (or 3) point "FLAT" routine, can give relatively undistorted spectra, capable of ordinate expansion. The use of these facilities is demonstrated on the spectrum produced from a micro disc, with a 1 mm aperture, of a paint sample dispersed in potassium iodide (Figure 13), placed directly in the sample

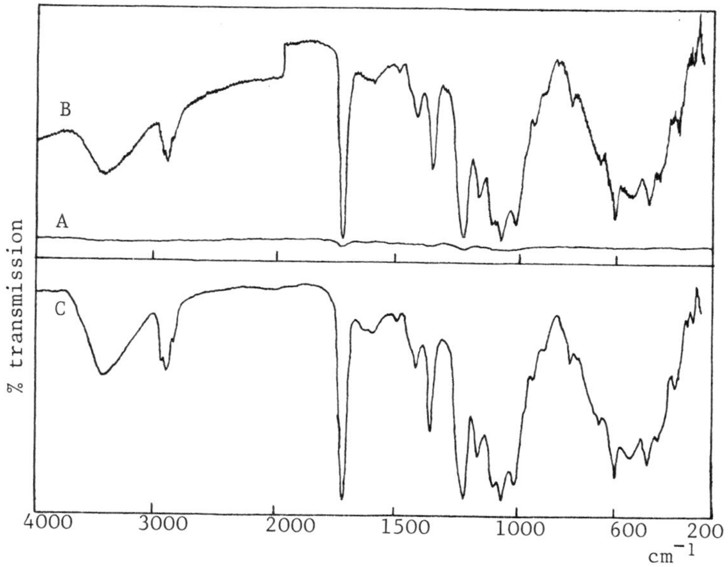

Figure 13. A, paint sample in KI micro disc (1 mm); B, spectrum A with full scale expansion; C, spectrum B after baseline adjustment.

beam - i.e. no beam condensing optics. A spectrum with a low transmittance (approximately 3% T) is obtained (Figure 13A) and this gives a distorted spectrum after computer expansion, Figure 13B. These distortions have been removed in Figure 13C by the FLAT routine and a baseline shift between 4000 cm^{-1} and 1980 cm^{-1}. This type of problem is frequently encountered in the examination of forensic and medical samples. With careful data manipulation, good quality results could be obtained from a very low sample loading.

Future trends in computer extended trace analysis

There are two main areas of development to improve spectrum quality and lower detection limits:

(i) The modification of standard sample preparation techniques,
(ii) The extension of the computer software to include more sophisticated routines.

One of the main limitations of this work is the presence of impurities which often exist at a level far greater than the sample concentration. Impurities that are intrinsic in the matrix are not a direct problem since they may be eliminated by the software. Contaminants picked up by the matrix during sample preparation, however, are the main source of concern. These may arise from atmospheric vapours, pump oil, solvent residues or even miscellaneous contaminants such as fingerprints.

With future improvements in computer techniques and careful sample manipulation, an overall gain in sensitivity of several orders of magnitude will be possible.

8. REFERENCES

1. J.J. Fox and A.E. Martin, Trans. Faraday Soc., 36 (1940) 897.
2. L.H. Thomas, Brit. J. of Soc. and Prev. Medicine, 29 (1975) 82.
3. R.A. Allen, Anal. Chem., 41 (1969) 552.
4. A. Huang and D. Firestone, J. of the AOAC., 54 (1971) 1288.
5. S.Z. Toma and S.A. Goldberg, Anal. Chem., 44 (1972) 431.
6. A.H. Gillieson and D.M. Farrell, Can. J. Spectr., 16 (1971) 21.
7. R.W. Freedman, S.Z. Toma and H.W. Lang, Amer. Indust. Hygiene Assoc. Journal, (1974) 411.
8. S.T. Beckett, A.P. Middleton and J. Dodgson, Ann. Occup. Hyg., 18 (1975) 313.

CHAPTER 11

RESONANCE RAMAN SPECTROSCOPY

R.J.H. Clark

1. INTRODUCTION

Raman scattering is considered to be a coherent absorption-emission sequence [1]. The intermediate state in this process is usually represented by a linear combination of eigenstates of the scattering molecule, each eigenstate being weighted according to its nearness to resonance. Far from resonance, the number of contributing states is large and many electronic manifolds are involved. In this case all the energy denominators in the expression for an element $\alpha_{\sigma\rho}$, of the scattering tensor are large, and hence the weighting factors are small and non-selective. The consequence of this is that only symmetric tensor components make any significant contribution to the scattering, and overtones have very low intensities. These simplifications to the scattering tensor arise because the statistical superposition of many eigenstates leads to an intermediate state with neither symmetry nor geometry clearly defined [2].

This situation changes under resonance conditions because then only some or possibly only one of the energy denominators become small, leading to large weighting factors for the corresponding eigenstates. The intermediate state in the scattering process assumes the symmetry and geometry of these dominant states, leading to both Franck-Condon progressions and contributions from antisymmetric tensor components. The simplest resonance case is that in which the absorption spectrum of the molecule consists of sharp vibronic or rovibronic lines and the energy of the incident light beam coincides with that of one of these lines. In this situation, the resonance fluorescence limit, only one vibronic or rovibronic state contributes significantly to the intermediate state. On the other hand, if the absorption spectrum of the scattering molecule consists of broad overlapping bands, or is continuous, it is necessary to sum the contribution of many states to the scattering process, even under resonance conditions. The situation nevertheless remains fairly simple if all these states belong to the same electronic manifold, in which case the intermediate state may be said to coincide with an electronic rather than a vibronic state. Most resonance Raman spectra so far reported are thought to correspond to this special case [3,4] as most relate to scattering by molecules in condensed phases, in which fluorescence is normally quenched.

2. RESONANCE RAMAN EQUATIONS

In order to apply resonance Raman spectroscopy to chemical problems it is necessary to establish what types of vibrational modes are enhanced by resonance with what types of electronic transitions. The quantum-mechanical expression for a Raman tensor element, as developed by perturbation methods, has the form [5-7]

$$(\alpha_{\sigma\rho})_{gi,gj} = A' + A + B \tag{1}$$

where

$$A' = \frac{1}{\hbar}\sum_{ev} \frac{\langle g^o|u_\sigma|e^o\rangle\langle e^o|u_\rho|g^o\rangle\langle gj|ev\rangle\langle ev|gi\rangle}{\nu_{ev,gi} - \nu_o + i\Gamma_{ev}} + \frac{\sigma \leftrightarrow \rho}{\nu_{ev,gi} + \nu_s + i\Gamma_{ev}} \tag{2}$$

$$A = -\frac{1}{\hbar^2}\sum_{ev'}\sum_{ev} \frac{\langle g^o|u_\sigma|e^o\rangle\langle e^o|\partial H/\partial Q|e^o\rangle\langle e^o|u_\rho|g^o\rangle\langle gj|ev'\rangle\langle ev'|Q|ev\rangle\langle ev|gi\rangle}{(\nu_{ev',gi} - \nu_o + i\Gamma_{ev'})(\nu_{ev,gi} - \nu_o + i\Gamma_{ev})} + \frac{\sigma \leftrightarrow \rho}{(\nu_{ev',gi} + \nu_s + i\Gamma_{ev'})(\nu_{ev,gi} + \nu_s + i\Gamma_{ev})} \tag{3}$$

$$B = -\frac{1}{\hbar^2}\sum_{fv'}\sum_{ev} \frac{\langle g^o|u_\rho|f^o\rangle\langle f^o|\partial H/\partial Q|e^o\rangle\langle e^o|u_\rho|g^o\rangle\langle gj|fv'\rangle\langle fv'|Q|ev\rangle\langle ev|gi\rangle}{(\nu_{fv',gi} - \nu_o + i\Gamma_{fv'})(\nu_{ev,gi} - \nu_o + i\Gamma_{ev})} + \frac{\sigma \leftrightarrow \rho}{(\nu_{fv',gi} + \nu_s + i\Gamma_{fv'})(\nu_{ev,gi} + \nu_s + i\Gamma_{ev})} \tag{4}$$

The wavefunctions are separated into their electronic and vibrational parts (adiabatic approximation). Excited state electronic wavefunctions are denoted by e and f, while the ground state electronic wavefunction is denoted by g; e^o, f^o, g^o, denote wavefunctions evaluated at the equilibrium nuclear position, Q = O, of the ground state. Initial and final vibrational levels of the ground state are denoted by gi and gj respectively, while vibrational levels of the excited electronic states are denoted by ev, ev' and fv'. The frequencies of the incident and scattered photons are denoted by ν_o and ν_s respectively. The vibronic coupling operator $\partial H/\partial Q$ (H being the electronic Hamiltonian) which connects two vibronic states either in the same electronic state e (A term), or in two different electronic states e and f (B term) is evaluated at the equilibrium configuration of the ground state (hence the subscript zero). u_σ and u_ρ are electric dipole moment operators i.e. $u_\sigma = -\sum_k e(r_k)_\sigma$, where $(r_k)_\sigma$ is the σ th component of the position vector of the kth electron. Γ_{ev}, $\Gamma_{ev'}$ and $\Gamma_{fv'}$, are damping constants (measures of the bandwidths of the electronic transitions).

Resonance enhancement of a band is, according to the above analysis, to be associated with the occurrence (for certain values of ν_0) of a very small denominator in one of the terms for $\alpha_{\sigma\rho}$. The summation over all the electronic states may thus be relaxed except for the electronic state in resonance, because the magnitude of $\alpha_{\sigma\rho}$ is then a function primarily of this particular term; it is also a function of the oscillator strength (directly) and the half-bandwidth (inversely) of the resonant electronic transition. The greatest resonance effects are therefore to be expected for strong, sharp electronic bands.

The terms A' and A both involve coupling with a single electronic excited state. Theoretical analysis indicates that the A' term is important only when resonance is approached very closely (it is otherwise zero for Raman scattering), whereas the A term has favourable Franck-Condon overlaps for Raman scattering away from resonance. The A term (also the A' term) requires some nuclear displacement in the excited electronic state and is non-zero only for totally symmetric vibrations; this is because the integral $\langle e^o|\partial H/\partial Q|e^o\rangle$, whose value determines that of term A, is zero unless $\partial H/\partial Q$, and therefore Q, belong to the totally symmetric representation.

On the other hand, the B term involves vibronic coupling between the resonant state and other excited electronic states. The active vibration (i.e. the one which displays the resonance enhancement) may have any symmetry contained in the direct product representation of the two electronic states which it connects. The enhancement depends on the magnitude of $\partial H/\partial Q$ as well as on the oscillator strengths; thus efficient vibronic coupling of the two excited electronic states is necessary for B-term enhancement to be effective.

The Raman tensor element contains complex terms, the real parts arising from the dispersive forces and the imaginary parts from the damping forces associated with the scattering processes [8]. The Raman intensity is proportional to the Raman tensor times its complex conjugate and thus does not contain cross terms between the real and imaginary parts. If several terms contribute to an element of the tensor, interference may occur (constructively or destructively) either between the real or the imaginary parts [9].

3. EXPERIMENTAL TECHNIQUES

The main difficulties to be overcome when trying to measure resonance Raman spectra are [4,10]: (a) the elimination of thermal decomposition of the sample and of the thermal lens effect (which interferes with the effectiveness with which the laser beam can be focussed on the sample and therefore with the intensity of the scattering), (b) the making of corrections to allow for absorption processes occurring concurrently with the scattering processes and (c) the elimination of fluorescence and photolysis.

Difficulty (a) may be largely overcome by spinning the sample rapidly (at ca. 2000 r.p.m.) or by a rapid scanning of the surface by the laser beam. Difficulty (b) is not easily overcome except by reducing the path length of the scattered beam to the absolute minimum (accurate measurement of the path length being nearly impossible). Other more refined procedures include Raman difference spectroscopy with a rotating divided cell [10]. Photolytic effects seem also to be reduced by sample spinning procedures, but the effects of fluorescence can only be effectively removed by time-discrimination procedures. The technique involves excitation of the sample with a pulsed laser, the pulse duration being nano- or picoseconds, together with time-adjusted gate electronics to permit the temporal resolution of the resonance Raman spectrum from the fluorescence spectrum. Rejection of even short-lived fluorescence by more than three orders of magnitude from the Raman scattering has been achieved.

4. RESONANCE RAMAN SPECTRA

Resonance Raman spectra are characterised by an enormous enhancement to the intensity of one or more totally symmetric modes of vibration of the molecule, together with an apparent breakdown in the simple-harmonic-oscillator selection rules leading to the appearance of a long series of overtone bands of usually only one of the totally symmetric modes (A-term enhancement). In some cases the Raman band intensities of fundamentals of other symmetries are also enhanced (B-term enhancement), e.g. on irradiation of heme proteins with exciting lines within the α and β band contours [3]. The optimum concentration for solution studies obviously depends on the oscillator strengths of the resonant electronic transitions; in practice, concentrations of 10^{-4} - 10^{-2} mol dm^{-3} seem to be satisfactory in most cases. The half-bandwidths of each member of an overtone progression invariably increase monotonically with increase in the vibrational quantum number, partly due to environmental effects and partly due to cross terms in the potential function which modify slightly the frequencies of the overlapping hot bands. Other factors may also be involved.

It is the enormous enhancement to certain band intensities under resonance conditions which has so far been of most significance to chemists (the intensities of most Raman bands are attenuated by absorption). This feature allows scattering by molecules (e.g. biological materials) to be detected in aqueous solutions at very low concentrations (e.g. 3×10^{-6} mol dm^{-3}) and it brings about a major simplification of Raman spectra so that the latter consist primarily of bands arising from totally symmetric fundamentals if A-term effects are dominant, or primarily of bands arising from non-totally symmetric fundamentals if B-term enhancement is dominant. Typical resonance Raman spectra are given by TiI_4 [11], $AuBr_4^-$ [12], $Mo_2Cl_8^{4-}$ [13], $Re_2Cl_8^{2-}$ [14], $Re_2Br_8^{2-}$ [14], etc. An important point is that the bands which are resonance enhanced are those which arise from fundamentals which are localised in the chromophoric part of the molecule. Moreover, the greatest enhancement is found for those fundamentals which are most responsible for converting the molecule from its ground to excited state geometry. This selective enhancement leads to a considerable simplification of the observed spectrum. Thus the correct identification of the most resonantly enhanced vibration will aid in the assignment of the resonant electronic transition and vice versa; resonance Raman and electronic spectroscopy are accordingly complementary.

5. HARMONIC FREQUENCIES AND ANHARMONICITY CONSTANTS

The observation of a large number of overtones of a totally symmetric fundamental, ν_1, of a molecule under resonance Raman conditions makes it possible to determine the harmonic frequency and certain anharmonicity constants. In the case of a diatomic molecule, the vibrational term, G, is given by the expression [15]

$$G = \omega_e(v + \tfrac{1}{2}) - \omega_e x_e(v + \tfrac{1}{2})^2 \tag{5}$$

where ω_e is the harmonic frequency, v is the vibrational quantum number, and $\omega_e x_e$ is the anharmonicity constant. The observed wavenumber, $\nu(v)$, of any overtone of a diatomic oscillator is thus given by the expression

$$\begin{aligned}\nu(v) &= G(v) - G(0)\\ &= \omega_e v_e - \omega_e x_e(v^2 + v)\end{aligned} \tag{6}$$

A plot of $\nu(v)/v$ versus v should be a straight line with a slope of $-\omega_e x_e$ and an intercept of $\omega_e - \omega_e x_e$; accordingly, both ω_e and $\omega_e x_e$ may be deduced from such a plot. Values for these quantities have been established by resonance Raman studies on many diatomic molecules and ions, and these have been tabulated recently [4].

In the case of polyatomic molecules and ions, the vibrational term is given by the expression [16]

$$G = \sum_j \omega_j(v_j + \frac{d_j}{2}) + \sum_j \sum_{k\geq j} x_{jk}(v_j + \frac{d_j}{2})(v_k + \frac{d_k}{2}) \tag{7}$$

where ω_j is the harmonic frequency of the jth fundamental, v_j and d_j are the vibrational quantum number and degeneracy, respectively, of this fundamental, and x_{jk} is the jkth anharmonicity constant. The observed wavenumber, $\nu(v_1)$, of any overtone of a fundamental, ν_1, of a polyatomic molecule is given by the expression

$$\nu(v_1) = G(v_1, v_2, v_3, \ldots v_n) - G(0, v_2, v_3, \ldots v_n)$$
$$= \omega_1 v_1 + x_{11}(v_1^2 + v_1 d_1)$$
$$+ v_1[x_{12}(v_2 + \frac{d_2}{2}) + x_{13}(v_3 + \frac{d_3}{2}) + \ldots x_{1n}(v_n + \frac{d_n}{2})] \quad (8)$$

With the simplification that $d_1 = 1$ (as ν_1 is totally symmetric and therefore non-degenerate), a plot of $\nu(v_1)/v_1$ versus v_1 therefore has a slope of x_{11} and an intercept of

$$\omega_1 + x_{11} + [x_{12}(v_2 + \frac{d_2}{2}) + x_{13}(v_3 + \frac{d_3}{2}) + \ldots x_{1n}(v_n + \frac{d_n}{2})]$$

It is commonly assumed, with little justification, that the term inside the square brackets is approximately zero, and therefore that both ω_1 and x_{11} may be deduced from the slope and the intercept in a manner analogous to that appropriate to the diatomic case. It should be noted that the opposite sign convention for the anharmonicity constants x_{jk} has frequently been employed.

It is also possible to observe further progressions in a totally symmetric mode under resonance Raman conditions, progressions which are based on one quantum of another fundamental designated ν_2. This fundamental ν_2 may or may not be totally symmetric, presumably depending on whether A-term only or whether B- as well as A-term enhancement is important. The observed wavenumber, $\nu_2 + \nu(v_1)$, of any member of such a progression is given by the expression

$$\nu_2 + \nu(v_1) = G(v_1, v_2, v_3, \ldots v_n) - G(0, v_2 - 1, v_3, \ldots v_n)$$
$$= \omega_1 v_1 + \omega_2 + x_{11}(v_1^2 + v_1 d_1) + x_{22}(2v_2 - 1 + d_2)$$
$$+ x_{12}(v_1 v_2 + v_1 \frac{d_2}{2} + \frac{d_1}{2}) + x_{13} v_1 (v_3 + \frac{d_3}{2})$$
$$+ \ldots x_{1n} v_1 (v_n + \frac{d_n}{2}) + x_{23}(v_3 + \frac{d_3}{2})$$
$$+ \ldots x_{2n}(v_n + \frac{d_n}{2}) \quad (9)$$

The observed wavenumber of the ν_2 fundamental (and its hot bands) is given by the expression

$$\nu_2 = G(v_1, v_2, v_3, \ldots v_n) - G(v_1, v_2 - 1, v_3, \ldots v_n)$$
$$= \omega_2 + x_{22}(2v_2 - 1 + d_2) + x_{12}(v_1 + \frac{d_1}{2}) + x_{23}(v_3 + \frac{d_3}{2})$$
$$+ \ldots x_{2n}(v_n + \frac{d_n}{2}) \quad (10)$$

Thus if the observed wavenumber of ν_2 is subtracted from that of each member of a progression in ν_1, based on one quantum of ν_2, the following general expression is obtained

$$[\nu_2 + \nu(v_1)] - \nu_2 = \omega_1 v_1 + x_{11}(v_1^2 + v_1 d_1) + v_1[x_{12}(v_2 - 1 + \frac{d_2}{2}) + x_{13}(v_3 + \frac{d_3}{2}) + \ldots\ldots x_{1n}(v_n + \frac{d_n}{2}) \quad (11)$$

A plot of $\{[\nu_2 + \nu(v_1)] - \nu_2\}/v_1$ versus v_1 has exactly the same slope (x_{11}) as that of the plot of $\nu(v_1)/v_1$ versus v_1, but it differs in its intercept by precisely the quantity x_{12}.* Thus, if further progressions in ν_1 are observed in a resonance Raman spectrum, based on one quantum of another fundamental ν_n, it is possible to deduce a value for the anharmonicity constant x_{1n}. This has been done in several cases recently, but the accuracy of such data so far obtained has severely restricted the accuracy of the x_{1n} values (usually to ±100%). Nevertheless, these are the first anharmonicities ever deduced for vibrations of such molecules and, with further improvements in experimental techniques and apparatus (notably in the reliability of the wavenumber drives of Raman spectrometers) the procedure described may well develop to be an important one for helping to determine molecular potential functions.

A summary of the spectroscopic results obtained from the $\omega_1 v_1$ progressions for polyatomic molecules and ions consisting of more than three atoms is given in Table 1; the compounds found to display these resonance Raman progressions are invariably ones with bonds with a high degree of covalent character or with a bond order greater than one. Table 1(a) summarises the results on tetrahedral MO_4 and MS_4, Table 1(b) summarises the results on polyatomic halogeno species and Table 1(c) summarises the results on metal-metal bonded species. The x_{11} values are invariably very small, the x_{11}/ω_1 values being on average -1.2×10^{-3} for all the discrete species listed in the Table. The two chain polymers, Wolffram's red and Reihlen's green (see below), both give rise to rather higher values for the ratio x_{11}/ω_1 (-6.2×10^{-3} and -2.1×10^{-3} respectively). Thus the fundamentals found to display the resonance Raman effect are <u>mechanically</u> virtually harmonic, although <u>electrically</u> they are far from being harmonic.

The cross terms, x_{1n}, have been evaluated as described above in a number of cases recently in which progressions

---

* Small additional terms of the sort $\sum_j \sum_{k \geqslant j} g_{jk} l_j l_k$ contribute to the vibrational energy in the case of degenerate vibrations for which v_j, $v_k \geqslant 1$ (the g_{jk} are small constants and the l_j represent vibrational angular momenta about the molecular symmetry axis). These additional terms would make slight changes in the intercepts referred to, but would affect neither the slopes nor the differences between the intercepts of the two plots.

Table 1 (continued).
(c) Resonance Raman spectra of metal-metal bonded species

Species	State	λ_0/nm	ω_1/cm^{-1}	x_{11}/cm^{-1}	Progression	$I(2\nu_1)/I(\nu_1)$	Ref.
$Mo_2Cl_8^{4-}$	Cs^+ salt	514.5	342.1±0.3	-0.66±0.07	$11\nu_1$	0.55	n
$Mo_2Cl_4[P(C_4H_9)_3]_4$	solid	514.5	352.4±1.0	-1.2 ±0.5	$3\nu_1$	0.45	o
$Re_2Cl_8^{2-}$	n-Bu_4N^+ salt	615.0	272.6±0.4	-0.35±0.05	$4\nu_1$	0.22	p
$Re_2Br_8^{2-}$	n-Bu_4N^+ salt	647.1	276.2±0.5	-0.39±0.06	$4\nu_1$	0.31	p
Re_3Cl_9	polymeric solid (300 K)	514.5	252,4±0.6	-0.72±0.08	$4\nu_1$	0.24	q
Re_3Cl_9	matrix-isolated (Ar)	488.0	278.0±0.5	-0.5 ±0.1	$5\nu_1$	0.53	r
$Ru_2(O_2CCH_3)_4Cl$	solid	514.5	327.6±0.5	-0.13±0.02	$5\nu_1$	0.15	s
$Ru_2(O_2CC_3H_7)_4Cl$	solid	514.5	331.4±0.5	-0.27±0.03	$4\nu_1$	0.18	s
$Ru_2OCl_{10}^{4-}$	K^+ salt	496.5	256.5±0.6	-0.16±0.03	$8\nu_1$	0.24	t

Table 1.
(a) Resonance Raman spectra of polyatomic MO_4 and MS_4 species (ν_1 is the totally symmetric stretching mode).

Species	State	λ_0/nm	ω_1/cm^{-1}	x_{11}/cm^{-1}	Progression	$I(2\nu_1)/I(\nu_1)$	Ref.
MnO_4^-	K^+ salt	514.5	845.5±0.5	-1.1 ±0.2	$8\nu_1$	0.63	a
CrO_4^{2-}	K^+ salt	363.8	854.4±0.5	-0.71±0.1	$10\nu_1$	0.67	a
MoS_4^{2-}	H_2O soln.	465.8	454.0	≃0.0	$6\nu_1$	0.50	b
PS_4^{3-}	Cu^+ salt	496.5	393.5±0.6	-1.15±0.15	$4\nu_1$	0.03	c
(b) Resonance Raman spectra of polyatomic halogeno species (ν_1 is the totally symmetric stretching mode).							
$FeBr_4^-$	Et_4N^+ salt	476.5	202.0±0.2	-0.38±0.05	$7\nu_1$	0.45	d
$AuBr_4^-$	Et_4N^+ salt	457.9	213.4±0.5	-0.29±0.05	$9\nu_1$	0.84	e
$PdBr_4^{2-}$	K^+/KBr disc	325.0	190 ±2	-0.6 ±0.6	$5\nu_1$	0.55	f
TiI_4	C_6H_{12} soln.	514.5	161.5±0.2	-0.11±0.03	$13\nu_1$	0.61	g
SnI_4	C_6H_{12} soln.	363.8	150.1±0.5	-0.05±0.05	$11\nu_1$	≃0.85	h
$PtBr_6^{2-}$	n-Bu_4N^+ salt	325.0	209.1±0.5	-0.1 ±0.1	$7\nu_1$	≃0.5	i,j
PtI_6^{2-}	H_2O soln.	488.0	150.3±0.5	≃0	$3\nu_1$	0.2	i,j
$SbBr_6^-$	Et_4N^+ salt	457.9	191.6±0.2	-0.05±0.01	$9\nu_1$	0.19	k
$PtCl_2^{2+}$	W.r.* solid	514.5	319.5±0.6	-2.00±0.08	$9\nu_1$	0.55	l,m
$PtBr_2^{2+}$	R.g.* solid	627.0	179.6±0.6	-0.37±0.04	$7\nu_1$	0.30	m

Table 1 (continued).

a W. Kiefer and H.J. Bernstein, Mol. Phys., 23 (1972) 835.
b A. Ranade and M. Stockburger, Chem. Phys. Letters, 22 (1973) 257.
c O. Sala and M.L.A. Temperini, Chem. Phys. Letters, 36 (1975) 652.
d Ref. 20.
e Y.M. Bosworth and R.J.H. Clark, Chem. Phys. Letters, 28 (1974) 611; ref. 12.
f H. Hamaguchi, I. Harada and T. Shimanouchi, Chem. Lett., (1973) 1049.
g R.J.H. Clark and P.D. Mitchell, J. Raman Spectrosc., 2 (1974) 399; ref. 11.
h R.J.H. Clark and P.D. Mitchell, Chem. Comm., (1973) 762.
i Ref. 27.
j H. Hamaguchi, I. Harada and T. Shimanouchi, J. Raman Spectrosc., 2 (1974) 517.
k R.J.H. Clark and M.L. Duarte, J. Chem. Soc. Dalton, in press.
l Ref. 22.
m Ref. 23.
* W.r. = Wolffram's red; R.g. = Reihlen's green (see text).
n Ref. 13.
o J. San Filippo and H.J. Sniadoch, Inorg. Chem., 12 (1973) 2326.
p Ref. 14.
q R.J.H. Clark and M.L. Franks, unpublished work.
r W.F. Howard and L. Andrews, Inorg. Chem., 14 (1975) 1727.
s R.J.H. Clark and M.L. Franks, J. Chem. Soc. Dalton, (1976) 1825.
t Ref. 18.

---

additional to the $v_1\nu_1$ progression have been observed. Thus for the $Mo_2Cl_8^{4-}$ ion [13] x_{12} is 0.5 cm^{-1} (ν_2 is the $a_{1g}\nu$(MoCl) mode), for the $AuBr_4^-$ ion [12] x_{14} is 0.55 cm^{-1} (ν_4 is the $b_{2g}\delta$(BrAuBr) mode), and for the S_3^- ion in ultramarine [17] x_{12} is 0.4 cm^{-1} (ν_2 is the degenerate bending mode). Thus, although the cross terms are very imprecisely known, they appear to be comparable in magnitude to the x_{11} values but opposite in sign. Substantial developments in this area are to be expected.

The richest resonance Raman spectrum so far encountered is that of the linear $[Ru_2OCl_{10}]^{4-}$ ion, for which nine separate progressions have been observed, in seven of which the progression-forming mode is the Ru-O-Ru symmetric stretching fundamental at 256 cm^{-1} [18]. Each Ru-O bond is double, and the Ru-O-Ru interbond angle is 180°.

Another resonance Raman spectrum which is very rich in features is that of matrix-isolated Re_3Cl_9 molecular units at low concentration (Table 1(c)). The orange argon matrix displays several progressions, in which there are two progression-forming modes ν_1 and ν_2 which are assigned to ν(ReRe) and the symmetric Cl-Re-Cl angle-bending mode. Two progression-forming modes have also been observed in the resonance Raman spectrum of the $SbBr_6^-$ ion, $\nu_1(a_{1g})$ and $\nu_2(e_g)$, to $v_1 = 8$ and $v_2 = 4$.

6. EXCITATION PROFILES

Plots of the frequency dependence of the Raman scattered intensity of a fundamental or its overtones (excitation profiles) are beginning to assume an important role in resonance Raman studies. For asymmetric modes responsible for vibronic coupling of two excited states, two conditions of resonance are predicted [8,19] from the equations for $\alpha_{\sigma\rho}$, viz. $\nu_0 \simeq \nu_{00}$ and ν_{10} (Stokes bands) and $\nu_0 \simeq \nu_{00}$ and ν_{01} (anti-Stokes bands), i.e. for incident radiation at the pure electronic $0 \leftarrow 0$ frequency and at the $1 \leftarrow 0$ vibronic sideband for Stokes bands, and at $0 \leftarrow 0$ and $0 \leftarrow 1$ frequencies for anti-Stokes bands. This has been demonstrated for haemoglobin and related molecules [3] on irradiation into the visible (α-β) absorption band at ca. 550 nm. For totally symmetric modes, the exciting frequency for maximum resonance enhancement of the Raman intensity need not coincide with either ν_{00} or ν_{10}; indeed for the species studied so far it seems to correspond closely with the electronic band maximum or to a frequency which differs therefrom by one excited-state quantum of the vibrational frequency under study [4,11,20].

It should be noted that if no vibronic coupling is assumed, as in Placzek's polarisability model, only pure electric dipole moments appear in the expression for an element of the polarisability tensor, and only a single maximum at ν_{00} would be expected in the excitation profile of a fundamental, whether it is totally symmetric or not.

The plotting of excitation profiles has only really become possible since the introduction of tunable dye lasers and a variety of effective dyes. Only in this way has it become possible to obtain a sufficient number of points on an excitation profile so as to be able to map it closely. The most commonly used and most efficient dye for this purpose is rhodamine 6G (tuning range 575-640 nm, maximum output 4 W at 600 nm for a 21 W argon ion pump, operating all-wave), but other dyes are being used frequently now viz. rhodamine B (615-665 nm), rhodamine 110 (550-590 nm), sodium fluorescein, incorporating cyclooctatetraene (540-570 nm), and a variety of coumarins (430-550 nm). The drawback in the case of the coumarin dyes is that many of them are not only very short-lived (ca. 48 h) but highly expensive. Raman band intensities are determined as band areas relative to the band area of a standard which shows no significant resonance enhancement in the frequency range under study (e.g. the a_1 bands of the ClO_4^- or SO_4^{2-} ions).

Excitation profiles for several different fundamentals of a number of molecules and ions have been obtained, e.g. TiI_4 [4, 11], $FeBr_4^-$ [20], $[Fe(bipy)_3]^{2+}$ [21], $[Fe(phen)_3]^{2+}$ [21], Wolffram's red [22], Reihlen's green [23]. The latter two compounds are examples of chain polymers of the sort $[Pt^{II}(C_2H_5NH_2)_4][Pt^{IV}(C_2H_5NH_2)_4X_2].X_4.4H_2O$, where X = Cl or Br respectively (Figure 1). They are therefore examples of class II mixed-valence compounds [24]. The resonant electronic transition is known to be polarised parallel with the chain axis and, significantly, of all the fundamentals of the compound, it is an axial one which is found to display the resonance Raman effect viz. the X-Pt-X symmetric breathing mode in each case, Figure 2. A close relationship between resonance Raman and

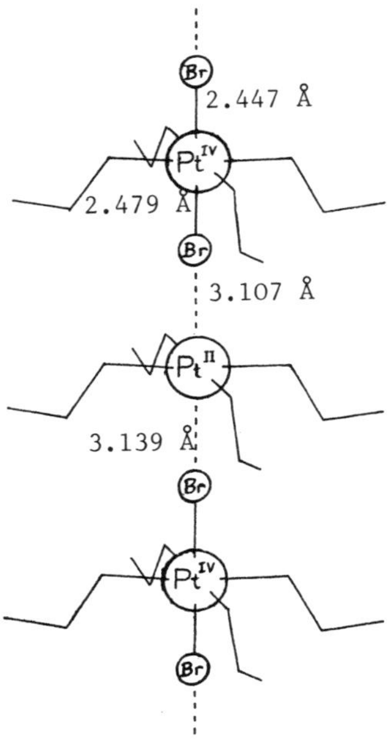

Figure 1. Structure of Reihlen's green.

electronic spectral assignments is thus implicit. Furthermore, the excitation profiles of these mixed-valence compounds illustrate the fact that very high spectral resolution of overlapping electronic bands can be achieved by resonance Raman studies, Figure 3.

A further important result of studies of excitation profiles is that for centrosymmetric ions, minima rather than maxima have been observed in the vicinity of electric dipole forbidden (vibronically allowed) ligand field bands [8,25]. This so-called Raman "antiresonance" is ascribed to destructive interference between scattering terms arising from the ligand-field transition at resonance and an electric-dipole-allowed charge-transfer transition at much higher energies. The requirement for interference is that the amplitudes of the scattering originating from the ligand field and from the charge-transfer states must be comparable. Both A and B terms probably contribute to the scattering from the latter, although the major contributor from the ligand-field state is likely to be a B term; this is because most of the intensity of ligand-field bands is vibronically induced [26]. Neither the A nor the A' term is likely to be important for ligand-field bands owing to the low transition moments of such transitions. These effects have been observed for the PdX_4^{2-} ions (X = Cl, Br or I), the $[Co(ethylenediamine)_3]^{3+}$, $PdCl_6^{2-}$, $PtBr_6^{2-}$, $RhCl_6^{3-}$ and $IrCl_6^{2-}$ ions [8,25,27]. Destructive interference effects may also occur between electric-dipole-allowed transitions.

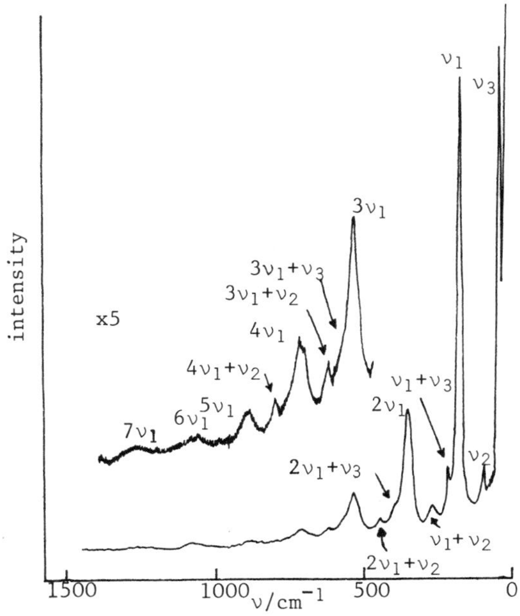

Figure 2. Resonance Raman spectrum of Reihlen's green obtained with 514.5 nm excitation (spectral slit width 1 cm^{-1}).

Constructive interference effects between electric-dipole-allowed and forbidden transitions have been suggested by Rimai et al. [28] to be responsible for local maxima observed in the excitation profiles of retinal, retinol and naphthalene at energies lower than that of the first allowed electronic transition of these molecules. Interference effects have also been discussed recently by Friedman and Hochstrasser [30], Mortensen [29] and Barron [8].

7. DEPOLARISATION RATIOS

Several elements of the scattering tensor may contribute to the intensity of a Raman band and, if resonance conditions are approached, the values of one or more of these elements may be enhanced (elements of the tensor which are zero in the non-resonance case remain zero in the resonance case). Thus in general $\alpha_{\sigma\rho} \neq \alpha_{\rho\sigma}$, and the tensor becomes asymmetric. Such a tensor can be decomposed into a symmetric part ($\alpha_{\sigma\rho}{}^{s} = \alpha_{\rho\sigma}{}^{s}$) and an antisymmetric part ($\alpha_{\sigma\rho}{}^{a} = -\alpha_{\rho\sigma}{}^{a}$). Modes for which $\alpha_{\sigma\rho} = -\alpha_{\rho\sigma}$ are inactive in normal Raman scattering, but become active under resonance conditions. Those for which $\frac{3}{4} < \rho_{\ell} < \infty$ are said to be anomalously polarised, while those for which $\rho_{\ell} = \infty$ are said to be inverse polarised.

The usual expression for the depolarisation ratio of a Raman band (linearly polarised radiation, 90° collection optics) of randomly oriented molecules is

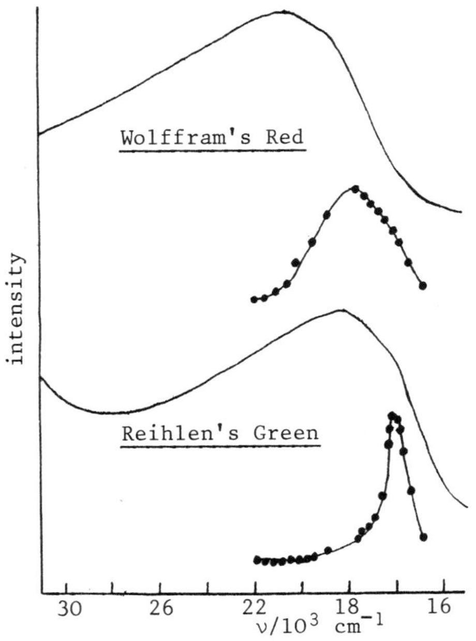

Figure 3. Diffuse reflectance spectra (-) of the mixed-valence compounds Wolffram's red and Reihlen's green together with excitation profiles (-•-•-) for the 316 cm^{-1} (Wolffram's red) and 179 cm^{-1} (Reihlen's green) bands observed in the Raman spectra of these compounds.

$$\rho_\ell = \frac{3\gamma_s'^2}{45\bar{\alpha}'^2 + 4\gamma_s'^2} \quad (12)$$

where

$$\bar{\alpha}' = 1/3(\alpha_{xx}' + \alpha_{yy}' + \alpha_{zz}')$$

and

$$\gamma_s'^2 = 1/2[(\alpha_{xx}' - \alpha_{yy}')^2 + (\alpha_{yy}' - \alpha_{zz}')^2 + (\alpha_{zz}' - \alpha_{xx}')^2 + 6(\alpha_{xy}'^2 + \alpha_{yz}'^2 + \alpha_{zx}'^2)]$$

This must be modified under resonance conditions (as realised by Placzek) to [31]

$$\rho_\ell = \frac{3\gamma_s'^2 + 5\gamma_a'^2}{45\bar{\alpha}'^2 + 4\gamma_s'^2} \quad (13)$$

where

$$\gamma_a'^2 = 3/4[(\alpha_{xy}' - \alpha_{yx}')^2 + (\alpha_{yz}' - \alpha_{zy}')^2 + (\alpha_{xz}' - \alpha_{zx}')^2]$$

The quantities $3\bar{\alpha}'$, γ_s' and γ_a' are referred to as the trace, symmetric anistropy, and antisymmetric anistropy respectively of the derived polarisability tensor. Clearly $\gamma_a' = 0$ if the tensor is symmetric, and then the second expression for ρ_ℓ collapses into the first, i.e. $0 < \rho_\ell \leq \frac{3}{4}$.

For a non-totally symmetric mode, $\bar{\alpha}' = 0$, and therefore

$$\rho = \tfrac{3}{4} + 5\gamma_a'^2/4\gamma_s'^2 \qquad (14)$$

If the scattering tensor is symmetric, i.e. if $\alpha_{\sigma\rho} = \alpha_{\rho\sigma}$, $\gamma_a' = 0$, $\rho_\ell = \frac{3}{4}$ and the depolarisation ratio shows no dispersion. If, through resonance enhancement, $\alpha_{\sigma\rho}$ becomes greater than $\alpha_{\rho\sigma}$, the scattering tensor becomes antisymmetric, and the depolarisation ratio must be anomalous. For non-totally symmetric modes of certain point groups, $\gamma_a'^2$ tends to become equal to $\gamma_s'^2$ with the consequence that ρ_ℓ approaches 2; in these cases, on passing fron non-resonance to resonance, the depolarisation ratio displays dispersion [2,19], and $\frac{3}{4} < \rho_\ell < 2$. For certain other point groups the species to which $(\alpha_{\sigma\rho} - \alpha_{\rho\sigma})$ belongs will become Raman active at resonance only, with $\alpha_{\sigma\rho} = -\alpha_{\rho\sigma}$ (i.e. $\gamma_s' = 0$, $\gamma_a' \neq 0$), leading to $\rho_\ell = \infty$ i.e. the depolarisation ratio of such a band shows no dispersion.

Totally symmetric modes may also give rise to anomalous polarisation, if $\gamma_a' \neq 0$, but only where $\gamma_a'^2 > 27\bar{\alpha}'^2/4$.

Such unusual polarisation phenomena have been observed for several bands of a number of large biological molecules such as haemoglobin and cytochrome c which have fundamentals belonging to the same irreducible representation as that of a pure rotation [3] (e.g. a_{2g} in D_{4h} or a_2 in C_{4v}). The actual ρ_ℓ value (estimated to exceed 100 in some cases) is considered to provide information on the loss of four-fold symmetry of the heme chromophore by steric or other effects.

Anomalous polarisation has also been observed for some cubic ions even though their vibrational representation lacks any mode which might normally be expected to display this phenomenon. The polarisation behaviour, e.g. for all fundamentals of the $IrCl_6^{2-}$ and $IrBr_6^{2-}$ ions [32-34], has been shown to be a consequence of the fact that the ground state of the ion is not totally symmetric but degenerate; this situation leads to contributions by antisymmetric components to the scattering tensor. A similar situation prevails for the $FeBr_4^-$ ion [20].

The principal use to which ρ_ℓ-values at resonance of totally symmetric modes are likely to be put in the future seems to be in the identification of the resonant electronic transition; if the latter is axially polarised, $\rho_\ell = 1/3$ at resonance, whereas if it is polarised perpendicular to the principal axis of the molecule, $\rho_\ell = 1/8$ at resonance. Thus for the iodine molecule $\rho_\ell = 0.35$ at resonance, consistent [1] with the assignment of the resonant electronic transition to ${}^3\Pi_{0u}{}^+ \leftarrow {}^1\Sigma_g{}^+$ (an axially polarised transition). Further, the fact that $\rho_\ell = 0.34$ at resonance for the ν_1(Ru-O-Ru) symmetric stretching mode of the $Ru_2OCl_{10}^{4-}$ ion leads unambiguously [18] to the assignment ${}^1A_{2u} \leftarrow {}^1A_{1g}$ for the resonant electronic transition of the ion (a z-axis polarised transition which is metal-localised).

8. CONCLUSIONS

The most characteristic feature of resonance Raman spectra is the enormous enhancement to the intensity of one or more totally symmetric modes, and the appearance of long overtone progressions in that totally symmetric fundamental which converts the molecule from the ground to the excited state; this fundamental is normally a stretching mode of the scattering molecule because it is much more common for electronic states to be displaced with respect to the ground state along a stretching than a bending coordinate. The extent of resonance for a fundamental, as determined by the amount of enhancement of a Raman band or by the length of the overtone progressions based thereon, reaches a maximum at or within one vibrational quantum of ν_{max} for a_1 bands. It is most effective when ν_0 comes into coincidence with a sharp, electric-dipole-allowed transition of high oscillator strength. The intensity enhancement is so great that the technique holds promise for the detection of only trace amounts of absorbing materials in any medium. The selective enhancement of bands arising from fundamentals belonging to some specific irreducible representation for large complicated molecules leads to an enormous simplification of the resonance Raman as compared with the Raman spectrum thereof. The observation of the long overtone progressions permits the calculation of the spectroscopic constants ω_1, x_{11}, x_{1n} and (rarely) x_{2n}, where the subscripts 1 and 2 relate to the progression-forming fundamentals. Finally the excitation profiles of resonantly enhanced (or "de-enhanced") Raman bands hold great promise as probes of the nature of excited electronic states and for spectral resolution of overlapping electronic bands.

9. REFERENCES

1. J. Behringer, in R.F. Barrow, D.A. Long and D.J. Millen (Editors), Molecular Spectroscopy, Vol. 3, Chemical Society, Specialist Periodical Reports, London, 1975, p. 163.
2. A.R. Gregory, W.H. Henneker, W. Siebrand and M. Zgierski, J. Chem. Phys., 63 (1975) 5475.
3. T.G. Spiro and T.M. Loehr, in R.J.H. Clark and R.E. Hester (Editors), Advances in Infrared and Raman Spectroscopy, Vol. 1, Heyden, London, 1975, p. 98.
4. R.J.H. Clark, in R.J.H. Clark and R.E. Hester (Editors), Advances in Infrared and Raman Spectroscopy, Vol. 1, Heyden, London, 1975, p. 143.
5. A.C. Albrecht, J. Chem. Phys., 34 (1961) 1476.
6. A.C. Albrecht and M.C. Hutley, J. Chem. Phys., 55 (1971) 4438.
7. J. Tang and A.C. Albrecht, in H.A. Szymanski (Editor), Raman Spectroscopy, Vol. 2, Plenum, New York, 1970, p. 33.
8. L.D. Barron, Mol. Phys., 31 (1976) 129.
9. P. Stein, V. Miskowski, W.H. Woodruff, J.P. Griffin, K.G. Werner, B.P. Gaber and T.G. Spiro, J. Chem. Phys., 64 (1976) 2159.
10. W. Kiefer, Appl. Spectrosc., 28 (1974) 115; R.J.H. Clark and R.E. Hester (Editors), Advances in Infrared and Raman Spectroscopy, Vol. 3, Heyden, London, 1977, p. 1.

11. R.J.H. Clark and P.D. Mitchell, J. Am. Chem. Soc., 95 (1973) 8300.
12. Y.M. Bosworth and R.J.H. Clark, J. Chem. Soc. Dalton, (1975) 381.
13. R.J.H. Clark and M.L. Franks, J. Am. Chem. Soc., 97 (1975) 2691.
14. R.J.H. Clark and M.L. Franks, J. Am. Chem. Soc., 98 (1976) 2763.
15. G. Herzberg, Spectra of Diatomic Molecules, Van Nostrand, Princeton, 1950, p. 95.
16. G. Herzberg, Infrared and Raman Spectra of Polyatomic Molecules, Van Nostrand, Princeton, 1945, p. 205.
17. R.J.H. Clark and M.L. Franks, Chem. Phys. Lett., 34 (1975) 69.
18. R.J.H. Clark and M.L. Franks, J. Amer. Chem. Soc., in press.
19. J.A. Koningstein and B.G. Jakubinek, J. Raman Spectrosc., 2 (1974) 317.
20. R.J.H. Clark and P.C. Turtle, J. Chem. Soc. Faraday II, 72 (1976) 1885.
21. R.J.H. Clark, P.C. Turtle, D.P. Strommen, B. Streusand, J. Kincaid and K. Nakamoto, Inorg. Chem., 16 (1977) 84.
22. R.J.H. Clark, M.L. Franks and W.R. Trumble, Chem. Phys. Lett., 41 (1976) 287.
23. R.J.H. Clark and M.L. Franks, J. Chem. Soc. Dalton, (1977) 198.
24. M.B. Robin and P. Day, in H.J. Eméleus and A.G. Sharpe (Editors), Advances in Inorganic Chemistry and Radiochemistry, Vol. 10, Academic Press, London, 1967, p. 247.
25. Y.M. Bosworth, R.J.H. Clark and P.C. Turtle, J. Chem. Soc. Dalton, (1975) 2026.
26. R. Dingle and C.J. Ballhausen, K. Dan. Vidensk, Selsk. Mat. Fys. Medd., 35 (1967) 26.
27. Y.M. Bosworth and R.J.H. Clark, J. Chem. Soc. Dalton, (1974) 1749.
28. L. Rimai, M.E. Hyde, H.C. Heller and D. Gill, Chem. Phys. Lett., 10 (1971) 207.
29. O.S. Mortensen, Chem. Phys. Lett., 30 (1975) 406.
30. J. Friedman and R.M. Hochstrasser, Chem. Phys. Lett., 32 (1975) 414.
31. T.G. Spiro and T.C. Strekas, Proc. Nat. Acad. Sci. U.S.A., 69 (1972) 2622.
32. Y.M. Bosworth and R.J.H. Clark, unpublished work, Jan. 1972; R.J.H. Clark and P.C. Turtle, to be published.
33. H. Hamaguchi, I. Harada and T. Shimanouchi, Chem. Phys. Lett., 32 (1975) 103.
34. H. Hamaguchi and T. Shimanouchi, Chem. Phys. Lett., 38 (1976) 370.

CHAPTER 12

ISOTOPIC SUBSTITUTION

A. Müller

1. INTRODUCTION

History

The first correct definition of isotopes was given by Soddy in 1910 from radioactivity studies. Thomson in 1912 found by mass separation in the magnetic and electric field that Ne contains two isotopes ^{20}Ne and ^{22}Ne [1]. Today, we know that roughly some 300 stable and 1000 radioactive isotopes exist. While some elements are isotopically pure (e.g. Co), some contain a lot of stable isotopes (e.g. Sn with 10 stable isotopes ranging in mass between 112 and 124 and Xe with 23 isotopes among which 9 are stable). Isotopes have been used in different fields of science and medicine, and also especially in vibrational spectroscopy.

The vibrational isotope effect was first observed in the rotation-vibration spectrum of HCl in 1920 [2] and showed the $^{35}Cl/^{37}Cl$ isotope splitting (in the $3 \leftarrow 0$ band, the vibrational shift is of the order 5.8 cm^{-1} whereas the shift in the rotational lines is roughly 0.2 cm^{-1}). Shortly after the discovery of heavy hydrogen by Urey et al. [3] in 1932, Dennison and co-workers [4] measured the i.r. spectrum of $D^{35}Cl$ present in ordinary HCl (first measurement of H/D isotope shift). Several rare isotopes such as ^{18}O were discovered from band spectra. A weak band due to $^{16}O^{18}O$ observed in the solar spectrum was the first proof for the existence of ^{18}O [5]. The isotopes ^{17}O, ^{13}C and ^{15}N were also first found from the isotope effect as observed in band spectra [6-8]. It is interesting to note in this connection, that the isotopic displacements of 0-0 bands, led to the recognition of the existence of zero-point vibrations.

In recent years, many investigations dealing with the use of isotopes in the study of molecular dynamics have been published. These studies have become feasible due to the fact that many stable isotopes are now commercially available and also due to the application of matrix isolation techniques.

Applications of isotope shifts

In vibrational spectroscopy, isotope shifts have been used for a number of purposes: (1) simple identification of bands (e.g. metal-ligand vibrations of coordination compounds) [9], (2) proof for rigorous and correct assignment of bands to the various irreducible representations [10,11], (3) rough estimation of anharmonicities (such as the one named after Dennison) [12], (4) correction of frequencies due to Fermi resonance [13], (5) determination of the structure of unstable species isolated

Table 1. (Continued).

Compounds	Advantages	Limitations	Examples	Ref.
(4) isotope substituted and matrix isolated	(a) no * or reduced † overlapping of bands (due to the abundance of several different isotopes)	difficulties as under (2)	(a)*$Ge^{35}Cl_4$ (see chapter 9, Figure 8),	d
			† $^{116}SnCl_4$	e
	(b) accurate isotope shifts as 3(b)			
	(c) in the case of small abundance of one isotope and small shifts		(c) $^{32/34}SPF_3$	f
	(d) very accurate measurement of central atom substitution data		(d) $Si^{35}Cl_4$	b
(5) ions in host lattices	(a) very sharp band as factor group splitting is avoided (b) best possibility for determination of exact force constants for ions	a suitable host has to be found	(a) $^{50/53}CrO_4^{2-}$ in K_2SO_4 (see Figure 3)	g

a R.S. McDowell, L.B. Asprey and L.C. Hoskins, J. Chem. Phys., 56 (1972) 5712.
b Ref. 17.
c Refs. 17 and 18.
d Ref. 19
e Ref. 20.
f Ref. 21.
g Ref. 22.

Table 1. Different methods for the measurement of isotope shifts.

Compounds	Advantages	Limitations	Examples	Ref.
(1) normal	very simple from the experimental point of view, cf. (2)	(a) for solids, liquids and solutions: only if the isotope shift is large and the percentage of the isotopes is high	(a) boron compounds	
		(b) gases: only in general for highly symmetrical compounds	(b) metal isotope splitting of the Q branch of RuO_4 (see Figure 1)	a
(2) isotope substituted	no partial overlapping of bands due to the abundance of different isotopes	(a) the corresponding isotope must be commercially available (highly enriched)		
		(b) some rare isotopes are extremely expensive	(b) ^{18}O, ^{34}S, ^{102}Ru	
		(c) difficulties of the preparation on a mg scale		
(3) matrix isolated	(a) no difficulties as under (2).	(a) abundance of too many isotopes with small mass differences*	(a*) $SnBr_4$ (see Figure 2)	b
	(b) accurate † isotope shifts (no hot band progressions and vib. rot.-interactions)	(b) matrix effects, no suitable matrix	(b†) RuO_4 (see Figure 1)	c

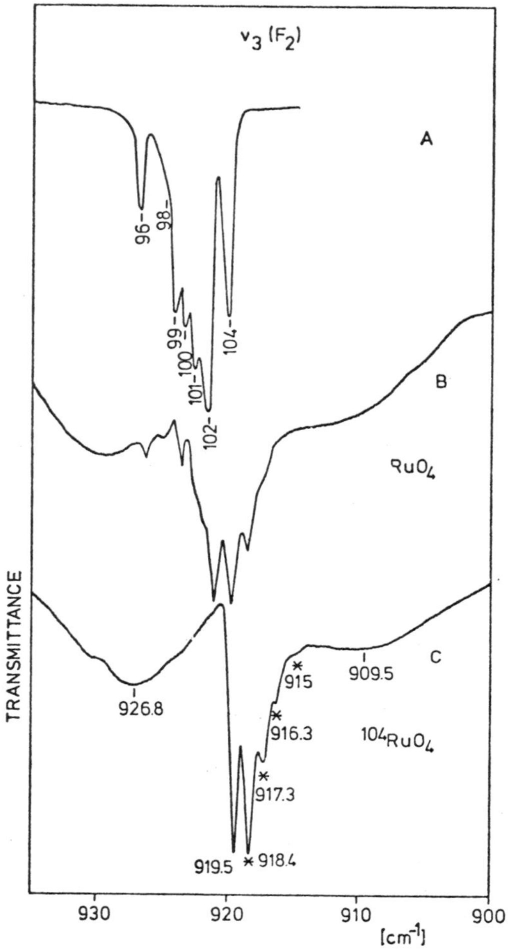

Figure 1. Matrix isolation (A), gas phase spectrum of normal (B) and $^{104}RuO_4$ (C) (reproduced with permission from ref. 18).

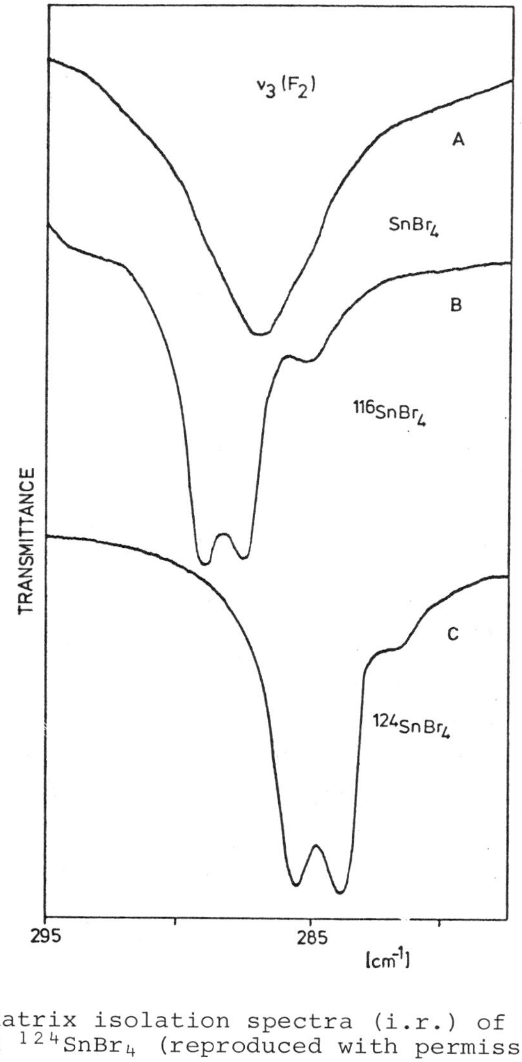

Figure 2. Matrix isolation spectra (i.r.) of $SnBr_4$, $^{116}SnBr_4$ and $^{124}SnBr_4$ (reproduced with permission from ref. 17)

in a matrix at low temperatures [14], (6) force constant refinements [10,15], (7) mathematical check on calculated values of molecular constants of isotopic molecules, (8) determination of the carrier of an observed band system (see introduction) [5], (9) estimation of unknown data for isotopic molecules [11,16] and (10) estimation of Cartesian displacement coordinates (see Equation (25) below).

2. EXPERIMENTAL ASPECTS

Measurement of isotope shifts

Methods used at present for the measurement of isotope shifts are summarised in Table 1. Several instructive examples which demonstrate the advantage of modern methods (e.g. the combined use of isotope substitution and matrix isolation techniques, which is in some cases the only possibility for measuring the isotope shifts) are presented, too. For more details, the reader should consult the original references [9, 14-23].

The use of isotopically pure compounds

In some cases the use of isotopically pure compounds represents the only possibility to solve problems complicated by the abundance of different isotopes (see Table 2).

3. CLASSICAL ISOTOPE RULES

Frequencies

Since within the Born-Oppenheimer approximation, the potential function remains invariant under isotopic substitution, one derives from the secular equation of Wilson [10], the well-known Teller-Redlich product rule (valid in the harmonic approximation; see [8,10]).

$$\prod \left(\frac{\lambda_k^i}{\lambda_k}\right) = \frac{|\underline{G}^i|}{|\underline{G}|} \qquad (1)$$

Deviations from Equation (1) are caused by the cubic, quartic and higher order terms in the potential energy [24].

Besides the above mentioned product rule, one can also derive sum rules of different orders, viz.

$$\sum_{i=0}^{m} (-1)^i F^i \phi^i_{k(p+1)} (a_\alpha a_\beta \ldots a_i) = 0 \qquad (2)$$

where

k designates the number of different factors (1,2,3,...n) see Equation (4) and Table 3.

i represents the number of substituted atoms a

F^i is the number of different functions $\phi^i_{k(p+1)}$ corresponding to the i-substituted molecules (determination see summary below)

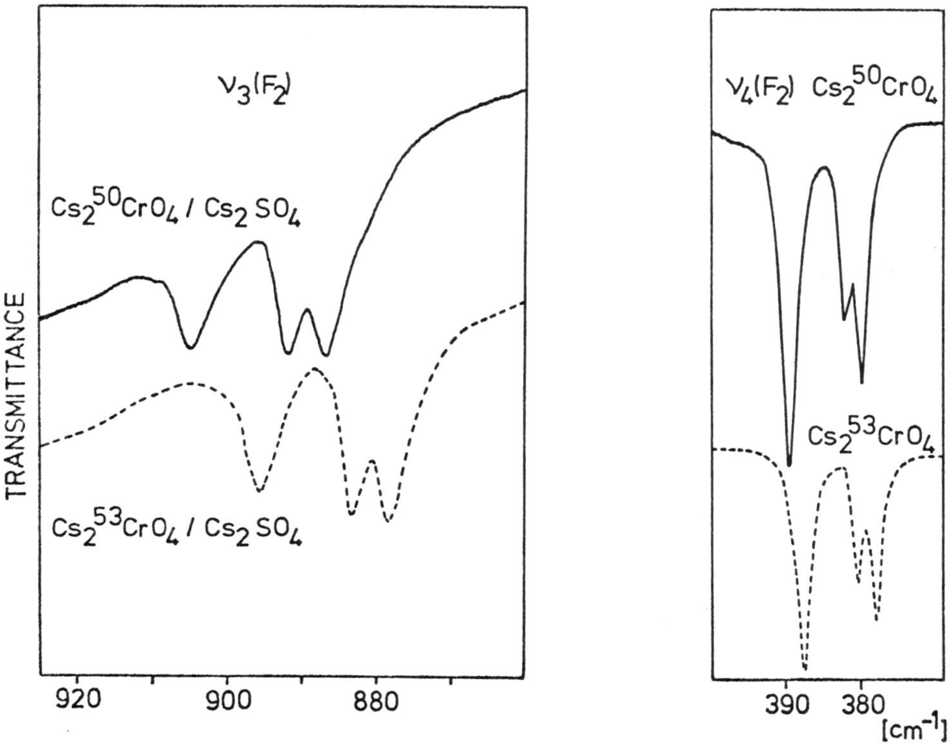

Figure 3. I.r. spectrum of $^{50}CrO_4^{2-}$ and $^{53}CrO_4^{2-}$ in Cs_2SO_4 (reproduced with permission from ref. 22).

ϕ_k^i equivalent to $c_{k(p+1)}$ in the secular equation of Wilson [10] written in an equivalent form to the characteristic equation for $(GF)^{p+1}$, viz.,

$$\lambda^n - c_{1(p+1)}\lambda^{n-1} + \ldots\ldots + (-1)^{n-1}c_{(n-1)(p+1)}\lambda + (-1)^n c_{n(p+1)} = 0 \quad (p = 0,1,2\ \ldots) \qquad (3)$$

with

$$c_{1(p+1)} = \sum_{\ell} \lambda_\ell^{p+1}$$

$$c_{2(p+1)} = \sum_{\ell,m} \lambda_\ell^{p+1}\lambda_m^{p+1} \qquad (4)$$

$$c_{n(p+1)} = \prod_{\ell} \lambda_\ell^{p+1}$$

The first and higher order sum rules follow as a special case of Equation (2) (see Table 3 and Refs. 11,25-30).

Table 2. The use of isotopically pure compounds

Applications	Advantages	Examples	Ref.
(1) band contour analysis	(a) distinction between vib.-rot. and isotope effects (b) distinction between bands of a hot band progression (also more accurate determination!) and those caused by isotopes	$H^{12}C^{35}Cl_3$	a
	(c) measurement of the position of PQR branches	$^{74}GeF_4$ (see Figure 4)	b
	(d) determination of ζ constants	$\zeta_3(Ge^{35}Cl_4)$ (see Figure 5)	c
(2) determination of accurate isotope shifts	in connection with matrix isolation technique often the only possibility	$\Delta\nu_3(F_2)$ $^{116}Sn/^{124}SnBr_4$	b

a K.H. Schmidt and A. Müller, J. Mol. Spectr., 50 (1974) 115; K.H. Schmidt, W. Hauswirth and A. Müller, J. Mol. Spectr., 57 (1975) 316.
b Ref. 17.
c Ref. 19.

(a) First order sum rule ($p+1 = 1$; $k = 1$)

Here we have

$$\phi^i_{k(p+1)}(a_\alpha a_\beta \ldots . \; a_i) = c^i_{11}(a_\alpha a_\beta \ldots . \; a_i)$$

$$= \sum_\ell \lambda_\ell (a_\alpha a_\beta \ldots . \; a_i) \quad (5)$$

Then it follows that

$$\sum_{i=0}^{m} (-1)^i F^i c^i_{11}(a_\alpha a_\beta \ldots . \; a_i) = 0 \quad (6)$$

(Sum rule of Decius-Wilson-Sverdlov [25])

or for the special case with two substituted atoms,

$$\sum \lambda_\ell [I] + \sum \lambda_\ell [IV(a_\alpha a_\beta)] = \sum \lambda_\ell [II(a_\alpha)] + \sum \lambda_\ell [III(a_\beta)] \quad (7)$$

As an example, applying the above rule to ethylene one gets

$$\sum \omega^2_\ell (C_2H_4) + \sum \omega^2_\ell (C_2D_4) = \sum \omega^2_\ell (C_2H_3D) + \sum \omega^2_\ell (C_2HD_3)$$

or for water,

$$\sum \omega^2_\ell (H_2O) + \sum \omega^2_\ell (D_2O) = 2\sum \omega^2_\ell (HDO)$$

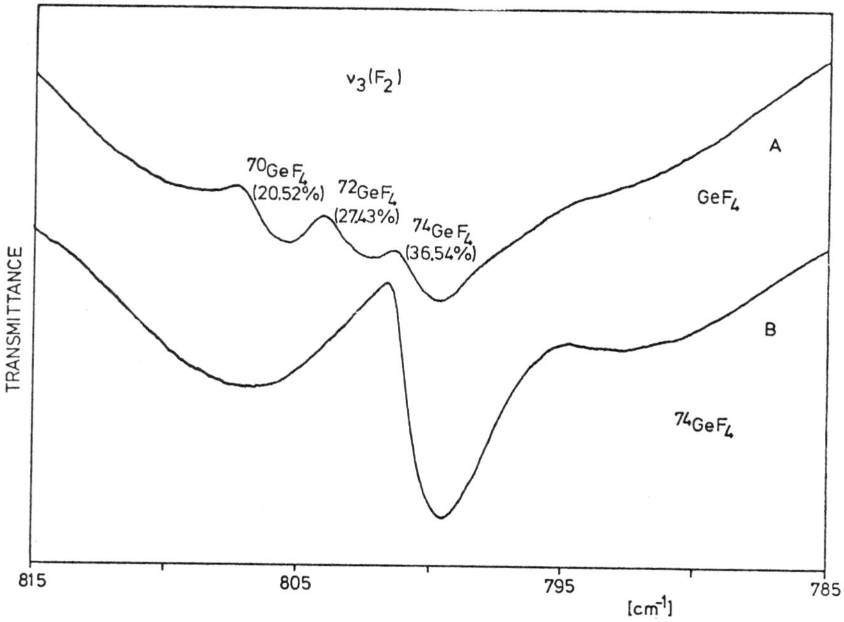

Figure 4. I.r. gas phase band contour of $\nu_3(F_2)$ of GeF_4 and $^{74}GeF_4$ (reproduced with permission from ref. 17).

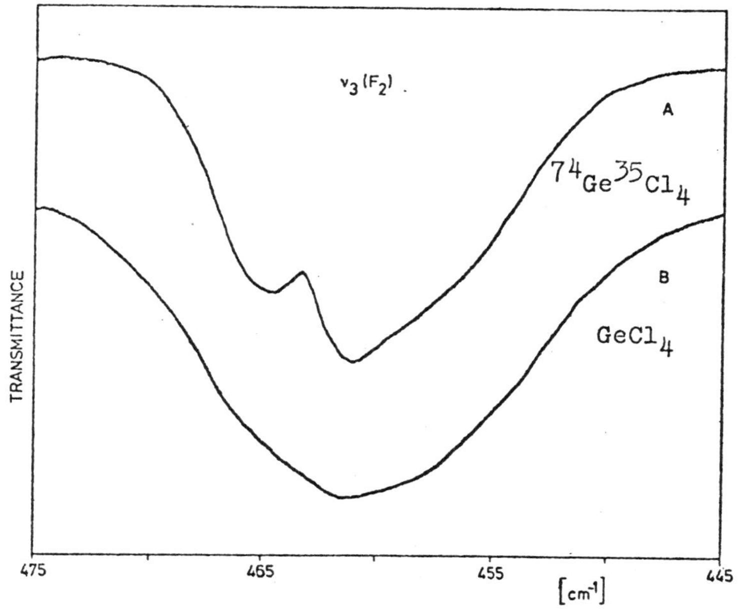

Figure 5. I.r. gas phase band contour of $\nu_3(F_2)$ of $GeCl_4$ and $^{74}Ge^{35}Cl_4$ (reproduced with permission from ref. 19).

(b) Higher-order sum rules

Higher order sum rules have also been obtained as for the powers of squared frequencies (Biegeleisen-Sverdlov) and products or products of powers of squared frequencies [26] (see Table 3).

Table 3. Presentation of the sum rules.

(p+1) \ k	1	2	n
1	$\sum_{\ell} \lambda_{\ell}$ [a]	$\sum_{\ell,m} \lambda_{\ell}\lambda_{m}$ [b]	$\prod_{\ell} \lambda_{\ell}$
2	$\sum_{\ell} \lambda_{\ell}^{2}$ [b]	$\sum_{\ell,m} \lambda_{\ell}^{2}\lambda_{m}^{2}$	$\prod_{\ell} \lambda_{\ell}^{2}$
3	$\sum_{\ell} \lambda_{\ell}^{3}$	$\sum_{\ell,m} \lambda_{\ell}^{3}\lambda_{m}^{3}$	$\prod_{\ell} \lambda_{\ell}^{3}$
(p+1)	$\sum_{\ell} \lambda_{\ell}^{p+1}$	$\sum_{\ell,m} \lambda_{\ell}^{p+1}\lambda_{m}^{p+1}$	$\prod_{\ell} \lambda_{\ell}^{p+1}$

a Corresponding to the rule of Decius-Wilson-Sverdlov [25].
b Corresponding to the rule of Biegeleisen-Sverdlov [26].

In addition, modifications of the sum rules [27], "a complete isotopic rule" (following the calculation of fundamental frequencies of one isotopic species from those of two others) [28], and a discussion of the accuracy of sum rules [29] have been published (for other refs. see [30]).

Infrared and Raman intensities

Important in checking the isotopic influence on infrared intensities, are the well known Crawford's sum rules [31], which in our new matrix notation [32] can be written as

$$\mathrm{Tr}(\underline{\mu}'\underline{G}) = k\mathrm{Tr}(\underline{A}) \quad (8)$$

$$\mathrm{Tr}(\underline{\mu}'\underline{C}) = k\mathrm{Tr}(\underline{A\sigma}) \quad (9)$$

where $\underline{A}$, whose diagonal elements are the observed integrated intensities, are defined as

$$\underline{A} = (k)^{-1}[\underline{L}^{t}\,\underline{\mu}'\underline{L}] \quad (10)$$

with $k = (3c^2/N_0\pi d)$ (d is the degeneracy of the irreducible representation and the other symbols have their usual meanings). In Equations (8) and (9), $\underline{\mu}'$ is a singular matrix containing the quadratic powers of the derivatives of the dipole moment with respect to a chosen set of coordinates [32] (e.g. internal or internal symmetry coordinates) and $\underline{C} = \underline{F}^{-1}$, $\underline{\sigma} = \underline{\Lambda}^{-1}$. When the molecule has no permanent dipole moment and/or when the species under consideration does not contain any component of rotation,

Equation (9) leads to the important isotopic rule [31]

$$\sum_i (A_i/\omega_i^2) = \text{invariant}$$

where A_i is the integrated intensity of the ith band.

Similar sum rules for Raman intensities and for the square of the derivatives of the trace of the polarisability tensor and its anisotropy with respect to normal coordinates are also valid under the condition that the static polarisability tensor has spherical symmetry and/or when the irreducible representation does not contain any rotations which shift the polarisability ellipsoid in space (see ref. 11).

Mean square amplitudes of vibration

Using the well known matrix notation [33]

$$\underline{\Sigma} = \underline{L\delta L}^t \tag{11}$$

for the mean square amplitudes ($\underline{\delta}$ is a diagonal matrix with elements $\delta_i = (h/8\pi^2\omega_i)\coth(h\omega_i/2kT)$; for the meaning of the symbols, see [33]) one can derive the isotopic product rule

$$\frac{|\underline{\Sigma}^i|}{|\underline{\Sigma}|} = \frac{|\underline{G}^i|}{|\underline{G}|} \prod_i (\delta_i^i/\delta_i) \tag{12}$$

which relates the mean square amplitudes of the parent molecule to those of the isotopically substituted one [33].

Centrifugal distortion constants

The centrifugal distortion constants $t_{\alpha\beta\gamma\delta}$ can be estimated from the relation [34-37]

$$t_{\alpha\beta\gamma\delta} = (\underline{J_{\alpha\beta}})^t \underline{F}^{-1} (\underline{J_{\gamma\delta}}) = \underline{A}_{\alpha\beta}\underline{G}^{-1}\underline{F}^{-1}\underline{G}^{-1}\underline{A}_{\gamma\delta} \tag{13}$$

The column matrix $J_{\alpha\beta}$ etc. contains the derivatives of the components of the inertia tensor with respect to the chosen set of coordinates and the elements $A_{\alpha\beta}$ etc. are related to the Cartesian components of the position vectors of the different atoms [36]. In Equation (13), $\underline{G}^{-1}$ and $\underline{F}^{-1}$ have their usual meanings. In cases where the directions of the principal axes of the inertia tensor are preserved, one obtains the simple isotopic product rule

$$\frac{|\underline{t}^{R^i}|}{|\underline{t}^R|} = \frac{|\underline{G}|^2}{|\underline{G}^i|^2} \tag{14}$$

directly from Equation (13) ($\underline{t}^R$ is a block matrix containing the different $t_{\alpha\beta\gamma\delta}$ constants; see [36]).

Also in cases where the directions of the axes of the principal moments of inertia tensor are altered by the isotopic substitution, one can get isotopic product rules. Isotopic sum rules involving linear combinations of the $t_{\alpha\beta\gamma\delta}$ constants have also been reported [36,37].

Coriolis coupling constants

For symmetric and spherical tops, the first order Coriolis constants ζ_i^α satisfy the linear relation [38]

$$\sum_i \zeta_i = C \tag{15}$$

where C is either a constant or a function of the components of the moment of inertia tensor. For spherical tops, the isotopic sum rule

$$\sum_i \zeta_i = \text{invariant} \tag{16}$$

follows from Equation (15). Using the familiar matrix notation [39]

$$\underline{\zeta}^\alpha = \underline{L}^{-1}\underline{C}^\alpha(\underline{L}^{-1})^t \tag{17}$$

one derives also the product rule

$$\frac{|\underline{\zeta}^{\alpha^i}|}{|\underline{\zeta}^\alpha|} = \frac{|\underline{G}|}{|\underline{G}^i|} \cdot \frac{|\underline{C}^{\alpha^i}|}{|\underline{C}^\alpha|} \tag{18}$$

More general isotopic sum rules involving quadratic forms of the zeta constants can also be derived and are applicable to any class of molecules (see [40] and references cited there).

Summary

Isotope rules for molecular constants have been briefly presented. These can be divided into sum and product rules. The product rules can be given in compact form for the matrix of molecular constant $\underline{O}$, viz.

$$\frac{|\underline{O}|}{|\underline{O}^i|}\left\{\frac{|\underline{G}^i|}{|\underline{G}|}\right\}^n = |\underline{X}| \tag{19}$$

(for the meaning of n, $\underline{O}$ and $\underline{X}$ see Table 4)

Also sum rules for frequencies and corresponding ones for i.r. as well as Raman intensities can be expressed with one equation [11,41] (equivalent to Equation (2)).

If all isotopic molecules have the same symmetry the sum rules are valid for modes of each irreducible representation. In other cases the correlation diagram relating the representations of the group (species with highest symmetry) and the subgroups (species with lower symmetry) has to be used [10]. The relations are valid for symmetry conserving types. For the determination of the coefficients, several methods have been proposed: (1) geometrically superimposing the molecules with

Table 4. Product rules (exact) for molecular constants ($\underline{O}$).

Molecular constant ($\underline{O}$) [a]	Matrix $\underline{X}$	n
$\underline{\Lambda}$	$\underline{E}$	+1
$\underline{\Sigma}$	$[(\underline{\delta}^i . \underline{\delta}^{-1})]^{-1}$	+1
$\underline{\zeta}^\alpha$	$[(\underline{C}^\alpha)^i . (\underline{C}^\alpha)^{-1}]^{-1}$	-1
$(\underline{t}^R)$ [b]	$\underline{E}$	-2

a The molecular constant is written in matrix form, the superscript i denotes the isotopic analogue.
b This is a matrix in block form (for spherical and symmetric tops) involving the centrifugal distortion constants $t_{\alpha\beta\gamma\delta}$. The equation given here (which holds for each block) is valid only when the directions of the principal axes are preserved under isotopic substitution.

appropriate signs so that the atoms vanish at all positions [10], (2) writing chemical exchange reactions in which all isotopic species are included e.g. $H_2O + D_2O \rightleftarrows 2HDO$ [42].

4. ISOTOPE RULES FOR HEAVY ATOM SUBSTITUTION

Perturbation theory (mass influence on molecular constants)

Considering heavy atom substitution, we are interested in studying the influence of small mass changes on molecular constants. If there is only a small percentage change in any molecular constant (as due to heavy atom isotopic substitution), the first order perturbation theroy [43] should give reasonable results.

For a secular equation

$$\underline{H}\underline{L}_o = \underline{L}_o\underline{E} \tag{20}$$

the change in the eigenvalue $\underline{E}$ due to a change $\underline{\Delta H}$ is

$$\underline{\Delta E}_k = [\underline{L}_o^{-1} \underline{\Delta H}\ \underline{L}_o]_{kk} \tag{21}$$

according to the first order perturbation theory (similar to the first order perturbation theory applied to a quantum mechanical problem in which the change in the eigenvalue due to a perturbation of an operator is given explicitly by the matrix element of the perturbation operator). One gets according to Equation (21), for instance, for the secular equation of Wilson,

$$\underline{L}_o^t \underline{F}\ \underline{\Delta G} (\underline{L}_o^t)^{-1} = \underline{\Delta\Lambda} \tag{22}$$

Table 5. Mass influence on molecular constants ($\underline{O}$)[a] studied with first order perturbation theory[b].

Molecular constant ($\underline{O}$)	Basic equation used to study the mass effect	Equation representing the isotopic variation of the molecular constant ($\underline{O}$)	Ref.
Vibrational frequency $(\underline{\Lambda})_i$	$\underline{G}\,\underline{F}\,\underline{L} = \underline{L}\,\underline{\Lambda}$	$(\underline{\Delta\Lambda}\,\underline{\Lambda}_o^{-1})_i = [\underline{L}_o^{-1}\underline{\Delta G}(\underline{L}_o^{-1})^t]_{ii}$	c
Coriolis constant ($\underline{\zeta}^{\alpha}$)	$\underline{L}^t\,\underline{F}\,\underline{K} = \underline{\Lambda}(\underline{E} - \underline{\zeta}^{\alpha})\underline{L}^t$	$\underline{\Delta\zeta}^{\alpha} = [\underline{\Delta\Lambda}\,\underline{\Lambda}_o^{-1}(\underline{E} - \underline{\zeta}_o^{\alpha}) - \underline{L}_o^{-1}\underline{\Delta K}(\underline{L}_o^{-1})^t]$	d
Mean square amplitude of vibration ($\underline{\Sigma}$)	$\underline{\Sigma}\,\underline{F}\,\underline{L} = \underline{L}\,\underline{\Lambda}\,\underline{\delta}$	$\underline{\Delta\Sigma} = [\underline{L}_o(\underline{\Delta\Lambda}\,\underline{\Lambda}_o^{-1}\underline{\delta}_o + \underline{\Delta\delta}).L_o^t]$	e
Integrated absorption intensity $(\underline{A})_i$	$\underline{A} = (N_o\pi d_i/3c^2).[\underline{L}^t\underline{\mu}'\underline{L}]$	$(\Delta A_i/A_i) = (\underline{\Delta\Lambda}\,\underline{\Lambda}_o^{-1})_i$	f

a The molecular constant is written in matrix form.
b For the meaning of the symbols, see text; the subscript o corresponds to the parent molecule.
c Ref. 10.
d A. Müller, K.H. Schmidt and N. Mohan, J. Chem. Phys., 57 (1972) 1752.
e A. Müller and N. Mohan, J. Chem. Phys., 58 (1973) 2994.
f Ref. 32.

Equation (22) leads to the result

$$\underline{\Delta\Lambda}\ \underline{\Lambda}^{-1} = [\underline{L}_o^{-1}\underline{\Delta G}(\underline{L}_o^{-1})^t] \qquad (23a)$$

or neglecting small off-diagonal elements,

$$\Delta\lambda_k/\lambda_k = [L_o^{-1}\underline{\Delta G}(\underline{L}_o^{-1})^t]_{kk} \qquad (23b)$$

The changes in other molecular constants (e.g. mean square amplitudes, Coriolis constants, and integrated infrared intensities) can also be derived in a similar way and are given in Table 5. In the case of infrared intensities and Coriolis coupling constants (for central atom substitution) one obtains a very simple equation relating the change in the constants directly to frequency shifts (Tables 5 and 6).

The matrix on the right hand side of Equation (23a) is not diagonal and it becomes necessary to employ off-diagonal elements coupling frequencies, if these are nearly equal (if due to symmetry lowering, two frequencies come under the same irreducible representation; see appendix).

Simple isotopic rule for frequencies

Simple isotopic sum rules valid for each constant separately (e.g. isotopic variation in each λ_k) can be derived by solving the vibrational problem in Cartesian coordinates in conjunction with the first order perturbation theory. Equation (23a) can be written in matrix form in Cartesian coordinates as [44] (subscript o not given)

$$\underline{\Delta\Lambda}\ .\ \underline{\Lambda}^{-1} = (\underline{L}^X)^t\underline{M}\ \underline{\Delta(M^{-1})}\ \underline{M}\ \underline{L}^X \qquad (24a)$$

or according to Equation (23b)

$$\Delta\lambda_k/\lambda_k = [(L^x_{pk})^2 + (L^y_{pk})^2 + (L^z_{pk})^2]m_p^2\Delta(1/m_p) \qquad (24b)$$

where $\underline{L}^X$ is the matrix of transformation between the Cartesian and the normal coordinates ($\underline{X} = \underline{L}^X\underline{Q}$), $\underline{M}$ is a diagonal (3N x 3N) matrix containing the masses of the various atoms and $\underline{\Delta(M^{-1})}$ is a matrix (naturally diagonal) representing the changes in the reciprocal masses $1/m_p$ due to isotopic substitution. Equation (24b) can be rewritten in a more general form as [45]

$$(\Delta\lambda_k/\lambda_k) = \sum_p[(L^x_{pk})^2 + (L^y_{pk})^2 + (L^z_{pk})^2].m_p^2\Delta(1/m_p) \qquad (25)$$

where L^x_{pk} etc. are the Cartesian components of the displacement of atom p during a vibration characterised by λ_k and the summation covers all substituted atoms m_p. It can be easily shown that $(L^x_{pk})^2 + (L^y_{pk})^2 + (L^z_{pk})^2 = A_p^2$ is identical for all atoms forming an equivalent set. This forms the basis of the isotopic rules derived below.

Equation (25) applied in conjunction with the normalisation condition

Table 6. Demonstration of the validity of the simple isotopic rule related to the absolute i.r. intensities ($\Gamma_i = A_i/\omega_i$).

	Vibration	Substitution	(Γ_i^+/Γ_i) (measured)	(ω_i^+/ω_i) (calculated)	Ref. for data
CF_4	ν_{as}(C-F)	$^{12}C/^{13}C$	1.01 ±0.02	0.97	a
$POCl_3/C_6H_{12}$	ν(P-O)	$^{16}O/^{18}O$	0.92 ±0.03	0.96	b
	ν_s(P-Cl)	$^{16}O/^{18}O$	0.97 ±0.03	0.99	b
	δ_s(PCl_3)	$^{16}O/^{18}O$	(0.82 ±0.16?)	0.99	b
benzoic acid	ν_s(C=O)	$^{12}C/^{13}C$	0.98 ±0.02	0.98	c
$Pd(N_2)_3$	ν_s(N-N)	$^{14}N/^{15}N$	0.91 (0.96)	0.97	d
$Ni(N_2)_4$	ν_{as}(N-N)	$^{14}N/^{15}N$	0.91 (0.96)	0.97	d

a I.W. Levin and T.P. Lewis, J. Chem. Phys., 52 (1970) 1608.
b I. Laulicht, S. Pinchas, D. Sadeh and D. Samuel, J. Chem. Phys., 41 (1964) 789.
c S. Pinchas, J. Phys. Chem., 69 (1965) 2256.
d Ref. 16. The experimental values of the relative absolute intensities given in brackets correspond to the data calculated using Cotton-Kraihanzel force field.

$$(\underline{L}^X)^t \underline{M}\ \underline{L}^X = \underline{E} \tag{26}$$

leads, for any molecular type, to the important result [45]

$$\sum_j (p/\alpha)_j (m_j^i/\Delta m_j)(\Delta\lambda_k^{(j)}/\lambda_k) = -1 \tag{27a}$$

where $\Delta\lambda_k^{(j)}$ is the shift in λ_k due to the isotopic substitution of α of the p atoms forming an equivalent set and the summation covers all sets of equivalent atoms (the superscript i corresponds to the isotope). Equation (27a) can be written in a very simple form for symmetric substitutions as

$$\sum_j m_j^{-1} \Delta\lambda_k^{(j)} = -\lambda_k \tag{27b}$$

$(\Delta m_j/m_j^i = m =$ relative mass change).

Thus, in AX_p type molecules, if we denote the shift due to the substitution $AX_p/{}^iAX_p$ by $\Delta\lambda^{(A)}$ and that due to the symmetrical substitution AX_p/A^iX_p by $\Delta\lambda^{(X)}$, we get from Equation (27):

$$(m_A^i/\Delta m_A)\Delta\lambda_k^{(A)} + (m_X^i/\Delta m_X)\Delta\lambda_k^{(X)} = -\lambda_k \tag{28}$$

Equation (27) shows that for a molecule with n different sets of atoms, only those shifts (in each λ_k) corresponding to the substitution of (n-1) sets of equivalent atoms are independent. Examples for the validity of these isotope rules are given in Table 7.

Equation (25) can be used to derive also isotopic rules valid for the substitution of any one set of equivalent atoms. Thus, for AX_p type molecules where A is either an atom or a group, we obtain for ligand substitution

$$\Delta\lambda_k^{(p)} = \text{constant} \tag{29}$$

where $\Delta\lambda_k^{(p)}$ corresponds to a successive substitution ($AX_p \rightarrow AX_{p-1}{}^iX$, $AX_{p-1}{}^iX \rightarrow AX_{p-2}{}^iX_2$ etc.). Table 8 demonstrates that these rules can be applied with success to heavy atom substitution. Correspondingly one obtains from Equation (25) for central atom substitution $AX_p \rightarrow {}^iAX_p$

$$\Delta\lambda_k/\Delta m_A \simeq \text{constant} \tag{30}$$

(assumption $m_A/m_A^i \simeq 1$), which means that the change in $\Delta\lambda$ is constant per unit mass change (see Table 9).

Considering the special case of symmetric substitution of the different sets of equivalent atoms, the simple isotopic additivity rule [45,46]

$$\sum_{a=1}^{p} \Delta\lambda_k^{(a)} = \Delta\lambda_k^{(pp)} \qquad (p \leqslant n) \tag{31}$$

Table 7. Isotopic rule related to central atom and ligand substitutions in A_nX_p type molecules[a] (all shifts in cm^{-1}).

A_nX_p	Mode	Central atom $\Delta\nu^{(A)}$ isotopic shift ($A_nX_p \rightarrow {}^iA_nX_p$) (meas).	Ligand isotopic shift $\Delta\nu^{(X)}$ ($A_nX_p \rightarrow A_n{}^iX_p$) (calc)	(meas)	Ref. for data
$^{16}O_3$	$\nu_2(A_1)$	7.1 ($^{16}O_3/^{18}O^{16}O_2$)	32.8	32.7 ($^{16}O_3/^{16}O^{18}O_2$)	c
$^{69}Ga^{35}Cl_3$	$\nu_3(E')$	2.8 ($^{69}Ga/^{71}Ga$)	7.4	7.5 ($^{35}Cl_3/^{37}Cl_3$)	d
$^{71}Ga^{35}Cl_3$	$\nu_3(E')$	-2.8 ($^{71}Ga/^{69}Ga$)	7.5	7.5 ($^{35}Cl_3/^{37}Cl_3$)	d
$^{28}Si^{35}Cl_4$	$\nu_3(F_2)$	13.1_5 ($^{28}Si/^{30}Si$)	6.0_5	5.6 ($^{35}Cl_4/^{37}Cl_4$) (6.2)[b]	e
$^{48}Ti^{35}Cl_4$	$\nu_3(F_2)$	5.6 ($^{48}Ti/^{46}Ti$)	6.6	6.6 ($^{35}Cl_4/^{37}Cl_4$)	f
$^{74}Ge^{35}Cl_4$	$\nu_3(F_2)$	5.0 ($^{74}Ge/^{70}Ge$)	7.8	7.8 ($^{35}Cl_4/^{37}Cl_4$)	f
$^{116}Sn^{35}Cl_4$	$\nu_3(F_2)$	3.6 ($^{116}Xn/^{124}Sn$)	8.2	8.0 ($^{35}Cl_4/^{37}Cl_4$)	g

a The shift associated with substitution $\Delta\lambda^{(X)}$ has been calculated from a knowledge of the shift due to central atom substitution $\Delta\lambda^{(A)}$ (see Equation (28)); the isotopic substitution associated with the shift is given in brackets following the numerical value.
b Calculated using the exact force field determined from the $^{28}Si/^{30}Si$ shift in $\nu_3(F_2)$ [17].
c L. Andrews and R.C. Spiker, J. Phys. Chem., 76 (1972) 3208.
d I.R. Beattie, H.E. Blayden, S.M. Hall, S.N. Jenny and J.S. Ogden, J. Chem. Soc. Dalton, (1976) 666.
e Ref. 17.
f Ref. 19.
g Ref. 20.

Table 8. Isotopic rule for stepwise substitution of equivalent atoms[a] (all shifts in cm^{-1}).

n	AX_n	Mode	$\Delta\nu^{(1)}$ AX_n ↓ $AX_{n-1}{}^iX$	$\Delta\nu^{(2)}$ $AX_{n-1}{}^iX$ ↓ $AX_{n-2}{}^iX_2$	$\Delta\nu^{(3)}$ $AX_{n-2}{}^iX_2$ ↓ $AX_{n-3}{}^iX_3$	$\Delta\nu^{(4)}$ $AX_{n-3}{}^iX_3$ ↓ A^iX_4	Ref. for data
4	CCl_4	$\nu_3(F_2)$	0.8[b]	0.9[b]	-	-	e
			(1.9)[c]	(0.9)[c]	-	-	
	$^{28}SiCl_4$	$\nu_3(F_2)$	1.4	1.5	1.5	1.3	f
	$^{74}GeCl_4$	$\nu_3(F_2)$	1.8	1.9	1.9	2.1	g
	$^{116}SnCl_4$	$\nu_1(A_1)$	2.8	2.1	2.1	2.5	h
		$\nu_2(E)$	0.7	0.7_5	0.7_5	0.6	h
		$\nu_3(F_2)$	2.0	2.0	2.1	2.1	h
	$^{124}SnCl_4$	$\nu_3(F_2)$	1.9	2.0	2.1	2.2	h
	$^{48}TiCl_4$	$\nu_3(F_2)$	1.5	1.6	1.6	1.5	g
3	PCl_3	$\nu_2(A_1)$	1.9	1.8	1.8		i
	$HCCl_3$	$\nu_2(A_1)$	1.6	1.6	1.5		j
		$\nu_5(E)$	1.1_5	1.3_5	1.2		j
2	CO_2	$\nu_1(\Sigma_g)$	39.1[d]	38.5[d]			k
		$\nu_2(\pi_u)$	5.1[d]	5.2[d]			k
	OCl_2	$\nu_1(A_1)$	2.9	2.9			ℓ

a See Equation (29). The weighted average of the components into which the F_2 and E modes for AX_4 type molecules, split in C_{3v} and C_{2v} type molecules was used in the calculations (see Equation (32)). The same procedure was adopted in the case of $HCCl_3$ (E mode).

b Corrected for Fermi resonance; see the reference under note e.

c The observed shifts without correcting for Fermi resonance.

d Harmonic shifts.

e H.J. Becher, Z. Physik. Chem. (Frankfurt), 81 (1972) 225.

f Ref. 17.

g Ref. 19.

h Ref. 20.

i A. Ruoff, H. Bürger, S. Biedermann and J. Cichon, Spectrochim. Acta, A, 28 (1972) 953.

j S. T. King, J. Chem. Phys., 49 (1968) 1321.

k Z. Cihla and A. Chedin, J. Mol. Spectr., 40 (1971) 337.

ℓ M.M. Rochkind and G.C. Pimentel, J. Chem. Phys., 42 (1965) 1361.

Table 9. Isotope shifts for central atom substitution $AX_p \rightarrow {}^{i}AX_p$ (example RuO_4)[a] [17,18].

Species	Abundance %	$\nu_3(F_2)$ (cm^{-1})	$\Delta\nu_3/\Delta m$ (cm^{-1}/amu)
$^{96}RuO_4$	5.51	921.8 ±0.1	
$^{98}RuO_4$	1.87	920 (sh)	0.9
$^{99}RuO_4$	12.72	919.2 ±0.1	0.8
$^{100}RuO_4$	17.62	918.3 ±0.1	0.9
$^{101}RuO_4$	17.02	917.5 ±0.1	0.8
$^{102}RuO_4$	31.61	916.8 ±0.1	0.7
$^{104}RuO_4$	10.58	915.2 ±0.1	0.8

a See Equation (30) Other examples:

$\Delta\nu_3(F_2)/\Delta m$ ($^{28/29/30}SiF_4$) = 8.6 ±0.2 cm^{-1}/amu;

$\Delta\nu_3(F_2)/\Delta m$ ($^{70/72/73/74/76}GeF_4$) = 1.4 ±0.1 cm^{-1}/amu.

can be derived easily from Equation (25) (the definition of the quantities is the same as in the general rule for molecular constants below). Some examples are given in Table 10.

In applying these isotopic rules, i.e. Equations (27) and (29), it is important to note that in cases where the degeneracy is removed by the isotopic substitution (i.e. asymmetric substitution), one has to estimate the frequencies corresponding to molecules of lower symmetry by the method of weighted average [47] according to which, we have

$$\lambda_k = [(r+s)^{-1}]\left[\sum_i^{r} \lambda_i \text{ (non-degenerate)} + s\lambda_j \text{ (degenerate)}\right] \quad (32)$$

where s is the degeneracy of the vibration corresponding to the molecule of lower symmetry. The examples in the Tables demonstrate that this is an excellent approximation in this case.

The appendix contains the application of the first order perturbation theory for cases exhibiting near accidental degeneracy (asymmetric substitutions in general).

A simple additive isotope rule for molecular constants

A simple additivity rule for the various molecular constants ($\Delta\lambda_i$, Σ_{ij}, $\zeta_{ii}^{(\alpha)}$, A_i, X_{ij} etc.) valid for heavy atom substitution can be derived from the perturbation theory [48,49]. The above mentioned molecular constants O_{ij} can be derived from one of the equations with the eigenvector matrix $\underline{L}$ (see Table 5) (m = +1)

$$\underline{O} = \underline{L}^m \underline{A} (\underline{L}^m)^t \qquad (\text{for } \underline{\Sigma},\ \underline{\zeta}^{\alpha}) \quad (33)$$

or

$$\underline{O} = (\underline{L}^m)^t \underline{A} \underline{L}^m \qquad (\text{for } \underline{A}) \quad (34)$$

Table 10. Additivity rule for different sets of atoms (see Equation (36)).

Species	Mode	$\Delta\nu^{(1)}$ (in cm^{-1})	$\Delta\nu^{(2)}$ (in cm^{-1})	$\Delta\nu^{(22)}$ (in cm^{-1}) calc.[a]	meas.	Ref.
		$^{16}O^{14}NF/^{16}O^{15}NF$	$^{16}O^{14}NF/^{18}O^{14}NF$	$^{16}O^{14}NF/^{18}O^{15}NF$		
ONF[b]	$\nu_1 = [\nu(NO)]$	32.9	49.7	83.5	83.4	c
	$\nu_2 = [\delta(ONF)]$	17.6	8.3	26.1	26.0	
	$\nu_3 = [\nu(NF)]$	2.5	8.8	11.1	10.7	
		$^{16}O^{14}N^{35}Cl/^{16}O^{15}N^{35}Cl$	$^{16}O^{14}N^{35}Cl/^{18}O^{14}N^{35}Cl$	$^{16}O^{14}N^{35}Cl/^{18}O^{15}N^{35}Cl$		
ONCl[b]	$\nu_1 = [\nu(NO)]$	32.0	49.5	82.4	82.2	d
	$\nu_2 = [\delta(ONCl)]$	14.4	7.7	22.3	22.5	
	$\nu_2 = [\nu(NCl)]$	2.1	7.1	9.2	9.1	
		$^{16}O^{14}N^{80}Br/^{16}O^{15}N^{80}Br$	$^{16}O^{14}N^{80}Br/^{18}O^{14}N^{80}Br$	$^{16}O^{14}N^{80}Br/^{18}O^{15}N^{80}Br$		
ONBr[b]	$\nu_1 = [\nu(ON)]$	31.9	49.3	82.1	82.4	e
		$^{14}N^{32}S^{35}Cl/^{14}N^{34}S^{35}Cl$	$^{14}N^{32}S^{35}Cl/^{15}N^{32}S^{35}Cl$	$^{14}N^{32}S^{35}Cl/^{15}N^{34}S^{35}Cl$		
NSCl	$\nu_1 = [\nu(SN)]$	11.8	30.7	42.8	43.0	f
		$^{14}N^{32}SBr/^{14}N^{34}SBr$	$^{14}N^{32}SBr/^{15}N^{32}SBr$	$^{14}N^{32}SBr/^{15}N^{34}SBr$		
NSBr	$\nu_1 = [\nu(SN)]$	11.9	30.2	42.4	42.4	f
		$^{116}Sn^{35}Cl_4/^{124}Sn^{35}Cl_4$	$^{116}Sn^{35}Cl_4/^{116}Sn^{37}Cl_4$	$^{116}Sn^{35}Cl_4/^{124}Sn^{37}Cl_4$		
$SnCl_4$	$\nu_3 = [\nu_{as}(SnCl)]$	3.6	8.0	11.7	11.7	g
		$H^{12}C^{35}Cl_3/H^{13}C^{35}Cl_3$	$H^{12}C^{35}Cl_3/H^{12}C^{37}Cl_3$	$H^{12}C^{35}Cl_3/H^{13}C^{37}Cl_3$		
$HCCl_3$	$\nu_2 = [\nu(CCl)]$	17.5	4.7	22.3	22.5	h

Table 10. (Continued).

a $\Delta\nu^{(22)} = [(\nu + \nu^{(1)})\Delta\nu^{(1)} + (\nu + \nu^{(2)})\Delta\nu^{(2)}]/[(\nu + \nu^{(22)}]$ where ν corresponds to the parent molecule, $\nu^{(1)}$ and $\nu^{(2)}$ to those involving monosubstitutions (i.e. where only one atom of the parent molecule is isotopically substituted) and $\nu^{(22)}$ to that involving multisubstitution.

b Harmonic frequencies.

c L.H. Jones, L.B. Asprey and R.R. Ryan, J. Chem. Phys., 47 (1967) 3371.

d L.H. Jones, R.R. Ryan and L.B. Asprey, J. Chem. Phys., 49 (1968) 581.

e J. Laane, L.H. Jones, R.R. Ryan and L.B. Asprey, J. Mol. Spectr., 30 (1969) 485.

f S.C. Peake and A.J. Downs, J. Chem. Soc. Dalton, (1974) 859.

g F. Königer and A. Müller, J. Mol. Spectr., 56 (1975) 200.

h K.H. Schmidt and A. Müller, J. Mol. Spectr., 50 (1974) 115.

or

$$\underline{O} = \underline{L}^{-m}\underline{AL}^{m} \qquad \text{(for } \underline{\lambda}) \tag{35}$$

Using the first order perturbation theory, the additivity rule

$$\sum_{a=1}^{p} (\Delta O^{(a)})_{ij} = (\Delta O^{(pp)})_{ij} \qquad (p \leqslant n) \tag{36}$$

is obtained ($\Delta O^{(a)}$ refers to mono-symmetric substitution in which only one set of equivalent atoms is substituted and $O^{(pp)}$ corresponds to the multi-symmetric substitution of p sets of equivalent atoms) by using the equation for the molecular constants together with the relation

$$\sum_{a=1}^{p} (\Delta L^{(a)})_{ij} = (\Delta L^{(pp)})_{ij} \tag{37}$$

The last equation (additivity rule) can be easily obtained by first order perturbation theory. The explicit relations involving the various molecular constants are given in Table 10. It should be noted that equations of the form (36) are valid for each constant of any one class (e.g. the shift in each λ_k, the shift in each Σ_{ij}, etc.) separately and hence serve as a very easy check on the experimental results. The above mentioned equation can be used to check anharmonicity constants of isotopic species, whereby excellent results have been obtained [48].

<u>Summary</u>

The methods of analysing theoretically small isotopic shifts of molecular constants have been covered. Whereas the classical isotope sum and product rules are valid within the harmonic approximation for all constants belonging to one irreducible representation, approximate equations valid for one individual isotope shift in heavy atom substitution data can be derived with first order perturbation theory.

These isotopic rules can be utilised in the analysis of complicated spectra as in deducing the effect of Fermi resonance on the isotope shifts or in the analysis of matrix spectra related to the stretching fundamentals, since in many cases the isotopic splitting pattern associated with the low frequency bending modes may not be known (in such cases, the classical isotope rules cannot be applied).

Acknowledgments

I thank Dr. N. Mohan for valuable discussions. Also, the financial help provided by the "Deutsche Forschungsgemeinschaft" and the "Fonds der Chemischen Industrie" and the NATO Scientific Affairs Division, is gratefully acknowledged.

5. REFERENCES

1. See A.F. Holleman and E. Wiberg, Lehrbuch der Anorganischen Chemie, de Gruyter, Berlin, 1976.
2. F.W. Loomis, Astrophys. J., 52 (1920) 248; A. Kratzer, Z. Physik, 3 (1920) 460; 4 (1921) 476.
3. H.C. Urey, F.C. Brickwedde and G.M. Murphy, Phys. Rev., 39 (1932) 164.
4. J.D. Hardy, E.F. Barker and D.M. Dennison, Phys. Rev., 42 (1932) 279.
5. G.H. Dieke and H.D. Babcock, Proc. Nat. Acad. U.S., 13 (1927) 670; W.F. Giauque and H.L. Johnston, Nature, 123 (1929) 318.
6. H.D. Babcock, Proc. Nat. Acad. U.S., 15 (1929) 471.
7. A.S. King and R.T. Birge, Astrophys. J., 72 (1930) 251.
8. S.M. Naudé, Phys. Rev., 36 (1930) 333.
9. N. Mohan, A. Müller and K. Nakamoto, Metal Isotope Effect on Molecular Vibrations, in R.J.H. Clark and R.E. Hester (Editors), Advances in Infrared and Raman Spectroscopy, Vol. 1, Heyden, London, 1975, p. 173.
10. E.B. Wilson, J.C. Decius and P.C. Cross, Molecular Vibrations, McGraw-Hill, New York, 1955.
11. L.M. Sverdlov, M.A. Kovner and E.P. Krainov, Vibrational Spectra of Polyatomic Molecules, John Wiley, New York, 1974.
12. D.M. Dennison, Rev. Mod. Phys., 12 (1940) 175.
13. For example, H.J. Becher, Z. Physik. Chem. (Frankfurt), 81 (1972) 225.
14. D.E. Milligan and M.E. Jacox, in D.A. Ramsay (Editor), MTP International Review of Science, Vol. 3, Butterworths, London, 1972, p. 1; H.E. Hallam (Editor), Vibrational Spectroscopy of Trapped Species, John Wiley, London, 1973.
15. L.H. Jones, Inorganic Vibrational Spectroscopy, Marcel Dekker, New York, 1971.
16. M. Moskovits and G.A. Ozin, in J.R. Durig (Editor), Vibrational Spectra and Structure, Vol. 4, Elsevier, Amsterdam, 1975.
17. F. Königer, A. Müller and W.J. Orville-Thomas, J. Mol. Struct., 37 (1977) 199; A. Müller and F. Königer, Can. J. Spectr., 21 (1976) 179.
18. F. Königer and A. Müller, J. Mol. Spectr., in press.

19. F. Königer, R.O. Carter and A. Müller, Spectrochim. Acta A, 32 (1976) 891.
20. F. Königer and A. Müller, J. Mol. Spectr., 56 (1975) 300.
21. F. Königer and A. Müller, Spectrochim. Acta A, in press.
22. A. Müller, F. Königer and N. Weinstock, Spectrochim. Acta A, 30 (1974) 641.
23. H.J. Becher, F. Friedrich and H. Willner, Z. Anorg. Allg. Chem., 395 (1973) 134.
24. I. Suzuki, Appl. Spectr. Rev., 9 (1975) 249.
25. J.C. Decius and E.B. Wilson, J. Chem. Phys., 19 (1951) 1409; L.M. Sverdlov, Doklady Akad. Nauk. SSSR, 78 (1951) 1115.
26. L.M. Sverdlov, Doklady Akad. Nauk. SSSR, 86 (1952) 513; 88 (1953) 249; Opt. i. Spektr., 8 (1960) 3; J. Biegeleisen, J. Chem. Phys., 28 (1958) 694.
27. T. Uno and K. Machida, Spectrochim. Acta, 18 (1962) 279; K. Machida, J. Chem. Phys., 38 (1963) 1360; J. Heiklen, J. Chem. Phys., 36 (1962) 721; J. Phys. Chem., 70 (1966) 989; L.M. Sverdlov, Doklady Akad. Nauk. SSSR, 94 (1954) 451.
28. S. Brodersen and A.A. Langseth, J. Mol. Spectr., 3 (1959) 114; S. Brodersen and T. Johannesen, J. Mol. Spectr., 23 (1967) 293.
29. N.L. Zhirnov, Opt. i. Spektr., 14 (1963) 580.
30. G. Strey, L. Cederbaum and K. Engelke, J. Chem. Phys., 54 (1971) 4402; D.L. Cummings and J.L. Wood, J. Mol. Struct., 26 (1975) 393; J. Wright and V.S. Ramachandra Rao, J. Mol. Spectr., 51 (1974) 520; N.T. Kyong and D.S. Bystrov, Opt. Spectr., 35 (1973) 613.
31. B.L. Crawford, J. Chem. Phys., 20 (1952) 977.
32. N. Mohan and A. Müller, J. Mol. Struct., 27 (1975) 255; N. Mohan, A. Müller and A. Alix, Mol. Phys., 33 (1977) 319.
33. S.J. Cyvin, Molecular Vibrations and Mean Square Amplitudes, Elsevier, Amsterdam, 1968.
34. D. Kivelson and E.B. Wilson, J. Chem. Phys., 20 (1952) 1575; 21 (1953) 1229.
35. M.R. Aliev, S.I. Subbotin and V.I. Tyulin, Opt. Spectr., 24 (1968) 47.
36. A.P. Alexandrov, M.R. Aliev and V.T. Alexanyan, Opt. Spectr., 29 (1970) 568.
37. A.P. Alexandrov and M.R. Aliev, J. Mol. Spectr., 47 (1973) 1.
38. E. Teller, Hand- und Jahrbuch der Chem. Physik, 9 (1934) 11, 43; M. Johnston and D.M. Dennison, Phys. Rev., 48 (1935) 868.
39. J.N. Meal and S.R. Polo, J. Chem. Phys., 24 (1956) 1119.
40. J.K.G. Watson, J. Mol. Spectr., 39 (1971) 364.
41. L.M. Sverdlov, Opt. i. Spectr., 8 (1960) 253.
42. L.M. Sverdlov, Opt. i. Spectr., 17 (1964) 947; S.I. Mizushima, Raman Effect, Handbuch der Physik, Bd. XXVI, Springer Verlag, Berlin, 1958.
43. E. Madelung, Die Mathematischen Hilfsmittel des Physikers, Springer Verlag, Berlin, 1964.
44. T. Miyazawa, J. Mol. Spectr., 13 (1964) 321.
45. N. Mohan and A. Müller, J. Chem. Phys., in press.
46. A. Müller, N. Mohan and F. Königer, J. Mol. Struct., 30 (1976) 297.
47. W. Lehmann, J. Mol. Spectr., 7 (1961) 261.
48. A. Müller, N. Mohan and A. Alix, to be published.
49. A.J.P. Alix and A. Müller, J. Chim. Phys., in press.

6. APPENDIX

The Green's function approach to the study of the isotope effect on vibrational frequencies

The Green's function approach suggested by Dewames and Wolfram [A1] is based on the fact that the dynamics of the "perturbed" or the "substituted" molecule is contained in the Green's function for the "unperturbed" or the "parent" molecule. Isotopic rules which do not involve a knowledge of the force field explicitly, follow from the relation

$$|\underline{G}(\omega^2)(\underline{\Delta M}\underline{M}^{-1}\omega^2 + \underline{E})| = 0 \qquad \text{(A1)}$$

where $\underline{G}(\omega^2)$ is the Green's function corresponding to the unperturbed molecule (in the spectral representation, $\underline{G}(\omega^2)$ is diagonal), $\underline{M}$ is a diagonal matrix (3N x 3N) whose elements are the masses of the various atoms forming the molecule and Δ relates to the isotopic variation. Equation (A1) is solved in the external symmetry coordinate representation where the normal coordinates Q_i are written as linear combinations (orthogonal) of the external symmetry coordinates $S_a^{(E)}$ through the matrix of mixing parameters ($\underline{A}$).

Although the isotopic rules (which are simply the ones involving the eigenvalues of the usual ($\underline{GF}$) matrix [A2]) derived from the Green's function approach are found to be the same as those derived classically [A1,A2], the procedure involved is a tedious one and further, this approach could be considered as a mere counterpart of the parametric study of molecular vibrations in internal vibrational coordinates [A3].

Perturbation theory for accidentally degenerate vibrations

In the study of the isotope effect on vibrational frequencies for heavy atom substitution, the usual first order perturbation theory is equivalent to the neglect of the cross (or non-diagonal) terms defined by Equation (23). This is not possible when two or more eigenvalues of the system (unperturbed) lie close together, so that accidental degeneracy occurs. In this case, one must take into consideration, the non-diagonal elements as well. When r of the eigenvalues of the unperturbed system are accidentally degenerate, one must obtain the corresponding r eigenvalues of the perturbed system by solving the appropriate secular equation of order r thus computing the eigenvalues.

Considering our isotope rules related to the vibrational frequencies, one has to take into account the cross terms $(\underline{L}_0^{-1})\Delta\underline{G}(\underline{L}_0^{-1})^t$ (or those of the product matrix $(\underline{L}^X)^t\underline{M\Delta}(\underline{M}^{-1}).\underline{M}(\underline{L}^X)$ in the Cartesian representation) in deriving the eigenvalues of the perturbed system (which is the isotopic analogue in our case) [A2]. Thus, the solution of the secular equation

$$\begin{vmatrix} H_{11} - \lambda & H_{12} & \dots & H_{1r} \\ H_{21} & H_{22} - \lambda & \dots & H_{2r} \\ H_{r1} & H_{r2} & & H_{rr} - \lambda \end{vmatrix} = 0 \qquad \text{(A2)}$$

of order r provides the r eigenvalues of the perturbed system which are nearly accidentally degenerate (in our case, $\underline{H}$ is the product matrix $(\underline{L}^X)^t\underline{M}\Delta(\underline{M}^{-1})\underline{M}(\underline{L}^X)$). Since the trace of a secular determinant is the sum of its diagonal elements (or the sum of the eigenvalues), it is clear that the sum of the r eigenvalues derived from Equation (A2) is the same as the sum of the corresponding r eigenvalues obtained from first order perturbation theory applied to non-degenerate systems. This fact, when taken into account, leads to the isotopic rules presented in Table A1 for nearly accidentally degenerate cases. The numerical results included in Table A2 give ample support to the validity of these isotopic rules.

Table A1. Isotopic rules following from the first order perturbation theory applied to r accidentally degenerate vibrations.

Nature of substitution	Isotopic rule	
All sets of atoms (symmetric or asymmetric)	$\sum_k^r \sum_j (p/\alpha)_j (m_j^i/\Delta m_j)(\Delta\lambda_k^{(j)}/\lambda_k) = (-1)r$ (corresponds to Equation (27a))	(A3)
All sets of atoms (symmetric)	$\sum_k^r \Delta\lambda_k^{(pp)}/\lambda_k = \sum_{a=1}^{p} \sum_k^r \Delta\lambda_k^{(a)}/\lambda_k$ (corresponds to Equation (31)) $(p \leqslant n)$*	(A4)
One set of equivalent atoms	$\sum_k^r (\Delta\lambda_k^{(p)}/\lambda_k) = \text{constant}$ (corresponds to Equation (29))	(A5)

* n is the number of sets of equivalent atoms.

Table A2. Verification of the isotopic rule (Equation (A5) in Table A1) following from perturbation theory (λ values in mdyn/amu Å)[a].

XY_2 (C_{2v})	$\Delta\lambda_1^{(p)}/\lambda_1$ ($X^{16}O_2/X^{16}O^{18}O$)	$\Delta\lambda_3^{(p)}/\lambda_3$ ($X^{16}O_2/X^{16}O^{18}O$)	$\Delta\lambda_1^{(p)}/\lambda_1$ ($X^{16}O^{18}O/X^{18}O_2$)	$\Delta\lambda_3^{(p)}/\lambda_3$ ($X^{16}O^{18}O/X^{18}O_2$)	$[\Delta\lambda_1^{(p)}/\lambda_1 + \Delta\lambda_3^{(p)}/\lambda_3]$ ($X^{16}O_2/X^{16}O^{18}O$)	$[\Delta\lambda_1^{(p)}/\lambda_1 + \Delta\lambda_3^{(p)}/\lambda_3]$ ($X^{16}O^{18}O/X^{18}O_2$)	Ref. for data
O_3	0.021 (12.3)	0.032 (17.6)	0.038 (21.7)	0.014 (7.4)	0.053 (29.9)	0.052 (29.2)	b
SO_2	0.049 (28.4)	0.028 (19.0)	0.040 (22.4$_5$)	0.036 (24.2)	0.077 (47.4)	0.076 (46.7)	c
UO_2	0.095 (37.0)	0.011 (4.4)	0.018 (6.8)	0.088 (34.6)	0.106 (41.4)	0.106 (41.4)	d

a The values given in brackets correspond to their respective isotope shifts in terms of the frequencies themselves (i.e. either $\Delta\nu_1^{(p)}$, $\Delta\nu_3^{(p)}$ or $[\Delta\nu_1^{(p)} + \Delta\nu_3^{(p)}]$. According to the perturbation theory for non-degenerate cases $\Delta\nu_k^{(p)}$ ($X^{16}O_2/X^{16}O^{18}O$) and $\Delta\nu_k^{(p)}$ ($X^{16}O^{18}O/X^{18}O_2$) should be nearly equal.

b A. Barbe, S. Secroun and P. Jouve, J. Mol. Spectr., 49 (1974) 171; the harmonic frequencies were employed in the calculations.

c M. Allavena, R. Rysnik, D. White, V. Calder and D.E. Mann, J. Chem. Phys., 50 (1969) 3399.

d S.D. Gabelnick, G.T. Reedy and M.G. Chasanov J. Chem. Phys., 58 (1973) 4468.

References

A1 R.E. Dewames and T. Wolfram, J. Chem. Phys., 40 (1964) 853; C.D. Bass, L. Lynds and T. Wolfram, J. Chem. Phys., 40 (1964) 3611; T. Wolfram, C.D. Bass, R.E. Dewames and L. Lynds, Bull. Chem. Soc. Japan, 39 (1966) 201.
A2 See ref. 10 above.
A3 N. Mohan and A. Müller, J. Mol. Spectr., 42 (1972) 400; A. Alix, N. Mohan and A. Müller, Z. Naturforsch. A, 28 (1973) 1158.

CHAPTER 13

INFRARED BAND INTENSITIES AND THE POLAR PROPERTIES OF MOLECULES

W.J. Orville-Thomas, S. Suzuki and G. Riley

1. INTRODUCTION

To a chemist the most useful concepts in describing chemical bonding are,

a) The state of hybridisation of the atomic orbitals used in constructing molecular orbitals, i.e., the valence states of the bonded atoms,
b) the extent to which the bond possesses ionic character, and
c) the presence of π-electrons.

For single bonds such as XH (X = C, N, O) (c) is considered to be absent. To these basic concepts must be added also the chemist's firm belief that a bond of a particular strength, as measured by bond energy, bond length, or bond stretching force constant, has a unique electronic structure. The chemist is therefore engaged in a never-ending search for a firm basis from which he can elucidate information which can then be interpreted to yield a description of the electronic structure of individual bonds in molecules couched in terms of the simple concepts outlined above.

In the majority of cases the chemist bases his quantitative ideas on valence bonding on experimental data such as vibrational frequencies, bond lengths, bond energies, spin-spin and nuclear quadrupole coupling constants. Studies based on these and other experimental data have provided an immense amount of structural information which is undoubtedly accurate so far as the grosser structural features are concerned.

A closer look at some of the results however indicates that this analysis is too simple and that other, more sensitive, experimental parameters should be used for bond structure analysis.

Bond stretching frequencies and spin-spin coupling constants

Bond stretching CH vibrations are relatively isolated because of their high frequency. Lehmann [1] has shown that the root-mean square value of the symmetric and asymmetric stretching frequencies of a methyl group is directly proportional to the force constant of the bond. Hence these rms values (Table 1) can be taken as giving a measure of the strength of the CH bond in the methyl group. Müller and Prichard [2] have shown that the C-H nuclear spin-spin coupling constant is proportional to the amount of 's' character in the bonding orbital. These authors showed in addition that this constant is almost independent of the ionic character of the

Table 1. Vibrational frequencies, electronegativities and spin-spin coupling constants, in CH_3X molecules.

Molecule	rms $\nu(CH)$ (cm^{-1})	(x)	$J(C^{13}-H)$ (Hz)
CH_3D	3014	2.15	
CH_3F	2992	3.95	149
CH_3Cl	3017	3.00	150
CH_3Br	3028	2.80	152
CH_3I	3032	2.55	151
CH_3CHO	2992	2.78	127
CH_3OH	2972	3.80	141
CH_3CN	2991	3.11	136
CH_3NC	2985	3.49	
CH_3COO^-	2985	2.92	

bond. Since the spin-spin coupling constant is almost invariant in the methyl halides (Table 1) it seems that the valence states of the carbon atoms in these compounds are almost identical: i.e. the hybridisation factor remains constant and is relatively unimportant in altering the CH bond strengths. Hence since the rms frequency varies from molecule to molecule the main factor causing this variation must be the change in the ionic character caused by the electrical effects induced by the neighbouring polar CX bonds.

The determination of the ionic character of a bond is difficult. However, it seems clear that the ionic character of the CH bond will depend, upon the electronegativity of X, $\chi(X)$. It seems therefore safe to claim that the frequency depends upon the ionic character of the CH bond which in turn is governed by the electronegativity of X. When the CH force constants are plotted against electronegativity difference in Figure 1, it is seen that the curves are concave. If a horizontal line corresponding to a definite value of the bond stretching force constant is drawn it will therefore cut the curves in more than one point. Since these points represent CH bonds of different electronic structure (or ionic character) the anomalous position is apparently reached of having more than one bond of the same strength but with different electronic structures [3].

Bond lengths and nuclear quadrupole coupling constants

In atoms and molecules nuclei are embedded in an electronic cloud. When the charge distribution outside a particular quadrupolar nucleus is non-spherical there is an interaction between the nuclear field and the external field. This interaction is responsible for hyperfine structure of rotational lines. A nuclear quadrupole coupling constant, eQq, can be calculated from this hyperfine structure when both Q and q are

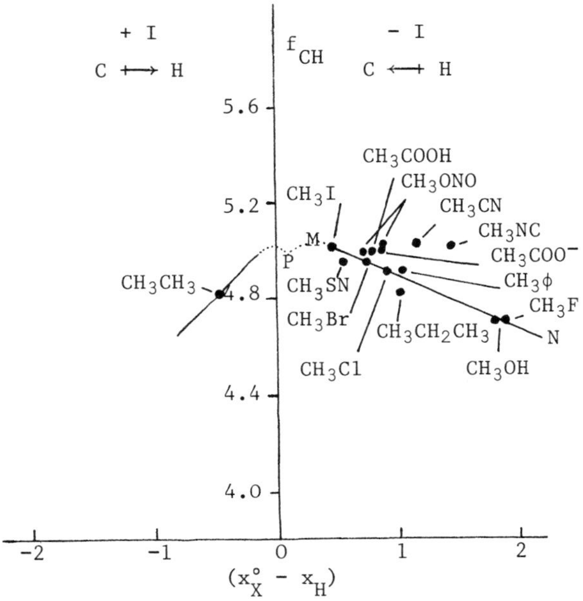

Figure 1. Variation of force constant with electronegativity factor for CH bonds.

finite. Effectively q, which is the field gradient at the nucleus is a measure of the departure from spherical symmetry of the charge distribution at the nucleus due to the electrons and other nuclei present in the same molecule. This quantity depends on the environment of the quadrupolar nucleus in the molecule, and is therefore intimately connected with the type of valence bonding surrounding the nucleus.

Dailey [4] pointed out that quadrupolar nuclei act as 'built in probes', but since their position remains fixed, they give information about the electron distribution at one point only.

Townes and Dailey [5] argued that since s electrons and filled inner shells have spherical symmetry, and since d and f electrons do not penetrate near the nucleus, the quadrupole coupling constant is largely due to the p electrons present in the valence shells.

A large number of microwave and infrared studies have been carried out on XCN molecules in which the distribution of bonding electrons can be pictured as, X-C ≡ N (⟷) i.e. the CN bond is considered to be triple in character and the lone-pair electrons form an atomic dipole. In the majority of cases a large number of isotopic modifications have been studied and the very accurate molecular parameters given in Table 2 obtained. The most striking feature about these data is that for the halogenated compounds the CN bond lengths are equivalent to within experimental error. If one takes this as an indication that

Table 2. Bond parameters in XCN molecules.

Molecule	r(CN) (Å)	eQq(^{14}N) (MHz)	f(CN) (mdyne $Å^{-1}$)
HCN	1.155	-4.58	18.68
ICN	1.159	-3.80	
BrCN	1.158	-3.83	18.51
ClCN	1.159	-3.63	17.92
FCN	1.159	-2.67	

the CN bonds have very similar electronic structures there seems no reason why the distribution of electrons in the atomic dipoles should differ. In effect one expects the same non-spherical distribution of electronic charge around the nitrogen nucleus in each of the four halogenated molecules. One is obviously led, therefore, to expect almost equal values for the nitrogen nuclear quadrupole coupling constants in the four molecules. Examination of Table 2 shows, however, that the value obtained for FCN is very different from the others. This again emphasises that bonds of equal length do not necessarily have the same electronic distribution. This possibility has also been pointed out by Tyler and Sheridan [6]. If, however, we insist on clinging to our chemical intuition and demanding that valence bonds with dissimilar electronic structures are different then it seems clear that the experimental data mentioned above are not particularly sensitive parameters in so far as determining the electronic structures of valence bonds is concerned. This raises the question as to whether some other quantity would more accurately mirror the electronic structure of molecules and in particular the structures of individual bonds. One such parameter would appear to be infrared vibrational band intensities.

Infrared band intensities as a molecular probe

Since ν(CH) or ν(CD) bond stretching vibrations are relatively isolated, it can be argued that if in a series of molecules the corresponding frequencies remain almost constant then the electronic structures of the bonds are very similar. A simple extension of this argument then leads one to expect comparable values for the CH(CD) band intensities.

Some intensities for the ν(CD) band of deuterated chloroform [7], in a variety of solvents, are summarised in Table 3. An examination of this table shows that although little or no frequency change occurs there is a very large variation in intensity as we go from one liquid system to another. Obviously part of this change [8] can be attributed to the different molecular environments of the absorbing molecule, and part to specific interactions of a charge transfer nature [9] between the solute and solvent. The experimental fact, however, that the intensity changes so markedly whilst the frequency stays almost unchanged indicates quite conclusively that band intensities are much more sensitive probes for structural and interaction studies than other parameters such as vibrational frequencies, bond lengths or nuclear quadrupole coupling constants.

Table 3. ν(CD) band intensities, A, in $CDCl_3$.

Solvent	$\Delta\nu$ (cm^{-1})*	A (10^3 cm $mole^{-1}$)	Ratio $\frac{A(solvent)}{A(CCl_4)}$
$CDCl_3$	0	62	1.24
CCl_4	0	50	1.00
Benzene	5	316	6.32
Nitrobenzene	4	300	6.00
Mesitylene	5	340	6.80
Acetone	0	540	10.8
n-Ethylacetamide	0	1700	34.0

* $\Delta\nu = \nu(CDCl_3) - \nu(solvent)$

2. BAND INTENSITIES

When infrared radiation is absorbed by a molecule the intensity of absorption is related to the way in which the electronic charges move during the molecular vibrations. During vibrations molecules are deformed and hence there must be associated changes in the electronic structure of individual bonds. It is clear therefore that infrared absorption band intensities should provide information on the electronic charge distributions in molecules and also how the electrons in the outermost valence shells redistribute themselves during molecular vibrations. In order to obtain this type of information for individual molecules it is clear that band intensities should be measured for molecules in the gas phase where intermolecular interactions are absent.

An infrared absorption band can be characterised by a number of parameters of which the frequency of the band centre, ν, and the absolute intensity of the band, A, are of great importance insofar as studies of molecular structure are concerned.

If frequency values can be assigned to all fundamental vibrations then a normal coordinate calculation, albeit of an approximate nature, is possible. This leads to information on the electronic structure of valence bonds, via the force constants, and also of the form of the normal vibrations, i.e. quantitative information is obtained concerning the normal coordinates, Q, that describe the motions of the individual atoms during the fundamental vibrations.

This knowledge of the normal coordinates can then be used in conjunction with measured band intensities to yield information on charge distributions in molecules and also on how the valence electrons (bonding and lone-pair) redistribute themselves during the deformations that occur during vibrations. These studies also provide some information on the equilibrium charge distributions and so provide a better insight into the polar nature of individual valence bonds.

Definition of band intensity

The basic theory of the absorption of radiation is dealt with in several texts and it has been shown that an explicit relationship exists between the band intensity Γ as defined by,

$$\Gamma_{n'',n'} = \frac{1}{n\ell} \int_{Band} \ln(I_0/I)\, d\ln\nu = \frac{8\pi^3 N}{3hc} <0|p|1>^2 \quad (1)$$

and the derivative of the molecular dipole moment of the normal coordinate. This relation is given by,

$$\Gamma_i = \frac{N\pi}{3c^2\omega_i} \left(\frac{\partial p}{\partial Q_i}\right)^2 \quad (2)$$

In this relation Γ_i is the integrated intensity of the fundamental absorption band, and ω is the harmonic frequency of the ith mode.

Intensity has also been measured in a different basic unit, defined as,

$$A = \frac{1}{c\ell} \int \ln(I_0/I)\, d\nu \quad (3)$$

From the definitions of Γ and A it can be seen that they are approximately related as follows,

$$A = \Gamma\nu_0 \quad (4)$$

where ν_0 is the frequency of the band centre in cm^{-1}. Band intensities have been reported in many units and relations between these units have been given by Mills [10].

3. BAND INTENSITIES AND THE POLAR PROPERTIES OF CHEMICAL BONDS

Using the best possible techniques it is now possible to measure experimental band intensities with an estimated probable error of ±1-2% and the list of reliable experimental results on small molecules is now quite considerable [10,11].

Once reasonably accurate gas-phase band intensities are known, we are in a position to obtain information on the polar properties of the chemical bonds constituting the molecule.

The main aim in studying infrared absorption intensities is to obtain an understanding of how the electrons move when a molecule changes its shape on vibration away from the equilibrium configuration. To achieve this goal it is necessary to create a model from which easily visualised polar parameters associated with individual bonds can be derived.

Three direct steps are involved in analysing band intensities to yield information on the polar characteristics of molecules. They are,

(i) the calculation of molecular dipole gradients, $\partial\mu/\partial Q$, for

each fundamental mode of vibration;
(ii) the transformation of the $\partial\mu(M)/\partial Q$'s to molecular dipole gradients with respect to symmetry coordinates, $\partial\mu/\partial S$ [12,13]; and (iii) the application of bond moment theory to yield bond moment or electrooptical constants from the set of $\partial\mu/\partial S$ values [13-21].

These will be dealt with in turn.

Calculation of $\partial\mu/\partial Q$ values from band intensities

The electric dipole moment, μ, for a molecule in a given electronic state is a function of the vibrational coordinates and, when the change in dipole moment resulting from vibration is small, it is permissible to develop μ as a power series in the normal coordinates, Q_i,

$$\mu = \mu_o + \Sigma\left[\frac{\partial\mu}{\partial Q_i}\right]_0 Q_i \tag{5}$$

where the subscript zero indicates one is working in the neighbourhood of the equilibrium position and μ_o is the static molecular dipole moment. In practice only the first two terms are retained so that a quantitative description of $\mu(M)$ requires therefore a knowledge of the values of μ_o and the various parameters $\partial\mu/\partial Q_i$.

Several relations connecting band intensities with $\partial\mu/\partial Q$ values have been derived for isolated molecules in the gas phase [22-24]. These form a special case of the general relation derived by Ratajczak and Orville-Thomas [25] which applies to molecules in any phase. This can be written as

$$A_i(\nu) = \frac{g_i \pi N}{3c^2 n(\nu)} \left(\frac{\nu_i}{\omega_i}\right)\left(\frac{\partial\mu}{\partial Q_i}\right)^2 \frac{E_i^2(\nu)}{E^2(\nu)} \tag{6}$$

$E_i(\nu)$ is the effective electrical field (the internal field) in a macroscopic field of strength $E(\nu)$ in a dielectric of refractive index $n(\nu)$, where ν is the frequency of the applied field; g_i is the degree of degeneracy of the vibration. In dilute gases this relation reduces to

$$A_i^G(\nu) = \frac{g_i \pi N}{3c^2} \left(\frac{\nu_i}{\omega_i}\right)\left(\frac{\partial\mu}{\partial Q_i}\right)^2 \tag{7}$$

identical with that derived by Overend [24].

Using relation (7), values for the quantities $\partial\mu/\partial Q_i$ can be calculated for each fundamental mode of vibration from the band intensities.

Since $(\partial\mu/\partial Q_i)$ appears as a squared term, its algebraic sign is unknown. The zero-order frequencies resulting from anharmonic corrections of ν_i are evaluated by assuming that the corrections are the same in both liquid and gaseous phases. The (ν_i/ω_i) ratios were used where possible [26]. For the degenerate bending vibrations, g_i takes the value 2.

Typical values for $\partial\mu/\partial Q_i$ are listed in Table 4 (taken from references cited in J. Mol. Structure, 4 (1969) 163).

Table 4. Intensity (A_i) and $|\partial\mu/\partial Q|$ values.

Molecule	Vibration	A_i (10^3 darks)			$\lvert\partial\mu/\partial Q\rvert$ ($cm^{3/2}sec^{-1}$)		
		Vapour	Liquid	Solid	Vapour	Liquid	Solid
YXY							
OCO	ν_3	5.4		7.7	62.2		64.6
	ν_2	66.5		45.9	308		206
SCS	ν_3	0.5	0.56	0.85	18.9	16.6	26.9
	ν_2	56.7	71.4	80.0	238	268	280
SeCSe	ν_3		0.86			20.6	
	ν_2	29.0	52.4		201	221.3	
YXZ							
OCS	ν_2	0.8	1.11	0.74	32.2	34.4	29.6
	ν_3	0.29	1.0	0.76	14.3	22.5	19.2
	ν_1	59.0	63.8	68.0	288	264	284
SCSe	ν_2		0.35			18.2	
	ν_3		1.46			26.2	
	ν_1		60.2			238	

Since the quantity $(\partial\mu/\partial Q)^2$ bears a direct relationship to the intensity A_i of an absorption band the results in Table 4 are not really surprising in that some very high values of $\partial\mu/\partial Q$ have been derived. If $\partial\mu/\partial Q$ is considered in its approximate form - $\Delta\mu/\Delta Q$ - it is evident in the case of a high $\partial\mu/\partial Q$ value, that either a large dipole moment change is associated with a small change in normal coordinate or a small $\Delta\mu$ associated with an extremely small ΔQ is in operation. The converse is true in the case of very small $\partial\mu/\partial Q$ values. It seems likely that the dipole moment change is the most likely phenomenon in the case of the present study to cause intense bands, i.e. large $\partial\mu/\partial Q$ values. Contributions which make up $\Delta\mu$ are complex and the bond moments (of which the overall dipole moment of the molecule is a resultant) contain contributions from [27]: (i) the asymmetry in the charge distribution of the bonding electrons; (ii) inequalities of size of the atoms (homopolar dipole); (iii) the asymmetry of the hybrid atomic orbitals, (iv) the asymmetry of charge associated with the lone pair orbitals (atomic dipoles).

Most of these contributions are expected to change during the vibration and thus contribute to $\partial\mu/\partial Q_i$. Moreover, electronic rearrangements may occur in bonds close to the vibrating one, even though they are not coupled vibrationally. Large contributions to the intensity may arise from this source particularly in conjugated systems. Hence the expression for $\partial\mu/\partial Q_i$ comprises a number of terms which may be of the same or opposite sign and in general it is difficult to determine which of the terms is dominant and what would be the influence of a change of environment on each term.

Since normal coordinates are complex functions of the changes that occur in bond lengths and bond angles during vibrations the quantities $\partial\mu/\partial Q_i$ do not have a simple physical meaning such as those associated with the terms ionic character and bond dipole constants. In order to obtain information useful to chemists the $\partial\mu/\partial Q$ values have to be transformed to quantities directly related to the polar properties of individual valence bonds. Examples of such parameters, which are known as electrooptical parameters (Russian school [15]) or bond moment constants [13,16-21], are bond moments, μ_b, and bond dipole gradients, $\partial\mu_b/\partial r_b$.

The problem of determining bond moment constants can be tackled either experimentally via infrared band intensities or theoretically using wave mechanical methods. These approaches will be discussed in turn.

<u>Transformation of $\partial\mu/\partial Q$ values to dipole moment gradients associated with symmetry coordinates, $\partial\mu/\partial S$.</u>

The measurement of vibrational band intensities enables one to calculate values of $\partial\mu/\partial Q_i$, i.e. the derivative of the molecular dipole moment with respect to a normal coordinate. The sign of this quantity however, is indeterminate. Next in order to obtain information on molecular properties it is necessary to transform this quantity, which has no simple physical meaning, into related quantities associated with the derivative of the molecular dipole moment with respect to a symmetry coordinate, S. This has been described several times [28,29]. The derivatives of the molecular dipole moment with respect to symmetry coordinates are given by,

$$\frac{\partial\mu}{\partial S_j} = \sum_i \frac{\partial\mu}{\partial Q_i}\frac{\partial Q_i}{\partial S_j} \tag{8}$$

To carry out this transformation it is necessary to calculate the values of the coefficients $\partial Q/\partial S$. This can be done as follows. The transformation relating symmetry coordinates to normal coordinates is,

$$S = LQ$$

On differentiating the inverse of this equation with respect to S one obtains the coefficients $\partial Q/\partial S$ which on substitution in Equation (8) gives the desired relations

$$\frac{\partial\mu}{\partial S_j} = \sum_i (L^{-1})_{ij}\frac{\partial\mu}{\partial Q_i} \tag{9}$$

Examination of Equation (9) shows immediately that in order to calculate a value for any particular $\partial\mu/\partial S_j$ values for all non-vanishing $\partial\mu/\partial Q$'s are required. In favourable cases where the molecule has symmetry the L^{-1} matrix factors into smaller blocks and the problem becomes tractable. In practice the necessity for obtaining all values of $\partial\mu/\partial Q$ for vibrations within a particular symmetry species in order to obtain $\partial\mu/\partial S$ for a particular symmetry coordinate leads to difficulties.

Further difficulties occur in the calculation of the coefficients occurring in the L^{-1} matrix. These coefficients can be calculated from the force constants occurring in the potential function. This emphasises the point that reliable values of $\partial\mu/\partial S$ can only be obtained for molecules for which an accurate force constant calculation has been carried out. It is important to note that the sign of each $\partial\mu/\partial Q$ is indeterminate and this is also true for the elements of the L^{-1} matrix. In practice then Equation (9) does not give a unique value for the $\partial\mu/\partial S$ parameters but in general yields 2^{n-1} different solutions, where n is the number of normal vibrations in the symmetry class being considered. Only one of these solutions can be physically correct.

The polar parameters $\partial\mu/\partial S$ are derived exactly within the double harmonic approximation but their accuracy depends critically on the accuracy with which one can measure the band intensities and the accuracy with which the potential function can be estimated.

4. BOND MOMENT THEORY

The molecular polar parameter $\partial\mu/\partial S$ is derived exactly within the double harmonic approximation. These quantities however apply to the molecule as a whole and for the general purposes of chemists it is necessary to manipulate these data to obtain polar parameters associated with individual chemical bonds. These constants can be visualised physically and hopefully should eventually become a 'bond' parameter and transferable from molecule to molecule in exactly the same way as bond energies and force constants can be. In the Russian literature bond moment theory is referred to as electrooptical theory. We prefer the more descriptive term "bond moment theory". By electrooptical or bond moment constants is meant bond dipole moments, μ_n, and the derivatives of bond dipole moments with respect to the bond-stretching or bond-bending internal coordinates, $\partial\mu_n/\partial r_n$, $\partial\mu_n/\partial r_m$ and $\partial\mu_k/\partial\gamma_s$.

In order to obtain bond moment constants it is necessary to carry out a normal coordinate calculation and then to provide values of $\partial\mu/\partial S_j$ from the measured band intensities (Equation 9) and then to interpret these polar parameters in terms of a bond moment theory which manipulates the $\partial\mu/\partial S_j$ data to yield bond moment constants.

Normal coordinate calculation

It has been shown (Equation 9) that the following relation holds

$$\frac{\partial\mu}{\partial S_j} = \sum_i (L^{-1})_{ij} \frac{\partial\mu}{\partial Q_i} \qquad (9)$$

and hence to obtain values for the polar parameters $\partial\mu/\partial S_j$ from band intensities it is necessary to evaluate the elements of the L^{-1} amplitude matrix. This is done as follows.

Linear triatomic molecules have four possible modes of vibration of which the bond-bending mode is doubly-degenerate. For CO_2, CS_2 and CSe_2 there are two bond-stretching modes of which the asymmetric mode only is infrared-active; for the YXZ molecules both the stretching modes are infrared-active.

The symmetric and the general linear triatomic molecules are illustrated in Figure 2.

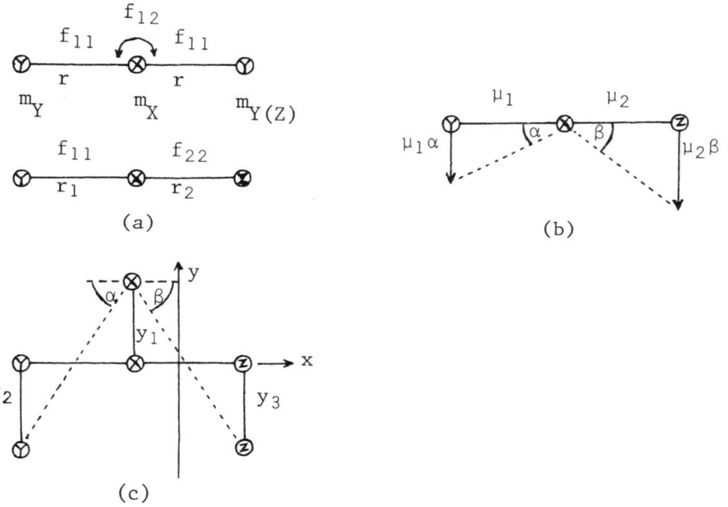

Figure 2. General linear triatomic molecule:
(a) molecular parameters; (b) polar parameters;
(c) coordinates.

Normal-coordinate calculations have been described many times (for example see refs. 28-30). A brief description only is included here.

Normal coordinates, Q_i, are complex linear functions of internal displacement coordinates. In terms of symmetry coordinates, S, the normal coordinate transformation matrix is

$$\mathbf{S} = \mathbf{LQ} \quad \text{whence} \quad \mathbf{Q} = \mathbf{L}^{-1}\mathbf{S} \qquad (10)$$

For a symmetry species of two vibrations Equation (10) is equivalent to,

$$Q_1 = L^{-1}_{11}S_1 + L^{-1}_{12}S_2 \; ; \quad Q_2 = L^{-1}_{21}S_1 + L^{-1}_{22}S_1 \qquad (11)$$

The secular equation can be written as $|\mathbf{GF} - \mathbf{E}\lambda| = 0$, and for this form the homogeneous equations for the amplitudes can be written in the form

$$(FG - E\lambda_k)(L^{-1})'_k = 0$$

or written out for a symmetry class containing two vibrations,

$$[(FG)_{11} - \lambda_k]L^{-1}_{k1} + (FG)_{12}L^{-1}_{k2} = 0 \tag{12}$$

$$(FG)_{21}L^{-1}_{k1} + [(FG)_{22} - \lambda_k]L^{-1}_{k2} = 0 \tag{13}$$

where k = 1 or 2 in turn for the two vibrations considered.

These equations are homogeneous and lead therefore only to the ratios for the coefficients of the L^{-1} reciprocal matrix, for a given value of λ_k.

From (12)

$$L^{-1}_{k2}/L^{-1}_{k1} = -[\{(FG)_{11} - \lambda_k\}/(FG)_{12}] \qquad k = 1,2 \tag{14}$$

To obtain separate values for the coefficients, it is necessary to normalise. That is, it is necessary to multiply the unnormalised L^{-1} coefficients by the appropriate normalisation constant in order to obtain the absolute values needed [28].

The normalisation condition is that,

$$\mathbf{LL'} = \mathbf{G} \quad \text{or} \quad \mathbf{L^{-1}G(L^{-1})'} = \mathbf{E}$$

The second condition is the easier condition to apply in practice.

Let $(L_0^{-1})_{kt}$ be an unnormalised value and K_k the normalising constant, then

$$K_k = \sum_{tt}[(L_0^{-1})_{kt}(L_0^{-1})_{kt'}G_{tt'}]^{-\frac{1}{2}}$$

where t = 1,2...n, t' = 1,2...n, for a vibrational species containing n vibrations.

For the trivial case of n = 1,

$$K_k = [(L_0^{-1})^2_{kk}G_{kk}]^{-\frac{1}{2}}$$

Putting $(L_0^{-1})_{kk} = 1$, the normalised value is obtained directly as

$$L^{-1}_{kk} = K_k = (G_{kk})^{-\frac{1}{2}} \tag{15}$$

For n = 2, putting $(L^{-1}_0)_{kt} = 1$, and t,t' = 1,2

$$K_k = [G_{11} + 2L^{-1}_{k2}G_{12} + (L^{-1}_{k2})^2G_{22}]^{-\frac{1}{2}} \tag{16}$$

where k = 1,2 in turn.

The general linear triatomic molecule is shown in Figure 2: force constants are designated by the symbols f_{11} and f_{22} for bond-stretching, $f(\alpha)$ for bond-bending whilst the bond-bond interaction constant is f_{12}. The force constants used are given in Table 5 which also includes the bond lengths.

The fundamental vibrations of linear XY_2 molecules (point

Table 5. Molecular parameters for linear X∴∴Y∴∴Z or X═Y═Z molecules.

Molecule	r(XY) (Å)	r(YZ) (Å)	Force constants (mdyne Å^{-1}) f(XY)	f(YZ)
OCO	1.160	1.160	16.02	16.02
OCS	1.161	1.560	15.95	7.53
OCSe	1.159	1.709	15.92	5.50
SCS	1.554	1.554	7.83	7.83
SCSe	1.557	1.709	7.80	5.82
SCTe	1.557	1.904	7.46	4.48
SeCSe[a]	1.711	1.711	5.94	5.94

[a] Value assumed for bond length.

group $D_{\infty h}$) are in different symmetry classes,

$$\Gamma = \Sigma_g^+ + \Sigma_u^+ + \Pi_u$$

and hence can be dealt with as three separate one-dimensional problems.

Suitable symmetry coordinates are,

$$\text{class } \Sigma_g^+ \quad S_1 = (\Delta r_{12} + \Delta r_{23})/\sqrt{2}$$

$$\text{class } \Sigma_u^+ \quad S_2 = (\Delta r_{12} - \Delta r_{23})/\sqrt{2}$$

$$\text{class } \Pi_u \quad S_3 = r\Delta\theta$$

In these cases the appropriate L^{-1} matrix elements (Equation 15) are

$$L_{11}^{-1} = (G_{11})^{-\frac{1}{2}} = \mu_y^{-\frac{1}{2}} \; ; \; L_{22}^{-1} = (G_{22})^{-\frac{1}{2}} = (2\mu_x + \mu_y)^{-\frac{1}{2}}$$

$$L_{33}^{-1} = (G_{33})^{-\frac{1}{2}} = (4\mu_x + 2\mu_y)^{-\frac{1}{2}}$$

where $1/m_x = \mu_x$ etc.

The two bond-stretching vibrations of linear YXZ molecules (point group $C_{\infty v}$) are in the same symmetry class whilst the bond-bending mode is in a symmetry class of its own, i.e. $\Gamma = 2\Sigma^+ + \Pi$.

Suitable symmetry coordinates are,

$$\text{Class } \Sigma^+ \begin{cases} S_1 = \Delta r_1 & \text{for } \nu_a \text{ vibrations} \\ S_2 = \Delta r_2 & \text{for } \nu_s \text{ vibrations} \end{cases}$$

$$\text{Class } \Pi \quad S_3 = (r_1 r_2)^{\frac{1}{2}}\Delta\theta$$

For the two class Σ^+ vibrations the ratios of the L^{-1}

matrix elements have been given by Equation (8). Using the standard methods it is found that,

$$(FG)_{11} = f_{11}(\mu_x + \mu_y) - f_{12}\mu_x$$

$$(FG)_{12} = -f_{11}\mu_x + f_{12}(\mu_x + \mu_z)$$

Substitution of the appropriate values for λ_k, $(FG)_{11}$ and $(FG)_{12}$ in Equation (14) followed by normalisation (Equation 16) leads to the L^{-1} matrix elements listed in Table 6.

Table 6. L^{-1} matrix elements.

Molecule	L^{-1}_{11}	L^{-1}_{12}	L^{-1}_{21}	L^{-1}_{22}	L^{-1}_{33}
		$(10^{-12}g^{\frac{1}{2}})$			$(10^{-20}g^{\frac{1}{2}})$
OCS	4.931	3.332	3.008	-0.601	
SCSe	1.144	1.246	2.717	-1.419	
OCO				1.905	2.210
SCS				2.050	3.185
SeCSe				2.154	3.685

The bond-bending vibration represents a one-dimensional problem and hence

$$L^{-1}_{33} = (G_{33})^{-\frac{1}{2}}$$

where

$$G_{33} = \mu_1/r_1^2 + \mu_3/r_2^2 + \mu_2(1/r_1 + 1/r_2)^2$$

These relations lead to the values given in Table 6.

When the normalised L^{-1} coefficients have been obtained it is then possible to obtain the $\partial\mu/\partial S_j$ values from Equation (9). This relation shows that L^{-1} coefficients give the contribution of the ith $\partial\mu/\partial Q$ to the jth $\partial\mu/\partial S$ value.

There is some ambiguity in the determination of the $\partial\mu/\partial S_j$ however, since the sign of $\partial\mu/\partial Q_i$ is unknown and since only the relative signs of the L^{-1} coefficients for a particular vibration are known. A sign ambiguity in the L^{-1} coefficients arises because the sign of the normalisation constant K_k is undetermined but this does not contribute to a further sign ambiguity in $\partial\mu/\partial S_j$. The L^{-1} coefficients enable us to tell what motion of the atoms is involved in each normal coordinate and the sign of L^{-1} therefore indicates which way the atoms are moving with respect to a given defined positive direction.

The only assumption involved in obtaining values of $\partial\mu/\partial S_j$ from $\partial\mu/\partial Q_i$ values by Equation (9) is that the quantities $\partial\mu/\partial S_j$ are constant and additive for any vibration of a given symmetry class. The extent to which $\partial\mu/\partial S_j$ varies among different sets of values obtained depends to a considerable extent on the values of the L^{-1} coefficients. Clearly, however, the values of some of the $\partial\mu/\partial S_j$ parameters involve only small contributions from all but one of the vibrations (i.e. all the L^{-1} coefficients except one are small for a highly localised vibration). This is generally accepted to be true, for example, for an OH bond-stretching vibration ν(OH) with its particularly high frequency. However if the assumption mentioned above is invalid then the values of $\partial\mu/\partial S_j$ which are mainly derived from localised vibrations are expected to give more meaningful bond moment constants when the $\partial\mu/\partial S_j$ values are interpreted further by means of bond moment theory than those obtained by analysing the data for vibrations which are delocalised.

Zero-order bond moment theory

It has been pointed out by Hornig and McKean [14] that so far as bond polar properties are concerned, the maximum information which can be derived from absolute intensity measurements without any major additional assumptions is the change in each component of the molecular dipole moment with a change in the appropriate symmetry coordinate, i.e. $\partial\mu/\partial S_j$. They also emphasised that there are considerable difficulties involved in obtaining accurate values for these parameters. This is because there are inherent errors involved in the measured intensities and in the $(L^{-1})_{ij}$ coefficients calculated from the force field (or force constants) of the molecule. It is also found that, unless the molecule has considerable symmetry, several different alternative values of $\partial\mu/\partial S_j$ are obtained (owing to sign ambiguities). Furthermore, in attempting to reduce the $\partial\mu/\partial S_j$ values to properties characteristic of individual bonds it has been customary, using zero-order bond moment theory, to make three further basic assumptions. These are,

a) when a bond is stretched by an amount dr, a moment $(\partial\mu/\partial r)dr$ is produced in the direction of the bond,
b) when a bond is bent through an angle $d\alpha$, a moment $\mu d\alpha$, where μ is the "effective" bond moment, is produced in the plane of bending and perpendicular to the direction of the bond,
c) when any one bond is bent or stretched no moments are produced in the other bonds (zero-order approximation of bond moment theory).

The dipole moment μ(XY) (from bending vibrations) and $\partial\mu(XY)/\partial r(XY)$ (from stretching vibrations) can then be related to $\partial\mu/\partial S$ by considering the geometry of the molecule in a particular symmetry class and using the above assumptions.

From general considerations it can be argued that the values of the bond moment constants μ(XY) and $\partial\mu(XY)/\partial r(XY)$ obtained by applying a particular bond moment theory should be approximately the same for similar bonds in different molecules. In addition they should be the same when derived from vibrations in different symmetry classes in the same molecule. Furthermore, the effective bond moments measured should be near the values obtained from "static" dipole moment measurements; it follows from this that the vector sum of the bond moments measured should be equal to the "static" molecular dipole moment.

Calculation of the bond moment constants, $\partial\mu/\partial r$

The rule for the differentiation of a function of several variables can be used to obtain the bond moment constants, $\partial\mu/\partial r$, from the known quantities $\partial\mu/\partial S$.

For the bond-stretching vibrations of linear YXY molecules one finds that,

$$\partial\mu/\partial r = (\partial\mu/\partial Q_2)(\partial Q_2/\partial r) \tag{17}$$

where Q_2 is the normal coordinate defining the ν_a(YXY) vibration which is infrared-active. Since this vibration is in a symmetry class of its own,

$$\partial Q_2/\partial r = L_{22}^{-1} = (G_{22}^{-1})^{-\frac{1}{2}} \tag{18}$$

and therefore on substituting Equation (13) in (12) one obtains the desired relation,

$$\partial\mu/\partial r = L_{22}^{-1}(\partial\mu/\partial Q_2) = (2\mu_x + \mu_y)^{-\frac{1}{2}}(\partial\mu/\partial Q_2) \tag{19}$$

For linear YXZ molecules both bond-stretching vibrations belong to the same symmetry class and one has,

$$\partial\mu/\partial r_i = (\partial\mu/\partial Q_1)(\partial Q_1/\partial r_i) + (\partial\mu/\partial Q_2)(\partial Q_2/\partial r_i) \tag{20}$$

For this symmetry species of two vibrations the normal coordinates are given by Equation (11). Suitable symmetry coordinates are $S_1 = \Delta r_1$ and $S_2 = \Delta r_2$ which on substitution in Equations (11) lead to relations between the normal and internal coordinates of the form,

$$Q_i = L_{i1}^{-1}\Delta r_1 + L_{i2}^{-1}\Delta r_2 \qquad (i = 1,2) \tag{21}$$

Differentiation of Equation (21) with respect to r_1 and r_2 in turn leads to values for the quantities $\partial Q/\partial r$ which on substitution in relation (20) yields the desired relations

$$\partial\mu/\partial r_i = L_{1i}^{-1}(\partial\mu/\partial Q_1) + L_{2i}^{-1}(\partial\mu/\partial Q_2) \tag{22}$$

Using Equations (19) and (22) the values for $\partial\mu/\partial r$ given in Table 7 were calculated.

Calculation of μ_0 (bond) from the bond bending band intensity

Linear triatomic molecules have a single doubly-degenerate bending mode in the Π class, ν_3, whose intensity can be used together with the molecular dipole moment to determine the bond moments. The procedure is as follows. From Equation (11) one has for ν_3,

$$\partial\mu/\partial S_3 = L_{33}^{-1}\partial\mu/\partial Q_3$$

The symmetry coordinate is defined as $S_3 = r\Delta\theta$, where θ is the total angle of bend and $r = \sqrt{r_1 r_2}$. From these relations it follows that,

Table 7. Values of $\partial\mu/\partial r$ for linear triatomic molecules (D Å^{-1}).

Molecule	$\partial\mu_1/\partial r_1$ Vapour	Liquid	Solid	$\partial\mu_2/\partial r_2$ Vapour	Liquid	Solid
OCO	6.0		4.C			
SCS	5.8[a]	5.5[b]	5.7			
SeCSe	4.3	4.8[b]				
OCS	7.1	6.2[b]	7.1	4.1	2.7[b]	3.8
SCSe		6.3[c]			3.6[b]	
		6.7[d]			3.1[b]	

a Bond moment derivatives calculated from gas-phase intensities given in ref. 31.
b This work.
c This work. Different signs for $\partial\mu/\partial Q$.
d This work. Similar signs for $\partial\mu/\partial Q$.

$$\partial\mu/\partial S_3 = r\partial\mu/\partial\theta = L_{33}^{-1}\partial\mu/\partial Q_3 \quad (23)$$

and hence $\partial\mu/\partial\theta$ can be calculated from the other known quantities.

It is now necessary to obtain a relationship between $\partial\mu/\partial S_3$ and the bond moments. This can be done using the principle of conservation of momentum.

Let us consider the general linear triatomic molecule shown in Figure 2. If bond XY of moment $\boldsymbol{\mu}_1$ bends through an angle α from the equilibrium, a moment of $\boldsymbol{\mu}_1\boldsymbol{\alpha}$ is produced perpendicular to the bond as shown and similarly for bond YZ. The total moment produced is therefore

$$\Delta\mu = \boldsymbol{\mu}_1\boldsymbol{\alpha} + \boldsymbol{\mu}_2\boldsymbol{\beta} \quad (24)$$

where the total angle of bending is,

$$\theta = \alpha + \beta \quad (25)$$

It follows from Equations (23) and (24) that

$$\partial\mu/\partial\theta = (L_{33}^{-1}/r)(\partial\mu/\partial Q_3) = \mu_1(\partial\alpha/\partial\theta) + \mu_2(\partial\beta/\partial\theta) \quad (26)$$

This equation contains four unknowns on the right-hand side but it is also known of course, that

$$\boldsymbol{\mu}(\text{molecule}) = \boldsymbol{\mu}_1 + \boldsymbol{\mu}_2 \quad (27)$$

Hence in order to solve for the bond moments it is necessary to determine $\partial\alpha/\partial\theta$ and $\partial\beta/\partial\theta$. This is done as follows.

On differentiating Equation (25) with respect to θ one obtains,

$$\partial\alpha/\partial\theta + \partial\beta/\partial\theta = 1 \tag{28}$$

A second relation involving these quantities can be found by making use of the law of conservation of momentum.

As reference take the coordinate system of Figure 2c with the centre-of-mass as origin in the equilibrium state. If the atoms vibrate parallel to the y-axis during the bond-bending vibration it can be shown that

$$m_2\dot{y}_1 - m_1\dot{y}_2 - m_3\dot{y}_3 = 0 \quad (\dot{y} = \partial y/\partial t) \tag{29}$$

and

$$-m_3r_2\dot{y}_3 + m_1r_1\dot{y}_2 = 0 \tag{30}$$

Again if the amplitudes of vibration are small

$$\tan\alpha = (y_1 + y_2)/r_1 \sim \alpha \; ; \; \tan\beta = (y_1 + y_2)/r_2 \sim \beta$$

On differentiating these with respect to time and substituting for $\dot{y}_1$ from Equation (29) one obtains,

$$r_1\dot{\alpha} = (m_1\dot{y}_2 + m_3\dot{y}_3 + m_2\dot{y}_2)/m_2 \tag{31}$$

and

$$r_2\dot{\beta} = (m_1\dot{y}_2 + m_3\dot{y}_3 + m_2\dot{y}_3)/m_2 \tag{32}$$

but $\dot{y} = (\partial y/\partial\theta)(\partial\theta/\partial t)$ and hence, Equations (31) and (32) become,

$$m_2r_1\alpha' = m_1y_2' + m_3y_3' + m_2y_2' \quad (y' = \partial y/\partial\theta)$$

and

$$m_2r_2\beta' = m_1y_2' + m_3y_3' + m_2y_3' \quad (\alpha' = \partial\alpha/\partial\theta)$$

These can be solved simultaneously for y_3' and y_2'. Substitution for these quantities in Equation (30), allowing for the fact that $\dot{y} = (\partial y/\partial\theta)(\partial\theta/\partial t)$ leads to

$$-(\partial\beta/\partial\theta)m_2r_2[r_2(m_1 + m_2) + m_1r_1]$$
$$+ (\partial\alpha/\partial\theta)m_1r_1[r_1(m_2 + m_3) + m_3r_2] = 0 \tag{33}$$

Equations (28) and (33) can then be solved simultaneously to determine individual values for $\partial\alpha/\partial\theta$ and $\partial\beta/\partial\theta$ which can be substituted in Equation (26). This procedure leads to a relation between $\partial\mu/\partial\theta$, which is known, and $\boldsymbol{\mu}_1$ and $\boldsymbol{\mu}_2$, the bond moments which are unknown. In order to obtain individual values for the bond moments use is made of relation (27) between the bond moments and the known values for the molecular dipole moments. The bond moment values obtained are given in Table 8.

The calculations leading to the bond moment constants given in Tables 7 and 8 have been described fully since detailed accounts are very seldom given in the literature.

Table 8. Bond moments.

Molecule	Bond moment μ (D)					
	μ_1			μ_2		
	Vapour	Liquid	Solid	Vapour	Liquid	Solid
OCO	1.33[a]		1.38[a]			
SCS	0.6[b]	0.53	0.85[c]			
SeCSe		0.76				
OCS	0.64[d]	0.75	0.77[d]	-0.07[d]	-0.04	±0.05[d]
SCSe		±0.89			1.01 or -0.77	

a H. Yamada and W.B. Person, J. Chem. Phys., 41 (1964) 2478.
b Calculated from intensity data given in ref. 31.
c Calculated from intensity data given in ref. 32.
d H. Yamada and W.B. Person, J. Chem. Phys., 43 (1965) 2519.

It is now clear, however, that zero order bond moment theory is inadequate to the extent that in many cases it leads to very different values for bond moments when vibrations of different symmetry classes of a particular molecule are concerned. Some typical values for bond moment constants using zero-order theory are given in Table 9.

Table 9. Bond moment constants for CH bonds.

Molecule	Symmetry class	$\partial\mu/\partial r$ (D Å^{-1})	μ(CH) (D)	Ref.
CH_4	f_2	±0.83	∓0.37	33
CH_3D, CH_2D_2, CD_3H	a_1,e	-0.61	0.33	34
C_2H_4, $C_2H_2D_2$	b_{2u}	±0.26	∓0.42	35
C_2D_4	b_{3u}	0.23	∓0.60	
	b_{1u}	0.67		
C_2H_2	Σ_u^+	0.80		36
	π_u		1.05	
C_2D_2	Σ_u^+	0.78		37
	π_u		0.89	
C_2H_6	a_{2u}	±1.24	∓0.23	38
	e_{1u}	±0.75	∓0.26	
C_6H_6	e_{1u}	+0.45	-0.31	39
	a_{2u}		-0.61	
$CH_3C{\equiv}CCH_3$	a_2''	±1.2	±0.09	40
	e'	±0.4	0.4	

An examination of Table 9 shows that the values for bond moments and gradients in different molecules differ quite considerably. It is not surprising that the bond moment constants differ for CH bonds in which the carbon atoms are in different states of hybridisation since this factor has a considerable effect on the electronic structure of the bonds. It is unacceptable however when these constants differ markedly for bonds in which carbon is, more or less, in a fixed hybridisation state.

The use of zero-order bond moment theory has been severely criticised. In particular it is very disturbing to find that different values for a particular bond moment constant are often obtained from a consideration of different symmetry species of the same molecule.

For example μ(CH) in benzene has been calculated from in-plane (e_{1u}) and out-of-plane (a_{2u}) bending modes. The μ(CH) value obtained for the a_{2u} mode is 0.61 D which is twice that of 0.31 D obtained for the e_{1u} vibration.

It is therefore obvious that zero-order bond moment order has to be modified. The position can be improved by admitting that bond/bond interactions must occur during vibration and consequently it is advisable to include cross-terms in the bond moment function. The analysis then becomes more difficult since the number of unknown bond moment constants is increased. Except in favourable cases where molecules have a high degree of symmetry the introduction of cross-terms into the bond moment theory quickly leads to a situation where an explicit solution is not possible since the number of unknown bond moment constants exceeds the number of known factors consisting of the molecular dipole moment, μ_o, and infrared band intensities, A_i.

Modified bond moment theory

In zero-order bond moment theory it is assumed that the dipole moment of a given bond is not affected by changes in other bonds and bond angles caused by the molecular vibration. Hence at this level of approximation each valence bond is characterised by a bond dipole moment μ_n and the derivative of this bond moment with respect to bond length $\partial\mu_n/\partial r_n$. In first order bond moment theory it is admitted that interactions occur between the individual bonds in the molecule and that the derivatives characterising a particular bond moment as a result of changes in the lengths of other bonds and bond angles in the molecule, are not zero. That is, $\partial\mu_n/\partial\mu_m \neq 0$ and $\partial\mu_k/\partial\gamma_s \neq 0$. Hence in first order bond moment theory each valence bond is characterised by bond dipole moment μ_n and the following derivatives of the bond moment in respect of changes in bond length and bond angle, $\partial\mu_n/\partial r_m$ and $\partial\mu_k/\partial\gamma_s$.

The calculation of bond moment parameters from infrared band intensities is exactly parallel to the calculation of force constants from vibrational frequencies. In each case except in most favourable circumstances, the number of unknown parameters, be they bond moment constants or force constants, is in excess of the number of known factors, such as band intensities, or vibrational frequencies, Coriolis and nuclear quadrupole coupling constants and data from isotopically modified species.

The Russian school has done a great deal of work on the band intensity problem. The first contributions to electro-optical (bond moment) theory were made by Vol'kenshtein and El'yashevich [41-43] and a general scheme put forward by Sverdlov and his co-workers [44]. These approaches have been extended and formulated in a comprehensive fashion by Gribov [15].

This work has been paralleled by considerable efforts made by small groups working in the U.S.A. and the U.K. These contributions are well described in comprehensive reviews on vibrational band intensities by Steele [45,47] and by Person and Steele [46].

First-order bond moment theory

Using zero-order bond moment theory, the strict additivity of molecular properties is assumed which here involves the premise that an individual bond dipole moment depends on the stretching of that bond but not on the coordinates which involve movements of the other atoms in the molecule. In other words, derivatives of bond dipole moments with respect to coordinates other than the corresponding bond stretching ones are put equal to zero. Taking the linear YXZ molecule as an example, one then has

$$\boldsymbol{\mu} = \boldsymbol{\mu}_1 + \boldsymbol{\mu}_2 + \{\partial\boldsymbol{\mu}_1/\partial\mathbf{r}_1\}\Delta\mathbf{r}_1 + \{\partial\boldsymbol{\mu}_2/\partial\mathbf{r}_2\}\Delta\mathbf{r}_2 + \boldsymbol{\mu}_1\Delta\boldsymbol{\alpha} + \boldsymbol{\mu}_2\Delta\boldsymbol{\beta} \tag{34}$$

where $\Delta\mathbf{r}_1$, $\Delta\mathbf{r}_2$, $\Delta\boldsymbol{\alpha}$ and $\Delta\boldsymbol{\beta}$ represent increments in the bond lengths and bond angle as the molecule vibrates (Figure 2).

Intuitively, however, one realises that as one bond in a polyatomic molecule is stretched, the polar properties of the other bonds present must be altered. The effect of bond interaction can then be taken into account by introducing cross-terms such as $\partial\boldsymbol{\mu}_1/\partial r_2$ and $\partial\boldsymbol{\mu}_2/\partial r_1$ into the bond moment function (34). As a consequence, however, the analysis is more difficult since the number of unknown bond polar constants has increased. The increase in complexity of the problem is indicated in Table 10.

Table 10. Observed data and bond polar constants in linear YXZ molecules.

Observables	Bond polar parameters		
	Zero order theory	First order theory	
μ	μ_1	μ_1	$\partial\mu_1/\partial r_2$
A_1	μ_2	μ_2	$\partial\mu_2/\partial r_1$
A_2	$\partial\mu_1/\partial r_1$	$\partial\mu_1/\partial r_1$	$\partial\mu_1/\partial\theta$
A_3	$\partial\mu_2/\partial r_2$	$\partial\mu_2/\partial r_2$	$\partial\mu_2/\partial\theta$

Earlier it has been shown how the bond polar parameters obtained from zero-order theory could be calculated for linear triatomic molecules. The corresponding first-order treatment will now be developed using Gribov's approach [15]. The determination of $\partial\mu/\partial Q$ values and the normal coordinate calculation, both essential steps in this type of work, have been described above.

General relations

As a starting point it is assumed that the moleculer dipole moment can be written as a sum of bond moments, each directed along a chemical bond of the molecule

$$\boldsymbol{\mu} = \sum_k \mathbf{e}_k \mu_k \tag{35}$$

where the μ_k denote the absolute values of the bond moments and the $\mathbf{e}_k$ are unit vectors parallel to the bond. In the case of the linear YXZ molecule, $\mathbf{e}_1 . \mathbf{e}_2 = 1$ (Figure 2) since the unit vectors are parallel and chosen to have the same sense. It is important to note that the relative signs of the bond unit vectors may be chosen arbitrarily, but that once chosen they are crucial in determining the "sense" of the bond moments. This is so since the contribution of a particular bond moment to the total molecular dipole moment (Equation 35) is given by the product $\mathbf{e}\mu$. If the sign of one of the $\mathbf{e}_k$ vectors is altered then it is necessary to change the sign of the associated scalar μ_k so that the entire sum in Equation (35) remains invariant. In practice this means that care has to be taken when comparing bond polar parameters obtained for different molecules since the sign convention for the $\mathbf{e}$ vectors might not be consistent: this stricture also applies when comparing results obtained by zero-order and first-order theory for the same molecule.

From Equation (35) for the molecular dipole moment, the derivative with respect to the normal coordinate Q_i is then

$$\frac{\partial \boldsymbol{\mu}}{\partial Q_i} = \sum_j \frac{\partial \boldsymbol{\mu}}{\partial S_j} . \frac{\partial S_j}{\partial Q_i} = \sum_j \frac{\partial (\sum_k \mu_k \mathbf{e}_k)}{\partial S_j} L_{ji} \tag{36}$$

where S_j represent the jth symmetry coordinate, and $\partial S_j/\partial Q_i = L_{ji}$ because $S_j = \sum_i L_{ji} Q_i$.

The stretching and bending coordinates in Equation (36) can be separated giving

$$\frac{\partial \boldsymbol{\mu}}{\partial Q_i} = \sum_\ell \frac{\partial (\sum_k \mu_k \mathbf{e}_k)}{\partial S_\ell} L_{\ell i} + \sum_n \frac{\partial (\sum_k \mu_k \mathbf{e}_k)}{\partial \theta_n} L_{ni} \tag{37}$$

where S_ℓ represents a bond stretching and θ_n a bond bending symmetry coordinate.

Now, since both μ_k and $\mathbf{e}_k$ are functions of the symmetry coordinate, we have

$$\frac{\partial(\sum_k \mu_k \mathbf{e}_k)}{\partial S_\ell} = \sum_k \mathbf{e}_k \frac{\partial \mu_k}{\partial S_\ell} + \sum_k \mu_k \frac{\partial \mathbf{e}_k}{\partial S_\ell} \qquad (38)$$

A similar expression is obtained for the derivatives with respect to the bending coordinates.

Substituting Equation (38) into Equation (37) and formulating the set of equations into matrix notation one has

$$\mathbf{D} = \{\mathbf{e} \left| \frac{\partial \mu}{\partial S} \,\vdots\, \frac{\partial \mu}{\partial \theta} \right| + \mu \left| \frac{\partial \mathbf{e}}{\partial S} \,\vdots\, \frac{\partial \mathbf{e}}{\partial \theta} \right| \} \left| \frac{L_s}{L_\theta} \right| \qquad (39)$$

where $\mathbf{D}$ denotes the row vector consisting of the $\partial \boldsymbol{\mu}/\partial Q_i$ elements, $\mathbf{e}$ and μ denote row vectors constructed from the direction vectors and bond moments, respectively, while

$$\left| \frac{\partial \mu}{\partial S} \,\vdots\, \frac{\partial \mu}{\partial \theta} \right| \quad \text{and} \quad \left| \frac{\partial \mathbf{e}}{\partial S} \,\vdots\, \frac{\partial \mathbf{e}}{\partial \theta} \right|$$

stand for matrices containing the derivatives of bond moments and unit vectors with respect to the stretching and bending symmetry coordinates.

The elements of the latter matrix are determined by the molecular geometry and their evaluation is described in detail by Gribov [15]. In Equation (39) the elements of the vector $\mathbf{D}$ can be obtained from infrared intensity measurements; the matrix L can be determined by a normal coordinate analysis; the vector $\boldsymbol{\mu}$ and the matrix

$$\left| \frac{\partial \mu}{\partial S} \,\vdots\, \frac{\partial \mu}{\partial \theta} \right|$$

contain the unknown parameters to be determined. If there are sufficient experimental data, these parameters can be evaluated by solving Equation (39). In general, however, this is not possible since the number of unknown parameters usually exceeds that of the experimental data, and only certain linear combinations of the parameters can be evaluated. Further progress to yield unique values for the bond polar parameters can then only be achieved by intuition based on chemical considerations or by an independent quantum mechanical calculation of some of the individual bond moments and their derivatives (see section 5).

Application to linear triatomic molecules

These molecules belong to the point group $C_{\infty v}$ and the bond stretching vibrations are in a different symmetry class to the doubly-degenerate bending mode: therefore they can be treated separately.

As symmetry coordinates one can choose,

$S_1 = \Delta r_1$

$S_2 = \Delta r_2$ and

$S_3 = r\Delta\theta$

(a) Bond bending vibration

On omitting the terms that do not involve bond bending, Equation (39) becomes

$$\mathbf{D}_\theta = \{\mathbf{e} \left|\frac{\partial\mu}{\partial\theta}\right| + \mu \left|\frac{\partial\mathbf{e}}{\partial\theta}\right|\}L_\theta \tag{40}$$

In the zero-order treatment the first term was neglected, i.e. it was assumed that the absolute magnitudes of bond moments remained invariant during bond bending vibrations.

For a linear YXZ molecule Equation (40) becomes [16]

$$\partial\boldsymbol{\mu}/\partial Q_3 = \{\mathbf{e}_1(\partial\mu_1/\partial\theta) + \mathbf{e}_2(\partial\mu_2/\partial\theta) + \mu_1(\partial\mathbf{e}_1/\partial\theta) + \mu_2(\partial\mathbf{e}_2/\partial\theta)\}L_{33} \tag{41}$$

The terms $\partial\mathbf{e}_i/\partial\theta$, which depend only on the geometry of the molecule and the atomic masses, have been evaluated above and detailed examples are given in Gribov's book [15]. Equation (41) contains four unknown bond moment parameters whilst the experimentally determined quantities number two, viz. $\partial\boldsymbol{\mu}/\partial Q_3$ from the band intensity and $\boldsymbol{\mu}$, the total dipole moment.

One can reduce the number of unknowns to three by introducing the composite term

$$\Delta \equiv [\partial\mu_1/\partial\theta + \partial\mu_2/\partial\theta]$$

which in the case of YCN is equivalent to

$$\Delta \equiv [\partial\mu(CY)/\partial S_3 + \partial\mu(CN)/\partial S_3]$$

This composite term can be regarded as giving an estimate of the extent to which rehybridisation of the orbitals associated with the central carbon atom takes place during the bending vibration.

By varying μ(CN), as a parameter, around the value obtained from the zero order approximation the equations can be solved for μ(CY) and Δ. The ranges of values obtained are given in Table 11.

Quantum mechanical calculations (see section 5) which reproduce the molecular dipole moment fairly well indicate, however, that even this first order theory is unable to interpret the dipole moment changes during the bending vibration, since the CN bond moment contains a reasonable contribution from the nitrogen atomic dipole, which is not necessarily parallel to the CN bond in the distorted molecule. Hence the bond moments calculated by first order theory need further careful investigation.

One cannot solve the problem explicitly even by assuming that the three unknown parameters are the same in HCN and DCN because the intensities of the bending vibrations and the molecular dipole moment are related to each other through an isotopic rule.

Table 11. Possible sets of values for μ(CN), μ(CY) and $\pm[\partial\mu(CY)/\partial S_3 + \partial\mu(CN)/\partial S_3]$.

μ(CN)		HCN	DCN	ClCN	BrCN
1.35	μ(CY)	1.58	1.58	1.45	1.59
	Δ	-0.44	-0.40	-0.10	-0.01
1.45	μ(CY)	1.48	1.48	1.35	1.49
	Δ	-0.34	-0.30	0	0.09
1.55	μ(CY)	1.38	1.38	1.25	1.39
	Δ	-0.24	-0.20	0.10	0.19
1.65	μ(CY)	1.28	1.28	1.15	1.29
	Δ	-0.14	-0.10	0.20	0.29
1.75	μ(CY)	1.18	1.18	1.05	1.19
	Δ	-0.04	0	0.30	0.39
1.85	μ(CY)	1.08	1.08	0.95	1.09
	Δ	0.06	0.10	0.40	0.49
μ(YCN)		2.93	2.93	2.80	2.94

(b) Stretching vibrations

Since pure stretching vibrations do not alter the directions of bonds, the second term in Equation (39) can be omitted, as well as the bending part from the first term. Equation (39) then reduces to,

$$\mathbf{D}_s = \mathbf{e}\left|\frac{\partial\mu}{\partial S}\right| L_s \qquad (42)$$

For the special case of a linear YXZ molecule Equation (42) becomes

$$|\partial\boldsymbol{\mu}/\partial Q_1, \partial\boldsymbol{\mu}/\partial Q_2| = |1,1| \begin{vmatrix} \partial\mu_1/\partial r_1 & \partial\mu_1/\partial r_2 \\ \partial\mu_2/\partial r_1 & \partial\mu_2/\partial r_2 \end{vmatrix} \begin{vmatrix} L_{11} & L_{12} \\ L_{21} & L_{22} \end{vmatrix} \qquad (43)$$

leading to the equations,

$$\partial\boldsymbol{\mu}/\partial Q_i = (\partial\mu_1/\partial r_1 + \partial\mu_2/\partial r_1)L_{1i} + (\partial\mu_1/\partial r_2 + \partial\mu_2/\partial r_2)L_{2i} \qquad (i = 1,2) \qquad (44)$$

From the known values of the L matrix elements and the $\partial\boldsymbol{\mu}/\partial Q$'s calculated from band intensities [50] the composite parameters in brackets in Equation (44) can be evaluated: these are listed in Table 12. Since the signs of the $\partial\boldsymbol{\mu}/\partial Q$ derivatives are unknown, two sets of values are obtained for the alternative possibilities when the $\partial\boldsymbol{\mu}/\partial Q_i$ quantities have the same sign and the opposite sign, respectively.

A comparison between HCN and acetylene is valuable since it enables an estimate of the individual derivatives in the composite terms listed in Table 12 to be made for HCN. The data listed in Table 13 emphasise the close similarity between the CH bonds in these two molecules. From this comparison it can be argued that the derivative $\partial\mu(CH)/\partial r(CH)$ for the two molecules must have almost equal values. During the CH infrared active vibration in acetylene, as one CH bond lengthens the

Table 12. Bond dipole moment derivatives calculated from infrared intensities by first order bond moment theory.

Molecule	Ratio positive		Ratio negative	
	$\frac{\partial\mu_{CY}}{\partial r_{CY}} + \frac{\partial\mu_{CN}}{\partial r_{CY}}$	$\frac{\partial\mu_{CY}}{\partial r_{CN}} + \frac{\partial\mu_{CN}}{\partial r_{CN}}$	$\frac{\partial\mu_{CY}}{\partial r_{CY}} + \frac{\partial\mu_{CN}}{\partial r_{CY}}$	$\frac{\partial\mu_{CY}}{\partial r_{CN}} + \frac{\partial\mu_{CN}}{\partial r_{CN}}$
HCN	1.088	-0.479	1.041	-0.784
DCN	1.075	-0.637	0.524	-1.846
ClCN	1.012	-1.085	-0.605	-2.041
(0% anh.)	1.820	-0.608	-1.414	-2.519
(1% anh.)	1.936	-0.539	-1.529	-2.587
BrCN	0.417	-0.861	-0.290	-1.255
(0% anh.)	0.594	-0.763	-0.466	-1.354
(1% anh.)	0.727	-0.689	-0.599	-1.428

Table 13. Comparison of CH bond parameters in HCN and H_2C_2

	r(CH) (Å)	μ(CH) (D)	ν(CH) (cm^{-1})
HCN	1.06	1.15	3312
HCCH	1.10	1.10	3287(C_2H_2) 3335(C_2HD)

other contracts. Effectively this then means that $\partial\mu(CC)/\partial r(CH) = 0$. A first-order bond moment analysis [15] of the available data therefore leads to a value of 0.90 D $Å^{-1}$ for $\partial\mu(CH)/\partial r(CH)$, in acetylene. Substitution of this value into the composite term listed in Table 12 for HCN leads to a value of 0.188 D $Å^{-1}$ for $\partial\mu(CN)/\partial r(CH)$.

5. QUANTUM MECHANICAL CALCULATION OF POLAR PARAMETERS

So far this account has dealt with the efforts made to calculate bond moment functions from experimental data using theory based on the concepts of bond moments and the derivatives of bond moments with respect to vibrational coordinates. In this approach it is unnecessary to have any explanation in structural terms of the various cross-terms such as $\partial\mu_n/\partial r_m$ that occur in the mathematical formulation. This represents a physicist's approach to the problem but a chemist needs more. He needs a model, preferably simple, which can be easily visualised and which accounts for changes in the electronic structure of chemical bonds as the molecule vibrates.

The analysis of infrared intensities obtained experimentally is not the only method for calculating bond moment constants. Since the absorption intensity due to a transition between two states is

$$\Gamma = \frac{8\pi^3 Ng}{3hc} |\langle\psi'|p|\psi''\rangle|^2 \tag{45}$$

where ψ' and ψ'' are the wavefunctions of the states. Using the Born-Oppenheimer approximation and separating the vibrational and rotational wavefunctions expression (45) leads to

$$\Gamma_i = \frac{\pi Ng}{3c\nu_i} \left(\frac{\partial\mu}{\partial Q_i}\right)^2 \tag{46}$$

provided that the double harmonic approximation (mechanical and electrical) is used. A comparison of Equations (45) and (46) shows how quantum mechanical calculations give information on the molecular charge distribution. Since the charge distribution is directly related to the polar properties of molecules information can be obtained about molecular dipole moments and also the bond moment constants occurring in the dipole moment function for a vibrating molecule.

In practice the calculation of the vibrational transition moment then leads to theoretical estimates of the polar parameters $\partial\mu/\partial Q_i$ which can be transformed to values for $\partial\mu/\partial S_j$.

Again in principle it is possible to calculate the dipole moment for different values of the spatial coordinates of the atoms in a molecule. If small displacements of the atoms are chosen to conform with an internal or symmetry coordinate then an approximate value for $\partial\mu/\partial r_k$ can be obtained as follows,

$$\frac{\partial\mu}{\partial r_k} \simeq \frac{\mu(r) - \mu(r_e)}{r - r_e} = \left[\frac{\langle 0|\mu|0\rangle_r - \langle 0|\mu|0\rangle_{r_e}}{r - r_e}\right] \tag{47}$$

where r is the value of r_k after a small displacement from its equilibrium value r_e along the appropriate coordinate.

The first attempt to calculate theoretical estimates of infrared intensities was made by Coulson and his co-workers [51]. The results were not particularly encouraging since the calculations were made using rather inaccurate wavefunctions. This early work is very valuable, however, since it provided some guide-lines as to the importance of changes in hybridisation and of lone-pair contributions to infrared intensities.

Considerable progress in the use of theoretical methods has been made in the past ten years and the value of semi-empirical methods has been adequately demonstrated.

Bond moment constants from molecular orbital theory

With a ready supply of programmes* and the increased

---

* Appropriate computer programmes are available from the Quantum Chemistry Programme Exchange, University of Indiana, Bloomington, Indiana, U.S.A.

availability of computers spectacular advances have been made in the 'ab initio' calculation of dipole moment functions. Such calculations are expensive, however, and most workers use the Complete Neglect of Differential Overlap (CNDO) method as developed by Pople and his school [52]. These CNDO methods have been shown to give reliable values for dipole moments for molecules in equilibrium configurations. These calculations have been extended to molecules in distorted configurations [53] and hence it is possible to calculate theoretical values for the polar parameters $\partial\mu/\partial S$.

Bond moment parameters in molecular orbital terms

If the most general electronic wavefunction of the molecule is ψ, then the Cartesian component of the dipole moment is given by

$$\mu_x = -e\int \psi^* (\sum_i x_i) \psi \, d\nu + e\sum_A Z_A X_A \qquad (48)$$

Where x_i is the time-averaged x-component of the radius vector of the ith electron, Z_A is the charge of nucleus A, and X_A is the nuclear coordinate; similar expressions hold for μ_y and μ_z.

Using the LCAO MO approach it has been shown [17] that for closed shell molecules, described by a single determinantal wavefunction, relation (48) becomes

$$\mu_x = -e\sum (\sum_\nu^A P_{\nu\nu} X_{\nu\nu} - X_A Z_A) - e\sum_{A\neq B} \sum_\mu^A \sum_\nu^B P_{\mu\nu} X_{\mu\nu} - e\sum \sum_{\mu\neq\nu}^A P_{\mu\nu} X_{\mu\nu} \qquad (49)$$

where $P_{\mu\nu}$ are the usual bond orders

$$P_{\mu\nu} = 2\sum_i c_{i\mu} c_{i\nu}$$

and

$$X_{\mu\nu} = \int \phi_\mu x \phi_\nu d\nu$$

ϕ_μ being an atomic orbital.

The first term in Equation (49) corresponds to the approximation in which a molecule is regarded as a system of point charges at nuclei provided a basis of pure s, p and d orbitals is used [17]. The second term in Equation (49) arises from the overlap density of orbitals on different atoms. This contribution represents the homopolar moments. The two-centre dipole integrals appearing here are given by Coulson and Rogers [54] for Slater type orbitals. The contribution to the total molecular dipole moment due to the homopolar moments is considerable, as shown by Giessner-Prettre and Pullman [55]. Therefore, a term to include this must be introduced into bond moment theory. It is important to point out that the value of the homopolar moment is not very sensitive to the accurate value of the hybridisation parameter for the orbitals belonging to a given atom [54].

The static bond moment

A number of factors contribute to determine the values of dipole moments. Four such basic factors have been identified. They are, (a) charge asymmetry arising from the unequal sharing of bonding electrons due to a difference in electronegativity between the bound atoms, (b) the homopolar dipole (or overlap moment) due to unequal sizes of the bonding atomic orbitals used by the two bound atoms, (c) charge asymmetry introduced by the use of hybrid character of atomic orbitals and (d) the presence of lone-pair atomic dipoles. Each of these factors will be affected by the atomic displacements occurring during molecular vibrations and hence contribute to the vibrational band intensity. It is possible to identify these factors with terms occurring in Equation (49).

The dipole moment contributions from the first two terms of relation (49) are bond-directed vectors. The homopolar moment, as was mentioned above, is not sensitive to the accurate value of the hybridisation parameter for the orbitals centred on a given atom. Therefore, the rehybridisation phenomena taking place during the small vibrational displacements are not likely to affect its magnitude, especially for bending vibrations. As far as the dipole moment contributions arising from the net atomic charges are concerned, the situation is different. Bruns and Person [56] showed that in certain cases during the vibrational motion a substantial flow of electronic charge from one atom to others takes place, which obviously is reflected in the band intensities. These authors divide the contribution to the infrared intensities due to the movement of electronic charges (first term in Equation (49)) into two parts, first, changes in μ during vibration due to the motion of equilibrium charges ($\mu_{q_{eq.}}$) and secondly changes of μ due to the flow of electronic charge ($\mu_{\Delta q}$) induced by the vibrational motion.

On this basis the static bond moment can be defined as composed of two contributions, the dipole moment due to the equilibrium atomic net charges and the homopolar moment of the bond.

Hybridisation and rehybridisation moments

The contribution arising from the third term in Equation (49) is the hybridisation moment, μ_h. In general it is not a bond directed vector. Neither can one predict without calculation the change in its direction during a vibrational motion and hence its contribution to the intensity. It is not related to the bond moments since it represents a sum of atomic moments [17,18]. As the molecule vibrates the state of hybridisation of the various atoms will change as the molecule is distorted from its equilibrium configuration. This means that the hybridisation moment itself changes during vibration, i.e. $\partial\mu_h/\partial R \neq 0$, where R represents an internal coordinate. This quantity can be calculated using molecular orbital theory for different molecular distortions and a 'rehybridisation' moment defined as

$$(\partial\mu_h/\partial R)_{R\to 0} = (\partial\mu_h/\partial R)_o \qquad (50)$$

can then be obtained by extrapolation.

In general the rehybridisation moment for a given molecular vibration is defined [18] as the derivative of the molecular hybrid moment with respect to the appropriate symmetry coordinate $(\partial\mu_h/\partial S_i)_0$. A detailed description of its role in determining infrared band intensities is available [17,18,53].

Modification of zero-order bond moment theory

Basically when zero-order theory is used the analysis of infrared band intensities leads to relations of the form

$$\partial\mu/\partial S_j = f[\Sigma\mu_n, \Sigma(\partial\mu_n/\partial r_n)] \quad (51)$$

where f stands for 'a function of'. Since this theory often leads to different values for a particular bond moment constant when vibrations from different symmetry classes are considered it is plainly inadequate.

On the other hand, the use of first-order bond moment theory which leads to expressions of the form [15],

$$\partial\mu/\partial S_j = f[\Sigma\mu_n, \Sigma(\partial\mu_n/\partial r_m), \Sigma(\partial\mu_n/\partial\theta_\ell)] \quad (52)$$

often leads to situations where an explicit solution is not possible since the number of unknown bond moment constants exceeds the number of known experimental polar factors such as molecular dipole moment and infrared band intensities.

A comparison of Equations (51) and (52) shows that the modification of zero-order to give first-order theory results is achieved by introducing cross-terms as correction factors to the right-hand side (RHS) of the relations. An alternative way of improving the situation is to leave the RHS of the relations unchanged and to use molecular orbital theory to introduce correction terms to the LHS of the equations so that one obtains consistent values for bond moment constants. Using this approach it is hoped to eliminate some of the discrepancies of zero-order theory. This approach amounts to an effort to redefine the term 'bond moment' so that it becomes a true transferable bond property.

The first attempt [17] along these lines introduced a correction term arising from the rehybridisation phenomenon into the basic relation (51). Subsequently a second correction term taking into account the dipole moment component due to changes of the effective electronic charges associated with the atoms which are induced by the vibrational motion was introduced [21, 56]. These terms modified the basic relation from

$$\partial\mu/\partial S_j = \sum_k(\partial\mu/\partial r_k)(\partial r_k/\partial S_j) \quad (53)$$

to

$$\frac{\partial\mu}{\partial S_j} - \frac{\partial\mu_h}{\partial S_j} - \frac{\partial\mu_{\Delta q}}{\partial S_j} = \sum_k(\partial\mu'/\partial r_k)(\partial r_k/\partial S_j) \quad (54)$$

The derivatives $(\partial\mu'/\partial r_k)$ according to the bond moment hypothesis are the bond moments, μ_b, and their derivatives with respect to the corresponding bond lengths, $(\partial\mu_b/\partial R)$. These quantities in Equation (54) differ from the corresponding $(\partial\mu/\partial r_k)$ derivatives in Equation (53) since as has been shown [17,18,56] the latter are not directly related to the polarity of the chemical bonds because of unknown contributions from the $(\partial\mu_h/\partial S_j)$ and $(\partial\mu_{\Delta q}/\partial S_j)$.

With the addition of the two correction terms the LHS of Equation (54) is in accord with the basic assumptions of bond moment theory. The first term in Equation (54) is obtained from experimental intensities, the second and third terms can be calculated theoretically by the molecular orbital method. It should be pointed out that although the accuracy of the theoretical terms may be low compared with the experimental quantities, even a correction of $(\partial\mu/\partial S_j)$ in the right direction (+ or -) will substantially improve the consistency of the results when the bond moment scheme is applied. In addition the method provides a possibility for the multiple checking of values obtained for bond polar parameters by carrying out calculations on different vibrations in a molecule, and on different molecules, containing similar types of chemical bonds as well.

In the following sections the application of the procedure outlined above is applied to the bending vibrations of ethylene.

Determination of the CH bond moment in ethylene

Ethylene possesses three infrared active bending vibrations. Suitable symmetry coordinates to describe these vibrations are

B_{1u} (out-of-plane) $\quad S_7 = \frac{1}{\sqrt{2}} r_0 \cos(\alpha/2)(\Delta\gamma_1 + \Delta\gamma_2)$

B_{2u} (in-plane) $\quad S_{10} = \frac{1}{\sqrt{2}} r_0 (\Delta\beta_1 + \Delta\beta_2)$

B_{3u} (in-plane) $\quad S_{12} = \frac{1}{\sqrt{2}} r_0 (\Delta\alpha_1 - \Delta\alpha_2)$

where α_i = HCH angle; β_i = angle between C-C bond and the bisector of C_iH_2 angle; γ_i = angle between C-C and the C_iH_2 plane, and r_0 is the equilibrium C-H bond distance. The geometric parameters taken [57] are r_{CH} = 1.085 Å, r_{CC} = 1.339 Å and $\angle$ HCH = 117°50'.

The experimental intensity data used in this study are those determined by Golike, Mills, Person and Crawford [58]. In calculating the $(\partial\mu/\partial Q_i)$ derivatives from the observed band areas [58] anharmonicity corrections for all vibrations were applied using the anharmonicity data given in ref. 59. The values obtained are shown in Table 14. Using the force field determined by Duncan, McKean and Mallinson [59] the sets of $\partial\mu/\partial S_j$ derivatives given in Table 15 are calculated. It can be seen that the sign combinations (+-) or (-+) for both B_{2u} and B_{3u} classes give best agreement between the isotopes.

In this study the different contributions to the total dipole moment were calculated using the CNDO/2 method [60,61]. The calculations were carried out for different nuclear configurations in accord with the symmetry coordinates S_7, S_{10}

Table 14. Values of $(\partial\mu/\partial Q_i)$/e.s.u. $cm^{-1}g^{-1}$.

Vibration		C_2H_4	C_2D_4
B_1	7	±107.7	±77.7
B_2	9	±60.8	±42.4
	10	±8.8	±2.5
B_3	11	±44.8	±33.4
	12	±37.7	±27.4

Table 15. Values of $(\partial\mu/\partial S_j)$/D $Å^{-1}$* for the infrared active vibrations of C_2H_4 and C_2D_4.

		C_2H_4		C_2D_4	
		$(+)^a$		(+)	
B_1	$(\partial\mu/\partial S_7)$	1.82		1.74	
		(+ +)	(+ −)	(+ +)	(+ −)
B_2	$(\partial\mu/\partial S_9)$	0.751	0.729	0.698	0.684
	$(\partial\mu/\partial S_{10})$	0.119	-0.197	0.023	-0.102
		(+ +)	(+ −)	(+ +)	(+ −)
B_3	$(\partial\mu/\partial S_{11})$	0.593	0.542	0.648	0.523
	$(\partial\mu/\partial S_{12})$	0.332	-0.320	0.320	-0.321

a Sign choice for the $(\partial\mu/\partial Q_i)$ derivatives. By multiplying each row by (-1) the remaining values of $(\partial\mu/\partial S_j)$ corresponding to the alternative sign choice for $(\partial\mu/\partial Q_i)$ can be obtained.
* 1 D ≡ 3.3 10^{-30} C m, 1 Å ≡ 10^{-10} m.

and S_{12}. In determining the $(\partial\mu_h/\partial S_j)$ and $(\partial\mu_{\Delta q}/\partial S_j)$ derivatives the method followed was that suggested by Segal and Klein [53] who calculated theoretical values for the total dipole moment derivatives $(\partial\mu/\partial S_j)$ for several molecules including ethylene. The large values obtained for the derivative $(\partial\mu_h/\partial S_j)$ and $(\partial\mu_{\Delta q}/\partial S_j)$ (see Table 16) show clearly the importance of these phenomena in determining infrared intensities.

Zero-order bond moment theory leads to the following relations:

$$\partial\mu/\partial S_7 = 2\sqrt{2}\mu'(CH)/r(CH)$$

$$\partial\mu/\partial S_{10} = -2\sqrt{2}\cos(\alpha/2)\mu'(CH)/r(CH) \qquad (55)$$

$$\partial\mu/\partial S_{12} = \sqrt{2}\sin(\alpha/2)\mu'(CH)/r(CH)$$

Substitution of the various possible choices for $(\partial\mu/\partial S_j)$ in Equation (55) leads to the set of values for μ(CH) given in the penultimate column of Table 16. Examination shows that they are inconsistent to the extent that no single value emerges

Table 16. Experimental and theoretical bond moment parameters for ethylene.

Vibration	(a)	$(\partial\mu/\partial S_j)$ /D Å^{-1} (b)	$(\partial\mu_h/\partial S_j)$ /D Å^{-1} (c)	$(\partial\mu_{\Delta q}/\partial S_j)$ /D Å^{-1} (d)	μ(CH)/D (e)	μ(CH)/D (f)
ν_7	+	1.82	-1.13	0	0.69	-1.12
	-	-1.82			-0.69	0.26
ν_{10}	+ +	0.116	-0.59	0.292	-0.09	-0.31
	- -	-0.116			0.09	-0.14
	+ -	-0.195			0.15	-0.08
	- +	0.195			-0.15	-0.37
ν_{12}	+ +	0.3288	0.613	-0.628	0.29	0.31
	+ -	0.3167			0.28	0.29
	- -	-0.3288			-0.29	-0.28
	- +	-0.3167			-0.28	-0.27

(a) Sign choice for the $(\partial\mu/\partial Q_i)$ derivatives; (b) calculated from the experimental intensity data given in ref. 58; (c), (d) theoretical values using CNDO/2 theory; (e) using Equation (55); (f) using Equation (56).

from the various possibilities for all symmetry classes. When the dipole moment contribution due to the rehybridisation and the change of electronic charges are taken into account, as indicated in Equation (54), Equations (55) have to be modified to expressions of the form

$$\partial\mu/\partial S_7 - \partial\mu_h/\partial S_7 - \partial\mu_{\Delta q}/\partial S_7 = 2\sqrt{2}\mu_b(\mathrm{CH})/r(\mathrm{CH}) \qquad (56)$$

Numerical values for the left-hand-side of Equation (56) can then be obtained by subtracting from the $(\partial\mu/\partial S_j)$ parameters, given in Table 15, the values for $(\partial\mu_h/\partial S_j)$ and $(\partial\mu_{\Delta q}/\partial S_j)$ theoretically determined in this work. Substitution of these values in Equation (56) leads to the set of μ_b(CH) values given in the last column of Table 16. This time a common value of (0.7±0.4) D with sense (C^+H^-) is found. This is the averaged value for μ_b(C-H) from the following sign choices: B_{1u}(+), B_{2u}(-+) and B_{3u}(-+). Taking into account the errors involved in both experimental and theoretical values used in the calculations one must not expect perfect agreement between the bond moments obtained from different symmetry classes. It can be said that the consistency of the values for the bond moments obtained using the procedure described here provides a supplementary criterion for the correct choice of the relative and absolute signs of the $(\partial\mu/\partial Q_i)$ derivatives for the ν_7 and ν_{10} vibrations. We have preferred sign combinations (-+)B_{2u} and (-+)B_{3u} instead of (++)B_{2u} and (--)B_{3u} which also give satisfactory agreement for μ_b(C-H) because this is indicated by the results presented in Table 15 and discussed above. The direction of the C-H bond moment (C^+H^-) is in accord with recent theoretical calculations [62]. The magnitude is much lower compared with the purely theoretical value and this fact can be attributed to a great extent to the difference in the definitions of the bond moments.

It should be pointed out that this approach can also be used to modify first-order bond moment theory.

6. POSTSCRIPT

Many factors contribute to the values of the derivatives $\partial\mu/\partial S_j$ obtained from experimentally determined infrared band intensities. Some of these factors lead to increments in the effective 'static' bond moments, others do not. Clearly, then, if one hopes to obtain consistent values for bond moments which remain constant as one goes from a study of normal vibrations in one symmetry class to another then it is imperative to remove from the experimentally determined $\partial\mu/\partial S_j$ value those increments which do not have any bearing on the value of the static bond moment μ_b which will then be a true bond constant capable of being transferred from one molecule to others with closely similar bonds.

That is one has to modify the basic zero-order bond moment theory relations which are of the form

$$\partial\mu/\partial S_j \quad = \quad f(\mu_n,\ \partial\mu_n/\partial r_n) \qquad (n = 1,2...)$$

to corrected relations of the form

$$\partial\mu/\partial S_j \ - \ \sum_i \partial\mu_i/\partial S_j \quad = \quad f(\mu_b, \partial\mu_b/\partial r_n)$$

Some of these correction terms have already been identified. They arise from the rehybridisation phenomenon [17,49] and from changes in the effective atomic charges which occur during vibrations [56].

One factor which has not been considered hitherto is the increment due to the homopolar (overlap) moment of the bond - this arises because the atoms use atomic orbitals of different size for bonding purposes. As vibration proceeds the overlap integral changes in value and this will affect the value of μ_s (s for size), the homopolar moment. Hence this effect will provide a contribution of $(\partial\mu_s/\partial S_j)_{S_j=0}$ to the total value of $\partial\mu/\partial S_j$.

This effect has been investigated for the first time by Riley and Orville-Thomas [63] in the case of the ν_{10} rocking mode of ethylene.

The contributions to $\partial\mu/\partial S_{10}$ were calculated, within the Hartree-Fock approximation, using optimised double zeta basis sets [64]. Each of these Slater functions was expanded as a linear combination of four Gaussian functions with coefficients and exponents as given as Stewart [65]. The equilibrium geometry of the ethylene molecule was determined from an electron diffraction study [66]. The results obtained are given in Table 17.

Table 17. Experimental and theoretical polar parameters (in D $Å^{-1}$) and the μ(CH) bond moment (in D) for ethylene.

Vibration	(a)	$\partial\mu/\partial S_{10}$ (b)	$\partial\mu_h/\partial S_{10}$ (c)	$\partial\mu_{\Delta q}/\partial S_{10}$ (c)	$\partial\mu_s/\partial S_{10}$ (c)	μ(CH) (d)	μ(CH) (e)
	+ +	0.119				-0.35	-0.31
ν_{10}	- -	-0.119	-2.87	1.99	0.60	-0.16	-0.13
	+ -	-0.196				-0.11	-0.08
	- +	0.196				-0.41	-0.37

(a) Sign choice for the $\partial\mu/\partial Q_i$ gradients; (b) experimental values; (c) theoretical values; (d) Hartree-Fock values; (e) CNDO/2 values from ref. 21.

The CH bond moment was calculated from Equation (55). The possible values are given in Table 17. Irrespective of the absolute value the sense of the bond moment must be $\mu(C \leftrightarrow H)$. It is interesting to note that as previously assumed the effect due to the homopolar moment is small. Studies proceeding on the other deformation modes of ethylene should enable the absolute value for μ(CH) to be determined.

7. REFERENCES

1. W.J. Lehmann, J. Mol. Spectr., 7 (1961) 261.
2. N. Müller and D.E. Prichard, J. Chem. Phys., 31 (1959) 768, 1471.
3. W.J. Orville-Thomas, J. Chem. Phys., 43 (1965) 5244.
4. B.P. Dailey, J. Phys. Chem., 57 (1953) 490.
5. C.H. Townes and B.P. Dailey, J. Chem. Phys., 17 (1949) 782.
6. J.K. Tyler and J. Sheridan, Trans. Faraday Soc., 59 (1963) 2661.
7. C.M. Huggins and G.C. Pimentel, J. Chem. Phys., 23 (1955) 896.
8. H. Ratajczak and W.J. Orville-Thomas, Trans. Faraday Soc., 61 (1965) 2603.
9. H. Ratajczak and W.J. Orville-Thomas, J. Mol. Struct., 14 (1972) 155; H. Ratajczak, W.J. Orville-Thomas and C.N.R. Rao, Chem. Phys., 17 (1976) 197.
10. I.M. Mills, Ann. Rep. Progr. Chem., Chem. Soc., London, LV (1958) 56.
11. W.B. Person and D. Steele, Molecular Spectroscopy, Vol. 2, Chem. Soc. Specialist Periodical Report, (1974) 357.
12. E.B. Wilson, J.G. Decius and P.C. Cross, Molecular Vibrations, McGraw-Hill, New York, 1955.
13. P.R. Davies and W.J. Orville-Thomas, J. Mol. Struct., 4 (1969) 163.
14. D.F. Hornig and D.C. McKean, J. Phys. Chem., 59 (1955) 1133.
15. L.A. Gribov, Intensity theory for infrared spectra of polyatomic molecules, Consultants Bureau, New York, 1964.
16. Gloria A. Thomas, G. Jalsovszky, J.A. Ladd and W.J. Orville-Thomas, J. Mol. Struct., 8 (1971) 1.
17. G. Jalsovszky and W.J. Orville-Thomas, Trans. Faraday Soc., 67 (1971) 1894.

18. B. Galabov and W.J. Orville-Thomas, J. Chem. Soc. Faraday II,68 (1972) 1778.
19. B. Galabov and W.J. Orville-Thomas, J. Mol. Struct., 18 (1973) 169.
20. Helen Stoeckli-Evans, A.J. Barnes and W.J. Orville-Thomas, J. Mol. Struct., 24 (1975) 73.
21. B. Galabov, S. Suzuki and W.J. Orville-Thomas, J. Chem. Soc. Faraday II, 71 (1975) 162.
22. A.N. Thorndike, A.J. Wells and E. Bright Wilson, J. Chem. Phys., 15 (1947) 157.
23. G.M. Barrow, J. Phys. Chem., 59 (1955) 1129.
24. J. Overend, in M.M. Davies (Editor), Infrared Spectroscopy and Molecular Structure, Elsevier, Amsterdam, Chapter 10, 1963.
25. H. Ratajczak and W.J. Orville-Thomas, Trans. Faraday Soc., 61 (1965) 2603.
26. T. Wentink, J. Chem. Phys., 30 (1959) 105.
27. C.A. Coulson, in E. Thornton and H.W. Thompson (Editors), Molecular Spectroscopy, Pergamon Press, London, 1959, p. 183.
28. E.B. Wilson, J.G. Decius and P.C. Cross, Molecular Vibrations, McGraw-Hill, New York, 1955.
29. S. Califano, Vibrational States, J. Wiley and Sons, London, 1976.
30. S.J. Cyvin, Molecular Vibrations and Mean Square Amplitudes, Elsevier, Amsterdam, 1968.
31. D.Z. Robinson, J. Chem. Phys., 19 (1951) 881.
32. H. Yamada and W.B. Person, J. Chem. Phys., 40 (1964) 309.
33. I.M. Mills, Mol. Phys., 1 (1958) 107.
34. R.E. Hiller and J.W. Straley, J. Mol. Spectr., 5 (1960) 24.
35. R.C. Golike, I.M. Mills, W.B. Person and B.L. Crawford, J. Chem. Phys., 25 (1956) 1266.
36. R.C. Kelly, R. Rollefson and B.S. Schurin, J. Chem. Phys., 19 (1951) 1595.
37. D.F. Eggers, I.C. Hisatsune and I. Van Alten, J. Phys. Chem., 59 (1955) 1124.
38. I.M. Nyquist, I.M. Mills, W.B. Person and B.L. Crawford, J. Chem. Phys., 26 (1957) 552.
39. H. Spedding and D.H. Whiffen, Proc. Roy. Soc. London, A, 238 (1956) 245.
40. I.M. Mills and H.W. Thompson, Proc. Roy. Soc. London, A, 228 (1955) 287.
41. M.V. Vol'kenshtein, Doklady Akad. Nauk. SSSR, 32 (1941) 185.
42. M.V. Vol'kenshtein, ZhETF, 11 (1941) 642.
43. M.V. Vol'kenshtein and M.A. El'yashevich, Doklady Akad. Nauk. SSSR, 41 (1943) 380; 43 (1944) 56; ZhETF, 15 (1945) 124.
44. L.M. Sverdlov, M.A. Kovner and E.P. Krainov, Vibrational Spectra of Polyatomic Molecules, J. Wiley and Sons, London, 1974.
45. D. Steele, Q. Rev. Chem. Soc., 18 (1964) 21.
46. W.B. Person and D. Steele, Molecular Spectroscopy, Vol. 2, Specialist Periodical Reports, Chem. Soc., London, 1974.
47. D. Steele, in R.J.H. Clark and R.E. Hester (Editors), Adv. Infrared and Raman Spectroscopy, Heyden and Son Ltd., London, 1975.
48. L. Burnelle and C.A. Coulson, Trans. Faraday Soc., 53 (1957) 403.

49. D. Steele and W. Wheatley, J. Mol. Spectr., 32 (1969) 265.
50. G.A. Thomas, J.A. Ladd and W.J. Orville-Thomas, J. Mol. Struct., 4 (1969) 179.
51. N.V. Cohan and C.A. Coulson, Trans. Faraday Soc., 53 (1956) 1160; C.A. Coulson and M.J. Stephan, Trans. Faraday Soc., 53 (1957) 272.
52. J.A. Pople and D.L. Beveridge, Approximate Molecular Orbital Theory, McGraw-Hill, New York, 1970.
53. G.A. Segal and M.L. Klein, J. Chem. Phys., 47 (1967) 4236.
54. C.A. Coulson and M.T. Rogers, J. Chem. Phys., 35 (1961) 593.
55. C. Giessner-Prettre and A. Pullman, Theor. Chim. Acta, 11 (1968) 159.
56. R.E. Bruns and W.B. Person, J. Chem. Phys., 57 (1972) 324.
57. J.L. Duncan, I.J. Wright and D. van Lerberghe, J. Mol. Spectr., 42 (1972) 463.
58. R.C. Golike, I.M. Mills, W.B. Person and B.L. Crawford, J. Chem. Phys., 25 (1956) 1266.
59. J.L. Duncan, D.C. McKean and P.D. Mallinson, J. Mol. Spectr., 45 (1973) 221.
60. J.A. Pople, D.P. Santry and G.A. Segal, J. Chem. Phys., 43 (1965) S 129.
61. J.A. Pople and G.A. Segal, J. Chem. Phys., 44 (1966) 3289.
62. R.H. Pritchard and C.W. Kern, J. Amer. Chem. Soc., 91 (1969) 1631; M.D. Newton, E. Switkes and W.N. Lipscomb, J. Chem. Phys., 53 (1970) 2645.
63. G. Riley and W.J. Orville-Thomas, unpublished data.
64. E. Clementi, J. Chem. Phys., 40 (1964) 1944.
65. R.F. Stewart, J. Chem. Phys., 52 (1970) 431.
66. L.S. Bartell, E.A. Roth, C.D. Hollowell, K. Kuchitzu and J.E. Young, Jr., J. Chem. Phys., 42 (1965) 2683.

CHAPTER 14

PREDICTION OF INFRARED AND RAMAN INTENSITIES BY PARAMETRIC METHODS

M. Gussoni, S. Abbate and G. Zerbi

1. INTRODUCTION

The vibrational (infrared and Raman) spectra provide two different sources of information in the molecular dynamics: the frequencies of the vibrational transitions and the intensities of the corresponding bands.

The first kind of data has been and is widely used as a tool for the characterisation of a compound both on empirical grounds and in association with normal coordinate calculations. In the latter case the experimental frequencies are used to compute potential energy parameters (force constants) which are essentially related to the strength of the chemical bond. If a set of force constants has been derived from a sufficiently large number of isotopically substituted molecules, it may be used to predict the frequencies of other molecules containing the same bonds and having similar electronic structure. A large amount of work in this direction has been done in the last twenty years. The force constants for several systems of atoms are well known and may be used to elucidate the structure and dynamical behaviour of very large and complicated molecules such as regular and disordered polymers, biological systems, etc. (see Chapter 24).

A parallel study should be possible on intensity data. The vibrational intensities may be represented as functions of parameters which are related to the fluctuations of charge distribution during the motion. These parameters should play the same role in the study of intensities as that of force constants in normal coordinate calculations.

We will briefly review here the more important kinds of parameters used so far in studying vibrational intensities and will underline the advantages of the so called electrooptical parameters (e.o.p.'s) with respect to other parameters. Infrared and Raman intensities will be considered in parallel.

The infrared intensity of a band due to the normal mode Q_i is given by

$$I_i^{IR} = K \left(\frac{\partial \vec{M}}{\partial Q_i}\right)^2 \qquad (1)$$

where K is a constant and $\vec{M}$ the molecular dipole moment. The Raman intensity of a band due to the normal mode Q_i is given by

$$I_i^R = K(\nu_i)\left[45\left(\frac{\partial \bar{\alpha}}{\partial Q_i}\right)^2 + 13\left(\frac{\partial \gamma}{\partial Q_i}\right)^2\right] \qquad (2)$$

where $K(\nu_i)$ is a known function of the fundamental frequency ν_i; $\bar{\alpha}$ and γ are the mean molecular polarisability and the molecular anisotropy respectively; the compound is assumed to be isotropic and to be irradiated with natural light, otherwise the numerical coefficients in Equation (2) have to be changed. Any infrared intensity determines uniquely $(\partial\vec{M}/\partial Q_i)^2$. Any Raman intensity is related to the two quantities $(\partial\bar{\alpha}/\partial Q_i)^2$ and $(\partial\gamma/\partial Q_i)^2$; for non-totally symmetric modes $(\partial\bar{\alpha}/\partial Q_i) = 0$ while for totally symmetric modes, if the polarisation

$$\rho_i = \frac{6\left(\frac{\partial\gamma}{\partial Q_i}\right)^2}{45\left(\frac{\partial\bar{\alpha}}{\partial Q_i}\right)^2 + 7\left(\frac{\partial\gamma}{\partial Q_i}\right)^2} \tag{3}$$

is experimentally available, one can achieve a unique determination of $(\partial\bar{\alpha}/\partial Q_i)^2$ and $(\partial\gamma/\partial Q_i)^2$.

The quantity $(\partial\vec{M}/\partial Q_i)$ may be expressed as a function of internal coordinates (see Chapter 13) as

$$\frac{\partial\vec{M}}{\partial Q_i} = \sum_m^M \left(\frac{\partial\vec{M}}{\partial R_m}\right) L^R_{mi} \tag{4}$$

where L^R_i is the internal eigenvector belonging to Q_i and M is the number of internal coordinates. The quantities $(\partial\vec{M}/\partial R_m)$, which we call molecular or total dipole moment derivatives, have often been assumed to be the more suitable parameters for the study of intensities. They represent the variation of the molecular dipole moment when the vibration R_m takes place. Care must be taken that the same R_m may be performed in different ways in the various isotopic derivatives in order to satisfy the different Eckart's conditions. This results in a possible difference between $(\partial\vec{M}/\partial R_m)^{(a)}$ and $(\partial\vec{M}/\partial R_m)^{(b)}$ (where (a) and (b) label two isotopic derivatives), even taking into account the fact that the electronic structure is the same. Crawford [1] has discussed this problem extensively and has given the relation among $(\partial\vec{M}/\partial R_m)^{(a)}$ and $(\partial\vec{M}/\partial R_m)^{(b)}$ as a function of the equilibrium dipole moment $\vec{M}°$; the two quantities are equal when the molecule is non-polar and/or when R_m is a symmetry coordinate belonging to a symmetry species which does not contain any rotation.

An expression analogous to that given in Equation (4) for $(\partial\vec{M}/\partial Q_i)$ can be written for $(\partial\alpha_{uv}/\partial Q_i)$, where u and v refer to the axes of the molecule fixed system:

$$\frac{\partial\alpha_{uv}}{\partial Q_i} = \sum_m^M \frac{\partial\alpha_{uv}}{\partial R_m} L^R_{mi} \tag{5}$$

The same considerations made for the infrared case apply also in this case: $(\partial\alpha_{uv}/\partial R_m)^{(a)}$ and $(\partial\alpha_{uv}/\partial R_m)^{(b)}$ need not be equal. Crawford [1] has shown that the difference is a function of the equilibrium polarisability tensor $((\alpha°))$, it vanishes when $((\alpha°))$ is spherical and/or when R_m belongs to a symmetry species where no rotation occurs.

Equations (4) and (5) allow the derivation of information from a simultaneous study of the intensities of a molecule and of its deuterated derivatives. However, only when $\vec{M}°$ vanishes (or when $((\alpha°))$ is spherical) is it possible to use the same $(\partial\vec{M}/\partial R_m)(\partial\alpha_{uv}/\partial R_m)$ for the isotopic derivatives. Otherwise it is necessary to apply rather cumbersome corrections [1-4] and these corrections make the physical meaning of the results less immediate. However all the earlier studies on intensities have been carried out in this frame especially in the case of infrared (see for instance all the works of the Minnesota School which have appeared in J. Chem. Phys. between 1952 and 1966 under the leading title: "Vibrational Intensities .... ", e.g. ref. 2).

The quantities $(\partial\vec{M}/\partial Q_i)$ and $(\partial\alpha_{uv}/\partial Q_i)$ may also be expressed as a function of Cartesian displacement coordinates, namely

$$\frac{\partial\vec{M}}{\partial Q_i} = \sum_n^{3N} \frac{\partial\vec{M}}{\partial x_n} L^x_{ni} \qquad (6)$$

$$\frac{\partial\alpha_{uv}}{\partial Q_i} = \sum_n^{3N} \frac{\partial\alpha_{uv}}{\partial x_n} L^x_{ni} \qquad (7)$$

where L^x_i is the Cartesian eigenvector belonging to Q_i and N the number of atoms in the molecule.* The quantities $(\partial\vec{M}/\partial x_n)$ and $(\partial\alpha_{uv}/\partial x_n)$ are independent of isotopic substitution [1]. This fact introduces a great advantage both in the calculations and in their physical interpretation. Recent studies [5,6] on infrared intensities make use of the $(\partial\vec{M}/\partial x_n)$'s. The derivatives of the molecular dipole moment components with respect to the three Cartesian coordinates of an atom are arranged into (3 x 3) matrices called polar tensors. Attempts are being made to find whether the polar tensors are characteristics of atoms and whether they can be transferred to other molecules containing the same atoms in a different surrounding [7,8]. Combinations of the $(\partial\vec{M}/\partial x_n)$'s and combinations of the $(\partial\alpha_{uv}/\partial x_n)$'s have been used by Mayants and his coworkers [9].

---

* It must be noticed that, while $(\partial\vec{M}/\partial R_m)$ and $(\partial\alpha_{uv}/\partial R_m)$ are purely vibrational quantities and are completely determined by the vibrational intensities, the quantities $(\partial\vec{M}/\partial x_n)$ and $(\partial\alpha_{uv}/\partial x_n)$ contain contributions from the rotations ρ_n of the molecule; therefore they are not completely determined by the knowledge of $(\partial\vec{M}/\partial Q_i)$ or $(\partial\alpha_{uv}/\partial Q_i)$, the knowledge of $(\partial\vec{M}/\partial\rho_n)$ and $(\partial\alpha_{uv}/\partial\rho_n)$ also being necessary.

2. ELECTROOPTICAL PARAMETERS

All the quantities appearing in Equations (4-7) refer to molecular dipole moments or polarisability tensors. It seems valuable to consider the possibility of relating the experimental intensities to bond dipole moments and to bond polarisability tensors. Following the original idea of Vol'kenshtein and El'yashevich [10] the assumption can be made that at any time during the motion the molecular dipole moment $\vec{M}$ can be written as a sum of bond dipole moments

$$\vec{M} = \sum_k^K \vec{\mu}^k = \sum_k^K \mu^k \vec{e}^k \tag{8}$$

and the molecular polarisability tensor can be written as a sum of bond polarisability tensors*

$$((\alpha)) = \sum_k^K ((\alpha^k)) = \sum_k^K \{\alpha^{kL}\vec{e}^{kL}\vec{e}^{kL} + \alpha^{kT}\vec{e}^{kT}\vec{e}^{kT} + \alpha^{kT'}\vec{e}^{kT'}\vec{e}^{kT'}\} \tag{9}$$

K is the number of bonds in the molecule. $\vec{e}^{kL} = \vec{e}^k$ is the unit vector in the instantaneous direction of the k-th bond; it is assumed that $\vec{\mu}^k$ and one of the principal axes of $((\alpha^k))$ lie in this direction. $\vec{e}^{kT}$ and $\vec{e}^{kT'}$ are the unit vectors in the direction of the other two principal axes of $((\alpha^k))$.

The quantity $(\partial\vec{M}/\partial Q_i)$ can now be written as [11]

$$\frac{\partial\vec{M}}{\partial Q_i} = \sum_m^M \left[\sum_k^K \left(\frac{\partial\mu^k}{\partial R_m}\right)^o \vec{e}^{ko} + \sum_k^K \mu^{ko}\left(\frac{\partial\vec{e}^k}{\partial R_m}\right)^o\right] L^R_{mi}$$

$$= [\tilde{\underline{\vec{e}}}^o \frac{\partial\mu}{\underline{\underline{\partial R}}} + \tilde{\underline{\mu}}^o \frac{\partial\vec{e}}{\underline{\underline{\partial R}}}] \underline{L}^R_i \tag{10}$$

The quantities included in $\underline{\vec{e}}^o$ and $(\partial\vec{e}/\underline{\underline{\partial R}})$ are known from molecular geometry or can be evaluated from the dynamics of the system [12,13]. The quantities included in $\underline{\mu}^o$ and $(\partial\mu/\underline{\underline{\partial R}})$ are called equilibrium electrooptical parameters and valence electrooptical parameters respectively; the coefficients which multiply the equilibrium electrooptical parameters are called deformation coefficients. The electrooptical parameters have been shown [12] to be the same for a molecule and its isotopic derivatives, thus representing only the fluctuations of charge distribution during the motion; the difference between $(\partial\vec{M}/\partial R_m)^{(a)}$ and $(\partial\vec{M}/\partial R_m)^{(b)}$ can be ascribed completely to the deformation part in Equation (10).

The quantity $(\partial\alpha_{uv}/\partial Q_i)$ can in an analogous way be expressed as [14]

---

* Dyadic notation for tensors is used.

$$\frac{\partial \alpha_{uv}}{\partial Q_i} = \sum_m^M \{ \sum_k^K [(\frac{\partial \alpha^{kL}}{\partial R_m})^\circ \delta_{uv} + (\frac{\partial \alpha^{kT}}{\partial R_m} - \frac{\partial \alpha^{kL}}{\partial R_m})^\circ (e_u^{kTO} e_v^{kTO}$$

$$+ e_u^{kT'O} e_v^{kT'O}) - (\alpha^{kLO} - \alpha^{kTO})(e_u^{kTO} \frac{\partial e_v^{kT}}{\partial R_m} + e_v^{kTO} \frac{\partial e_u^{kT}}{\partial R_m}$$

$$+ e_u^{kT'O} \frac{\partial e_v^{kT'}}{\partial R_m} + e_v^{kT'O} \frac{\partial e_u^{kT'}}{\partial R_m})]\} L_{mi}^R$$

$$= \{\delta_{uv} \underline{\tilde{1}} \underline{\underline{\frac{\partial \alpha}{\partial R}}} + (\widetilde{\underline{e_u^{LO} e_v^{LO}}} - 1/3 \delta_{uv} \underline{\tilde{1}}) \underline{\underline{\frac{\partial \gamma}{\partial R}}} + \underline{\tilde{\gamma}}^\circ \underline{\underline{\frac{\partial e_u^L e_v^L}{\partial R}}}\} \underline{L}_i^R \qquad (11)$$

where use has been made of the orthonormality relations among unit vectors and the assumption has also been made that the bond tensors are cylindrical at the equilibrium and during the motion.* The quantities $\partial\bar{\alpha}/\partial Q_i$ and $\partial\gamma/\partial Q_i$ which enter directly into the intensity expression because of Equation (11), become

$$(\frac{\partial \bar{\alpha}}{\partial Q_i})^2 = \underline{\tilde{L}}_i^R \underline{\underline{\frac{\partial \bar{\alpha}}{\partial R}}} \underline{1} \underline{\tilde{1}} \underline{\underline{\frac{\partial \bar{\alpha}}{\partial R}}} \underline{L}_i^R \qquad (12)$$

$$2(\frac{\partial \gamma}{\partial Q_i})^2 = \underline{\tilde{L}}_i^R \{ 3 \underline{\underline{\frac{\partial \tilde{\gamma}}{\partial R}}} \sum_{uv} [\underline{e_u^{LO} e_v^{LO}} \; \widetilde{\underline{e_u^{LO} e_v^{LO}}} - \frac{\delta_{uv}}{q} \underline{1} \underline{\tilde{1}}] \underline{\underline{\frac{\partial \gamma}{\partial R}}}$$

$$+ 6 \underline{\underline{\frac{\partial \tilde{\gamma}}{\partial R}}} \sum_{uv} \widetilde{\underline{e_u^{LO} e_v^{LO}}} \; \underline{\tilde{\gamma}}^\circ \underline{\underline{\frac{\partial e_u^L e_v^L}{\partial R}}} + 3 \sum_{uv} \widetilde{\underline{\underline{\frac{\partial e_u^L e_v^L}{\partial R}}}} \underline{\gamma}^\circ \underline{\tilde{\gamma}}^\circ \underline{\underline{\frac{\partial e_u^L e_v^L}{\partial R}}} \} \underline{L}_i^R \qquad (13)$$

$\partial\bar{\alpha}/\underline{\underline{\partial R}}$ contains the derived bond mean polarisabilities, $\partial\gamma/\underline{\underline{\partial R}}$ the derived bond anisotropies, γ^o the equilibrium bond anisotropies; the quantities in the first two matrices are the so-called valence electrooptical parameters and those in the latter matrix are the equilibrium electrooptical parameters. All the other quantities entering Equations (12) and (13) are known from the equilibrium geometry or from the dynamics of the molecule.

Also in the Raman case it can be shown [15] that the electrooptical parameters do not depend on geometrical distortions during the motion but only on the fluctuations of charge density and that they are independent of isotopic substitution. The so-called "bond moment model" [17] and "bond polarisability model" [18] are a simplified version of the e.o.p.'s theory.

---

* An expression similar to Equation (11) has been derived [15] also for the non-cylindrical case. It has also been shown [16] that, if a bond polar tensor is cylindrical at the equilibrium, it remains cylindrical during the motion.

The above models in their original formulations allow for the changes of the dipole moment and of the polarisability tensor only with the stretching of the same bond; all the other $\partial\mu/\partial R_m$ are taken to be zero. Because of these too drastic assumptions these models often lead to inconsistencies [17]. Nevertheless several steps towards the understanding of the intensity problem have been made in this frame [19,20]. Some further modifications of the bond moment model and of the bond polarisability model have been proposed more recently [21-23].

The electrooptical parameters seem thus to provide the more useful tool for studying intensities; they can be treated just as force constants in the study of frequencies; they can be refined by a simultaneous least squares adjustment ("overlay") on the data from a molecule and from its isotopic derivatives and they may be transferred from one molecule to another when some groups of atoms are the same [24,25]. The tranferability seems more realistic for e.o.p.'s than for the kinds of parameters described previously since the e.o.p.'s refer to bonds (and therefore to local characteristics), while the $\partial\vec{M}/\partial R_m$, $\partial\vec{M}/\partial x_n$, $\partial\alpha_{uv}/\partial R_m$, $\partial\alpha_{uv}/\partial x_n$ refer always to molecules (and therefore to overall characteristics).

Just as an illustration of the last concept let us discuss the interpretation of infrared intensities of methyl halides (Figure 1) [12]. The equilibrium e.o.p.'s are μ°_{CH} and μ°_{CX}; the

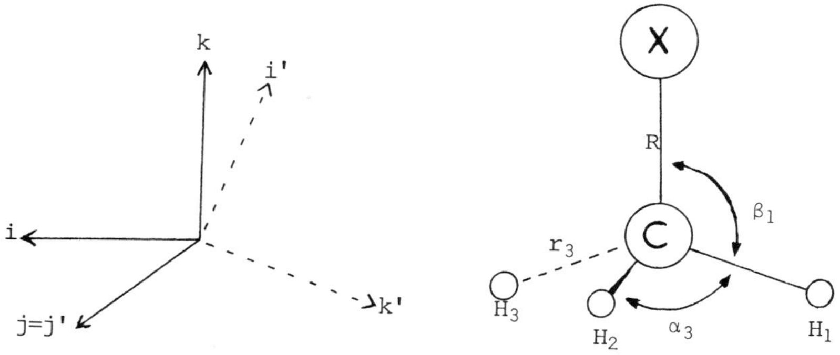

Figure 1. Reference system (i j k) for the polar tensor of the X atom (X = F,Cl,Br,I) and reference system (i' j' k') for the polar tensor of the H_1 atom.

valence e.o.p.'s are reported in Table 1. If the symmetry coordinates are chosen as follows

$$S_1 = \frac{1}{\sqrt{3}}(r_1 + r_2 + r_2)$$

$$S_2 = \frac{1}{\sqrt{6}}(\alpha_1 + \alpha_2 + \alpha_3 - \beta_1 - \beta_2 - \beta_3)$$

$$S_3 = R$$

Table 1. Valence electrooptical parameters; the internal coordinates are defined in Figure 1; m and μ are the bond moments of C-X and C-H respectively.

	R_{CX}	r_1	r_2	r_3	β_1	β_2	β_3	α_1	α_2	α_3
C-X	$\frac{\partial m}{\partial R}$	$\frac{\partial m}{\partial r}$	$\frac{\partial m}{\partial r}$	$\frac{\partial m}{\partial r}$	$\frac{\partial m}{\partial \beta}$	$\frac{\partial m}{\partial \beta}$	$\frac{\partial m}{\partial \beta}$	$\frac{\partial m}{\partial \alpha}$	$\frac{\partial m}{\partial \alpha}$	$\frac{\partial m}{\partial \alpha}$
$C-H_1$	$\frac{\partial \mu}{\partial R}$	$\frac{\partial \mu}{\partial r}$	$\frac{\partial \mu}{\partial r'}$	$\frac{\partial \mu}{\partial r'}$	$\frac{\partial \mu}{\partial \beta}$	$\frac{\partial \mu}{\partial \beta'}$	$\frac{\partial \mu}{\partial \beta'}$	$\frac{\partial \mu}{\partial \alpha}$	$\frac{\partial \mu}{\partial \alpha'}$	$\frac{\partial \mu}{\partial \alpha'}$
$C-H_2$	$\frac{\partial \mu}{\partial R}$	$\frac{\partial \mu}{\partial r'}$	$\frac{\partial \mu}{\partial r}$	$\frac{\partial \mu}{\partial r'}$	$\frac{\partial \mu}{\partial \beta'}$	$\frac{\partial \mu}{\partial \beta}$	$\frac{\partial \mu}{\partial \beta'}$	$\frac{\partial \mu}{\partial \alpha'}$	$\frac{\partial \mu}{\partial \alpha}$	$\frac{\partial \mu}{\partial \alpha'}$
$C-H_3$	$\frac{\partial \mu}{\partial R}$	$\frac{\partial \mu}{\partial r'}$	$\frac{\partial \mu}{\partial r'}$	$\frac{\partial \mu}{\partial r}$	$\frac{\partial \mu}{\partial \beta'}$	$\frac{\partial \mu}{\partial \beta'}$	$\frac{\partial \mu}{\partial \beta}$	$\frac{\partial \mu}{\partial \alpha'}$	$\frac{\partial \mu}{\partial \alpha'}$	$\frac{\partial \mu}{\partial \alpha}$

$$S_4 = \frac{1}{\sqrt{2}}(r_2 - r_3)$$

$$S_5 = \frac{1}{\sqrt{2}}(\alpha_2 - \alpha_3)$$

$$S_6 = \frac{1}{\sqrt{2}}(\beta_2 - \beta_3)$$

the derivatives of the molecular dipole moment with respect to the symmetry coordinates may be expressed as in Table 2; the

Table 2. Analytical expressions of the dipole moment derivatives as functions of the electrooptical parameters.

$$\frac{\partial M^z}{\partial S_1} = \sqrt{3}\,\frac{\partial m}{\partial r} - \frac{1}{\sqrt{3}}\left\{\frac{\partial \mu}{\partial r} + 2\,\frac{\partial \mu}{\partial r'}\right\}$$

$$\frac{\partial M^z}{\partial S_2} = \frac{1}{\sqrt{6}}\left\{3\left[\frac{\partial m}{\partial \alpha} - \frac{\partial m}{\partial \beta}\right] - \frac{\partial \mu}{\partial \alpha} - 2\,\frac{\partial \mu}{\partial \alpha'} + \frac{\partial \mu}{\partial \beta} + 2\,\frac{\partial \mu}{\partial \beta'} + a\mu^\circ_{CH}\right\}$$

$$\frac{\partial M^z}{\partial S_3} = \frac{\partial m}{\partial R} - \frac{\partial \mu}{\partial R}$$

$$\frac{\partial M^y}{\partial S_4} = \frac{2}{\sqrt{3}}\left\{\left[\frac{\partial \mu}{\partial r} - \frac{\partial \mu}{\partial r'}\right] - b\mu^\circ_{CH}\right\} + c\mu^\circ_{CX}$$

$$\frac{\partial M^y}{\partial S_5} = \frac{2}{\sqrt{3}}\left\{\frac{\partial \mu}{\partial \alpha} - \frac{\partial \mu}{\partial \alpha'}\right\} + d\mu^\circ_{CH} + e\mu^\circ_{CX}$$

$$\frac{\partial M^y}{\partial S_6} = \frac{2}{\sqrt{3}}\left\{\frac{\partial \mu}{\partial \beta} - \frac{\partial \mu}{\partial \beta'}\right\} - f\mu^\circ_{CH} - g\mu^\circ_{CX}$$

deformation coefficients are given in Table 3. The same experimental data could be used to derive numerical values either for the $\partial M/\partial s_t$'s or for the e.o.p.'s. But one sees that,

Table 3. Values of the deformation coefficients (Å^{-1}) defined in Table 2.

	F	Cl	Br	I
a	3.097	3.097	3.097	3.097
b	-0.023	-0.017	-0.015	-0.012
c	0.043	0.037	0.036	0.033
d	0.839	0.853	0.858	0.861
e	0.055	0.041	0.037	0.033
f	0.767	0.815	0.830	0.840
g	0.128	0.079	0.064	0.054

while the $\partial\vec{M}/\partial s_t$'s only show a general trend with electronegativity (see Table 4) several e.o.p.'s remain constant throughout the series. It can be noticed, for example, that the values of $\partial M^y/\partial S_5$ undergo very little variation: this can be justified by the fact that only C-H bond parameters enter its expression except for μ°_{CX}, which however enters with a very small coefficient.

Table 4. Experimental values [26] of the dipole moment derivatives expressed in D Å^{-1}.

	F	Cl	Br	I
$\frac{\partial M^z}{\partial S_1}$	0.5668	0.5927	0.5654	0.4772
$\frac{\partial M^z}{\partial S_2}$	0.1420	-0.1541	-0.3264	-0.4846
$\frac{\partial M^z}{\partial S_3}$	-4.3391	-2.2730	-1.4471	-0.5790
$\frac{\partial M^y}{\partial S_4}$	0.8744	0.3496	0.2848	0.2077
$\frac{\partial M^y}{\partial S_5}$	-0.2527	-0.2963	-0.3021	-0.2922
$\frac{\partial M^y}{\partial S_6}$	0.1506	0.1994	0.2870	0.3327

The same kind of comparison can be done in the case of polar tensors. The expressions of the polar tensors as a function of e.o.p.'s are reported in Table 5. The values of the polar tensors for atoms X and H_1 (see Figure 1) through the series are reported in Table 6. Again the same considerations made above apply here: while the polar tensors can only show a trend throughout the series, the e.o.p.'s allow a more subtle separation of the contributions of the various bonds to the variations of the molecular dipole moment. For example, the constancy of the element (2,2) of the polar tensor for H_1

(Table 6) can be predicted since its values are completely determined by C-H bond parameters (Table 5) which might be approximately the same. Moreover the trend among the values of the element (1,1) for the polar tensor of X:

$$\sqrt{2}\left(\frac{\partial\mu}{\partial\beta'} - \frac{\partial\mu}{\partial\beta}\right) + \mu^{\circ}_{CX} = -1.698\ (F);\ -2.001\ (Cl);\ -2.002\ (Br);\ -1.942\ (I)$$

reproduces the trend among the corresponding experimental values for $\vec{M}^{\circ}$:

$$M^{\circ} = -1.790\ (F);\ -1.869\ (Cl);\ -1.797\ (Br);\ -1.650\ (I)$$

showing again the transferability of CH parameters. These considerations and other parallel ones for the Raman case support the idea that e.o.p.'s should become a tool suitable for the study of infrared and Raman intensities and should throw some light on the behaviour of the bond charge distribution during the motion.

3. E.O.P.'S OF NORMAL PARAFFINS

In our laboratory attempts have been made and are being made to calculate a general set of infrared e.o.p.'s and Raman e.o.p.'s for normal paraffins. We report here only some preliminary results: (1) those of an overlay calculation on the infrared intensities of deuterated methanes and ethanes; (2) those of two separate "overlays" on the Raman intensities of deuterated methanes and of C_6H_{12}, C_6D_{12}. The results for case (1) are reported in Table 7. While the first set of parameters gives much better fitting because the bending valence e.o.p.'s are not constrained to be equal in methanes and ethanes, the second set of parameters is in our opinion more interesting because all the intensities of these molecules are satisfactorily predicted by only six parameters (Table 8) without any distinction between C-H parameters in methanes and ethanes. Further studies in this direction (overlay calculation of the methyl group parameters between ethanes and propanes) led to the results reported in Table 9. Some general comments may already be made at this stage: the stretching valence e.o.p.'s are always higher by an order of magnitude than the corresponding bending valence e.o.p.'s; there is an appreciable difference between valence e.o.p.'s in methyl and methylene groups; the equilibrium e.o.p.'s for C-H are roughly the same in methane, ethane and propane (methyl and methylene groups). The values reported in the second column of Table 9 have been used to predict the infrared intensities of polyethylene and perdeuteropolyethylene (see Chapter 24, Figures 8 and 9). To this purpose Equation (10) has been modified as follows [30]

$$\frac{\partial\vec{M}}{\partial Q_i(o)} = [\vec{e}^{\circ}\,\underline{\underline{\frac{\partial\mu}{\partial R}}}(o) + \underline{\tilde{\mu}}^{\circ}\,\underline{\underline{\frac{\partial\vec{e}}{\partial R}}}(o)]\,\underline{L}^{R}_{i}(o) \qquad (14)$$

where $Q_i(o)$ is a normal mode of the polymer for wavevector

Table 5. Analytical expression for the polar tensors of halogen X and hydrogen H_1 atoms as functions of the electrooptical parameters. The reference systems (i j k) and (i' j' k') are defined in Figure 1. s°_{CX} and s°_{CH} are the equilibrium length of C-X and C-H bonds, respectively.

	x_X	y_X	z_X
$\hat{i}$	$(\sqrt{2}[\frac{\partial\mu}{\partial\beta'} - \frac{\partial\mu}{\partial\beta}] + \mu^\circ_{CX})/s^\circ_{CX}$	0	0
$\hat{j}$	0	$(\sqrt{2}[\frac{\partial\mu}{\partial\beta'} - \frac{\partial\mu}{\partial\beta}] + \mu^\circ_{CX})/s^\circ_{CX}$	0
$\hat{k}$	0	0	$\frac{\partial m}{\partial R} - \frac{\partial\mu}{\partial R}$
	x'_{H_1}	y'_{H_1}	z'_{H_1}
$\hat{i}'$	$(\frac{2\sqrt{2}}{3}[\frac{\partial m}{\partial\alpha} - \frac{\partial m}{\partial\beta}] + \frac{\sqrt{2}}{3}[2\frac{\partial\mu}{\partial\beta'} - \frac{\partial\mu}{\partial\alpha} - \frac{\partial\mu}{\partial\alpha'}] + \mu^\circ_{CH})/s^\circ_{CH}$	0	$\frac{2\sqrt{2}}{3}[\frac{\partial m}{\partial r} - \frac{\partial\mu}{\partial r'}]$
$\hat{j}'$	0	$(\sqrt{2}[\frac{\partial\mu}{\partial\alpha} - \frac{\partial\mu}{\partial\alpha'}] + \mu^\circ_{CH})/s^\circ_{CH}$	0
$\hat{k}'$	$\frac{1}{3}[\frac{\partial m}{\partial\beta} - \frac{\partial m}{\partial\alpha}] - \frac{1}{3}[3\frac{\partial\mu}{\partial\beta} - 2\frac{\partial\mu}{\partial\beta'} + \frac{\partial\mu}{\partial\alpha} - 2\frac{\partial\mu}{\partial\alpha'}]$	0	$-\frac{1}{3}\frac{\partial m}{\partial r} + \frac{\partial\mu}{\partial r} - \frac{2}{3}\frac{\partial\mu}{\partial r'}$

Table 6. Experimental values [7] of the polar tensors of halogen atoms X = F,Cl,Br,I and of the hydrogen atom H_1. The units employed are D $Å^{-1}$.

	x_F	y_F	z_F	x_{Cl}	y_{Cl}	z_{Cl}	x_{Br}	y_{Br}	z_{Br}	x_I	y_I	z_I
$\hat{i}$	-1.225	0	0	-1.124	0	0	-1.033	0	0	-0.908	0	C
$\hat{j}$	0	-1.225	0	0	-1.124	0	0	-1.033	0	0	-0.908	0
$\hat{k}$	0	0	-4.481	0	0	-2.142	0	0	-1.436	0	0	-0.552
	$x_{H_1}^F$	$y_{H_1}^F$	$z_{H_1}^F$	$x_{H_1}^{Cl}$	$y_{H_1}^{Cl}$	$z_{H_1}^{Cl}$	$x_{H_1}^{Br}$	$y_{H_1}^{Br}$	$z_{H_1}^{Br}$	$x_{H_1}^I$	$y_{H_1}^I$	$z_{H_1}^I$
$\hat{i}'$	0.058	0	0.101	-0.173	0	0.269	-0.274	0	0.274	-0.317	0	0.264
$\hat{j}'$	0	0.351	0	0	0.351	0	0	0.327	0	0	0.288	0
$\hat{k}'$	-0.034	0	-0.687	-0.115	0	-0.293	-0.149	0	-0.202	-0.183	0	-0.124

Table 7. Overlay least squares refinement on intensity data for methanes [27] and ethanes [28].

	ν_{obs}		A_{obs} (10^3 cm mol^{-1})	A^{I}_{calc}	A^{II}_{calc}
CH_4	3019	F_2	6728	6784	7418
	1306	F_2	3364	3253	2960
CD_4	2259	F_2	2609	3066	3414
	996	F_2	1936	2012	1843
CH_2D_2	3013	B_1	3484	3372	3693
	2976	A_1			
	2234	B_2	1383	1487	1654
	2202	A_1			
	1436	A_1	-	409	372
	1234	B_2			
	1090	B_1	2146	2269	2073
	1033	A_1			
CH_3D	3021	E	4919	5027	5502
	2945	A_1			
	2200	A_1	628	728	810
	1471	E			
	1300	A_1	2923	2969	2706
	1155	E			
CD_3H	2993	A_1	1570	1680	1841
	2200	E	2093	2277	2534
	2142	A_1			
	1291	E	665	752	689
	1036	E	1525	1608	1471
	1003	A_1			
CH_3CH_3	2974	E_u	17800	17616	17388
	2915	A_{2u}			
	1460	E_u	1830	2010	1944
	1370	A_{2u}			
	822	E_u	630	582	1320
CD_3CD_3	2236	E_u	6440	6402	5079
	2095	A_{2u}	2410	3040	3528

Table 7. (Continued)

	ν_{obs}		A_{obs} (10^3 cm mol^{-1})	A^{I}_{calc}	A^{II}_{calc}
CD_3CD_3	1082	E_u	1200	1089	1291
	1077	A_{2u}			
	594	E_u	320	332	677
CH_3CD_3	2976	E	9260	8679	8588
	2912	A_1			
	2240	E	3030	3329	2604
	2090	A_1	1370	1520	1769
	1471	E	950	978	987
	1387	A_1			
	1122	A_1			
	1115	E	660	542	810
	1066	E			
	904	A_1	-	61	16
	678	E	420	398	816
CH_3CH_2D	2994	A"			
	2980	A'			
	2968	A"	14600	14629	14441
	2953	A'			
	2948	A'			
	2182	A'	1400	1624	1440
	1469	A"			
	1460	A'	1290	1199	1399
	1449	A'			
	1388	A'			
	1312	A"	350	628	445
	1310	A'			
	1159	A"	80	87	122
	1120	A'			
	978	A'	20	15	54
	804	A"	380	246	683
	713	A'	160	228	400
	273	A"	-	0	0

Table 8. Electrooptical parameters from overlay of infrared intensity data of methanes and ethanes.

e.o.p.'s	Set I	Set II	
$\frac{\partial\mu}{\partial r}$	.882	.934	D Å^{-1}
$\frac{\partial\mu}{\partial r'}$	.222	.248	D Å^{-1}
$\frac{\partial\mu}{\partial\alpha'} = -\frac{\partial\mu}{\partial\alpha}$ (CH_4)	.027	.022	D rad^{-1}
$\frac{\partial\mu}{\partial\alpha}$	.132	-.022	D rad^{-1}
$\frac{\partial\mu}{\partial\alpha'}$	-.154	.022	D rad^{-1}
$\frac{\partial\mu}{\partial\beta}$	.077	.022	D rad^{-1}
$\frac{\partial\mu}{\partial\beta'}$	.049	-.022	D rad^{-1}
$\frac{\partial M}{\partial r}$	-.068	-.054	D Å^{-1}
$\frac{\partial M}{\partial\beta}$	.029	-.082	D rad^{-1}

$\mu^{\circ}_{C-H} = -0.26$ D $\quad M^{\circ}_{C-C} = 0$ D

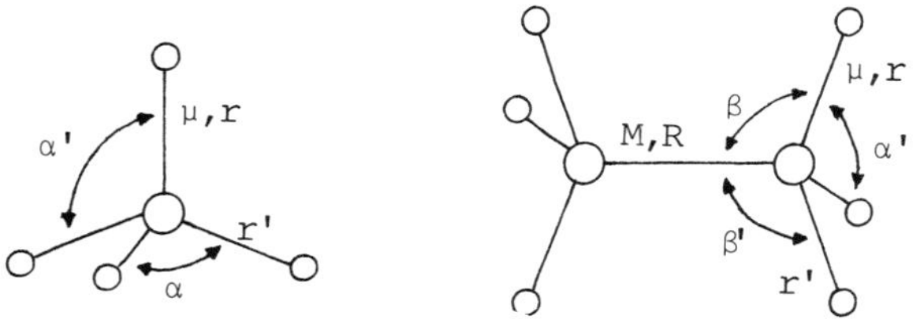

k = 0 and $\underline{L}_i^R(0)$ the corresponding internal eigenvector, while

$$\frac{\partial x}{\underline{\underline{\partial R}}}(0) = \sum_{s=-p}^{+p} \frac{\partial x}{\underline{\underline{\partial R(s)}}} \quad \text{for } x = \mu, \vec{e} \tag{15}$$

The last equation means that, for instance, the elements of the matrix $(\partial\mu/\underline{\underline{\partial R}})(0)$ depend on the variations of the bond dipole

Table 9. E.o.p.'s from overlay least squares refinement on infrared intensity data of ethanes and propanes [28].

$\frac{\partial\mu}{\partial r} = 0.918\ \text{D}\,\text{Å}^{-1}$	$\frac{\partial m}{\partial d} = 0.863\ \text{D}\,\text{Å}^{-1}$	$\frac{\partial M}{\partial r} = 0.017\ \text{D}\,\text{Å}^{-1}$
$\frac{\partial\mu}{\partial r'} = 0.138\ \text{D}\,\text{Å}^{-1}$	$\frac{\partial m}{\partial d'} = 0.063\ \text{D}\,\text{Å}^{-1}$	$\frac{\partial M}{\partial d} = 0.011\ \text{D}\,\text{Å}^{-1}$
$\frac{\partial\mu}{\partial\alpha} = -0.004\ \text{D rad}^{-1}$	$\frac{\partial m}{\partial\omega} = -0.022\ \text{D rad}^{-1}$	$\frac{\partial M}{\partial\beta} = 0.084\ \text{D rad}^{-1}$
$\frac{\partial\mu}{\partial\alpha'} = 0.029\ \text{D rad}^{-1}$	$\frac{\partial m}{\partial\gamma} = -0.022\ \text{D rad}^{-1}$	$\frac{\partial M}{\partial\omega} = 0.021\ \text{D rad}^{-1}$
$\frac{\partial\mu}{\partial\beta} = -0.006\ \text{D rad}^{-1}$	$\frac{\partial m}{\partial\gamma'} = 0.038\ \text{D rad}^{-1}$	$\frac{\partial M}{\partial\gamma} = -0.098\ \text{D rad}^{-1}$
$\frac{\partial\mu}{\partial R} = 0.422\ \text{D}\,\text{Å}^{-1}$	$\frac{\partial m}{\partial R} = 0.233\ \text{D}\,\text{Å}^{-1}$	$\frac{\partial M}{\partial\gamma'} = 0.061\ \text{D rad}^{-1}$

$$\mu^\circ_{C-H} = m^\circ_{C-H} = -0.245\ \text{D} \qquad M^\circ_{C-C} = 0\ \text{D}$$

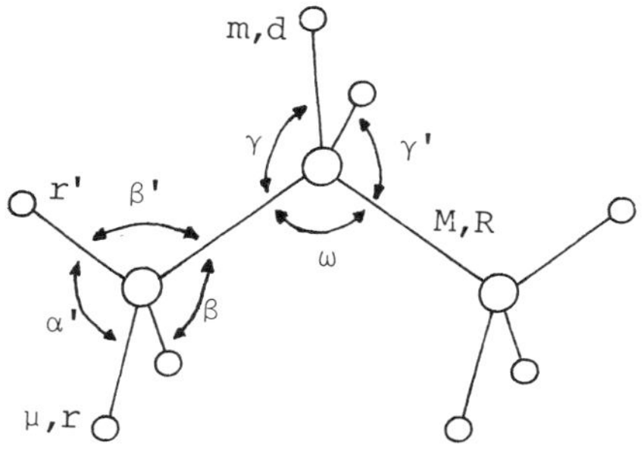

moment of a translational unit with the internal coordinates of the same unit (s=0) and of some neighbouring units (s= -1 , ... -p; s= +1 , ... +p). The above calculations for polyethylene have been made in the assumption that the electrooptical interactions extend only to p=1.

The results for case (2) are reported in Tables 10-12. The intensity data for cyclohexane are available only in the bending region, thus the constancy of $\partial\bar{\alpha}_{CH}/\partial r_{CH}$ and $\partial\gamma_{CH}/\partial r_{CH}$ in the two overlay calculations has little significance. On the contrary it seems to us very important that $\partial\gamma_{CH}/\partial\beta$, $\partial\gamma_{CH}/\partial\beta'$ and γ°_{CH} have retained almost the same value in the two calculations. The set of parameters derived from cyclohexanes (Table 12) have been used to predict the Raman intensities of polyethylene and deuteropolyethylene (Chapter 24, Figures 10 and 11). To this purpose the variation of the molecular

Table 10. E.o.p.'s combinations calculated from methanes data [29].

$\frac{\partial\bar{\alpha}}{\partial r} + 3\,\frac{\partial\bar{\alpha}}{\partial r'}$	$= 1.3\ \text{Å}^2$
$\frac{\partial\gamma}{\partial r} - \frac{\partial\gamma}{\partial r'}$	$= 2.22\ \text{Å}^2$
$\frac{\partial\gamma}{\partial\alpha'} = -\frac{\partial\gamma}{\partial\alpha}$	$= -0.099\ \text{Å}^3\ \text{rad}^{-1}$
γ°	$= 0.305\ \text{Å}^3$

polarisability with a normal mode has been written [30] as

$$\frac{\partial\alpha_{uv}}{\partial Q_i(o)} = \{\delta_{uv}\,\underline{\tilde{1}}\,\frac{\partial\alpha}{\underline{\underline{\partial R}}}(o) + (\underline{\widetilde{e_u^{LO}e_v^{LO}}} - \frac{1}{3}\delta_{uv}\,\underline{\tilde{1}})\,\frac{\partial\gamma}{\underline{\underline{\partial R}}}(o)$$

$$+ \tilde{\gamma}^\circ\,\frac{\partial e_u^L e_v^L}{\underline{\underline{\partial R}}}(o)\}\,L_i^R(o) \tag{16}$$

where Equation (15) has been used for $x = \bar{\alpha}$, γ, $e_u^L e_v^L$. Again p has been assumed to be equal to 1.

The results presented in this chapter, even if still incomplete, are rather encouraging; we hope therefore to be able to find a set of e.o.p.'s (in infrared and Raman) which is able to predict the intensity behaviour for all n-paraffins.

4. REFERENCES

1. B.L. Crawford, J. Chem. Phys., 20 (1952) 977.
2. A.D. Dickson, I.M. Mills and B. Crawford, J. Chem. Phys., 27 (1957) 445.
3. A.J. Van Straten and W.M.A. Smit, J. Mol. Spectr., 56 (1975) 484.
4. M. Gussoni and S. Abbate, J. Mol. Spectr., 62 (1976) 53.
5. F. Biarge, J. Herranz and J. Morcillo, An. R. Soc. Esp. Fis. Quim., A57 (1961) 81.
6. W.B. Person and J.H. Newton, J. Chem. Phys., 61 (1974) 1040.
7. J.H. Newton and W.B. Person, J. Chem. Phys., 64 (1976) 3036.
8. B.J. Krohn, W.B. Person and J. Overend, J. Chem. Phys., 65 (1976) 969.
9. L.S. Mayants and B.S. Averbukh, J. Mol. Spectr., 22 (1967) 197.
10. M.V. Vol'kenshtein , M.A. El'yashevich and B.I. Stepanov, Molecular Vibrations, Vol. 2, Gesudavst Izd. Tekh. i Teoret. Lit. Moskow i Leningrad, 1949.
11. L.A. Gribov, Intensity Theory for Infrared Spectra of Polyatomic Molecules, Consultants Bureau, New York, 1964.

Table 11. Calculated intensities for methane and deuterated derivatives. The intensity values are relative to the A_1 intensity of CH_4.

		ν_{obs}	I_{obs}	I_{obs} [29]	I_{calc}
CH_4	A_1	2915	1	1	1
	E	1534	0.21	0.213	0.144
	F_2	3019	1.26	0.945	0.932
	F_2	1306	-	0.003	0.002
CD_4	A_1	2108	-	0.855	0.848
	E	1092	-	0.034	0.110
	F_2	2259	-	0.847	0.814
	F_2	996	-	-	0.000
CH_2D_2	A_1	2976	-	0.461	0.635
	A_1	2202	-	0.522	0.580
	A_1	1436	-	-	0.048
	A_1	1033	-	-	0.017
	A_2	1329	-	-	0.065
	B_1	3013	-	0.085	0.311
	B_1	1090	-	0.021	0.000
	B_2	2234	-	0.042	0.271
	B_2	1234	-	0.021	0.000
CH_3D	A_1	2945	-	0.729	0.812
	A_1	2200	-	0.284	0.432
	A_1	1300	-	-	0.000
	E	3021	-	(0.127)	0.621
	E	1471	-	0.042	0.120
	E	1155	-	0.042	0.017
CHD_3	A_1	2993	-	0.258	0.469
	A_1	2143	-	0.626	0.543
	A_1	1003	-	-	0.001
	E	2265	-	(0.127)	0.719
	E	1297	-	0.013	0.084
	E	1036	-	0.013	0.033

Table 12. E.o.p.'s calculated from cyclohexane data.

	(a)	(b)		(a)	(b)
$\frac{\partial \bar{\alpha}^{CH}}{\partial r_{CH}}$	(1.300)	1.300	$\frac{\partial \gamma^{CH}}{\partial \beta'}$	0.109	0.099
$\frac{\partial \bar{\alpha}^{CH}}{\partial \omega}$	-0.017	-	$\frac{\partial \gamma^{CC}}{\partial R_{CC}}$	1.457	-
$\frac{\partial \bar{\alpha}^{CC}}{\partial R_{CC}}$	0.921	-	$\frac{\partial \gamma^{CC}}{\partial \beta}$	-0.049	-
$\frac{\partial \bar{\alpha}^{CC}}{\partial \omega}$	0.079	-	$\frac{\partial \gamma^{CC}}{\partial \beta'}$	0.018	-
$\frac{\partial \gamma^{CH}}{\partial r_{CH}}$	(2.200)	2.200	γ°_{CC}	(0.050)	-
$\frac{\partial \gamma^{CH}}{\partial \beta}$	-0.089	-0.099	γ°_{CH}	0.316	0.305

(a) Results from our refinement on cyclohexanes. The parameters in parentheses have not been refined; the first two because the data on stretching intensities are not available and the third because no intensity depends on it in a sufficient large way to affect it.
(b) Results from our refinement on methanes.

12. M. Gussoni and S. Abbate, J. Chem. Phys., 65 (1976) 3439.
13. M. Gussoni, G. Dellepiane and S. Abbate, J. Mol. Spectr., 57 (1975) 323.
14. M. Vol'kenshtein, L. Gribov, M. El'yashevich and Y. Stepanov, Kolebanya Molekul, Yzhdatelstvo Nauka, Moscow, 1972.
15. M. Gussoni, S. Abbate and G. Zerbi, to be published.
16. S. Abbate and M. Gussoni, to be published.
17. D.F. Hornig and D.C. McKean, J. Phys. Chem., 59 (1955) 1133.
18. D.A. Long, Proc. Roy. Soc. London, 217A (1953) 203.
19. M.J. Hopper, J.W. Russell and J. Overend, Spectrochim Acta A, 28 (1972) 1215.
20. T. Yoshino and H.J. Bernstein, J. Mol. Spectr., 2 (1958) 241.
21. W.J. Orville-Thomas, B. Galabov, G. Jalsovsky and H. Ratajczak, J. Mol. Struct., 19 (1973) 761.
22. R.G. Snyder, J. Mol. Spectr., 36 (1970) 222.
23. S. Montero and D. Bermejo, 5th Int. Conf. Raman Spectroscopy, Freiburg, 1976, p. 438.
24. S. Abbate, M. Gussoni and G. Zerbi, 31st Symp. Molecular Spectroscopy, Columbus, Ohio, 1976.
25. S. Abbate, M. Gussoni, G. Zerbi and G. Masetti, Proc. 5th Int. Conf. Raman Spectroscopy, Freiburg, 1976.
26. J.W. Russell, G.D. Needham and J. Overend, J. Chem. Phys., 45 (1966) 3383.
27. R.E. Hiller and J.W. Straley, J. Mol. Spectr., 3 (1959) 632.
28. S. Kondo and S. Saëki, Spectrochim. Acta A, 29 (1973) 735.
29. R.O. Kagel, Vibrational Intensity Studies, thesis, Univ. of Minnesota, 1964.
30. S. Abbate, M. Gussoni, G. Masetti and G. Zerbi, J. Chem. Phys., in press.

CHAPTER 15

BAND CONTOUR ANALYSIS

W.H. Fletcher

1. INTRODUCTION

The appearance of a vibration-rotation band (either infrared or Raman) for a molecule in the gas phase is a function of the selection rules, the transition probabilities, the moments of inertia, temperature, first order Coriolis coupling (if any), perturbations such as ℓ-type resonance, second order Coriolis coupling, Fermi resonance, and the resolution with which the spectrum is observed. One would like to observe bands with sufficient resolution that definite line assignments could be made for a rotational analysis, but this is not generally possible except for relatively small molecules. It is therefore often necessary to look at the grosser features of the band envelope, and from this try to obtain molecular parameters.

The use of band contours to extract information on molecular parameters from unresolved bands of non-linear molecules may be traced back to three classic papers. Placzek and Teller, in their discussion of rotational structure of Raman bands [1], included a discussion of the unresolved Raman band shapes for perpendicular vibrations of symmetric tops, varying from the oblate top (planar molecule) to the limiting case of a prolate top (the linear molecule). Gerhard and Dennison [2] considered the contours of unresolved parallel and perpendicular bands in the infrared, but did not include the effects of Coriolis coupling or other perturbations on the observed shapes. The contours of unresolved infrared bands of asymmetric tops were discussed by Badger and Zumwalt [3], but no perturbing effects were considered.

The effect of Coriolis coupling on band shapes in symmetric tops was first discussed by Teller [4], and Johnston and Dennison showed how first order Coriolis constants could be derived from the resolved line structure in the perpendicular bands [5]. The relationship of Coriolis coupling and band shapes in spherical tops and in oblate symmetric tops was later investigated by Edgell and Moynihan [6] and subsequently by Hoskins [7]. These authors showed how the separations of the P-R maxima in these bands depend on ζ, and they derived equations expressing this dependence for molecules with large moments of inertia. These methods and their variations have been applied to the determination of Coriolis constants for a large number of spherical and symmetric tops [8-17]. While these methods have been very valuable for molecules for which only smooth band contours are observable, and for which no serious perturbations were observed, users should be aware of certain serious limitations and the sources of error which exist.

First order Coriolis coupling is a result of the vibrational angular momentum which is developed between components of doubly degenerate vibrations in symmetric tops and between components of triply degenerate vibrations in spherical tops. The most prominent effect of this is a change in the spacing between adjacent lines. The coupling coefficients vary from -1.0 to 1.0, with the sign simply indicating the direction of the vibrational angular momentum. Second order Coriolis coupling may occur in molecules of any symmetry, the only necessary condition being that the direct product of the species of two interacting vibrations contains the species of one of the components of pure rotation. In symmetric tops an A x E coupling represents an interaction between the motions of atoms of a non-degenerate vibration (along the z axis) and the motions of atoms in one of the components (x or y) of a degenerate vibration. The effects are often small and negligible, but if the origins of two fundamentals are nearly coincident, this perturbation may become strong through a mixing of the wave functions of certain rovibrational levels. The result is a shift in energy levels and an anomalous intensity distribution in the observed lines.

2. SPHERICAL TOPS

The rotational energy levels for the ground state of a spherical top are given by the simple equation

$$F''(J) = B''J(J + 1) \qquad (1)$$

where B" is the rotational constant for the lower state and the effects of centrifugal distortion have been neglected. The latter effects are small enough to have no measurable effect on unresolved band contours. The rotational levels of the first excited state of triply degenerate bands are split into three levels by Coriolis coupling, and the energies are given in a low order of approximation by

$$F^{+}(J) = + B'J(J + 1) + 2B'\zeta(J + 1) \qquad (2)$$

$$F^{o}(J) = + B'J(J + 1) \qquad (3)$$

$$F^{-}(J) = + B'J(J + 1) - 2B'\zeta J \qquad (4)$$

In the infrared spectrum the selection rules permit the transitions for ΔJ = -1, 0 and +1, and these terminate only in the levels F^+, F^o, and F^-, respectively. This allows only one branch in each of the rotational wings of the band, and the separation of the P and R maxima can be expressed adequately by Equation (5), below. The computer simulation and utility of Raman band contours of spherical tops has been discussed by Masri and Fletcher [18].

The Raman selection rules are much less restrictive and each of the allowed transitions, ΔJ = 0, ±1, ±2, is allowed to terminate in each of the three upper states produced by the Coriolis splitting. This produces a total of fifteen sub-branches which we label Q^+, Q^o, Q^-, R^+, R^o, R^-, etc. If $\zeta_i > 0$, the R^+, R^o, R^-, S^+, S^o, S^-, and Q^+ sub-branches all contribute to the high frequency wing of the band, while the P^+, P^o, P^-, O^+, O^o, O^-, and Q^- components contribute to the low frequency wing. For $\zeta_i < 0$, the contributions to the low and high frequency wings are just reversed. Since the total band is a composite of many sub-branches, no simple expression for the separations of the OP-RS maxima exist, and for some values of zeta two sets of maxima occur [18].

Analytical methods

For the infrared bands of the triply degenerate species of point groups T_d and O_h, Edgell and Moynihan [6] and McDowell [19] showed that the Coriolis coupling constants can readily be evaluated from the separations of the P and R maxima of the unresolved bands by the relation

$$\Delta\nu_{PR}\ (\text{cm}^{-1}) = 4(BkT/hc)^{\frac{1}{2}}(1 - \zeta) \tag{5}$$

where $B = h/8\pi^2cI_B$.

This equation has been applied successfully to a large number of spherical top molecules, particularly tetrafluorides [6, 8-13], tetrachlorides [14,15] and hexafluorides [8,16].

In addition to the classic work of Placzek and Teller [1], the contours of Raman bands in spherical tops have been discussed by Masri and Fletcher [18], Clark and Rippon [20], and by Sportouch and Gaufrès [21]. Only a few spherical tops (CH_4, SiH_4, GeH_4, etc., their deuterated analogues, and corresponding fluorides) have easily resolvable rotational structure [22], so the molecular data are usually obtainable only from the band contours. The species E bands of point group T_d and the E_g bands of point group O_h involve no Coriolis coupling, and they are only Raman active. The separation of the maxima of the O-P and R-S branches of these bands has been shown by Clark and Rippon [20] to be

$$\Delta\nu_{OP-RS}\ (\text{cm}^{-1}) = 4.41\ (BkT/hc)^{\frac{1}{2}} \tag{6}$$

The validity of Equation (6) has been verified by extensive examinations of tetrahedral [20,21] and octahedral molecules [21,23]. The B values used in these studies have been obtained from high resolution infrared data on the triply degenerate fundamentals of the lighter molecules and calculated from electron diffraction or X-ray data for the heavier ones.

For the triply degenerate bands of spherical tops, the relationship between zeta and the OP-RS separations has been treated by two different, but essentially equivalent methods. Both methods involve the use of a reduced frequency parameter

$$\Delta\nu_{OP-RS}\ (\text{cm}^{-1}) = B^{\frac{1}{2}}\Delta X \tag{7}$$

which is valid only if $Bhc << kT$. This is equivalent to the one used by Placzek and Teller* in expressing the separation of maxima in the doubly degenerate bands. Clark and Rippon [20] have derived analytical expressions for the intensities of each of the sub-branches contributing to each wing of a band, summed the seven equations in each case and differentiated them to find the maxima. Some approximations are necessarily involved, but comparison of their results with observed separations in Raman bands of tetrahedral and octahedral molecules [20,21] demonstrate the validity of the method.

Raman band contour calculations

An alternative method used by Masri and Fletcher involves the calculation of the frequencies and intensities of all transitions from the appropriate energy levels, selection rules and matrix elements. The intensity of a given J line in a triply degenerate band of a spherical top is given to a sufficient approximation for this work by Herranz and Stoicheff [24].

$$I = C(2J + 1)g_\tau \exp[-hcF(J,\tau)/kT]\beta^2 \quad (8)$$

C is a constant for a given band, so it does not influence the contour, g_τ is the statistical weight due to nuclear spin and the vibration-rotation symmetry of the level, $F(J,\tau)$ is the rotational energy level of the ground state, and β^2 is the averaged anisotropy invariant that is similar to the more familiar Hönl London formula for infrared bands [25]. The method of calculating β^2 is described in reference [18], and the matrix elements and selection rules for all fifteen sub-branches are given. The calculation of the band contour is typical of many programs which exist for computing contours of Raman, infrared and electronic bands. The frequency of each spectral line is computed and stored as an element of an array representing appropriate frequency intervals. The corresponding intensity is calculated and stored in a parallel array. For calculating a band contour which is observed with slits that are too wide to observe any structure, it is sufficient to leave the intensity as a line function which is then convoluted with an appropriate slit function (usually triangular) to simulate what the spectrometer actually records as a spectral line. This distributes the observed intensity over a number of adjacent intervals. After all line positions have been computed and convoluted with the slit function, the summed intensities are then plotted to produce the spectrum.

One of the advantages of a computer program which will calculate and plot band contours is the facility to plot the sub-branches separately as well as the composite band. In certain cases prominent features of a band can be due almost entirely to a specific one of the seven sub-branches which contribute to each of the low and high frequency wings of a triply degenerate Raman band. This is illustrated by Figures 1-4, which show how band contours change as zeta varies from +0.9 to -0.9. The contributions of some of the different sub-branches can be

---

* Placzek and Teller's relation is $\Delta\nu = (BkT/hc)^{\frac{1}{2}}x$ [1]. This is the relation used by Clark and Rippon [20] and by Sportouch and Gaufrès [21].

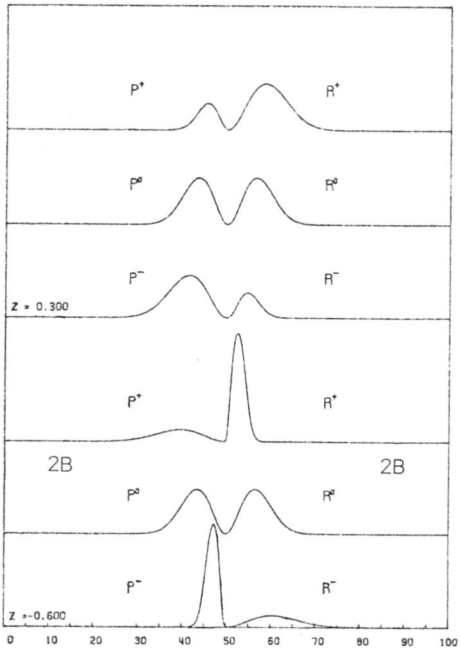

Figure 1. Sublevel components of P and R branches for ζ = -0.6 and ζ = +0.3.

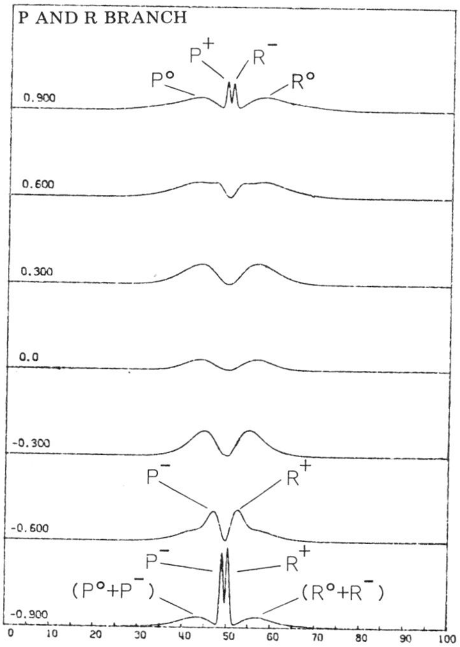

Figure 2. P and R branch components for different values of ζ.

traced to the composite band. These contours are computed with B' = B" = 0.05 cm^{-1}, a spectral slit width of 1.0 cm^{-1}, and a temperature of 300 K. Figure 1 shows the plus, minus, and zero components of the P and R branches for two typical values of zeta. Note that the P^--R^- and P^+-R^+ components are mirror images of each other in each case. Consequently, their composites (Figure 2) are symmetric, but the small changes in B that occur in the heavy symmetric tops would not alter these contours significantly, so these are adequate representations of real bands. In the P and R branches, where two distinguishable maxima occur for some values of zeta, the major sub-branch contributors are indicated. The Q branches in Figure 3 have a very simple structure because ΔJ = 0, but the relative positions of the Q^+ and Q^- sub-branches depend on the sign of zeta. Figures 2-4 were plotted on the same intensity scale so they may be compared directly. The S and O branches are shown in reference [18], but are not reproduced here because they have only weak and very broad contours. The strongest are for ζ = -0.9 and 0.6, and these have approximately the same intensity and shape as the P and R branches for ζ = +0.3 and 0.0, respectively. The sub-branches which make the major contributions to the most prominent features of the composite bands are indicated in Figure 4. These can be verified by comparison with the previous figures, and reference to Table 1, which gives the line frequencies of the 15 sub-branches as a function of J.

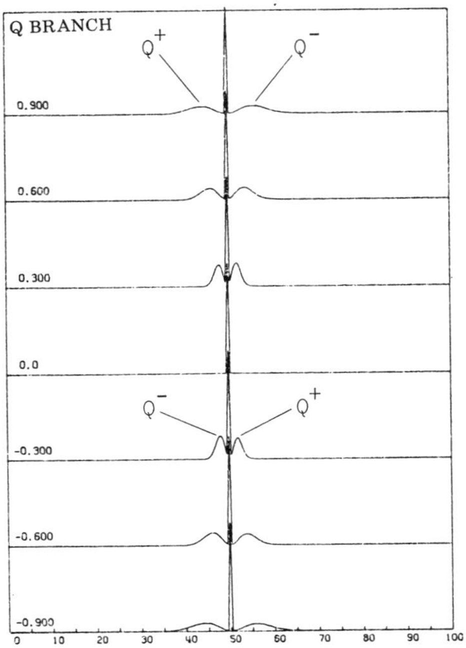

Figure 3. Q branch components for different values of ζ.

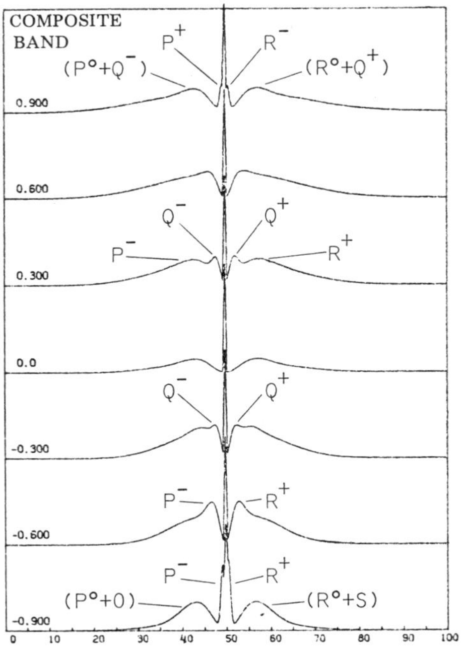

Figure 4. Composite band for different values of ζ.

(Figures 1-4 reproduced with permission from ref. 18.)

The separations of both of these maxima can be used to estimate zeta values [18], but there are regions where the separation is only weakly dependent on zeta, especially from -0.6 to -0.4 and +0.5 to +0.7. Furthermore, the OP-RS separations are not single-valued functions of zeta, so some additional information may be necessary to select the correct estimate of zeta.

Extensive observations of band contours of both tetrahedral and octahedral spherical tops have been made by Clark and Rippon [20], and by Bosworth, et al. [23]. They have determined the Coriolis coupling coefficients from the OP-RS separations using the analytical methods they have outlined and by using computer simulation of band contours as outlined by Masri and Fletcher [18]. Comparison of zeta values determined from Raman band contours with those obtained from infrared data of CF_4, SiF_4, and GeF_4 show excellent agreement.

The determination of zeta values from Raman band contours has also been discussed by Sportouch and Gaufrès [21], who observed the doubly and triply degenerate bands of CF_4, SiF_4 and SF_6. These authors have also used a computer program to calculate the band contours, and have taken into account the small changes in B which occur for CF_4 and SiF_4. A change of a few tenths of a percent in B for these molecules is enough to produce a distinct asymmetry, but slit widths of 2 to 4 cm^{-1}

Table 1. Frequency of displacements of sub-branch lines from the band origins, ν_0.

Sub-branch	$(\nu - \nu_0)$ (cm^{-1})
S^+	$6B(1 + \zeta_i) + 4B(1 + \zeta_i/2)J$
S^0	$6B + 4BJ$
S^-	$6B(1 - 2\zeta_i/3) + 4B(1 - \zeta_i/2)J$
R^+	$2B(1 + 2\zeta_i) + 2B(1 + \zeta_i)J$
R^0	$2B + 2BJ$
R^-	$2B(1 - \zeta_i) + 2B(1 - \zeta_i)J$
Q^+	$2B\zeta_i + 2B\zeta_i J$
Q^0	0
Q^-	$- 2B\zeta_i J$
P^+	$- 2B(1 - \zeta_i)J$
P^0	$- 2BJ$
P^-	$2B\zeta_i - 2B(1 - \zeta_i)J$
O^+	$2B(1 - \zeta_i) - 4B(1 - \zeta_i/2)J$
O^0	$2B - 4BJ$
O^-	$2B(1 + 2\zeta_i) - 4B(1 + \zeta_i/2)J$

were used in the observations, so no other detail was observed.

In the case of these smallest spherical tops, it would be worthwhile to observe the band contours with slit widths of less than 1 cm^{-1}. For instance, in ν_4 of SiF_4 a sharp splitting of the Q branch occurs and should be observed with slits of 0.8 cm^{-1} or less, and similar features should be observed in other bands. For changes in B as large as 0.5% some sub-branch heads begin to appear in computed bands.

While the measurement of zeta values from unresolved Raman band contours has produced encouraging results, caution should be used in taking the results at face value. The accuracy may be no better than ±0.05, and this can cause a serious error in calculation when the zeta values are used to help refine a force field. The problem is that measurement of OP-RS separations is not an accurate process in itself, and perturbations such as changes in B values, isotope effects, and the effects of hot bands are generally not taken into account. If the zeta sum rule is used because one of the bands cannot be observed, the uncertainty is simply compounded. McDowell and Asprey [26] have studied the effect of hot bands on the observed zeta values for WF_6 and find $\zeta_3 = 0.123$ by measuring OP-RS spacings at many temperatures from 190° to 310°K and using the fact that the spacing is a linear function of $T^{\frac{1}{2}}$. Two previously reported values for ζ_3 are 0.20 ±0.03 [11] and 0.12 ±0.01 [16].

3. SYMMETRIC TOPS

The band contours of symmetric top molecules are more difficult to handle than spherical tops because much more structure is generally observable, and there are more perturbations which produce effects that can be observed even at moderate resolution. This means that conclusions drawn from smooth band contours are much less reliable in symmetric tops, but as resolution is increased and more detail is observable, much more information can be obtained from fitting band contours, even though rotational assignments are not possible. Band contours of symmetric tops for infrared and their relationship to Coriolis constants were discussed first by Gerhard and Dennison [2], Edgell and Moynihan [27], Edgell and Valentine [28], and by Hoskins [7]. More recently the effects of strong perturbations such as the ℓ-type resonance and second order Coriolis coupling have been investigated and taken into account in contour calculations by Mills et al. [29-32], Masri and Blass [33-34], and Duncan et al. [35-37].

While the early work on band contours dealt with analytical techniques to obtain Coriolis constants from P-R separations, the development of computer programs has made it possible to calculate the complete band with no more than minor approximations. For a symmetric top the selection rules for a parallel band [25] are $\Delta K = 0$ and $\Delta J = \pm 1$ for $K = 0$, and $\Delta J = 0, \pm 1$ for $K \neq 0$. If $[(A' - B') - (A'' - B'')]$ is sufficiently small this produces a well resolved J structure at high resolution but no resolvable K structure. At very high resolution the K structure of the Q branch is observable [38,39]. The selection rules for a perpendicular band are $\Delta K = \pm 1$ and $\Delta J = 0, \pm 1$, and the main features of the band are the Q branches of the subbands. These positions are given by

$$\nu_o^{sub} = \nu_o + [A'(1 - 2\zeta) - B'] \pm 2[A'(1 - \zeta) - B']K + [(A' - B') - (A'' - B'')]K^2 \qquad (9)$$

where centrifugal distortion and perturbations other than first order Coriolis coupling have been neglected. These Q branches are broadened by the term $(B' - B'')J(J + 1)$ which contributes to the $\Delta J = 0$ transitions. The complete band is synthesised by calculating the positions and intensities of all lines which have a measurable intensity, applying an appropriate line shape function to each line, convoluting with a slit function corresponding to the conditions used in scanning the experimental spectrum, and plotting the summed results. In this way it is possible to simulate any observed spectrum with scanning conditions varying from the use of very narrow slits to achieve high resolution to the use of wide slits which obscure most of the spectral details. The effects of perturbations such as second order Coriolis coupling and ℓ-type resonance, which often produce strong intensity perturbations, can be included in such calculations. The fitting of a band contour normally requires a trial and error procedure, in which one begins with the best parameters available and alters them systematically to produce the best fit of the observed spectrum. Both frequency and intensity are used as criteria in judging the best fit.

The analysis of the band contours of the ν_3 and ν_4 fundamentals of BF_3 by Duncan [40] shows what can be achieved by fitting a partially resolved band contour. Using the ν_3 band of $^{10}BF_3$ observed by Dreska, et al. [41], Duncan found the best fit was achieved by using the value of B" = 0.3527 cm^{-1} determined by Nielsen [42] from ν_2, and assuming that A" = B"/2 and A' = B'/2. This left just two independent parameters to be used in fitting the observed band contour, (B" - B') and ζ_3. Refinement of the fitting showed the best results with (B" - B') = 0.0012 ±0.0002 and ζ_3 = +0.81. Comparison of Duncan's ζ_3 with the value of 0.7808 ±0.0036 found by Brown and Overend [43] from high resolution data for $2\nu_3$ might be taken as an indication of the reliability of the zeta value obtained by fitting the band contour. However, more recent high resolution work by Ginn et al. [44] on ν_4 of $^{10}BF_3$ and $^{11}BF_3$ yield values of ζ_4 = -0.809 and -0.789 for the two isotopic species, respectively. The zeta sum rule requires that $\zeta_3 = -\zeta_4$ in planar symmetric tops, so an explanation of the low value of ζ_3 obtained from $2\nu_3$ is needed. Ginn et al. [44] examined the data for the ν_4 bands carefully for signs of additional perturbations, such as $\ell(2,2)$ resonance and concluded that only negligible effects could be present. Therefore, the value of ζ_3 found from $2\nu_3$ is an effective value containing the effects of resonance with $\nu_1 + \nu_3 + \nu_4$.

The ν_4 fundamental of $^{11}BF_3$ observed at low resolution shows only a broad smooth band contour [45] and a satisfactory determination of zeta can hardly be made by measuring the P-R separation. However, the width of the Q branch is a rather sensitive function of zeta, and an examination of this gives a value of -0.81 ±0.03 for ζ_4 [40]. Good agreement such as in this case between zetas from band contours and high resolution data can occur fortuitously. It should be noted that the presence of the ^{10}B isotope can produce a significant broadening of the Q branch, although in this case it seems to be negligible.

The effects of second order Coriolis coupling have been observed and discussed for many molecules; i.e. formaldehyde [46,47], allene [48-51], methane [52,53], methyl cyanide [54-56], methyl acetylene [57], and methyl halides [58-62]. It is responsible for the appearance in the infrared of the forbidden ν_2 fundamental of methane [52-53] and germane [63]. The necessary condition for this to occur (Jahn's rule) is that the direct product of the species of two interacting levels contains the species of one of the rotations. The first thorough analysis of $A_1 \times E$ second order Coriolis coupling in symmetric tops by calculation of the observed band contours was carried out by di Lauro and Mills for the ν_2 and ν_5 fundamentals of CH_3F [29], which have band centres at 1460 and 1468 cm^{-1}, respectively. Jones et al. [58] observed this region with a resolution of 0.25 cm^{-1} and found intensity anomalies in the Q branches of ν_5. They also found it necessary to use two different sets of coefficients in the equations required to fit the ^RR lines and the ^PP lines in the sub-bands on the two sides of $^RQ_0$, and they attributed these anomalies to Coriolis interactions. Similar anomalies had been observed for lines in degenerate CH_3 deformations of other methyl halides [59-62] and methyl cyanide [54-56]. These perturbations were not satisfactorily accounted for until di Lauro and Mills [29] performed an exact calculation of line positions and intensities and compared computer calculated band contours with the observed spectrum. The results of their calculation are shown in Figure 5.

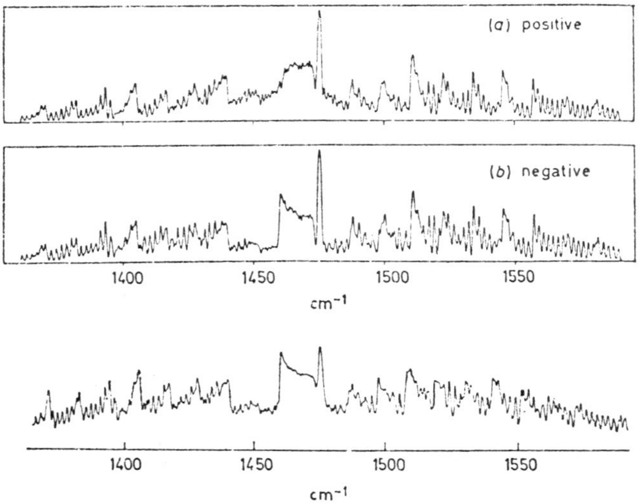

Figure 5. Observed and calculated spectra for the ν_2 and ν_5 region of CH_3F (reproduced with permission from ref. 29).

The perturbation is dominated by an off-diagonal element in the Hamiltonian,

$$H_{s,t} = 2^{\frac{1}{2}} B \zeta^{y}_{s,ta} [J(J+1) - k(k \pm 1)]^{\frac{1}{2}} \tag{10}$$

which connects the symmetric vibration, ν_2, with one component of the degenerate vibration, ν_5. The two vibrations are identified by s and t, respectively, and k is the signed quantum number of the component of angular momentum along the figure axis of the molecule. In addition to accounting for the appearance of the band, the magnitude of $\zeta^{y}_{2,5a}$ is established, and the relative sign of the dipole moment derivatives $(\partial\mu^z/\partial Q_2)$ and $(\partial\mu^x/\partial Q_{5a})$, is determined. The correct intensity perturbation is calculated when the two dipole moment derivatives have the same sign and $\zeta^{y}_{2,5a} = -0.602$. Perturbation by second order Coriolis coupling is possible only if Jahn's rule is satisfied, but in addition, for the effects to be observable, it is necessary that the two vibrational levels be nearly degenerate. That is $(\nu_s - \nu_t)$ must be small because it appears as a resonance denominator in a cross term of the Hamiltonian.

The occurrence of ℓ-type resonance, which is also a second order Coriolis effect, is responsible for a prominent and frequently observed intensity perturbation which produces a split Q branch in many symmetric tops. The theory was discussed thoroughly by Mills [32] and by Cartwright and Mills [30], who demonstrated several successful applications with cyclopropane bands. The ℓ-doubling of energy levels in a symmetric top arises from the vibrational angular momentum in a degenerate mode, and it is analogous (but more complex) to the familiar ℓ-doubling which occurs in linear molecules [64]. In linear molecules the ℓ-doubling effect increases with J and the splitting is expressed

$$\Delta\nu = \bar{q}J(J + 1) \tag{11}$$

but in symmetric tops it is observed only in the $K = 1 \leftarrow 0$ sub-band, since the splitting of the sub-levels decreases with increasing K. The splitting arises here because the $^RQ_0$ branch and the $^RP_0$ and $^RR_0$ branches terminate in different components of the ℓ-doubled levels. This splitting is observed only at very high resolution [65].

The effect of ℓ-type resonance becomes easily observable if $[A'(1 - \zeta^z) - B'] \simeq 0$ in prolate tops or if $[C'(1 - \zeta^z) - B'] \simeq 0$ in oblate tops. This causes the Q branches of a perpendicular band to be clustered together and give the appearance of a parallel band. Bands of this type are called "pseudoparallel" and they are observed most often in oblate tops [33,34]. The effect of the resonance is to cause a distinct "hole" near the centre of the Q branch region of these pseudoparallel bands, as seen in the ν_{11} band of cyclopropane [33-35] shown in Figure 6.

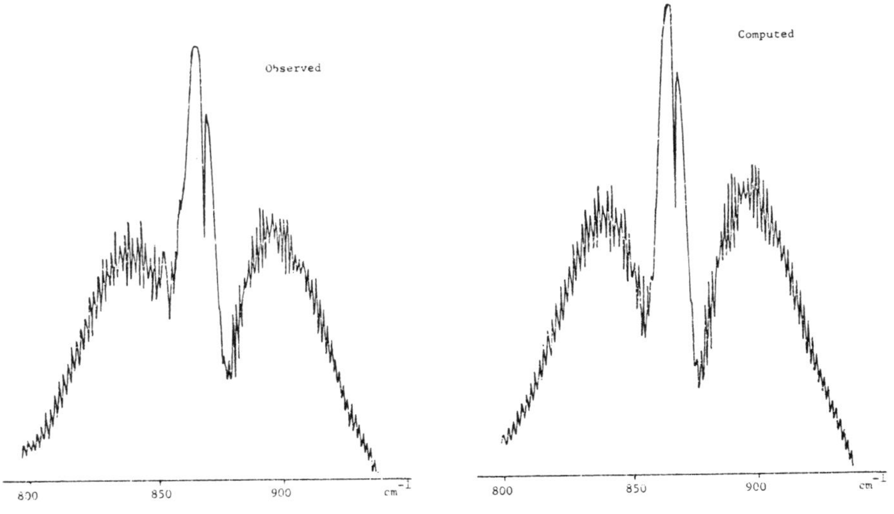

Figure 6. Observed and computed spectra for cyclopropane ν_{11} (reproduced with permission from ref. 33).

The interaction occurs between the rovibrational levels (v_t, ℓ_t, J, K) and (v_t, $\ell_t \pm 2$, J, $K \pm 2$) and is frequently called ℓ(2,2) resonance. The effect is observed in the species E' bands of cyclopropane [33,35,36], and in some of the E bands of NF_3 and CHF_3 [34].

Raman band contours

As in the case of spherical tops, the bands of symmetric tops are composites of many sub-bands. In some cases two distinct maxima appear on either side of the band centre, and in other cases only one or none appear. The theory of the energy levels, selection rules and line intensities for the doubly degenerate bands of symmetric tops has been given by Placzek and Teller [1] and this has been applied to the computer simulation of the Raman bands by Masri and Fletcher [66]. In these bands, the Raman selection rules for the rotational quantum number J are quite general. These are

$$\Delta J = 0, \pm 1, \pm 2, \quad \text{with} \quad J' + J'' > 2.$$

However, the Raman selection rules for the rotational quantum number K and the associated ones for the vibrational angular momentum quantum number ℓ may take one of the following forms (for C_3 type groups):

(i) $\Delta K = \pm 1$ with $\Delta\ell = \pm 1$ or $\Delta\ell = \mp 1$

(ii) $\Delta K = \pm 2$ with $\Delta\ell = \mp 1$ or $\Delta\ell = \pm 1$

(iii) $\Delta K = \pm 1$ with $\Delta\ell = \pm 1$ and

$\Delta K = \pm 2$ with $\Delta\ell = \mp 1$

depending upon the particular point group to which the molecule belongs. A complete tabulation, giving the correct form for each point group together with rules for associating the correct ΔK and $\Delta\ell$ selection rules in cases (i) and (ii), has been given by Mills [67]. In particular, case (i) applies to the E" bands of D_{3h}, case (ii) applies to the E' bands of D_{3h}, and case (iii) applies to the E species Raman bands of molecules of the point group C_{3v}. For these latter bands, both sets of sub-bands, corresponding to $\Delta K = \pm 1$ and to $\Delta K = \pm 2$, appear in the spectrum but their intensities are determined by different components of the polarisability tensor and so their relative intensity may vary over a wide range. The case of $\Delta K = \pm 1$ and $\Delta\ell = \pm 1$ will be recognised as the general selection rules for the degenerate infrared bands of symmetric tops of any point group. The cases where $\Delta K = \pm 1$ and $\Delta K = \pm 2$ occur separately will be discussed below.

The general expression for the intensity of a line in a doubly degenerate Raman band is

$$I = (C_{\Delta K})(\nu_0 - \nu_i)^4(2J + 1)g_{KJ}B_{KJ}\exp[-hcF(K,J)/kT] \quad (11)$$

except for a constant factor. $C_{\Delta K}$ is a constant giving the relative intensity of the $\Delta K = \pm 1$ to the $\Delta K = \pm 2$ sub-bands when both sets appear in the same band, such as for C_{3v} molecules. F(K,J) is the rotational energy term of the ground state, g_{KJ} is the statistical weight of the state, B_{KJ} is an intensity factor, and $(\nu_0 - \nu_i)$ is the frequency of the Raman radiation. If the influence of nuclear spin is neglected for our particular application to molecules with large moments of inertia then g_{KJ} becomes simply the K degeneracy of the levels, which is 1 for $K = 0$ and 2 for $K \neq 0$. Molecules with single 3-fold axis of symmetry will have a strong, weak, weak, strong intensity alternation, but if these are observable a more detailed analysis may be done without depending on the band shape analysis.

The factors B_{KJ} have been given in a condensed form by Placzek and Teller and have occasionally been misquoted. The correct B_{KJ} factors (including the statistical weight factor, $(2J+1)$), together with the full restrictions on K and J have been clearly and explicitly given by Masri and Williams [68]. The K and J restrictions are listed using Herzberg's notation, and they must be obeyed in addition to the general restriction $K \leqslant J$. The B_{KJ} factors have also been given by Gaufrès and Sportouch [69].

The rotational energies of the ground and first excited states of degenerate bands are given by

$$F''(J,K) = B''J(J+1) + (A'' - B'')K^2 \tag{13}$$

$$F'(J,K) = B'J(J+1) + (A' - B')K^2 + 2A'\zeta K \tag{14}$$

The selection rules $\Delta K = \pm 1$ and $\Delta K = \pm 2$ give two sets of Q branches (sub-band origins),

$$\nu_o^{R,P} = \nu_o + [A'(1-\zeta)^2 - B'] \pm 2[A(1-\zeta) - B']K + [(A' - A'') - (B' - B'')]K^2 \tag{15}$$

$$\nu_o^{S,O} = \nu_o + [A'(2+\zeta)^2 - 4B'] \pm 4[A'(1+\zeta/2) - B']K + [(A' - A'') - (B' - B'')]K^2 \tag{16}$$

The terms for the additional allowed transitions, $\Delta J = 0, \pm 1, \pm 2$, which are given by Masri and Fletcher [66], must be added to each of these equations to produce the complete sub-band. The band contours are calculated as described previously, with line shapes functions included to provide a more correct representation of the observed band if sufficient resolution is available to show appreciable structural detail. The R, P, S and O sub-bands are computed separately, as are their component sub-branches (i.e. ${}^RQ$, ${}^RR$, ${}^RP$, ${}^RS$ and ${}^RO$), so that each sub-branch, sub-band, or their composite may be plotted.

For C_{3v} molecules where $\Delta K = \pm 1, \pm 2$, each side of the band origin is a composite of 10 sub-branches, one for each combination of ΔK and $\Delta J = 0, \pm 1, \pm 2$. Since the R-P and S-O sub-bands depend on different components of the polarisability tensor, and these have unknown relative values, an examination was made of the cases where either $\Delta K = \pm 1$ or $\Delta K = \pm 2$. These correspond to the species E" and E' bands, respectively of point group D_{3h}. As in the case of spherical tops, a reduced frequency parameter is necessary to relate zeta in a general way to the moments of inertia and the OP-RS separation (cf. Equation (7)).

$$\Delta\nu\ (\mathrm{cm}^{-1}) = (\beta B)^{\frac{1}{2}} \Delta X \tag{17}$$

where $\beta = (A/B - 1)$, and again, this is valid only if $hc\beta B \ll kT$.

The OP-RS separation is a function of β and ζ. Calculations were made for values of zeta ranging from -1.0 to +1.0, for β varying from -0.5 to +3.0 and with slit widths of 2.0 cm^{-1}. For the bands with $\Delta K = \pm 2$, only one set of OP-RS maxima occur [66], and reliability of estimated zeta values is not very good except for oblate tops with negative, or at most small positive values of zeta. The bands with $\Delta K = \pm 1$ show two maxima in some cases and a much more complex behaviour with regard to the inner maxima. The sub-band origins are given by Equation (15), and in fact are identical with those for an infrared band. Examination of the data from the simulated bands [66] and Equation (15) shows that separation of the inner maxima is zero at two points for every molecule. When $\zeta = +1$, the coefficient of the K term in Equation (15) is $\pm 2B$, so the sub-band origins are so closely spaced that only a central maximum is observed. As zeta decreases a splitting occurs, which first has a negative sign, and approaches a maximum absolute value at some value of zeta, which depends upon β, and then decreases to zero. The splitting is necessarily zero when zeta reaches a value that causes $[A'(1 - \zeta) - B']$ to vanish. This accidental coincidence, or a near coincidence, of sub-band origins produces pseudoparallel bands. As zeta decreases from this point the OP-RS maxima separate again, but in a positive manner. The OP-RS separations have been used successfully to estimate zeta from unresolved band contours [66,70-72] but the reliability is about the same as for the corresponding infrared bands. For infrared inactive bands for which no rotational detail can be observed, the Raman band contour is the only source of zeta. The uncertainty in zetas determined in this way is unfortunate because they are needed as constraints in evaluating force fields.

The gas phase Raman spectra of cyclopropane (C_3H_6 and C_3D_6) have been examined by Daunt and Shurvell [73] with spectral slit widths ranging from 0.5 to 5 cm^{-1}. Five perpendicular bands of C_3H_6 and six perpendicular bands of C_3D_6 were analysed by computer simulation with excellent results. The derived molecular constants α^B and α^C, were in satisfactory agreement with values obtained from infrared data or other sources. While the numerical values deduced are not highly accurate, the comparison of observed contours is a valuable aid in deducing and confirming band assignments.

Clark and Rippon [70] examined the contours of POF_3, $POCl_3$, VOF_3, $VOCl_3$, $PSCl_3$ and $FClO_3$ in order to evaluate the Coriolis coefficients and, in addition, to estimate the relative contributions of the polarisability components of the $\Delta K = \pm 1$ and $\Delta K = \pm 2$ transitions to the entire band. This was done by selecting a value of zeta from the literature, either from infrared data or from force constant calculations, and then varying both zeta and the relative intensities of the $\Delta K = \pm 1$ and $\Delta K = \pm 2$ transitions until the best fit of the observed and computed band contour was obtained. The value of $\delta = (\Delta K = \pm 1)/(\Delta K = \pm 2)$ varied from 1.0 for POF_3 and $FClO_3$ to 1.5 for $VOCl_3$ and 2.0 for $PSCl_3$, and the uncertainties in zetas varied from ± 0.02 to ± 0.10. Complete data were not obtainable for all of these molecules, so these large uncertainties are not unexpected. The zetas are in acceptable agreement with infrared values and they appear to be of equal reliability, although some are not well established by either method.

A study of the Raman band contours of ^{12}C and ^{13}C isotopes of CH_3F and $CHCl_3$ was made by Clark et al. [72] in order to obtain additional information from partially resolved contours. Not all of the zeta constants compare well with those from high resolution infrared work, but the δ ratios seem to be well determined and very significant. The band intensity is very slightly dominated by the $\Delta K = \pm 2$ transitions in ν_4 of both CH_3F and $CHCl_3$, but the ν_5 and ν_6 are definitely dominated by the $\Delta K = \pm 1$ transitions in CH_3F. In $CHCl_3$, however, the intensities of ν_5 and ν_6 are primarily due to the $\Delta K = \pm 2$ transitions.

The first observation of a second order Coriolis resonance in a high resolution Raman spectrum has just been reported by Escribano, et al. [74]. The ν_2, ν_5 region, which is well known from infrared work [29], was recorded with slit widths of 0.15 to 0.20 cm^{-1}. The analysis was made using a computer program which handles strong Coriolis resonances, ℓ-doubling and ℓ-type resonance effects, and centrifugal distortions. Details of the results cannot be given here, but this surely represents a significant achievement in the observation and analysis of higher order perturbations in the Raman spectra of gases.

4. ASYMMETRIC TOPS

The band envelopes produced by the infrared active vibrations of asymmetric tops were first described and classified by Badger and Zumwalt [3]. Following Nielsen's notation, they defined type A bands as those produced by vibrations with dipole moment changes parallel to the axis of the smallest moment of inertia. Type B bands are produced by vibrations with dipole moment changes parallel to the axis of the intermediate moment of inertia, and type C bands represent vibrations with dipole moment changes parallel to the axis of the largest moment of inertia. As a general rule the type A bands may be described as having well formed P, Q and R branches, resembling parallel bands of symmetric tops. Type B bands have a central minimum with two distinct peaks on either side of ν_0, and they may exhibit shoulders or even distinct peaks on the extreme sides of the two central peaks. Type C bands typically have very strong Q branches and relatively weak P and R branches, which may appear only as narrow shoulders on a rather wide Q branch. However, what is actually seen depends upon the size and asymmetry of the molecule examined. Perturbations of the types described previously further complicate the observed envelopes.

Refinement of the calculation of P-R separations and other details in asymmetric tops have been described by McDowell [75], Seth-Paul and De Meyer [76], and by Seth-Paul and Dijkstra [77]. The latter authors have noted that the branch spacings of type A ($\|$), type B ($\perp$), and type C ($\perp$) bands, which correspond to prolate tops, are in general different from those found in type A ($\perp$), type B ($\perp$), and type C ($\|$), which are more closely related to oblate tops. They have applied this to a large number of molecules and given semi-empirical expressions for calculating the P-R separations. The Q-Q separations in type B bands has been discussed in detail by McDowell [75] and a semi-empirical equation has been applied to molecules with κ ranging from -0.963 to +0.670*. Seth-Paul and De Meyer [76] have also fitted the Q-Q and P-R separations to semi-empirical equations and applied them to oblate asymmetric tops. Seth-Paul [78] has reviewed classical and modern methods of calculating P-R separations in symmetric and asymmetric tops.

Band contour calculations

The complex nature of asymmetric top energy levels plus the numerous perturbations (second order Coriolis coupling, accidental resonances, changes in moments of inertia, etc.) make the use of computed band contours a necessary and powerful tool for analysis, but one which must be used with care. A number of programs for calculating bands for asymmetric tops have been described [81-84], and an excellent critical review of band contour methods has been given by Kidd and King [85].

Nakagawa, et al. [86] have modified the band contour program written by Ueda and Shimanouchi [84] to account for centrifugal distortion in the upper and lower states. The introduction of these small perturbations is necessary for a light molecule in which structure may be resolved. The ν_2 and ν_3 fundamentals of H_2CO were analysed, using spectra recorded with resolution varying from 0.3 to 0.6 cm^{-1}. Ground state and upper state rotational constants and centrifugal distortion constants from microwave spectra were used for simulating the observed spectrum. The identification of second order Coriolis interactions with ν_4 and ν_6 in these observed bands demonstrates the effectiveness of using computer simulation, especially when considerable structure is observed. In this case, to be sure, the resolution was not enough to show more than blends of many lines.

The first use of a computer synthesised spectrum of a Raman band of an asymmetric top was recently described by Duncan et al. [87]. They have observed the ν_5 fundamental of ethylene with a resolution of 0.3 cm^{-1}, and a program written by Hills [88] was used to simulate the band. The band centre of ν_5 was accurately relocated at 3083.19 $\pm$0.10 cm^{-1}. The symmetric top approximation was used for $K > 2$, and this seems to be justified for the resolution which was used.

Acknowledgment

The author would like to acknowledge his sincere appreciation to Dr. Steve Daunt for many helpful discussions and for his critical reading of the manuscript.

---

* $\kappa = (2B - A - C)/(A - C)$ is the commonly used asymmetry parameter. A, B and C are the rotational constants corresponding to the three principle axes of the molecule ($A > B > C$). κ varies from -1 to +1, with $\kappa = -1$ representing the limiting case of the prolate top and $\kappa = +1$ representing the oblate top [79,80].

5. REFERENCES

1. G. Placzek and E. Teller, Z. Physik. 81 (1933) 209.
2. S.L. Gerhard and D.M. Dennison, Phys. Rev., 43 (1933) 197.
3. R.M. Badger and L.R. Zumwalt, Phys. Rev., 6 (1938) 711.
4. E. Teller, Hand und Jahrbuch der Chemische Physik, Akademische Verlagsgesellshaft, Leipzig, 1934, Vol. 9, Ch. 2, p. 43.
5. M. Johnston and D.M. Dennison, Phys. Rev., 48 (1935) 868.
6. W.F. Edgell and R.E. Moynihan, J. Chem. Phys., 27 (1957) 155.
7. L.C. Hoskins, J. Chem. Phys., 45 (1966) 4594.
8. I.W. Levin and S. Abramowitz, J. Chem. Phys., 44 (1966) 2562.
9. I.W. Levin and S. Abramowitz, J. Chem. Phys., 50 (1969) 4860.
10. I.W. Levin and S. Abramowitz, Inorg. Chem., 5 (1966) 2024; 6 (1967) 538.
11. S. Abramowitz and I.W. Levin, J. Chem. Phys., 44 (1966) 3353.
12. D.C. McKean, Spectrochim. Acta, 22 (1966) 269.
13. J.L. Duncan and I.M. Mills, Spectrochim. Acta, 20 (1964) 1089.
14. H. Bürger and A. Ruoff, Spectrochim. Acta, 24A (1968) 1863.
15. A. Müller and B. Krebs, J. Mol. Spectr., 26 (1968) 136.
16. H. Kim, P.A. Souder and H.H. Claassen, J. Mol. Spectr., 26 (1968) 46.
17. T.D. Goldfarb, J. Chem. Phys., 39 (1963) 2860.
18. F.N. Masri and W.H. Fletcher, J. Chem. Phys., 52 (1970) 5759.
19. R.S. McDowell, J. Chem. Phys., 43 (1965) 319.
20. R.J.H. Clark and D.M. Rippon, J. Mol. Spectr., 44 (1972) 479.
21. S. Sportouch and R. Gaufrès, J. Chim. Phys., 69 (1972) 470.
22. A. Maki, E.K. Plyler and R. Thibault, J. Chem. Phys., 37 (1962) 1899.
23. Y.M. Bosworth, R.J.H. Clark and D.M. Rippon, J. Mol. Spectr., 46 (1973) 240.
24. J. Herranz and B.P. Stoicheff, J. Mol. Spectr., 10 (1963) 488.
25. G. Herzberg, Infrared and Raman Spectra of Polyatomic Molecules, D. Van Nostrand, New York, 1945, Ch. 4, pp. 416-426.
26. R.S. McDowell and L.B. Asprey, J. Mol. Spectr., 48 (1973) 254.
27. W.F. Edgell and R.E. Moynihan, J. Chem. Phys., 45 (1966) 1205.
28. R.W. Valentine, Ph.D. dissertation, Purdue University, 1957.
29. C. di Lauro and I.M. Mills, J. Mol. Spectr., 21 (1966) 386.
30. G.J. Cartwright and I.M. Mills, J. Mol. Spectr., 34 (1970) 415.
31. I.M. Mills, Pure and Appl. Chem., 11 (1965) 325.
32. I.M. Mills, Pure and Appl. Chem., 18 (1969) 285.
33. F.N. Masri and W.E. Blass, J. Mol. Spectr., 39 (1971) 21.
34. F.N. Masri and W.E. Blass, J. Mol. Spectr., 39 (1971) 98.
35. J.L. Duncan, J. Mol. Spectr., 25 (1968) 451.

36. J.L. Duncan and D. Ellis, J. Mol. Spectr., 28 (1968) 540.
37. J.L. Duncan and I.J. Wright, Mol. Phys., 20 (1971) 673.
38. M. Morrillion-Chapey and G. Graner, J. Mol. Spectr., 31 (1969) 155.
39. G. Graner, J. Mol. Spectr., 51 (1974) 238.
40. J.L. Duncan, J. Mol. Spectr., 22 (1967) 247.
41. N. Dreska, K.N. Rao and L.H. Jones, J. Mol. Spectr., 18 (1965) 404.
42. A.H. Nielsen, J. Chem. Phys., 22 (1954) 659.
43. C.W. Brown and J. Overend, Can. J. Phys., 46 (1968) 977.
44. S.G.W. Ginn, D. Johansen and J. Overend, J. Mol. Spectr., 36 (1970) 448.
45. I.W. Levin and S. Abramowitz, J. Chem. Phys., 43 (1965) 4213.
46. T. Nakagawa and Y. Morino, J. Mol. Spectr.,38 (1971) 84.
47. K. Yamada , T. Nakagawa, K. Kuchitsu and Y. Morino, J. Mol. Spectr., 38 (1971) 70.
48. I.M. Mills, W.L. Smith and J.L. Duncan, J. Mol. Spectr., 16 (1965) 349.
49. J. Overend and B.L. Crawford, J. Chem. Phys., 29 (1958) 1002.
50. J.L. Duncan, D. Ellis, I.J. Wright, J.M.R. Stone and I.M. Mills, J. Mol. Spectr., 38 (1971) 508.
51. H.H. Nielsen, Rev. Mod. Phys., 23 (1951) 90.
52. M. Dang Nhu, Ann. Phys. (Paris), 4 (1969) 273.
53. H.M. Kaylor and A.H.Nielsen, J. Chem. Phys., 23 (1955) 2139.
54. J.L. Duncan, D. Ellis and I.J. Wright, Mol. Phys., 20 (1971) 673.
55. F.N. Masri, J.L. Duncan and G.K. Speirs, J. Mol. Spectr., 47 (1973) 163.
56. S. Kondo and W.B. Person, J. Mol. Spectr., 52 (1974) 287.
57. J.L. Duncan, I.J. Wright and D. Ellis, J. Mol. Spectr., 37 (1971) 394.
58. E.W. Jones, R.J.L. Popplewell and H.W. Thompson, Proc. Roy. Soc., A290 (1966) 490.
59. E.W. Jones, R.J.L. Popplewell and H.W. Thompson, Spectrochim. Acta, 22 (1966) 647.
60. E.W. Jones, R.J.L. Popplewell and H.W. Thompson, Spectrochem. Acta, 22 (1966) 669.
61. T.M. Holladay and A.H. Nielsen, J. Mol. Spectr., 14 (1964) 371.
62. G. Amat and H.H. Nielsen, J. Mol. Spectr., 23 (1967) 359.
63. R.J. Corice, K. Fox and W.H. Fletcher, J. Mol. Spectr., 41 (1972) 95.
64. G. Herzberg, Infrared and Raman Spectra of Polyatomic Molecules, D. van Nostrand, New York, 1945, pp. 377, 451.
65. C. Betrencourt-Stirnemann and G. Graner, J. Mol. Spectr., 51 (1974) 216.
66. F.N. Masri and W.H. Fletcher, J. Raman Spectr., 1 (1973) 221.
67. I.M. Mills, Mol. Phys., 8 (1964) 363.
68. F.N. Masri and I.R. Williams, Computer Phys. Comm., 1 (1970) 349.
69. R. Gaufrès and S. Sportouch, J. Mol. Spectr., 39 (1971) 527.
70. R.J.H. Clark and D.M. Rippon, Mol. Phys., 28 (1974) 305.

71. R.J.H. Clark and O.H. Ellestad, J. Mol. Spectr., 56 (1975) 386.
72. R.J.H. Clark, O.H. Ellestad and R. Escribano, Mol. Phys., 31 (1976) 651.
73. S.J. Daunt and H.F. Shurvell, J. Raman Spectr., 2 (1974) 463.
74. R. Escribano, I.M. Mills and S. Brodersen, J. Mol. Spectr., 61 (1976) 249.
75. R.S. McDowell, Spectrochim. Acta, 30A (1974) 1271.
76. W.A. Seth-Paul and H. DeMeyer, Spectrochim. Acta, 25A (1969) 1671.
77. W.A. Seth-Paul and G. Dijkstra, Spectrochim. Acta, 23A (1967) 2861.
78. W.A. Seth-Paul, J. Mol. Structure, 3 (1969) 403.
79. G. Herzberg, Infrared and Raman Spectra of Polyatomic Molecules, D. van Nostrand, New York, 1945, Ch. 1, p. 44-50.
80. G.W. King, R.M. Hainer and P.C. Cross, J. Chem. Phys., 11 (1943) 27.
81. J.E. Parkin, J. Mol. Spectr., 15 (1965) 483.
82. J. Christoffersen, J.M. Hollas and G.H. Kirby, Proc. Roy. Soc., (London), A307 (1968) 97.
83. F.W. Birss, S.D. Colson and D.A. Ramsay, Can. J. Phys., 51 (1973) 1031.
84. T. Ueda and T. Shimanouchi, J. Mol. Spectr., 28 (1968) 350.
85. K.G. Kidd and G.W. King, J. Mol. Spectr., 40 (1971) 461.
86. T. Nakagawa, H. Kashiwagi, H. Kurihara and Y. Morino, J. Mol. Spectr., 31 (1969) 436.
87. J.L. Duncan, F. Hegelund, R.B. Foster, G.W. Hills and W.J.Jones, J. Mol. Spectr., 61 (1976) 470.
88. G.W. Hills, Ph.D. dissertation, University of Cambridge, 1974.

CHAPTER 16

SOME COMMENTS ON THE USE OF CONSTRAINTS AND ADDITIONAL DATA BESIDES FREQUENCIES IN FORCE CONSTANT CALCULATIONS

A. Müller and N. Mohan

1. INTRODUCTION

Numerous methods have been proposed in the past to calculate the force constants from one set of frequencies alone. These methods employ either reduced force field models (as the Urey-Bradley Force Field (UBFF), Orbital Valence Force Field (OVFF), Hybrid Orbital Force Field (HOFF), Hybrid Bond Force Field (HBFF), etc.) or assumptions about the General Quadratic Force Field (GQFF) so that the approximate GQFF constants can be computed in a straightforward manner. Some of these methods may be termed kinematic methods. A critical study of these methods has already been reported [1,2]. Although it is in general impossible to calculate a single set of $n(n+1)/2$ physically meaningful force constants from a set of n frequencies alone, the use of these methods has often been overstressed in the past. In order to obtain a physically meaningful set of force constants, one should employ as many additional data (such as isotope shifts, Coriolis constants, centrifugal distortion constants, mean amplitudes of vibration, inertia defect (for planar molecules), ℓ-type doubling constants (for symmetric and spherical tops), intensities of bands, etc.) as possible. However, since for many polyatomic molecules, it is almost impossible to get a unique solution for all the F-elements, one is forced to impose certain constraints on the force field. Such constraints should be viewed critically in that they should be physically reasonable and one should be able to estimate the magnitude of errors in the other force constants due to the use of the constraints. Basically, these methods involve constraints on the GQFF constants by (i) setting some $\mathbf{F}_{ij}$ elements equal to zero or assuming the values of some F_{ij} elements [3]; (ii) transferring the force constants for a group (as in "Local Symmetry Force Field" (LSFF)) [3]; (iii) reducing the order of the secular determinant by the separation of high and low frequencies (the usual "High-Low Frequency Separation" method (HLFS)) [4]; or (iv) assuming a reduced molecular model (as in the "Point Mass Model" (PMM) approach) [5,6]. In all these cases after a simplification (the error involved should have a negligible influence on the rest of the force constants), the secular equation can be solved to obtain a unique solution (or a limited number of solutions) by taking into account additional data. In this review, a critical discussion of the above methods as well as other model calculations based solely on a single set of frequencies (which in general cannot lead to physically meaningful results) is presented.

2. MODEL CALCULATIONS

By model calculations, we refer to those reduced potential models which contain as many independent potential energy constants as (or less than) the number of vibrational frequencies. The UBFF, OVFF, HOFF and HBFF calculations come under this category. A very brief elucidation of the basic principles of these models is presented below.

The UBFF [7] assumes that the off-diagonal elements of the GQFF are caused by non-bonded atoms interaction. In the simple form, the UBFF potential can be written as

$$V = \tfrac{1}{2}\sum K_{ij}(\Delta r_{ij})^2 + \tfrac{1}{2}\sum H_{ijk}(\Delta\Theta_{ijk}) + \tfrac{1}{2}\sum Y_{ijkl}(\Delta t_{ijkl})$$

$$+ \tfrac{1}{2}\sum F_{ij}(\Delta q_{ij})^2 \qquad (1)$$

where K_{ij}, H_{ijk}, Y_{ijkl} and F_{ij} are the bond stretching, bending, torsional and non-bonded interaction force constants respectively. The redundancy conditions could be used to write the potential in terms of independent coordinates. Although this model could, in many cases, be used to assign the vibrational frequencies and possibly to interpret the isotopic splittings of the fundamentals [8], the force constants thus derived do not have, strictly speaking, the same physical significance as that of the GQFF constants. Thus, for example, the UBFF stretching force constant K_{ij} cannot be equated with f_r of the GQFF.

The OVFF model [9] contains "bond bending" and the interaction among the non-bonded atoms in the potential function. It assumes that the distortions (due to vibration) can be explained entirely by "orbital following" and by the overlapping of the orbitals. Modifications of this model which take into account the redundancies, have been proposed by Claassen and co-workers [10] who have also set up the model in Cartesian coordinates [11]. Although for some selected types of molecules, this force field is found to be reasonable [12] (with respect to an interpretation of the symmetry force constants), the criticism listed under the UBFF is valid in this case, too.

The HOFF model uses the concept of "hybridisation changes" to correlate some of the off-diagonal GQFF constants and in this way attempts to reduce the number of independent GQFF constants. This model was applied successfully to the methyl halides (CH_3F, CH_3Cl, etc.) [13], but the rules derived from this model are not obeyed in all cases [14]. Further, this model can be applied only to those molecules which involve solely sp^n hybridisation (i.e. d or higher orbitals are not involved in the bond formation).

The HBFF approach [15] is very similar to HOFF and is found to be reasonable in the case of selected sets of molecules.

3. ADDITIONAL DATA AND THEIR SENSITIVITY

It is well known that the use of additional data (mentioned in the introduction) increases the reliability of the force constants, thus computed. However, the force constants derived from different additional data (employed either separately or together) cannot be characterised by the same degree of precision. This is due to the facts that (i) the accuracy of the experimental values of the various molecular constants is not the same and (ii) the sensitivity of the force constants to these additional data varies. Hence, it is worth studying the expressions which give information on the sensitivity of the force constants to additional data.

Small frequency shifts due to isotopic substitution

As shown by Wilson et al. [4] (see also ref. 16) the small frequency shifts involved in the substitution of heavy and very heavy atoms could be predicted reasonably accurately by the first order perturbation theory. Using this formalism, it can be shown [4,16] that

$$\Delta_{ii} = (\Delta\lambda_i/\lambda_i) = [\underline{L}^{-1}\underline{\Delta G}(\underline{L}^{-1})^t]_{ii} \qquad (2)$$

when all the frequencies lie far apart (so that the nondegenerate perturbation theory could be applied [17]). In this case, one derives the Jacobians [18]

$$\frac{\partial\Delta_{ii}}{\partial F_{ab}} = \frac{\partial(\Delta\lambda_i/\lambda_i)}{\partial F_{ab}} = 2\sum_{j\neq i}\Delta_{ij}L_{ai}L_{bj}(\lambda_i - \lambda_j)^{-1} \qquad (3)$$

where $\Delta_{ij} = [\underline{L}^{-1}\underline{\Delta G}(\underline{L}^{-1})^t]_{ij}$. Another form of the Jacobian, given by

$$\frac{\partial(\lambda_i^+/\lambda_i)}{\partial F_{ab}} = 2\sum_{j\neq i}A_{ai}A_{bj}L_{ai}L_{bj}\lambda_i^{-1} \qquad (4)$$

could also be derived from the nondegenerate first order perturbation theory ($\underline{A} = \underline{L}^{-1}\underline{L}^+$ with + being used to denote the quantities related to the isotopic analogue) [19]. Using Equations (3) and (4), it was shown [19,20] that the force constants are best fixed by the isotopic shifts when the vibrational fundamental in question has a low value. For XY_n type molecules exhibiting small mass coupling, the approximate error in F_{aa} (i.e. δF_{aa}) can be computed from the relation [20]

$$\frac{\delta F_{aa}}{\delta(\Delta\lambda_i/\lambda_i)} \simeq \frac{\lambda_i - \lambda_j}{\Delta G_{aa}} \qquad (5)$$

Equations (3)-(5) show that for transition metal complexes for which the vibrational frequencies corresponding to metal-ligand vibrations are low, the force constants could be fixed with reasonable precision even if the isotope shifts carry large uncertainties (e.g. $\simeq 3 \pm 1$ cm^{-1}) [19,20] when the corresponding change in G is not very small (which is the case with the first and second row transition elements). Large frequency shifts due to H/D substitution, on the other hand, are virtually useless because of the uncertainties arising out of anharmonicity effects [21].

Coriolis coupling constants

Since the equation involving the Coriolis coupling constants is similar to the one pertaining to small frequency shifts [3,22], it is clear that the Jacobians in the two cases should also be similar. Using nondegenerate first order perturbation theory, it can be shown that [23]

$$\frac{\partial \zeta_{ii}}{\partial F_{ab}} = 2 \sum_{j \neq i} \zeta_{ij} L_{ai} L_{bj} (\lambda_i - \lambda_j)^{-1} \quad (6)$$

Equation (6) once again indicates that the force constants are best constrained by the Coriolis coupling constants, when the ζ_{ij} elements are large and when simultaneously the vibrational frequencies are low and lie close together [24].

The published results so far, have demonstrated that both the isotopic shifts (small shifts) and the Coriolis constants (both measured with the current accuracy) are in general quite effective in fixing the precise values of the force constants. However, it is not easy to obtain accurate values of the isotope shifts when the bands are broad (as in the case of solid state spectra) or when a Fermi resonance is present. Similarly, the Coriolis constants may not be determined accurately when the band contour is disturbed by hot band progressions or by overlapping due to the presence of a number of isotopes (also in those cases where no rotational fine structure may be observed).

Centrifugal distortion constants

The centrifugal distortion constants $-\tau_{\alpha\beta r\delta}$ can be determined from the relation [25]

$$-\tau_{\alpha\beta r\delta} = (\underline{J}_{\alpha\beta})^t \underline{F}^{-1} (\underline{J}_{r\delta}) / I_{\alpha\alpha} I_{\beta\beta} I_{rr} I_{\delta\delta} \quad (7)$$

where $I_{\alpha\alpha}$ etc. are the Cartesian components of the principal moment of inertia tensor, $\underline{F}^{-1}$ is the matrix inverse of $\underline{F}$, and $J_{\alpha\beta}$ contains the elements which are the derivatives of the $\alpha\beta$th component of I with respect to the chosen set of vibrational coordinates. From the nondegenerate first order perturbation theory, one obtains [3]

$$\frac{\partial \tau_{\alpha\beta r\delta}}{\partial F_{ab}} = \tfrac{1}{2} \sum_k \sum_m J_{\alpha\beta}^{(k)} (\underline{F}^{-1})_{ka} (\underline{F}^{-1})_{bm} J_{r\delta}^{(m)} / I_{\alpha\alpha} I_{\beta\beta} I_{rr} I_{\delta\delta} \quad (8)$$

A study of the numerical results for different molecules has indicated [26,27] that the τ elements are in general fixed well by the diagonal force constants alone. This means that in general, the off-diagonal elements of $\underline{F}$ are not controlled well by the centrifugal distortion constants unless these constants are known with a high degree of precision.

Mean square amplitudes of vibration

The matrix of mean square amplitudes $\underline{\Sigma}$ can be generated from the relation [28,29]

$$\underline{\Sigma} = \underline{L}\ \underline{\delta}\ L^{t} \tag{9}$$

where the diagonal matrix $\underline{\delta}$ has the usual meaning [28]. The Jacobians $(\partial\Sigma_{ij}/\partial F_{ab})$ were shown to be [29]

$$\frac{\partial\Sigma_{ij}}{\partial F_{ab}} = \sum_{k} L_{ik}L_{jk}H(\partial\lambda_k/\partial F_{ab}) + (1 - \delta_{ab}/2). \sum_{k}\sum_{m}(L_{ik}L_{jm} + L_{im}L_{jk}).(L_{ak}L_{bm} + L_{am}L_{bk}).(\delta_k - \delta_m).(\lambda_k - \lambda_m)^{-1} \tag{10}$$

where

$$H = -\frac{16.85748}{2\lambda_k}[(\omega_k \tanh(0.719399\omega_k/T))^{-1} + (0.719399/T).(\sinh^2(0.719399\omega_k/T))^{-1}] \tag{11}$$

Numerical calculations in the past have indicated that in general, the mean square amplitudes do not serve as an effective constraint on the force field.

Inertia defect in planar molecules

For planar molecules, the inertia defect Δ (which reflects the deviation from planarity) is a function of the harmonic force constants alone [30,31] and can be computed from the relation [32]

$$\Delta = (h/\pi^2 c)\mathrm{Tr}(\underline{V}\ \underline{\Omega}\ \underline{X}) \tag{12}$$

where $\underline{V}$ is a diagonal matrix containing the vibrational quantum numbers and the degeneracies, $\underline{\Omega}$ a square matrix whose elements are functions of the squares of the vibrational frequencies, and $\underline{X}$ another square matrix comprising the first and second order Coriolis constants. Since the inertia defect in any vibrational state is related directly not to the force constants, but to the Coriolis constants [32], it is clear that the Jacobians $(\partial\Delta/\partial F_{ab})$ would involve the quantities $(\partial\Delta/\partial\zeta_{ij})$. These expressions are complicated [33] and hence are not reported here. It is generally found that the ground state constant $\Delta(0)$ is not a sensitive function of the force constants [34,35].

Integrated infrared intensities

The integrated infrared intensities A_i of the vibrational bands can, in principle, be used to compute the force constants when the dipole moment derivatives $(\partial\mu/\partial S_i)$ are known [36,37]. The Jacobians $(\partial A_i/\partial F_{ab})$ are given by [36]

$$\frac{\partial A_i}{\partial F_{ab}} = 2k_i \sum_{j \neq i} A_{ij} L_{ai} L_{bj} (\lambda_i - \lambda_j)^{-1} \tag{13}$$

where $k_i = (N_0 \pi d_i / 3c^2)$ with d_i denoting the degeneracy of the mode and the other symbols have their usual meanings (A_{ij} is fixed by the relation $A_{ij} = \pm(A_i A_j)^{\frac{1}{2}}$ for all i and j [37]). Equation (13) shows that the intensities serve to constrain the force constants effectively when they have a large value and when the frequencies are low.

<u>ℓ-type doubling constants</u>

ℓ-type doubling occurs as a direct consequence of vibration-rotation interaction [38,39] and in the first approximation, the separation $\Delta\nu$ of the ℓ-doublets is given by

$$(\Delta\nu)_\ell = \bar{q}_i(v_i + 1)J(J + 1) \tag{14}$$

where $\bar{q}_i$ is the ℓ-type doubling constant associated with the ith degenerate mode at the equilibrium configuration and v and J are the vibrational and the rotational quantum numbers respectively [38]. For linear molecules for which $\bar{q}_i$ is a function of the quadratic force constants alone [39] (this is in general not true in the case of all low frequency bending vibrations; see the results for $\bar{q}_7$ in C_3O_2 [40]), the Jacobians involve $(\partial\zeta_{ij}/\partial F_{ab})$ and one gets explicitly (from the general relation presented in ref. 39)

$$\frac{\partial \bar{q}_i}{\partial F_{ab}} = (4B_e^2/\omega_i) \sum_j \zeta_{ij} (\partial\zeta_{ij}/\partial F_{ab})(3\omega_i^2 + \omega_j^2)(\omega_j^2 - \omega_i^2)^{-1} \tag{15}$$

where B_e is the rotational constant at the equilibrium configuration. As seen from the above equation, the effectiveness of $\bar{q}_i$ in fixing the accurate values of the force constants increases with increasing value of B_e^2 and decreasing values of $(\omega_j^2 - \omega_i^2)$ (for all j).

Considering the Jacobians (i.e. Equations (3), (4), (6), (8), (10), (13) and (15)), it should be noted that they have been derived for an independent set of force constants. It can be shown [41] that the form of the Jacobians is unaltered by removal of the redundancies.

<u>Extremal properties and sensitivity*</u>

With regard to the sensitivity of the force constants to the various additional data, one important point is to be noted. When the molecular constants are extremal, they are almost ineffective in controlling the force constants within a reasonable range, unless the molecular constants are known with

* A complete study of the extremal properties of the various molecular constants and their relative effectiveness in fixing the precise values of the force constants is presented in ref. 42.

a high degree of accuracy. This can be deduced easily from the Jacobians given above. A careful study has shown [34,43] that the mean square amplitudes in general and the centrifugal distortion constants as well as the ℓ-type doubling constant for XYZ ($C_{\infty v}$) type molecules and the inertia defect for XY_2 (C_{2v}) type molecules are, with some exceptions (e.g. $\bar{q}_2$ for ICN; see the experimental results in ref. 44) not very useful in fixing the force constants. This may be attributed to the fact that while the experimental values of the bonded mean square amplitudes for most of the molecules are close to minima, those of the centrifugal distortion constant and the inertia defect $\Delta(0)$ lie in the vicinity of their maxima for XYZ ($C_{\infty v}$) and XY_2 (C_{2v}) type molecules respectively (for XYZ type molecules $\bar{q}_2$ is also close to minima). The isotopic shifts due to the substitution of heavy atoms and the Coriolis coupling constants are in general not extremal and hence they can, in general, be used as effective constraints on the force field. For the sake of easy comparision, the relative effectiveness of the different molecular constants on fixing the force field is characterised in Table 1.

4. APPROXIMATIONS USED IN CONJUNCTION WITH ADDITIONAL DATA

Approximations have to be employed when the additional data are not sufficient to fix a unique force field. Some physical observations subject to certain tests are often quoted to justify such approximations. The force field obtained from additional data along with some plausible approximations might be termed "pseudo-exact" in the sense that such a force field is presumably close to the one that would have been obtained, if enough data are available for solving the problem completely without approximations. Admittedly, the validity of such approximations differs from case to case.

The approximation F_{ab} = constant ($a \neq b$)

In this approximation, some force constants (practically always the off-diagonal ones) are constrained at a particular value so that the other force constants can be computed using the available additional data. In this connection, it becomes important to study the errors in the values of the other force constants due to such approximations and hence to see beforehand, in which cases it is valid to make such an approximation.

We are interested in assessing the magnitude of errors in the diagonal force constants F_{aa} due to those involved in the off-diagonal ones F_{ab}. Let $\underline{F}_0$ and ($\underline{F}_0$ + $\underline{\Delta F}$) be two force fields reproducing any particular assignment for a molecule. In our case, we identify $\underline{F}_0$ and ($\underline{F}_0$ + $\underline{\Delta F}$) with the true force field and the one obtained by constraining the values of some of the off-diagonal elements, respectively. It can be shown [23] that in this process $\underline{L}_0$ changes to $\underline{L}_0\underline{A}$ where $\underline{A}$ is an orthogonal matrix. We obtain the result

$$\underline{A}^t\underline{\Lambda}\;\underline{A} + \underline{A}^t\underline{L}_0^t\underline{\Delta F}\;\underline{L}_0\underline{A} = \underline{\Lambda} \qquad (16)$$

Table 1. (Continued)

Data	Accuracy of measurement	Limitations for the calculation of force constants	Sensitivity of data to the force constants	Example of application	Ref.
ℓ-type doubling constants $\bar{q}_t$	high accuracy in general	exist only for symmetrical molecules with degenerate fundamentals (for non-linear molecules, anharmonic force constants should be known)	often not very good (e.g. XYZ ($C_{\infty v}$) type molecules)	OCS OCSe	h h
Infrared intensities Γ_i or A_i	not very accurate	the dipole moment derivatives should be known; multiplicity of solutions is unavoidable			
Relative Raman intensities (I_i/I_j)	not very accurate	the force constants can be calculated only with the help of approximate bond polarisability theory; multiplicity of solutions is unavoidable		$SnCl_4$	i
Mean amplitudes Σ	not very accurate	the usefulness is limited by accuracy and sensitivity; Σ can be constructed fully from the electron diffraction data for simple molecular types only	generally insensitive to the force constants	$SnCl_4$ OsO_4	i j

Table 1. Additional data (besides frequencies) used for the calculation of exact force constants.

Data	Accuracy of measurement	Limitations for the calculation of force constants	Sensitivity of data to the force constants	Example of application	Ref.
Large frequency shifts (e.g. H/D) $\Delta\nu$	good	the shifts must be corrected for anharmonicity; further, the shifts might be perturbed by Fermi resonance	in general poor	CF_3H (H/D)	a
Small frequency shifts (heavy atom substitution) $\Delta\nu$	good, but in general not so in condensed phases	anharmonicity corrections necessary in some cases; perturbation of the shifts by Fermi resonance might occur	in general good	RuO_4 ($^{16}O/^{18}O$); NSCl ($^{14}N/^{15}N$; $^{32}S/^{34}S$; $^{35}Cl/^{37}Cl$)	b c
Coriolis coupling constants ζ_{ii}	often good, if the rotational fine structure can be resolved; otherwise a band contour analysis is necessary	can be measured accurately for simple molecules only; pure isotopes might be needed to study the band contours accurately	in general good, with exceptions	WF_6 $SnCl_4$ $GeCl_4$	d e f
Centrifugal distortion constants τ	very high in general, but analysis of a large number of rotational transitions under high resolution is necessary	measured only in the gas phase; very high resolution data needed to fix the τ_s precisely	often not very good (e.g. XYZ ($C_{\infty v}$) type molecules)	NSCl PSF_3	c g

Table 1. (Continued)

Data	Accuracy of measurement	Limitations for the calculation of force constants	Sensitivity of data to the force constants	Example of application	Ref.
Inertia defect Δ	the measurement errors are usually difficult to estimate; but, claimed to be accurate to ±0.005 amu Å^2	applicable for planar molecules only; the ground state constant is in general not very sensitive to the force field and hence those in excited states are needed	the ground state constant Δ(0) is often insensitive to the force constants	NSCl ONCl	c k

a R.W. Kirk and P.M. Wilt, J. Mol. Spectr., 58 (1975) 102.
b R.S. McDowell, L.B. Asprey and L.C. Hoskins, J. Chem. Phys., 5 (1972) 5712.
c Ref. 35.
d R.S. McDowell and L.B. Asprey, J. Mol. Spectr., 48 (1973) 254.
e F. Königer and A. Müller, J. Mol. Spectr., 56 (1975) 200.
f F. Königer, R.O. Carter and A. Müller, Spectrochim. Acta, A, 32 (1976) 891.
g F. Königer and A. Müller, Spectrochim. Acta, A, in press.
h Y. Morino and C. Matsumura, Bull. Chem. Soc. Japan, 40 (1967) 1095, 1101.
i H. Fujii and M. Kimura, J. Mol. Spectr., 37 (1971) 517.
j A. Müller, B. Krebs and S.J. Cyvin, Acta Chem. Scand.,21 (1967) 2399.
k G. Cazzoli, R. Cervellati and A.M. Mirri, J. Mol. Spectr., 56 (1975) 422.

from the standard secular equation of molecular vibrations [4]. From Equation (16) it follows that

$$\mathrm{Tr}(\underline{L}_0^t \ \underline{\Delta F} \ \underline{L}_0) = \mathrm{Tr}(\underline{G} \ \underline{\Delta F}) = 0 \tag{17}$$

which leads to the result

$$\delta F_{aa} = -(G_{aa})^{-1}(\sum_{b \neq a} G_{bb}\delta F_{bb} + 2\sum_{b} G_{ab}\delta F_{ab}) \tag{18}$$

The influence of the error in any one F_{ab} (i.e. ∂F_{ab}) on that associated with any F_{aa} (i.e. ∂F_{aa}) is given by the Jacobian $(\partial F_{aa}/\partial F_{ab})$ and from Equation (18), we obtain the result

$$\frac{\partial F_{aa}}{\partial F_{ab}} = -2G_{ab}/G_{aa} \tag{19}$$

Thus, it can be noted that the error in the diagonal force constant due to constraining the off-diagonal force constant, is a minimum when the factor (G_{ab}/G_{aa}) is negligible. This is the case for molecules containing a heavy metal and light ligands. Equation (19) provides the explanation as to why the Modified Valence Force Field (MVFF) with $F_{ab} = 0$ is a good approximation for coordination compounds with a heavy central atom (this is also true of the various approximation methods).

A typical example of the above approximation is $Ni(CO)_4$ where some of the off-diagonal force constants (those pertaining to the 2 x 2 E species and the 4 x 4 F_2 species) were constrained at particular values. It was shown [45] that an error of even 100% in the values of the off-diagonal force constants has very little influence on the values of the diagonal force constants.

Local symmetry force field (LSFF)

This model assumes the transferability of the force constants related to different functional groups in big molecules having similar structure. In this way, the number of unknown force constants is reduced, thus permitting the solution of the complete force field using the available additional data. In this approach, the interaction force constants pertaining to the different functional groups in a molecule are generally neglected (this assumption, presumably introduces only small errors if the coupling of the vibrations is small). The symmetry coordinates for the whole molecule are constructed in terms of the "local" symmetry coordinates for the various functional groups and hence the name "Local Symmetry Force Field" (LSFF) is used. The functional group corresponding to which the transferability of the force constants is generally assumed, is mainly CH_n. A complete review of this model has already been presented elsewhere [3,46] and these works should be consulted for more details.

Since no guarantee for the transferability of the force constants and for the correctness of assumptions about the next neighbour interactions (i.e. among the different functional groups) could be given, the LSFF model cannot be applied to all cases uniformly to deduce a physically meaningful set of force constants. This model is less applicable for pure inorganic compounds compared to organic ones having groups in nearly the same chemical environment.

High-low frequency separation method (HLFS)

In the HLFS method [4,47], the separation of the high and low frequencies is achieved by dropping the corresponding frequencies, thereby truncating the order of the original secular determinant. In this approximation, the high frequency vibrations are studied by merely dropping the rows and columns of $\underline{G}$ corresponding to the low frequency vibrations (LFS) whereas the low frequency vibrations are studied by using the modified elements of $\underline{G}$ given by [4,47]

$$G^0_{tt'} = G_{tt'} - \sum_{ss'} G_{ts} X_{ss'} G_{s't'} \qquad (20)$$

corresponding to the low frequency vibrations (HFS).

Müller et al. [24,47,48] suggested the use of Coriolis coupling constants and isotopic shifts in solving the truncated problem exactly, thus obtaining a set of "pseudo-exact" force constants. They showed that the force constants thus derived are close to the exact ones in cases where the condition for the separability (see below) is fulfilled. In applying additional data to solve the truncated problem, the Teller-Redlich product rule [4] (for the frequencies) and the ζ sum rule [22, 28] (for the Coriolis constants) should be satisfied as closely as possible in order that the HLFS method yields reasonable results [24,48] (this serves as experimental proof for the validity of the approximation). Under this condition, the error inherent in the use of this model is also expected to be rather small [19,24,48].

Recently, Cotton and Kraihanzel [49,50] proposed a model to study the high frequency vibrations (mainly in metal carbonyls). This method was used later by several authors [51,52] to obtain the force constants corresponding to the reduced block from the isotopic data and to calculate the absorption intensities for the corresponding bands. If $(AB)_n$ is the functional group, all the corresponding $\underline{G}$ elements in this model are given by $\mu_A + \mu_B$ where μ stands for the reciprocal mass. If it is recognised that in the classical LFS method [4] this is also the expression for the $\underline{G}$ elements related to the high frequency vibrations (when the different (AB) groups are not bonded together as in the case of the molecules considered), it is clear that the two models are identical. Interesting in the latter case [51,52] is the use of relative infrared intensities for the identification of the different isotopic species. In this model, the dipole moment components are assumed to be directed along the bonds and a simple vectorial addition is used to construct the derivatives ($\partial\mu/\partial S_i$) with respect to the various bond stretching symmetry coordinates (this method gives the relative values only). Using the Crawford's rules, the relative intensities could be calculated. These authors [50-52] showed that this model works well in many cases. The converse can also be deduced from these observations which is that the intensity data (relative) together with the HLFS method can lead to good results for the force constants when the Crawford's rules [53] are obeyed for the reduced block. It might be very dangerous to use the Cotton and Kraihanzel force field [49,50] to study the vibrations of other types of ligands without testing beforehand, the validity of the approximations (e.g. N-N vibrations in Pd and Ni complexes; see Huber et al. [52]).

Point mass model (PMM)

In this approach, the XH_n group is treated as a point with the aggregate mass ($m_X + nm_H$) and this results in considerable reduction of the order of the original secular determinant (for the whole molecule). This model has been employed in the past for the determination of force constants pertaining to skeletal vibrations using the isotope shifts, Coriolis coupling and centrifugal distortion constants [5,6,24].

In applying this procedure to several hexa- and tetra-ammine complexes, it was found [19] that the values and the error limits (the error limits correspond only to the experimental uncertainities of the molecular constants used as additional data) of the skeletal force constants derived from the PMM and the HLFS approaches separately, are very nearly the same (for the same force constants). This appears to justify the use of the PMM approach in the case of the ammine complexes (for which, it can be shown that the mixing between the ligand internal and the skeletal vibrations is negligible).

The limitations of this method are obvious. This model can be applied only in the case of molecules containing XH_n (other groups cannot be treated as a point mass). Further, this approach is valid only when the coupling between the vibrations of the XH_n group and the rest of the molecule, is negligible. This means that the PMM approach cannot be applied to molecules with MXH_n group having a lighter M atom since in this case a high degree of mixing between $\rho(XH)_n$ and $\nu(MX)$ might be expected. The validity of the Teller-Redlich product [4] and ζ sum rules [22,28], and the one related to the centrifugal distortion constants [54] pertaining to the reduced molecular model, have to be tested for the reliability of the PMM approach. It was shown [24,48] that if these conditions are fulfilled, reliable force constants for several molecules could be obtained (in this case the pseudo-exact force constants show good agreement with the exact ones). In order to compare the relative effectiveness of the different methods mentioned above, we have presented the limitations and the conditions for the applicability of these methods in Table 2. This table is expected to serve as an easy means of judging the reliability of the force constants obtained using the various approximations.

5. OTHER GENERAL PROBLEMS ASSOCIATED WITH FORCE CONSTANT SOLUTIONS

It is well known [46,55] that although in some cases, the additional data are sufficient in number and in accuracy to determine a set of force constants, problems related to the

Table 2. Conditions for the validity of the approximation methods used in conjunction with additional data[a].

Method	Condition for the validity of the method [b]
F_{ab} $(a \neq b)$ = constant	$\frac{\partial F_{aa}}{\partial F_{ab}} = -2 \frac{G_{ab}}{G_{aa}}$ The above equation indicates that this approximation is a good one for molecules containing a heavy central atom and light ligands (where $G_{ab} << G_{aa}$).
HLFS and PMM with	
(a) isotope shifts	$\frac{\prod \lambda_i^{iso}}{\prod \lambda_i} = \frac{\lvert \underline{G}^{iso} \rvert}{\lvert \underline{G} \rvert}$
(b) Coriolis constants	$Tr(\underline{\zeta}^{\alpha}) = Tr(\underline{G}^{-1}.\underline{C}^{\alpha})$
(c) infrared intensities	$Tr(\underline{\mu}'\underline{G}) = (3c^2/N_o \pi d) Tr\underline{A}$
PMM with centrifugal distortion constants	$\frac{\lvert (t^R)^{iso} \rvert}{\lvert (t^R) \rvert} = \frac{\lvert (\underline{G}^{-1})^{iso} \rvert^2}{\lvert (\underline{G}^{-1}) \rvert^2} \cdot \frac{\lvert \underline{A}^{iso} \rvert^2}{\lvert \underline{A} \rvert^2}$

a For the meaning of the symbols and equations, see text.
b The equations presented here, can be found in refs. 24, 36, 37, 48, 54. In the case of HLFS and PMM approaches, these equations pertain to the reduced block (see text) and these should be satisfied as closely as possible for the validity of the approaches.

multiplicity of the solutions and to the sensitivity of the force constants to the additional data do remain. In this section, a brief survey of these aspects is reported.

Multiple solutions

A unique solution for the force constants is possible only when there exist as many equations which are all independent and linear as the number of independent force constants. But, unfortunately, one is often forced to use non-linear equations to solve the force constants problem and hence the multiplicity of solutions appear as a stumbling block in the calculation of exact force constants. In n=2 cases, it was shown [34,43,56] that the isotope shifts due to asymmetric substitution, the mean square amplitudes of vibration and the inertia defect (for XY_2 type molecules) can be used to choose a unique set of force constants.

A classic example of the above feature is the E species force field for NF_3 [55] where the $^{14}N/^{15}N$ shifts, Coriolis constants and the centrifugal distortion constants were all shown to be ineffective in fixing a unique force field (two solutions are obtained). It was later shown [56] that the mean amplitude related to the bonded atom pair, could be used to choose a unique set. The problem of multiple solutions in higher order cases, has also been discussed in the literature [57] where it was shown that the isotope shifts due to asymmetric substitution are indispensable in fixing a unique set of force constants.

Lack of sensitivity of the data to the force constants (mainly off-diagonal)

In some cases, it is found (see the results for CF_3H and NSCl [35,58]) that some of the off-diagonal force constants are not fixed well by the data (even though the data are rather accurate). Thus in the case of CF_3H and NSCl, the stretch-stretch interaction constant F_{12} is found to be 0.52 ±0.86 and 1.18 ±0.40 mdyn $Å^{-1}$ respectively even though the data used (isotope shifts, centrifugal distortion constants, etc.) are defined within a narrow range. In order to limit the uncertainties associated with the force constants, one needs such pieces of data which are rather sensitive to the corresponding force constants; but, this may not be possible in all cases.

Acknowledgments

The financial support provided by the "Deutsche Forschungsgemeinschaft", the "Fonds der Chemischen Industrie" and the NATO Scientific Affairs Division, is gratefully acknowledged.

6. REFERENCES

1. Yu.N. Panchenko, R.A. Munoz, N.F. Stepanov and G.S. Koptev, J. Mol. Struct., 12 (1972) 289; A.J.P. Alix, H.H. Eysel, B. Jordanov, R. Kebabcioglu, N. Mohan and A. Müller, J. Mol. Struct., 27 (1975) 1; H.H. Eysel, W.J. Lehmann, K. Lucas, A. Müller and K.H. Schmidt, Z. Naturforsch., A, 29 (1974) 332.
2. P. Gans, in R.J.H. Clark and R.E. Hester (Editors), Advances in Infrared and Raman Spectroscopy, Vol. 2, Heyden, London, 1976.
3. T. Shimanouchi, in H. Eyring, D. Henderson and W. Jost (Editors), Physical Chemistry, Vol. 4, Academic Press, New York, 1970, p. 233.
4. E.B. Wilson, J.C. Decius and P.C. Cross, Molecular Vibrations, McGraw-Hill, New York, 1955.
5. A. Müller, K.H. Schmidt and G. Vandrish, Spectrochim. Acta, A, 30 (1974) 651; K.H. Schmidt and A. Müller, J. Mol. Struct., 22 (1974) 343; K.H. Schmidt, W. Hauswirth and A. Müller, J. Chem. Soc. Dalton, (1975) 2199; K.H. Schmidt and A. Müller, Inorg. Chem., 14 (1975) 2183.
6. R.L. Cook, J. Mol. Struct., 26 (1975) 126.
7. H.C. Urey and C.A. Bradley, Phys. Rev., 38 (1931) 1969; T. Shimanouchi, J. Chem. Phys., 17 (1949) 245, 743, 848;
8. S.T. King, J. Chem. Phys., 49 (1968) 1321; P.D. Mallinson, D.C. McKean, J.H. Holloway and I.A. Oxton, Spectrochim. Acta, A, 31 (1975) 143.

9. D.F. Heath and J.W. Linnett, Trans. Faraday Soc., 44 (1948) 873.
10. H. Kim, P.A. Souder and H.H. Claassen, J. Mol. Spectr., 26 (1968) 46.
11. J. Tyson, H.H. Claassen and H. Kim, J. Chem. Phys., 54 (1971) 3142.
12. N. Mohan, K.H. Schmidt and A. Müller, J. Mol. Struct., 13 (1972) 155.
13. I.M. Mills, Spectrochim. Acta, 19 (1963) 1585; W.H. Fletcher and W.T. Thompson, J. Mol. Spectr., 25 (1968) 240.
14. For example, J.L. Duncan, D.C. McKean and G.D. Nivellini, J. Mol. Struct., 32 (1976) 255; the rule $F_{12} = -(F_{67} - F_{68})/2(2)^{\frac{1}{2}}$ is not obeyed in this case.
15. W.T. King, J. Chem. Phys., 36 (1962) 165.
16. A. Müller, K.H. Schmidt and N. Mohan, J. Chem. Phys., 57 (1972) 1752.
17. H. Eyring, J. Walter and G.E. Kimball, Quantum Chemistry, John Wiley, New York, 1944, p. 92-96.
18. M. Tsuboi, J. Mol. Spectr., 19 (1966) 4.
19. N. Mohan, S.J. Cyvin and A. Müller, Coord. Chem. Rev., 21 (1976) 221.
20. A. Müller, N. Mohan, F. Königer and M.C. Chakravorti, Spectrochim. Acta, A, 31 (1975) 107; A. Müller, N. Mohan and F. Königer, J. Mol. Struct., 30 (1976) 297.
21. A.A. Chalmers and D.C. McKean, Spectrochim. Acta, A, 22 (1966) 251.
22. J.H. Meal and S.R. Polo, J. Chem. Phys., 24 (1956) 1119, 1126.
23. I.M. Mills, J. Mol. Spectr., 5 (1960) 334.
24. A. Müller and S.N. Rai, J. Mol. Struct., 24 (1975) 59.
25. D. Kivelson and E.B. Wilson, J. Chem. Phys., 20 (1952) 1575; 21 (1953) 1229.
26. I.M. Mills, in D.R. Lide and M.A. Paul (Editors), Critical Evaluation of Chemical and Physical Structural Information, National Academy of Sciences, Washington, D.C., 1974, p. 269.
27. J. Hog and T. Pedersen, J. Mol. Spectr., 61 (1976) 243.
28. S.J. Cyvin, Molecular Vibrations and Mean Square Amplitudes, Elsevier, Amsterdam, 1968.
29. S.J. Cyvin and G. Hagen, in S.J. Cyvin (Editor), Molecular Structures and Vibrations, Elsevier, Amsterdam, 1972.
30. T. Oka and Y. Morino, J. Mol. Spectr., 6 (1961) 472; 8 (1962) 9, 300.
31. M.U. Chan and P.M. Parker, J. Mol. Spectr., 42 (1972) 449.
32. E.I. Kredentser and L.M. Sverdlov, Opt. Spectr., 29 (1970) 272.
33. E.I. Kredentser and L.M. Sverdlov, Opt. Spectr., 26 (1969) 464.
34. A. Müller, N. Mohan, S.N. Rai, A. Alix and L. Bernard, J. Chim. Phys., 72 (1975) 158.
35. See A. Müller, N. Mohan, S.J. Cyvin, N. Weinstock and O. Glemser, J. Mol. Spectr., 59 (1976) 161.
36. N. Mohan and A. Müller, J. Mol. Struct., 27 (1975) 255.
37. N. Mohan, A.J.P. Alix and A. Müller, Mol. Phys., in press.
38. G. Herzberg, Infrared and Raman Spectra of Polyatomic Molecules, Van Nostrand, New York, 1945.

39. G.J. Cartwright and I.M. Mills, J. Mol. Spectr., 34 (1970) 415.
40. J.A. Duckett, I.M. Mills and A.G. Robiette, J. Mol. Spectr., 63 (1976) 249.
41. G. Strey, J. Mol. Spectr., 17 (1965) 265; 19 (1966) 229.
42. N. Mohan and A. Müller, in J.R. Durig (Editor), Vibrational Spectra and Structure, Vol. 6, Elsevier, Amsterdam, in press.
43. A. Müller and N. Mohan, J. Chem. Phys., 58 (1973) 2994; A. Müller, N. Mohan and A. Alix, J. Chem. Phys., 59 (1973) 6112; A. Müller, N. Mohan and S.N. Rai, J. Chem. Phys., 60 (1974) 3958.
44. J.B. Simpson, J.G. Smith and D.H. Whiffen, J. Mol. Spectr., 44 (1972) 558.
45. L.H. Jones, R.S. McDowell and M. Goldblatt, J. Chem. Phys., 48 (1968) 2663.
46. J. L. Duncan, in Molecular Spectroscopy, Vol. 3, Chemical Society, London, 1975.
47. A. Müller, N. Mohan, K.H. Schmidt and I.W. Levin, Chem. Phys. Lett., 15 (1972) 127.
48. K.H. Schmidt and A. Müller, J. Mol. Struct., 18 (1973) 135.
49. F.A. Cotton and C.S. Kraihanzel, J. Amer. Chem. Soc., 84 (1962) 4432; C.S. Kraihanzel and F.A. Cotton, Inorg. Chem., 2 (1963) 533; F.A. Cotton, Inorg. Chem., 3 (1964) 702.
50. M. Moskovits and G.A. Ozin, in J.R. Durig (Editor), Vibrational Spectra and Structure, Vol. 4, Elsevier, Amsterdam, 1975, p. 187.
51. H. Haas and R.K. Sheline, J. Chem. Phys., 47 (1967) 2996; G. Bor, J. Organo-metallic Chem., 10 (1967) 343; J.H. Darling and J. S. Ogden, J. Chem. Soc. Dalton, (1972) 2496.
52. H. Huber, E.P. Kündig, M. Moskovits and G.A. Ozin, J. Amer. Chem. Soc., 95 (1973) 332.
53. B.L. Crawford, J. Chem. Phys., 20 (1952) 977.
54. A.P. Aleksandrov and M.R. Aliev, J. Mol. Spectr., 47 (1973) 1; A.P. Aleksandrov, M.R. Aliev and V.T. Aleksanyan, Opt. Spectr., 29 (1970) 568.
55. A. Allan, J.L. Duncan, J.H. Holloway and D.C. McKean, J. Mol. Spectr., 31 (1969) 368.
56. A.R. Hoy, J.M.R. Stone and J.K.G. Watson, J. Mol. Spectr., 42 (1972) 393.
57. D.C. McKean and J.L. Duncan, Spectrochim. Acta, A, 27 (1971) 1879.
58. R.D. Cunha, J. Mol. Spectr., 43 (1972) 282.

CHAPTER 17

LIMITATIONS OF FORCE CONSTANT CALCULATIONS FOR LARGE MOLECULES

Giuseppe Zerbi

1. INTRODUCTION

Much human energy and computing time has been spent in the last two decades on the calculation of vibrational force constants for organic and inorganic molecules as isolated entities. More recently a parallel, even if relatively weaker, effort has been made to calculate inter and intramolecular vibrational force constants for molecules in the crystalline state. Such calculations have been greatly helped by the availability of larger and faster computers and of almost fully automatic computing programs which, once the molecule is suitably described, carry out the whole calculation.

The justifications for doing these calculations can be briefly summarised as follows. The first, more basic aim is the attempt to achieve a detailed knowledge of the interatomic potential which depends on the electronic structures of the vibrating molecules. The problem of the interaction potential between submicroscopic particles is of fundamental importance in physics and chemistry. Since in the vibrational spectrum normal frequencies give direct information on the potential, a great effort was made to understand the vibrational spectra of model molecules and to derive (in conjunction with the study of other physical properties influenced by molecular vibrations) general information on the origin and nature of the interatomic forces.

Just in the same way as diagnostic correlative studies between group frequencies provided the chemist with detailed information on the chemical structure of the molecule [1-3], chemists entered the field of vibrational molecular force constants with the hope of obtaining a detailed description of the nature of the chemical bond and its variation with the chemical or physical surroundings. This is particularly true in the field of inorganic chemistry [3]. Vibrational spectroscopists were also interested to analyse the vibrational spectrum and to derive sets of parameters associated with suitably chosen models of the interatomic potential which could be transferred for the prediction of the spectrum of a chemically similar molecule in terms of its known or unknown structure. Much structural work has been done by using parameters determined for a given model of the potential to predict the unknown structure of a compound by comparison between the calculated and experimental spectrum [4].

Finally, since molecular systems vibrate in all phases, several physical properties depend to a small or large extent on the atomic vibrations. Vibronic spectroscopy, vibrorotational spectroscopy, rotational spectroscopy, electron, X-ray and neutron diffraction studies, thermal, magnetic and dielectric properties, etc. contain a vibrational contribution which must be accounted for. Workers specialising in each field tried to determine the vibrational dependency of the physical properties encountered in their work and to provide numbers to describe them. The validity of such numbers in describing the same phenomenon, the atomic vibrations, depends on the model adopted and on the accuracy and nature of the physical measurements.

In an excellent article, Duncan [5] recently discussed the state of the art of force constant calculations in molecules. Many references to previous papers or books are given for both theory and numerical calculations. This article, however, deals in great detail with relatively small molecules and does not consider the field of "large" polyatomic molecules for which a great effort has been made by many workers. The vibrations of small and large molecules have been recently treated by Califano [6]. In this chapter we shall try, from a personal viewpoint, to assess the progress made in the field of the calculations of force constants of large molecules and to critically evaluate the validity of some of the general results achieved. The definition of "large" comprises molecules such as, say, methanol up to organic polymers or even biological systems. Most of the time the symmetry of these systems is rather low. Many compromises must then be made in the calculations. The purpose of this article is to discuss some of these limitations and the validity of the approximations adopted. We will mainly consider the case of intramolecular interactions in isolated molecules and only when necessary reference will be made to intermolecular interactions in crystalline systems.

2. CHOICE OF THE VIBRATIONAL COORDINATES

We assume that the reader is familiar with the classical mechanics of molecular vibrations as well as with the quantum mechanical treatment of them [7]. We concentrate mainly on the interatomic potential which enters the equation of motion and which in its turn determines the position of the vibrational levels and the normal modes.

Let the unknown interatomic potential (function only of the displacements of the atoms) be expanded in terms of Cartesian coordinates about the equilibrium geometry (which, for the time being, is assumed to be known).

$$2V = 2V_o + 2\sum_{i=1}^{3N}\left(\frac{\partial V}{\partial x_i}\right)_o x_i + \sum_{i,j=1}^{3N}\left(\frac{\partial^2 V}{\partial x_i \partial x_j}\right)_o x_i x_j + \dots \qquad (1)$$

The term V_o is removed by a suitable shifting of the reference: the linear term is zero since for independent coordinates at the equilibrium each $(\partial V/\partial x_i) = 0$. If one wishes to describe the interatomic potential by semiempirical or "ab initio" methods, before carrying out any vibrational calculation, the equilibrium geometry compatible with the theoretical model chosen must be found. Once the choice of the semiempirical atom-atom potential or of the basis set of atomic wavefunctions has been made, a calculation of the energy minimum is required: only in such a case is $(\partial V/\partial x_i) = 0$. This is the basis of the so-called "force method" by Pulay [8].

The Taylor expansion of the potential is always truncated at the second order term. Attempts to account for higher order terms have been made for a few small molecules in the gas phase for which vibrorotational spectra and rotational spectra could be fully analysed [9]. Any comment on these is beyond the purpose of this article since most of the large molecules either do not exist in the gas phase or (if they can be vaporised) their vibrorotational spectrum is useless for these purposes. Throughout this article we will discuss small amplitude oscillations in a rigorously harmonic potential.

The use of Cartesian coordinates as in Equation (1) has been generally avoided in molecular problems since on this basis the physical interpretation of force constants is difficult. A discussion of the usefulness of the force constant tensor has been recently presented by King [10]. Cartesian displacements have received new attention in the past few years when the vibrations of one-dimensional or three-dimensional crystals had to be considered [11] (see Chapter 24 of this book).

The most common set of vibrational coordinates is that described as Internal Displacement Coordinates and originally defined by Wilson [7]. Let $\underline{R} = \underline{\underline{B}}\underline{x}$ be the transformation (written in matrix notation) from Cartesian to internal coordinates: the corresponding vibrational potential takes the form

$$2V = 2\sum_i \left(\frac{\partial V}{\partial R_i}\right)_o R_i + \sum_{ij} \left(\frac{\partial^2 V}{\partial R_i \partial R_j}\right)_o R_i R_j = \underline{R}\,\underline{\underline{F}}^R\,\underline{R} \qquad (2)$$

where $F^R_{ij} = \left(\frac{\partial^2 V}{\partial R_i \partial R_j}\right)_o$. The advantage of the force constants in terms of internal coordinates is that the F^R_{ij}'s offer a more direct physical interpretation which is very appealing to the chemist seeking information on the properties of the bonds in the molecule.

Chemical spectroscopists in vibrational assignments like to use the so-called "group coordinates" defined by the linear transformation*

$$\underline{\mathfrak{R}} = \underline{\underline{U}}^{\mathfrak{R}}\,\underline{R} \qquad (3)$$

$\mathfrak{R}$ coordinates for several groups of atoms are closer to the qualitative description of the group vibrations adopted in

---

* Use is made here of the symbol $\mathfrak{R}$ for group coordinates while the same symbol is used by Overend et al. [9] in another context for the definition of curvilinear coordinates.

diagnostic spectroscopy correlations (e.g. CH_2 rocking, twisting, wagging, scissors) [1,2]. Group force constants $F_{ij}^{\mathcal{R}}$ or "local symmetry force constants" have often been used by the Japanese School [12].

The use of internal coordinates may introduce one or more redundancies in the vibrational problem. This case occurs when the number of internal coordinates is larger than the number of vibrational degrees of freedom. It is generally unwise to remove the redundancies by simply ignoring some of the internal coordinates for at least the following two reasons:

(i) only a complete set of internal coordinates forms a basis for a completely reduced representation of the symmetry point group of the molecule; therefore the use of a symmetrically incomplete set of internal coordinates often prevents advantage being taken of group theory.
(ii) the lack of one or more coordinates gives a description of the potential energy in terms of parameters which become difficult to transfer to other molecules because they contain also the contributions of the missing coordinates.

The proper handling of the redundancy has been the subject of controversy in the literature [13-15]. However, it seems well established that the redundancy relation among Wilson's coordinates is linear.

In the solution of the vibrational problem on the basis of redundant coordinates, a number of vanishing frequencies equal to the number of redundancies will appear. Since the existence of redundancies generates many singularities in the matrices involved in the calculations, it is more advisable to remove such redundancies. The explicit form of the redundancy condition can be found either by analytical procedures [16,17] or by computer [18,19]. The latter procedure is more suitable for big molecules where the number of redundancies may be large and analytically intractable. In this case use is made of the diagonalisation

$$\underline{\underline{G}}_R \underline{\underline{D}} = \underline{\underline{D}}\,\underline{\underline{\Gamma}} \tag{4}$$

of the inverse kinetic energy matrix $\underline{\underline{G}}_R$. Among the eigenvalues Γ_i of $\underline{\underline{G}}_R$ there are as many zeroes as redundancies exist among the internal coordinates:

$$\underline{\tilde{D}}_i\, \underline{R} = 0 \tag{5}$$

Once the explicit form has been derived, one can then build a set of linear combinations of internal coordinates orthogonal to the redundancy conditions.

Symmetry coordinates S (when symmetry exists) can finally be defined by another linear transformation from R to $\mathcal{R}$ or from a non-redundant set of coordinates. When S are chosen as basis, the final $\underline{\underline{G}}_S$ and $\underline{\underline{F}}_S$ matrices are factored to their maximum extent thus making calculations easier.

Methods for the construction of symmetry coordinates are clearly described in ref. 7. However, since the $\underline{\underline{G}}_R$ matrix contains in itself all the information on the symmetry of the problem [20], the eigenvectors obtained by numerical diagonalisation of $\underline{\underline{G}}_R$ provide directly the coefficients for the construction of a set of symmetry coordinates [19]. It has been shown [19] that the new set of coordinates

$$\underline{\Sigma} = \tilde{\underline{\underline{D}}}\,\underline{R} \tag{6}$$

obtained using the unitary matrix $\underline{\underline{D}}$ defined in Equation (4) gives a completely reduced representation of the symmetry point group of the molecule and that therefore the Σ are symmetry coordinates. Moreover a computer can be taught to recognise the symmetry species of each set of eigenvectors thus allowing the whole group theory of a vibrational problem to be treated directly by computer [21,22]. The numerical handling of group theory is particularly useful when large symmetric molecules (e.g. adamantane) must be handled.

3. SECULAR DETERMINANT, EIGENVALUES AND EIGENVECTORS

Once the vibrational coordinates have been chosen and redundancies eventually removed, the construction of the dynamical matrix can be easily done with suitable computer programs. In the Wilson technique the eigenvalue equation to be solved is of the type

$$\underline{\underline{G}}\,\underline{\underline{F}}\,\underline{\underline{L}} = \underline{\underline{L}}\,\underline{\underline{\Lambda}} \tag{7}$$

where $\underline{\underline{G}}$ contains all the information on the geometry and masses of the molecule, $\underline{\underline{F}}$ is the matrix of the force constants in the harmonic approximation, $\underline{\underline{L}}$ is the matrix of the eigenvectors which provide the relative amplitudes of oscillation, $\underline{\underline{\Lambda}}$ is the diagonal matrix of the eigenvalues which provide the normal or fundamental frequencies. The choice of the vibrational coordinates will determine the numerical values of the $\underline{\underline{G}}$, $\underline{\underline{F}}$ and $\underline{\underline{L}}$ matrices appearing in Equation (7). Since linear transformations relate one basis set to another, $\underline{\underline{G}}$ and $\underline{\underline{F}}$ matrices transform under similarity transformations, thus the characteristic frequencies do not change.

The numerical calculations of the eigenvalues and eigenvectors are easily performed by computer programs routinely available in all laboratories. Attention should be paid to numerical techniques which must provide good and acceptable accuracies both for eigenvalues and eigenvectors even in the case of truly degenerate or quasi-degenerate eigenvalues.

The method which at present ensures the most satisfactory solutions [23] is the so-called "Jacoby diagonalisation". Such a method can handle only symmetrical matrices. Since the matrix $\underline{\underline{GF}}$ of Equation (7) is not symmetrical, manipulations are required.

Using Equation (4), Equation (7) can be rewritten in the form

$$\underline{\underline{D}}\,\underline{\underline{\Gamma}}^{\frac{1}{2}}\,\underline{\underline{\Gamma}}^{\frac{1}{2}}\,\tilde{\underline{\underline{D}}}\,\underline{\underline{F}}\,\underline{\underline{D}}\,\underline{\underline{\Gamma}}^{\frac{1}{2}}\,\underline{\underline{\Gamma}}^{-\frac{1}{2}}\,\tilde{\underline{\underline{D}}}\,\underline{\underline{L}} = \underline{\underline{L}}\,\underline{\underline{\Lambda}}$$

or

$$(\underline{\underline{\Gamma}}^{\frac{1}{2}}\ \tilde{\underline{\underline{D}}}\ \underline{\underline{F}}\ \underline{\underline{D}}\ \underline{\underline{\Gamma}}^{\frac{1}{2}})\ \underline{\underline{\alpha}} = \underline{\underline{\alpha}}\ \underline{\underline{\Lambda}} \tag{8}$$

The matrix in brackets is now symmetrical. Since it is related to GF by a similarity transformation, the eigenvalues remain the same.

If the vanishing eigenvalues of $\underline{\underline{G}}$ appearing in $\underline{\underline{\Gamma}}$ and the corresponding eigenvectors are left out*, Equation (8) has the dimension of the vibrational space, i.e. redundancies have been removed. The amplitudes of vibration $\underline{\underline{\alpha}}$ are defined in the (3N-6) dimensional space

$$\underline{\overline{\Sigma}} = \underline{\underline{\Gamma}}^{-\frac{1}{2}}\ \tilde{\underline{D}}\ \underline{R} = \underline{\underline{\Gamma}}^{-\frac{1}{2}}\ \underline{\Sigma} \tag{9}$$

There are now several possibilities: one can express the solution either in the $\overline{\Sigma}$ or in the Σ space (redundancies are removed), or one can return to the R space, i.e.

$$\underline{\underline{L}} = \underline{\underline{D}}\ \underline{\underline{\Gamma}}^{\frac{1}{2}}\ \underline{\underline{\alpha}}$$

where $\underline{\underline{\Gamma}}$ is the non-singular (3N-6) x (3N-6) part of the diagonalised $\underline{\underline{G}}$ and $\underline{\underline{D}}$ is the corresponding Mx(3N-6) matrix. Of course $\tilde{\underline{D}}_i\ \underline{\underline{L}} = 0$ when $\underline{D}_i$ belongs to $\gamma_i = 0$.

For large molecules symmetry factoring of the secular equation (when symmetry exists) is mandatory for the reduction of the computing time. Very often, however, the number of degrees of freedom is very large and secular equations of the order of 80 or 100 must be solved. The problem becomes worse when large polymeric systems must be treated. For polymers which possess translational symmetry (one-dimensional crystals) the wave-vector dependent dynamical matrix may be of high order and, if dispersion curves $\omega(k)$ must be calculated [11], the number of high order secular equations to be solved becomes large. When the polymer lacks any symmetry, as in disordered systems, the dynamical matrix reaches the order of thousands. Numerical techniques are however available for calculating both eigenvalues and eigenvectors [24]. The application of the so-called Negative Eigenvalue Theorem [25] allows the histogram of the density of vibrational states $g(\omega)$ to be plotted, which represents the number of frequencies occurring in a chosen frequency range. The resolving power of this computer experiment can be chosen at will and the steps of the histogram taken very narrow so as to reach a single exact eigenvalue. The application of the inverse iteration method [26] allows the calculation of the approximate eigenvectors corresponding to an approximate eigenvalue. The more exact the eigenvalue the more exact is the corresponding eigenvector. These numerical techniques have been widely used in treating problems in polymer systems [27-29].

---

* Notice that $\underline{\underline{G}} = \underline{\underline{D}}\ \underline{\underline{\Gamma}}^{\frac{1}{2}}\ \underline{\underline{\Gamma}}^{\frac{1}{2}}\ \tilde{\underline{\underline{D}}}$ has the same dimensions and the same values also if the vanishing eigenvalues and corresponding eigenvectors are omitted in the matrix multiplication. Confusion on this point can be found in the literature.

4. CALCULATIONS OF FORCE CONSTANTS

Let us, for the time being, restrict ourselves to the space of internal coordinates R. The largest effort of many workers has been made to calculate quadratic force constants from experimental frequencies in the hope of determining transferable and physically meaningful sets of parameters.

It is well known that even for very simple molecules the number of symmetrically nonequivalent and independent force constants in a symmetric n-dimensional block is $n(n+1)/2$. This number is much larger than that of the experimental frequencies available from a single molecule. Attempts have been made to reduce the number of unknown parameters either by defining suitable models of the potential or by introducing additional experimental data in order to reach the condition "number of experimental data > number of parameters to be calculated". As usual least squares procedures had to be introduced.

The basic concepts which allow the development of least squares calculations are the following. It can be shown [30] that the relation between the eigenfrequencies λ_i and the force constants F_{ij} can be expanded in a Taylor series

$$\Delta\lambda_i = \sum\left(\frac{\partial\lambda_i}{\partial F_{jk}}\right)\Delta F_{jk} + \tfrac{1}{2}\sum\sum\left(\frac{\partial^2\lambda_i}{\partial F_{jk}\partial F_{\ell m}}\right)\partial F_{jk}\partial F_{\ell m} + \ldots \tag{10}$$

Only the first term of Equation (10) is retained. From a known $\underline{\underline{G}}$ matrix and an initial $\underline{\underline{F}}^\circ$ matrix the eigenvalue equation can be solved

$$\underline{\underline{G}}\ \underline{\underline{F}}^\circ\ \underline{\underline{L}}^\circ = \underline{\underline{L}}^\circ\ \underline{\underline{\Lambda}}^\circ \tag{11}$$

$\underline{\underline{\Lambda}}^\circ$ is the diagonal matrix of the calculated frequencies resulting from the starting guess of $\underline{\underline{F}}^\circ$. $\underline{\underline{L}}^\circ$ are the corresponding eigenvectors. Using Equation (10) it can be shown that

$$\Delta\lambda_i = \sum_j[(L^\circ_{ji})^2\Delta F_{jj}] + 2\sum_{j<i}[L^\circ_{ji}L^\circ_{ki}\Delta F_{jk}] \tag{12}$$

which can be written in matrix notation as

$$\underline{\Delta\lambda} = \underline{\underline{J}}\ \underline{\Delta F} \tag{13}$$

where $\underline{\Delta\lambda}$ is a column matrix of $\Delta\lambda_i$, $\underline{\Delta F}$ is a column matrix of the elements ΔF_{ij} and $\underline{\underline{J}}$ is a rectangular matrix containing the products of the L°_{ij} elements, calculated in Equation (11). One moves on to least squares calculations in trying to compute from Equation (13) the ΔF corrections to $\underline{\underline{F}}^\circ$ such that the errors $\Delta\lambda_i = \lambda_i^{\circ bs} - \lambda_i^{\circ cal}$ are minimised. From Equation (13) one has

$$\tilde{\underline{\underline{J}}}\ \underline{\underline{P}}\ \underline{\Delta\lambda} = (\tilde{\underline{\underline{J}}}\ \underline{\underline{P}}\ \underline{\underline{J}})\ \underline{\Delta F} \tag{14}$$

$$\underline{\Delta F} = (\tilde{\underline{\underline{J}}}\ \underline{\underline{P}}\ \underline{\underline{J}})^{-1}\ \tilde{\underline{\underline{J}}}\ \underline{\underline{P}}\ \underline{\Delta\lambda} \tag{15}$$

and one calculates the corrections to be applied to $\underline{\underline{F}}^{\circ}$ such that the sum of weighted squares of the residuals is minimised. A new force constant matrix $\underline{\underline{F}} = \underline{\underline{F}}^{\circ} + \underline{\underline{\Delta F}}$ is then generated and the cycle is repeated several times until (if) $\underline{\Delta F}$ becomes very small. $\underline{\underline{P}}$ is a diagonal matrix comprising the statistical weights which may be given to the observed quantities λ_i^{obs}.

The column matrix of the force constants $\underline{F}$ can be expressed in terms of a chosen set of quadratic force constants ϕ by the relation

$$F_{jk} = \sum_{\ell} Z_{jk\ell}\phi_{\ell} \tag{16}$$

or in matrix notation

$$\underline{F} = \underline{\underline{Z}}\ \underline{\phi} \tag{17}$$

$\underline{\underline{Z}}$ is a rectangular matrix of dimensions n x m where n is the number of force constants in the $\underline{\underline{F}}$ matrix and m is the number of parameters in $\underline{\phi}$. One can then write

$$\underline{\Delta F} = \underline{\underline{Z}}\ \underline{\Delta\phi} \tag{18}$$

and substitution of Equation (18) in Equation (14)

$$\underline{\underline{\tilde{J}}}\ \underline{\underline{P}}\ \underline{\Delta\lambda} = (\underline{\underline{\tilde{J}}}\ \underline{\underline{P}}\ \underline{\underline{J}})\ \underline{\underline{Z}}\ \underline{\Delta\phi} \tag{19}$$

gives, by left multiplication by $\underline{\underline{Z}}$, the normal equation

$$\underline{\underline{\tilde{Z}}}\ \underline{\underline{\tilde{J}}}\ \underline{\underline{P}}\ \underline{\Delta\lambda} = (\underline{\underline{\tilde{Z}}}\ \underline{\underline{\tilde{J}}}\ \underline{\underline{P}}\ \underline{\underline{J}}\ \underline{\underline{Z}})\ \underline{\Delta\phi} \tag{20}$$

from which

$$\underline{\Delta\phi} = (\underline{\underline{\tilde{Z}}}\ \underline{\underline{\tilde{J}}}\ \underline{\underline{P}}\ \underline{\underline{J}}\ \underline{\underline{Z}})^{-1}\ \underline{\underline{\tilde{Z}}}\ \underline{\underline{\tilde{J}}}\ \underline{\underline{P}}\ \underline{\Delta\lambda} \tag{21}$$

Equations (17) and (21) are of basic importance in the field of force constant calculations. Indeed, the efforts made by several workers to overcome the problem of the determination of quadratic force constants are reflected in and make large use of such equations.

Choice of the interatomic potential model

Since the potential is not known, several models have been proposed in the literature and have been applied with varying degree of success to the study of the normal vibrations of molecules and vice versa. Among those we note the Central-Force-Field (CFF) and the so-called Urey-Bradley-Force-Field (UBFF), renamed also Shimanouchi-Urey-Bradley-Force-Field (SUBFF) since Shimanouchi has revived [31] in the case of organic molecules the model proposed by Urey and Bradley. A Simple-Valence-Force-Field (SVFF) with only diagonal terms was also proposed and later modified with arbitrary introduction of several interaction terms*. For improving the success in

---

* Only a diagonal SVFF would have number of force constants < number of experimental frequencies; however, such a field does not reproduce frequencies.

reproducing frequencies authors started introducing additional constants both in the SUBFF (variously named as Modified-Shimanouchi-Urey-Bradley-Force-Field (MSUBFF), etc.) or in the Valence-Force-Field. Efforts have been made (and have reached a reasonable success) to reach the General-Force-Field (GFF) for the case of small molecules where all elements F_{ij} are evaluated. For large molecules a GFF has never been reached.

The main concept behind a potential model in the case of large molecules is to reduce the number of parameters in a systematic and impersonal way, once the recipe is adopted for all the molecules studied. The definition of the model determines the matrices Z and ϕ of Equation (17). SUBFF and MSUBFF have reached a high degree of success and application in the sixties mainly by the Japanese School, which has also tried to give some physical meaning to the calculated quantities [31] in terms of non-bonded atom-atom interactions somewhat related to the Lennard Jones potential. The need to improve the results has however forced the authors to introduce additional valence type interactions. With the availability of fast computers which has allowed a large degree of freedom in the calculations, workers smoothly turned their attention again to a valence force field where several (even if arbitrary) interaction terms were introduced. At present for most large organic molecules a valence force field is used.

In a way parallel to that developed for small polyatomic molecules in the gas phase, attempts have been recently made for large molecules (generally in the solid state) to correlate several physical properties which contain a vibrational contribution in order to improve the set of experimental data for the determination by a "grand least squares refinement" of the parameters of an "overall force field". Generally valence terms have been included together with atom-atom interactions of some sort in order to reproduce the vibrational spectrum together with lattice constants, heat of sublimation, specific heat and derived thermodynamic quantities, etc. [32]. This kind of potential is very popular among workers in the field of conformational analysis or when crystal structures of large molecules must be predicted. The popularity of these force fields is much less among spectroscopists since the predicted vibrational frequencies (frequencies and symmetry species) are very rough and generally much too far from the general fitting accepted by specialists in normal coordinate calculations.

Use of experimental frequencies from isotopic derivatives

Since the potential remains the same from one isotopic species to another, the number of experimental data can be greatly increased if isotopic derivatives are studied. The least squares calculation outlined above can be extended in order to take into account data from isotopic molecules. The transferability of the force field from one isotopic molecule to another is in this case mathematically exact. Equation (13) can be rewritten as

$$\underline{\Delta\lambda} = \underline{\underline{J}}\ \underline{\underline{Z}}\ \underline{\Delta\phi} \qquad (22)$$

and can be applied for a series of molecules in the following way

$$\begin{bmatrix} \underline{\Delta\lambda}^{(1)} \\ \underline{\Delta\lambda}^{(2)} \\ \cdots \\ \underline{\Delta\lambda}^{(n)} \end{bmatrix} = \begin{bmatrix} \underline{\underline{J}}^{(1)} & \underline{\underline{Z}}^{(1)} \\ \underline{\underline{J}}^{(2)} & \underline{\underline{Z}}^{(2)} \\ \cdots & \\ \underline{\underline{J}}^{(n)} & \underline{\underline{Z}}^{(n)} \end{bmatrix} \underline{\Delta\phi} \qquad (23)$$

where $\underline{\underline{J}}^{(i)}$ $\underline{\underline{Z}}^{(i)}$ are the matrices for a single molecule (or a symmetry block, if symmetry is operative). The final normal equation matrix can then be written as

$$\sum_{i=1}^{n} \widetilde{(\underline{\underline{J}}^{(i)}\underline{\underline{Z}}^{(i)})}\underline{\underline{P}}^{(i)}\underline{\Delta\lambda}^{(i)} = \sum_{i=1}^{n} \widetilde{(\underline{\underline{J}}^{(i)}\underline{\underline{Z}}^{(i)})}\underline{\underline{P}}^{(i)}(\underline{\underline{J}}^{(i)}\underline{\underline{Z}}^{(i)})\underline{\Delta\phi} \qquad (24)$$

It is thus possible to calculate from many frequencies of isotopic molecules a few force constants ϕ. These kinds of calculation have been and are routinely performed on all classes of molecules.

Frequencies from isotopic species, however, do not solve all the problems of force constants calculations. A careful choice of the isotopic atoms as well as of their positions in the molecule is required. Since the least squares procedure is based on frequency shifts they must be larger than the intrinsic uncertainties of the least squares calculations. For this reason, while H/D substitution is very useful since in general it generates large frequency shifts, other substitutions such as $^{12}C/^{13}C$, $^{14}N/^{15}N$, etc. become useful only in special cases which must be properly and carefully chosen. For instance they may become useful if added as special constraints in the least squares calculations [33-35]. Even H/D substitution may not solve the problem when the motions of heavy atoms are considered. In such a case H and D motions may be little coupled with the motions of the heavy atoms thus causing little or no shift.

Overlay with frequencies of similar molecules

Equation (22) offers another way to calculate force constants from experimental data. As already stated the transferability of force constants to isotopic derivatives is strictly mathematically correct. A mathematically non-rigorous, but chemically reasonable, way is to assume that the concept of transferability of force constants between chemically similar molecules may be applied. In such a case the same set of ϕ can be adopted in Equation (23) where each (i) does not refer to frequencies from isotopic molecules but from chemically similar molecules. In other words, with the help of chemical feeling, it is assumed that bonds in a chemical class of compounds are similar and their motions can be described by the same set of parameters. Such a method is arbitrary in the sense that chemical intuition and its validity are only verified by the success of the calculations in reproducing frequencies.

Based on this assumption, the so-called "overlay" least squares refinement has been worked out and translated in computing programs widely used [30,36]. When the fitting between experimental and calculated frequencies is not satisfactory the chemical intuition of the worker has to be modified, or, in other words, the chemical bonds shown not to be the same within the class of compounds chosen.

Even when all these devices are used, the number of force constants to be calculated from experimental frequencies is still too large for most large molecules. To make the problem solvable some (or many) interaction terms are arbitrarily forced to be equal or to be zero. In calculations of this type the degree of freedom of each researcher becomes practically unlimited and each author takes his own decision even if based on logical (but philosophical) grounds. The general procedure in an overlay calculation on many molecules is to test the validity of a given arbitrary choice of interaction constants by looking at the rate of convergence, at the statistical dispersion of the chosen constants and at the numerical values of the force constants which must be "reasonable" within the logical frame built up by the previous experience of the spectroscopist. Unfortunately "ab initio" methods in general are still too approximate to give even an indication of the order of magnitude of the interaction constants for reasonably large molecules. Systematic numerical procedures provide the way to test numerically the effectiveness of a given interaction constant in the calculations (see the "regression" program in ref. 30).

The main validity of a set of force constants obtained by the methods just described is given by their capability of reproducing the spectra of the compounds included in the refinement and of predicting the spectra of similar compounds. Before expressing a final, even if personal, judgement on the validity of such types of calculation, additional factors should be taken into account.

Uncertainity of the experimental frequencies

The true value of the experimental fundamental frequency to be introduced say in Equation (13) can never be known with certainty and fairly large uncertainties must be allowed in the case of large molecules (say $\sim 10\ cm^{-1}$) because of practical reasons.

First the accuracy and the resolving power of the spectrometers used makes the reading of the frequencies inaccurate (sometimes data are taken from the infrared, sometimes from the Raman spectra). Secondly, for most large molecules the experimental frequencies are obtained from the compounds in different phases (gas, liquid, solution in different solvents, solid). Mechanical and electrical effects make the fundamental frequencies phase dependent without the possibility of introducing reasonable corrections.

Another important source of uncertainty in the location of the fundamentals is the obvious difficulty in the accurate identification of all or part of the fundamentals in such large molecules. We assume in this article that the basic vibrational assignment has been carried out in the most critical way by using all the evidence which experiments can provide [37]. For such large molecules the identification of the fundamentals and of their symmetry species is sometimes very difficult because of strong overlapping, intrinsic weakness of the transition, existence of a large population of overtones and combinations, etc. Fermi resonance is not mentioned here since, even if it exists, it is hard to account for in complex cases.

All these uncertainties should be somehow translated into the weighting matrix $\underline{\underline{P}}$ of Equation (15), which should tell the computer our knowledge or ignorance on each observed frequency given to the least squares calculations. Actually the evaluation of the uncertainty is impossible and any attempt to change the weight P_i of frequency ν_i is very often a game with numbers. When many large molecules are introduced in the least squares adjustment it is preferable to leave some of the frequencies out of the calculation if their identification is not so satisfactory.

Use of anharmonic frequencies

The fundamental frequencies which are read on the experimental spectrum derive from the motions of the atoms in an anharmonic potential. Normal coordinate calculations are instead based on the approximation of a harmonic potential. The experimental frequencies to be used in the calculation of a harmonic potential should be suitably corrected by taking into account the anharmonic contributions. Methods have been proposed for such corrections [37,38] which have seldom been applied even in the case of small polyatomic molecules. These corrections are generally large for X-H stretching motions (where X = C,N,O,S,P, etc.) and are still not negligible for bending motions when H or D atoms are involved. For heavier atoms corrections are generally negligible. In the absence of enough data for a reliable calculation of the corrections for anharmonicity most of the force constant calculations on large molecules neglect the anharmonicity and use the experimental value as read from the spectrum. With such an approximation all force constant calculations on the stretching of C-H, O-H, N-H, etc. have only an indicative value; most times C-H stretchings are even left out of the least squares calculations. Another consequence of this approximation is the following: if a set of force constants for C-H motions (including the bending and torsional motions) is derived from an "overlay" calculation on only perhydrogenated molecules, it will turn out to be different if the calculations are made with an "overlay" only on perdeuterated molecules. Hence force constants derived from only hydrogenated molecules may not be transferable to the same set of perdeuterated molecules. This fact is not a contradiction to the exact transferability of force constants among isotopic species discussed above, but rather a consequence of the approximations and compromises just discussed in the calculation of force constants of large molecules.

Problems of a least squares refinement of quadratic force constants

A satisfactory convergence of the least square refinement towards a set of force constants which minimises the weighted sum of the squares of the residuals is not always reached. First the inversion of the matrix ($\tilde{\underline{\underline{Z}}}$ $\tilde{\underline{\underline{J}}}$ $\underline{\underline{P}}$ $\underline{\underline{J}}$ $\underline{\underline{Z}}$) appearing in Equation (21) may not be feasible. The most common reasons are the following:

(i) the matrix contains one or more singularities and cannot be inverted. The singularities can occur if one or more force constants are linearly related because of redundancies among internal coordinates or if the number of experimental data is smaller than the number of parameters we wish to calculate.
(ii) the matrix is quasi singular, i.e. it contains one or more very small eigenvalues such that its determinant becomes extremely small. The inversion of the matrix becomes numerically unstable. In such a case one or more force constants we are trying to adjust are ill-determined, i.e. they are not sufficiently determined by the experimental frequencies used in the calculation.

Analysis of the eigenvalues and eigenvectors of the normal equation matrix allows us to find which force constants are poorly defined in the calculations. Generally these force constants appear with a very large statistical uncertainty in the calculation [30].

Least squares calculations based on the drastic approximation of a linear relationship between force constants and frequencies (Equation (13)) by subsequent iterations may converge to a set of force constants which is certainly not the unique set. The existence of a multiplicity of solutions has been discussed by several authors [39,40] but very few practical suggestions to avoid the problem have been actually offered.

One can also reach different minima by using a different starting set of zero order parameters $\underline{\underline{F}}^{\circ}$ in Equation (11). Another possibility is that the refinement process oscillates between different minima.

If the problem of multiple solutions does exist for a least squares determination of a GFF for small molecules, the problem is even worse in the case of large molecules. This fact has to be clearly kept in mind when using published tables of quadratic force constants. The set one decides to adopt is "one of the possible sets" which reproduce the frequencies of the molecules considered and is able to predict the spectrum of other similar molecules. A typical example can be found in the well tested case of paraffins [41]. On the basis of much experimental evidence the B_{2g} Raman active CH_2 wagging mode of polyethylene (which is the limiting mode for the wagging-bending dispersion curve of paraffins) had to be reassigned from a band at 1415 cm^{-1} to another at 1385 cm^{-1}. A new least squares refinement of the force constants generated a set of parameters not dramatically different from the one which originally fitted the 1415 cm^{-1} band.

In the opinion of the author a few practical suggestions can be offered. The sensitivity of the calculation to the above problems must be tested carefully during the actual calculation, i.e. the stability of the problem must be verified. One way to assure a reasonably narrow convergence towards a statistically well defined set of parameters (i.e. parameters with small uncertainties) is to do the calculations only when the ratio of "number of experimental data/number of parameters" approximates 4 or 3. This suggestion has by no means any theoretical justification but it is dictated only by practical experience with calculations in the case of organic molecules.

In the final set of force constants to be accepted each parameter must have a small statistical uncertainty. These uncertainties must always be reported.

Calculation of force constants for redundant sets of coordinates

In the expression of the potential energy given in Equation (2) the linear term can be omitted in the assumption that the internal coordinates R_i are independent. On the contrary, if a redundancy exists the forces acting in the molecule on each R_i at the equilibrium do not vanish. The elimination of the linear term may be accomplished by the use of undetermined Lagrange multipliers.

The redundancies among the Wilson's coordinates are linear (Equation (5)):

$$\sum_i D_{ij} R_i = 0 \qquad (25)$$

Each of them can be multiplied by a parameter λ_j and added to the r.h.s. of Equation (2). The conditions for V to be stationary are then

$$\frac{\partial V}{\partial R_i} + \sum_j \lambda_j D_{ij} = 0 \qquad i=1,M;\ j=1,M-(3N-6) \qquad (26)$$

Such equations determine all the forces at the equilibrium as a parametric function of the Lagrange multipliers λ_j. Because of Equation (26), Equation (2) reduces to

$$2V = \sum_{ij} F^R_{ij} R_i R_j \qquad (27)$$

also when the R_i are independent.

The λ_j are also called "intramolecular tensions" because they describe the tensions in the springs at the equilibrium due to the existence of constraints. The tensions are certainly non-vanishing but they do not influence the vibrational frequencies, as it can be seen in Equation (27). This statement is correct if the whole vibrational problem is stated and solved in the linear space of Wilson's internal coordinates [14,42].

However, Mills [15] has pointed out that it may be desirable sometimes to express the potential energy in terms of curvilinear coordinates ρ. If the ρ coordinates are dependent, the redundancy among them may be non-linear:

$$\sum_i \Delta_{ij} \rho_i + \sum_{ik} \Delta_{ikj} \rho_i \rho_k = 0 \qquad (28)$$

The removal of the linear term in the potential energy

$$2V = 2\sum_i \left(\frac{\partial V}{\partial \rho_i}\right)_o \rho_i + \sum_{i,k} \left(\frac{\partial^2 V}{\partial \rho_i \partial \rho_k}\right)_o \rho_i \rho_k \tag{29}$$

leads then to

$$2V = \sum_{i,k} \left[\left(\frac{\partial^2 V}{\partial \rho_i \partial \rho_k}\right)_o + \sum_j (\lambda_j \Delta_{ikj})\right] \rho_i \rho_k \tag{30}$$

In this case the "intramolecular tensions" affect the vibrational frequencies. Since the secular equation can be solved only in the linear space [43], before inserting the potential energy into the secular equation, a transformation to Wilson's internal coordinates (linear) must be performed:

$$\rho_i = \sum_k t_{ik} R_k + \sum_{km} t_{ikm} R_k R_m \tag{31}$$

If the ρ coordinates are chosen in such a way that $t_{ik} = \delta_{ik}$, the harmonic potential in the linear space is given by

$$2V = \sum_{ik} \left[\left(\frac{\partial^2}{\partial \rho_i \partial \rho_k}\right)_o + \sum_j \lambda_j \Delta_{ikj}\right] R_i R_k \tag{32}$$

The $\underline{\underline{F}}_R$ matrix given by Equation (32) can then be multiplied by the usual $\underline{\underline{G}}_R$ matrix and the vibrational problem solved: the λ_j play there the same role as the force constants.

The use of intramolecular tensions is legitimate every time one feels it necessary to make explicit reference to a potential described in curvilinear coordinates. In this chapter, however, we deal mainly with the problem of finding a suitable set of parameters which can reproduce the frequencies and elucidate the structure and dynamical behaviour of chemically similar systems of large molecules. In this case we think it unnecessary and unwise to introduce additional parameters as the intramolecular tensions making reference to a curvilinear potential when also the determination of the "linear" force field may be hard. As a matter of fact, the more widely used force fields for large systems of molecules never use intramolecular tensions (see for instance ref. 41).

We mentioned earlier some computational problems connected with the singularity of the normal equation matrix. A peculiar problem arises in the calculation of F when one or more redundancies exist. As mentioned above the matrix ($\tilde{\underline{\underline{Z}}}$ $\tilde{\underline{\underline{J}}}$ $\underline{\underline{P}}$ $\underline{\underline{J}}$ $\underline{\underline{Z}}$) appearing in Equation (21) cannot be inverted if some force constants are linearly related because of redundancies. The simple example of a diatomic molecule clarifies the problem. If the only degree of vibrational freedom of such a molecule is described by two stretching coordinates R_1 and R_2 with corresponding force constants k_1 and k_2, from one frequency we can only determine one force constant $K = k_1 + k_2$. The splitting of the value of K into k_1 and k_2 is obviously arbitrary. One can well set $k_1 = K$ and $k_2 = 0$. A similar problem (or a more complex one when many relationships exist among diagonal and off-diagonal elements) is always found in calculations of large molecules, where local or cyclic redundancies always occur.

It is beyond the purpose of this chapter to discuss in detail the various difficulties introduced in the dynamical problem when such cases occur. We wish here only to discuss a few points we feel relevant for the problems treated in this chapter. Authors have dealt with these difficulties in various ways:

(i) One way is to treat the whole dynamical problem in terms of 3N-6 independent symmetry coordinates and to report only "symmetrised force constants". This corresponds, in the example above, to reporting only K. Such a procedure is mathematically rigorous and exact and is adopted by several authors in order to avoid the breaking down of each symmetrised force constant into the various F^R_{ij} when the R are redundant. The use of symmetrised force constants may be useful when comparing trends of values in sets of structurally similar simple and very symmetrical molecules. A possible improvement would require the use of sets of standardised symmetry coordinates on which all authors should agree [43]. The symmetrised force constants are of no use in the study of transferability to large molecules.

(ii) Most authors decide instead to break the combinations of force constants F^S into one of the possible sets of F^R and report a chosen set of values analogous to one of the possible pairs k_1 and k_2 of the diatomic molecule taken as example. Such a procedure is completely arbitrary and adds another decision to the many already taken in force constants calculations. Such arbitrariness involves both diagonal and off-diagonal elements. It has then to be pointed out that in the process of least squares refinement of force constants in terms of internal coordinates, while a few constants may not be numerically determinable because of lack of experimental data, others are not numerically determinable because of redundancies.

Workers in this field should be aware of the facts just discussed either when a choice of the parameters of the potential is made or when data of different authors are compared. The decisions taken should be clearly stated in reporting the calculated numbers.

When a fitting can be considered satisfactory

When all the complicating factors discussed above are taken into account a dogmatic statement on when a fitting can be considered satisfactory becomes practically impossible. Unbelievably successful frequency fittings have been reported in the literature for rather complex molecules: on the other hand conclusions on structural properties or on vibrational assignments have been reported based on an average frequency fitting of ± 50 cm^{-1}. Each author takes his own viewpoint depending on the aim of his work, on the computing facilities at his disposal and on the type of spectra from which the experimental data were taken.

The most important factor in judging the frequency fitting is again the ratio between the number of experimental data and the number of parameters considered in the calculations. The general rule that "by adding new constants one can fit everything" holds very seriously in this case and throws strong doubts on some sets of force constants published.

When force constants are calculated on the basis of a very large set of experimental frequencies, one possible good way to judge the frequency fit is to plot the histogram of the frequency differences for all frequencies considered. If the distribution of frequency errors is centred around zero and is smoothly and symmetrically distributed with a narrow half width ($\Delta\nu_{\frac{1}{2}} \simeq 10\ cm^{-1}$), calculations can be considered "satisfactory". Examples of these plots can be found in refs. 44 and 45.

5. PHYSICAL MEANING OF QUADRATIC FORCE CONSTANTS

At the beginning of this article it was mentioned that one of the main hopes of the chemist is to learn about the nature of the chemical bond in molecules from vibrational frequencies and from the derived force constants. The discussion so far presented in the case of large molecules clearly points out that the number of approximations adopted and the number of compromises with the reality of the spectra or with the difficulties of numerical calculations make the physical meaning of the calculated constants very doubtful.

Mention must be made here to the very interesting attempts made by Bernstein [46] and by McKean and Duncan [47] to derive bond properties from C-H stretching frequencies and force constants. These correlations are made on small molecules whose spectra have been accurately analysed. Qualitative correlations between force constants and the nature of the bond have been reviewed by Shimanouchi [31].

In the opinion of the writer calculated force constants are simply a set of vibrational parameters "carefully and critically" determined, whose physical meaning is practically nil, but which are very valuable for predicting spectra of similar molecules or for providing data for the approximate prediction of other physical properties related to vibrations. Obvious overall approximate trends or order of magnitude differences can be found. Force constants for C-C, C=C and C≡C stretchings show a logical trend [7]; it has to be remembered, however, that the stretching of the C=O group in acetone could be satisfactorily reproduced with a force constant for the C=O in the range 8 to 12 mdyne $Å^{-1}$ i.e. from a quasi single bond to a quasi triple bond!

The limit of force constant calculations and their related physical meaning is clearly shown in Table 1, which is given as an example for this discussion. In Table 1 we report the results of calculations of quadratic force constants of a network of C-C bonds in a set of molecules for which chemical intuition would tell us that they are very similar, if not identical. Calculations were performed from data derived from two different systems (diamond and normal paraffins) and from two completely different physical techniques: data for diamond are derived from phonon dispersion curves obtained from neutron scattering experiments [48]; data for normal paraffins are derived from spectroscopic data (infrared and Raman) [45]. The

Table 1. Valence force constants ϕ_i for C-C bonds and uncertainties $\sigma(\phi_i)$ derived from phonon dispersion curves of diamond and from the infrared and Raman spectra of normal paraffins[a].

	Diamond[b]		n-Paraffins[c]	
	ϕ_i	$\sigma(\phi_i)$	ϕ_i	$\sigma(\phi_i)$
K_R	3.831	0.023	4.532	0.041
H	0.872	0.121	1.302	0.085
F_R	0.164	0.017	0.083	0.027
F_H	0.392	0.012	0.303	0.018
$F_{\Lambda'}$	-0.015	0.010	-	-
$F_{\Lambda'''}$	0.173	0.043	{ 0.097[d]	0.018
			-0.005[d]	0.021

a Units: stretch and stretch-stretch in mdyne Å^{-1}, bend and bend-bend in mdyne Å rad^{-2}, stretch-bend in mdyne rad^{-1}.
b Ref. 48.
c Ref. 45
d In ref. 45 values are reported for trans and gauche conformations, f^t_ω and f^g_ω (symbols as in ref. 45)

number of experimental frequencies in both cases is very large, moreover frequencies for diamond refer only to motions of carbon atoms while in the case of paraffins the number of frequencies referring to the motions of carbon atoms is still large. In the light of our previous discussion the following comments can be made: (i) the ratio of number of experimental frequencies to number of parameters is large; (ii) force constants are statistically well determined; (iii) calculations show a good convergence toward a stable solution for the set of F^R chosen; (iv) the uniqueness of the solution for hydrocarbons is not so strongly established even if the various sets found do not dramatically differ especially for the constants related to C-C bonds.

In the light of the present discussion Table 1 shows that great caution must be taken in describing the nature of the C-C bond from its quadratic force constants even if so carefully determined.

The types of information chemists would like to derive from force constant calculations can be described with two examples which the authors and his collaborators have carried out just with the philosophy discussed in this chapter. The first case is that of monohalogeno-diacetylenes H-C≡C-C≡C-X (X = Cl,Br,I) [49]. A GVFF for these molecules should contain 15 independent constants for the $5\Sigma^+$ parallel modes and 10 independent constants for the 4Π perpendicular modes. Since no data from isotopic species were available we have carried out an "overlay" calculation on the three molecules for which careful experimental data by Klaboe et al. [50] were available. The question we wished to answer was the following: are force constants a good way to reveal differences in the electronic structure of similar molecules where only the halogen atom is changed? Parallel and perpendicular modes could be treated independently and the potential functions adopted after several attempts were the following:

$$2V\ (\text{bond stretching}) = [f(C{-}H)\Delta R^2 + f'(C{\equiv}C)\Delta l_1^2 + f(C{-}C)\Delta r^2 + f''(C{\equiv}C)\Delta l_2^2 + f'(C{\equiv}C/C{-}C)\Delta l_1 \Delta r]_{\text{overlay}} + [f(C{-}X)\Delta t^2 + f''(C{-}C/C{\equiv}C)\Delta r \Delta l_2]$$

$$2V\ (\text{angle bending}) = [\phi(H{-}C{\equiv}C)\Delta_1^2 + \phi'(C{\equiv}C{-}C)\Delta\phi^2 + \phi(H{-}C{\equiv}C/C{\equiv}C{-}C)\Delta\phi_1\Delta\phi_2]_{\text{overlay}} + \phi''(C{-}C{\equiv}C)\Delta\phi_3^2 + \phi(C{\equiv}C{-}X)\Delta\phi_4^2$$

The frequency fitting obtained is satisfactory, force constants are well determined. From Table 2 one can try to answer the

Table 2. "Stretching" and "bending" force constants and statistical dispersions of monohalodiacetylenes from an overlay calculation.

	Chloride		Bromide		Iodide	
	$\bar{\phi}_i$	$\sigma(\bar{\phi}_i)$	$\bar{\phi}_i$	$\sigma(\bar{\phi}_i)$	$\bar{\phi}_i$	$\sigma(\bar{\phi}_i)$
f(C-H) [a]	5.91	-	5.91	-	5.91	-
f'(C≡C)	15.97	0.38	15.97	0.38	15.97	0.38
f(C-C)	7.20	0.15	7.20	0.15	7.20	0.15
f"(C≡C)	14.11	0.32	14.11	0.32	14.11	0.32
f'(C≡C/C-C)	0.87	0.27	0.87	0.27	0.87	0.27
f(C-X)	5.19	0.12	4.53	0.13	3.59	0.11
f"(C-C/C≡C)	0.51	0.29	0.64	0.30	0.92	0.32
ϕ(C≡C-H)	0.132	0.0057	0.132	0.0057	0.132	0.0057
ϕ'(C≡C-C)	0.37	0.086	0.37	0.086	0.37	0.086
ϕ"(C-C≡C)	0.41	0.040	0.31	0.054	0.27	0.057
ϕ(C≡C-X)	0.36	0.06	0.57	0.11	0.68	0.13
ϕ(C≡C-H/C≡C-C)	0.126	0.036	0.126	0.036	0.126	0.036

a Force constant held fixed.

original question. Diagonal stretching force constants for each internal coordinate remain the same for the three molecules, with the obvious exception of f(C-X). While f'(C≡C/C-C) remains the same, f"(C-C/C≡C) does change from Cl to I. It is then concluded from the Σ^+ modes that the -C≡C-H portion of the molecule is electronically the same for the three molecules. The perturbation by the halogen atom goes as far as the C≡C to which it is attached. Support for this conclusion comes from the similar behaviour of the data from the Π modes.

Analogous types of informations were previously derived from the series of propargyl halides $CH_2X-C \equiv C-H$ (X = F,Cl,Br, I) [51]. In such a case a correlation between the electronegativity of the X atom with the splitting of in and out-of-plane $-C \equiv C-H$ deformation modes suggested on the basis of frequency correlations [52] has been shown not to be true, but only dependent on the kinetic effect because of the different mass of the halogen.

A great caution, however, has to be applied in deriving such types of information. Finer studies which take into account vibrorotational interactions, Fermi resonances, anharmonic potential, etc. may well destroy the picture of the molecules so nicely derived from quadratic force constants.

6. SOME POSITIVE ASPECTS OF FORCE CONSTANT CALCULATIONS OF LARGE MOLECULES

We wish to report here some constructive contributions to molecular spectroscopy, molecular structure and dynamics given by quadratic force constants derived from large molecules in spite of all the limitations and approximations discussed before.

Structural Analysis

Most organic molecules in the gas and liquid phase exist in several conformational states since no intermolecular forces force them in one of the conformational minima they generally choose when packed in a crystal. The case of structural disorder in the solid state is less common. The practical possibility of determining the existence and the type of conformers in a conformational mixture strongly relies on vibrational spectroscopy interpreted via normal coordinate calculations. The types of conformers in the liquid and solid states of normal paraffins [53] and ethers [4] have been determined in just this way. The force field derived from normal paraffins [54] has allowed the prediction of the spectrum of the planar zigzag structure of syndiotactic polypropylene before that modification of the polymer was actually obtained [55]. The fit between the predicted and experimental spectrum turned out to be very satisfactory. The application of the force field of refs. 54 and 4 has provided the way to understand the structure of polyolefins [56] and polyethers [4] respectively.

Vibrational assignment

The difficulties in locating the fundamentals in the experimental (infrared and Raman) spectra of large polyatomic molecules is common knowledge and many authors have tried to support their analysis of the experimental data with normal coordinate calculations. The approach to this type of work varies and can be generally described as follows:

(i) Zeroth order calculations: A set of force constants well determined on similar molecules is applied for the prediction of the spectrum of the molecule under study and the fitting is judged without further refinement. When this is possible it is certainly the best unbiased way to support the vibrational

assignment [57,58].
(ii) Least squares calculations: The technical aspects of the problems have been discussed in this chapter. If some frequencies have not been located in the spectrum "overlay" calculations can give an indication where such bands should occur. In the author's opinion only a few frequencies should be left to be located by calculations [59].
(iii) Force constant adjustment on only one molecule: The modification of the force field to fit our own choice of the fundamentals should be avoided since most of the time a "reasonable" set of force constants which fits our assignment will always be found. Such a calculation is not a proof of the validity of the proposed vibrational assignment or a reasonable suggestion of reliable force constants.

Lattice dynamics, phonon dispersion curves and neutron scattering for chain molecules

Since intermolecular forces in organic polymers are at least one order of magnitude weaker than intramolecular forces, the dynamical properties of chain molecules can be studied on the basis of the "single chain model" where only intramolecular forces are acting. The knowledge of intramolecular force constants derived for large molecules allows several experiments which depend on the vibrations of organic chain molecules to be accounted for. As discussed in Chapter 24 of this book, phonon dispersion curves $\omega(k)$ and densities of vibrational states $g(\omega)$ can be experimentally obtained from neutron coherent and incoherent scattering experiments respectively. The same quantities can be predicted using quadratic force constants derived from least squares refinements [60-65]. Sometimes experimental phonon dispersion curves have provided additional vibrational experimental data to be inserted in a least squares calculation. The known cases are those of polytetrafluoroethylene [66], hexagonal ice [67], diamond, silicon, germanium and tin [48].

Elastic constants and chain length of organic polymers

The knowledge of intramolecular force constants of chain molecules allows the calculation of the elastic modulus of several polymeric chains and theoretical values have been compared with the experimental ones [68].

The recent availability of experimental or theoretical phonon dispersion curves in the low energy region has opened up the possibility of determining the elastic constants of infinite polymer chains from the slope of the longitudinal acoustic branch at zero wave vector [69,70]. For long, but not infinite, chain molecules the lowest (but non-zero) frequency along the longitudinal acoustic branch corresponds to the so-called "accordion motion" [69] for $k \simeq 0$. The observation of the accordion motion for finite chain molecules allows the determination of either the length of the chain [71] or its Young's modulus.

The knowledge of a good and reliable intramolecular force field is of basic importance for the determination of the exact slope of the acoustical longitudinal branch and of the location of the accordion motion.

7. REFERENCES

1. R.N. Jones and C. Sandorfy, in A. Weissberger (Editor), Chemical Applications of Spectroscopy, Techniques of Organic Chemistry, Vol. IX, Interscience, New York, 1956.
2. L. Bellamy, The Infrared Spectra of Complex Molecules, Wiley, New York, 1958.
3. K. Nakamoto, Infrared Spectra of Inorganic and Coordination Compounds, Wiley, New York, 1963.
4. See for example, R.G. Snyder and G. Zerbi, Spectrochim. Acta A, 23 (1967) 391.
5. J.L. Duncan, Force Constant Calculations in Molecules, Specialist Report, Chemical Society, Vol. 3, 1975, Chap. 2.
6. S. Califano, Vibrational States, Wiley-Interscience, New York, 1976.
7. E.B. Wilson, J.C. Decius and P.C. Cross, Molecular Vibrations, McGraw-Hill, 1955.
8. See for instance, P. Pulay and W. Meyers, Mol. Phys., 27 (1974) 473.
9. I. Suzuki, M.A. Pariseau and J. Overend, J. Chem. Phys., 44 (1966) 3561.
10. W.T. King, J. Chem. Phys., 61 (1974) 4026.
11. G. Zerbi, Molecular Vibrations of High Polymers, in J. Brame (Editor), Applied Spectroscopy Reviews, 2 (1969) 193.
12. See for instance, H. Matsuura, I. Harada and T. Shimanouchi, Proc. 5th Int. Symp. on Raman Spectroscopy, Freiburg, 1976.
13. B. Crawford and J. Overend, J. Mol. Spectry, 42 (1964) 307.
14. M. Gussoni and G. Zerbi, Chem. Phys. Letters, 2 (1968) 145; M. Gussoni and G. Zerbi, Rend. Accad. Nazionale Lincei, serie VIII, vol. XL, fasc. 5 (1966).
15. I.M. Mills, Chem. Phys. Letters, 3 (1969) 267.
16. S. Califano and B. Crawford, Z. Elektrochem., 64 (1960) 571.
17. T. Shimanouchi, J. Chem. Phys., 17 (1949) 245.
18. M. Gussoni and G. Zerbi, Rend. Accad. Nazionale Lincei, serie VIII, vol. XL, fasc. 6 (1966).
19. M. Gussoni and G. Zerbi, J. Mol. Spectr., 26 (1968) 485.
20. M. Gussoni, G. Dellepiane and G. Zerbi, J. Chem. Phys., 53 (1970) 3450.
21. M. Gussoni, G. Dellepiane and G. Zerbi, Vibrational Symmetry from Computers, in S.J. Cyvin (Editor), Molecular Structure and Vibrations, Elsevier, Amsterdam, 1972.
22. M. Gussoni and G. Dellepiane, Chem. Phys. Letters, 10 (1971) 559.
23. P.A. White and R.R. Brown, Math. Computation, July (1964) 457.
24. G. Zerbi, Defects in Organic Crystals: Numerical Methods, in S. Califano (Editor), Dynamics and Intermolecular Forces, Academic Press, New York, 1975.
25. P. Dean, Rev. Mod. Phys., 44 (1972) 126.
26. J.H. Wilkinson, The Algebraic Eigenvalues Problem, Clarendon Press, Oxford, 1965.
27. M. Gussoni and G. Zerbi, J. Chem. Phys., 60 (1974) 4862.
28. A. Rubcic and G. Zerbi, Macromolecules, 7 (1974) 754, 759; Chem. Phys. Letters, 34 (1975) 343.
29. G. Zerbi and M. Sacchi, Macromolecules, 6 (1973) 692.
30. J.H. Schachtschneider, Shell Report, No. 57-65 (1966).

31. For a review see, T. Shimanouchi, in H. Eyring, D. Anderson and W. Jost (Editors), Physical Chemistry, An Advanced Treatise, Vol. IV, Academic Press, New York, 1970.
32. A. Warshel, in G. Segal (Editor), Modern Theoretical Chemistry, Plenum Press, New York, 1977, Chap. 7.
33. D.C. McKean and J.L. Duncan, Spectrochim. Acta A, 27 (1971) 1879.
34. D.C. McKean, Spectrochim. Acta, 22 (1966) 269.
35. M. Tsuboi, J. Mol. Spectry, 19 (1966) 4.
36. T. Shimanouchi, Computer Programs for Molecular Vibrations, 1975.
37. See for example, G.H. Herzberg, Molecular Spectra and Molecular Structure, Vol. 2, Van Nostrand, New York, 1950.
38. R.L. Arnett and B. Crawford, J. Chem. Phys., 18 (1950) 118.
39. J. Aldous and I.M. Mills, Spectrochim. Acta, 18 (1962) 1073.
40. D.A. Long, R.B. Gravenor and M. Woodger, Spectrochim. Acta, 19 (1963) 937.
41. R.G. Snyder, J. Mol. Spectry, 23 (1967) 224.
42. M.V. Vohlkenstein, L.A. Gribov, M.A. Eliashevich and L.I. Stepanov, Kolebaniya Molekul, Moscow, 1972.
43. S.J. Cyvin in S.J. Cyvin (Editor), Molecular Structures and Vibrations, Elsevier, Amsterdam, 1972.
44. J.R. Scherer, Spectrochim. Acta, 20 (1964) 345; ibid., 24A (1967) 1489.
45. R.G. Snyder, J. Chem. Phys., 47 (1967) 1316.
46. H. Bernstein, Spectrochim. Acta, 18 (1962) 161.
47. D.C. McKenn, Chem. Comm., (1971) 1373; D.C. McKean, J.L. Duncan and L. Batt, Spectrochim. Acta A, 29 (1973)1037.
48. R. Tubino, L. Piseri and G. Zerbi, J. Chem. Phys., 56 (1972) 1022.
49. B. Minasso and G. Zerbi, J. Mol. Struct., 7 (1971) 59.
50. P. Klaboe, E. Kloster-Jenses and S.J. Cyvin, Spectrochim. Acta A, 23 (1967) 2733.
51. G. Zerbi and M. Gussoni, J. Chem. Phys., 41 (1964) 456.
52. J.C. Evans and R.A. Nyquist, Spectrochim. Acta, 19 (1963) 1153.
53. R.G. Snyder, J. Chem. Phys., 47 (1967) 1316.
54. J.H. Schachtschneider and R.G. Snyder, Spectrochim. Acta, 21 (1965) 1527.
55. M. Peraldo and M. Cambini, Spectrochim. Acta, 21 (1965) 1509.
56. G. Zerbi and M. Gussoni, Spectrochim. Acta, 22 (1966) 2111.
57. G. Dellepiane and G. Zerbi, J. Chem. Phys., 48 (1968) 3573.
58. G. Zerbi and S. Sandroni, Spectrochim. Acta A, 26 (1970) 1951.
59. E.L. Wu, G. Zerbi, S. Califano and B. Crawford, J. Chem. Phys., 35 (1961) 2060.
60. M. Tasumi, T. Shimanouchi and T. Miyazawa, J. Mol. Spectry, 9 (1962) 261.
61. L. Piseri and G. Zerbi, J. Chem. Phys., 48 (1968) 3561.
62. G. Zerbi and L. Piseri, J. Chem. Phys., 49 (1968) 3840.
63. G. Zerbi and M. Sacchi, Macromolecules, 6 (1973) 692.
64. R. Tubino and G. Zerbi, J. Chem. Phys., 53 (1970) 1428.
65. A.B. Dempster and G. Zerbi, J. Chem. Phys., 54 (1971) 3600.
66. L. Piseri, B.M. Powell and G. Dolling, J. Chem. Phys., 57 (1973) 158.
67. P. Bosi, R. Tubino and G. Zerbi, J. Chem. Phys., 59 (1973) 4578.

68. T. Shimanouchi, A. Asahina and S. Enomoto, J. Polymer Sci., 59 (1962) 93.
69. R.F. Shaufele and T. Shimanouchi, J. Chem. Phys., 47 (1967) 3605.
70. J.W. White in K.J. Ivin (Editor), Structural Studies of Macromolecules by Spectroscopic Methods, Wiley, New York, 1976.
71. H.C. Olf, A. Peterlin and W.L. Peticolas, J. Polymer Sci., Polymer Phys., 12 (1974) 359.

CHAPTER 18

ATOM-ATOM AND DIPOLE-DIPOLE INTERMOLECULAR POTENTIALS IN THE LATTICE DYNAMICS OF MOLECULAR CRYSTALS

S. Califano

1. THE DYNAMICS OF SIMPLE LATTICES

The dynamics of simple lattices can be easily handled in terms of the treatment given many years ago by Born and Huang [1]. In the Born-Huang theory, a basis of Cartesian displacements of the atoms from their equilibrium positions is used to set up the equations of motion for a crystal. The potential and the kinetic energy are written in the form

$$V = \sum_{a\mu\alpha}\left(\frac{\partial V}{\partial U_{\alpha}^{a\mu}}\right)_{o} U_{\alpha}^{a\mu} + \tfrac{1}{2}\sum_{ab}\sum_{\mu\nu}\sum_{\alpha\beta}\left(\frac{\partial^{2}V}{\partial U_{\alpha}^{a\mu}\partial U_{\beta}^{b\nu}}\right)_{o} U_{\alpha}^{a\mu}U_{\beta}^{b\nu} + \ldots \tag{1}$$

$$T = \tfrac{1}{2}\sum_{a\mu\alpha} m_{\mu}(\dot{U}_{\alpha}^{a\mu})^{2} \tag{2}$$

where $U_{\alpha}^{a\mu}$ represents the Cartesian displacement in the α-th direction of an atom at the site μ in the unit cell a. Since the crystal is stress-free when the atoms are at rest, we have the equilibrium conditions

$$\left(\frac{\partial V}{\partial U_{\alpha}^{a\mu}}\right)_{o} = 0 \tag{3}$$

We notice that the Cartesian displacements utilised above are referred to a crystal-fixed reference system. If $X_{\alpha}^{a\mu}$ and $\bar{X}_{\alpha}^{a\mu}$ represent the instantaneous Cartesian coordinate of atom $a\mu$ and the corresponding equilibrium value, the displacement coordinate is defined as

$$U_{\alpha}^{a\mu} = X_{\alpha}^{a\mu} - \bar{X}_{\alpha}^{a\mu} \tag{4}$$

By substitution of (1) and (2) in the equations of motion, we obtain

$$m_{\mu}\ddot{U}_{\alpha}^{a\mu} + \sum_{b\nu\beta} f_{\alpha\beta}\binom{ab}{\mu\nu} U_{\beta}^{b\nu} = 0 \tag{5}$$

where

$$f_{\alpha\beta}\binom{a\ \ b}{\mu\ \ \nu} = \left(\frac{\partial^2 V}{\partial U_\alpha^{a\mu}\,\partial U_\beta^{b\nu}}\right)_o \tag{6}$$

The translational symmetry of the lattice imposes some important restrictions on the force constants (6). Because of the periodicity of the lattice the force constants cannot depend upon the individual values of the unit cell indices a and b, but only upon their difference [2] i.e. only upon the relative position of the unit cells. Thus we have

$$f_{\alpha\mu}\binom{a\ \ b}{\mu\ \ \nu} = f_{\alpha\beta}\binom{a\ \ -b}{\mu\ \ \nu} = f_{\alpha\beta}\binom{o\ \ b-a}{\mu\ \ \nu} \tag{7}$$

where o indicates a unit cell chosen as origin.

For an infinite crystal we have obviously an infinite number of equations of the type (5). The translational symmetry permits however reduction of the number of equations to a finite one [2]. If we consider corresponding atoms in different unit cells, i.e. atoms of different cells at the same site, we see that because of the translational symmetry their displacements must be identical in amplitude and may differ only in phase. We look therefore for solutions of (5) of the type

$$U_\alpha^{a\mu} = (m_\mu)^{-\frac{1}{2}} \epsilon_\alpha^\mu(\vec{k})\, e^{2\pi i\vec{k}.\vec{r}_a} e^{-i\omega t} \tag{8}$$

which correspond to waves propagating in the crystal with a phase difference between equivalent atoms in different unit cells, controlled by the vector $\vec{k}$. The vector $\vec{r}_a$ is the position vector of unit cell a in the space-fixed system.

By substitution of (8) in (5) we obtain, using (7)

$$\omega^2 \epsilon_\alpha^\mu(\vec{k}) = \sum_{\nu}\sum_{\beta} D_{\alpha\beta}^{\mu\nu}(\vec{k})\, \epsilon_\beta^\nu(\vec{k}) \tag{9}$$

where

$$D_{\alpha\beta}^{\mu\nu}(\vec{k}) = (m_\mu m_\nu)^{-\frac{1}{2}} \sum_b f_{\alpha\beta}\binom{o\ \ b}{\mu\ \ \nu} e^{-2\pi i\vec{k}.\vec{r}_b} \tag{10}$$

The number of equations (9) is now 3Z where Z is the number of sites in each unit cell. The infinite set of equations (5) have thus been reduced to a small and finite number of equations through the choice of solutions of the type (8). The quantities $D_{\alpha\beta}^{\mu\nu}(\vec{k})$ defined in (10) are the elements of a 3Z x 3Z matrix $D(\vec{k})$, called the dynamical matrix. Equation (9) shows that the quantities $\epsilon_\beta^\nu(\vec{k})$ are the eigenvectors of the dynamical matrix whereas ω^2 are the eigenvalues. The vibrational frequencies of the lattice are thus found by solving the determinantal equation

$$\left| D_{\alpha\beta}^{\mu\nu}(\vec{k}) - \omega^2 \delta_{\alpha\beta}\delta_{\mu\nu} \right| = 0 \tag{11}$$

for each value of the vector $\vec{k}$.

We notice that for each value of $\vec{k}$ we have 3Z solutions for ω^2 that we shall call $\omega_i^2(\vec{k})$ with i = 1,2, ... 3Z. The infinite number of vibrations of an infinite crystal is thus grouped into 3Z branches, each branch collecting the values of one $\omega_i^2(\vec{k})$ for all allowed values of $\vec{k}$. The plot of $\omega_i^2(\vec{k})$ as a function of $\vec{k}$ is called the dispersion curve of $\omega_i^2(\vec{k})$.

From the definition (10) of the elements $D_{\alpha\beta}^{\mu\nu}(\vec{k})$, we see that the dynamical matrix is Hermitian since

$$D_{\alpha\beta}^{\mu\nu}(\vec{k}) = D_{\beta\alpha}^{\nu\mu}(\vec{k})^* \quad (12)$$

and therefore the eigenvalues of $\omega_i^2(\vec{k})$ are real. A further property of the dynamical matrix is that

$$D_{\alpha\beta}^{\mu\nu}(-\vec{k}) = D_{\alpha\beta}^{\mu\nu}(\vec{k})^* \quad (13)$$

The frequencies $\omega_i(0)$ for $\vec{k} = 0$ which correspond to crystal vibrations in which all equivalent atoms move in phase are of great importance. The vibrations at $\vec{k} = 0$ are the only ones that can interact with electromagnetic radiation because of momentum conservation. Therefore only these vibrations will give rise to bands in the infrared and Raman spectra of a crystal. Three of these frequencies will be zero since they correspond to the three translations of the crystal. As $\vec{k}$ varies from zero to its maximum value in a given direction, the frequencies of these vibrations change from zero to some finite values. These three branches are called acoustic branches since the corresponding motions represent acoustic waves propagating in the crystal. The remaining 3Z-3 branches are called optical branches.

The vector $\vec{k}$ introduced in equation (8) is called the wavevector and is a vector in the reciprocal space since the exponential in (8) must be dimensionless. If $\vec{t}_1$, $\vec{t}_2$ and $\vec{t}_3$ are the primitive translation vectors of the lattice, the basic vectors of the reciprocal space are defined as [2,3]

$$\vec{\tau}_1 = \frac{1}{v}(\vec{t}_2 \times \vec{t}_3) \qquad \vec{\tau}_2 = \frac{1}{v}(\vec{t}_3 \times \vec{t}_1) \qquad \vec{\tau}_3 = \frac{1}{v}(\vec{t}_1 \times \vec{t}_2) \quad (14)$$

where v is the volume of the unit cell. The vectors $\vec{\tau}$ satisfy the relation

$$\vec{\tau}_i . \vec{t}_j = \delta_{ij} \quad (15)$$

The vector $\vec{k}$ can be expressed in terms of the basic vectors of the reciprocal space in the form

$$\vec{k} = \frac{n_1}{L_1}\vec{\tau}_1 + \frac{n_2}{L_2}\vec{\tau}_2 + \frac{n_3}{L_3}\vec{\tau}_3$$

where n_i are integers and L_i is the number of unit cells in the i direction.

2. MOLECULAR CRYSTALS

The treatment of the previous section is easily extended to molecular crystals, provided suitable expressions are found to obtain the Cartesian force constants (6) from intermolecular and intramolecular potentials.

We recall that in a molecular crystal we have molecules at each crystal site and that the forces which hold together the atoms in a molecule are much stronger than the forces which hold together the molecules in the crystal. This situation is clearly shown in the vibrational spectrum through a well defined separation between molecular or internal vibrations, with frequencies close to those of the vapour spectrum, and lattice or external modes of very low frequencies. Some internal vibrations of very low frequency, such as torsions, can however occur for molecular systems in the region of the lattice vibrations and in this case considerable mixing of such modes with the external vibrations will take place.

A convenient form of the potential for molecular crystal is of the type [4]

$$V = V_M + V_I \tag{16}$$

where V_M is the intramolecular potential of all molecules in the crystal and V_I is the intermolecular potential which contains all the interactions between different molecules. By using again the labels a and b for unit cells and the labels μ and ν for sites in the unit cell, we have

$$V_M = \sum_{a\mu} V_M^{a\mu} \tag{17}$$

where $V^{a\mu}$ is the intramolecular potential of a single molecule at the site μ in the a-th unit cell. Information on this part of the potential is obtained from the internal spectrum in much the same way as the calculation of force constants of isolated molecules. Since no analytical forms are available for the intramolecular potential except for diatomic and for a few triatomic molecules, the potential is normally expressed as a power series in terms of internal coordinates. By using the labels p and q for the internal degrees of freedom of a molecule, we have then

$$V_M^{a\mu} = \sum_p \left(\frac{\partial V_M^{a\mu}}{\partial S_p^{a\mu}}\right)_o S_p^{a\mu} + \tfrac{1}{2} \sum_{pq} \left(\frac{\partial^2 V_M^{a\mu}}{\partial S_p^{a\mu} \partial S^{a\mu}}\right)_o S_p^{a\mu} S_q^{a\mu} \tag{18}$$

For an isolated molecule the equilibrium conditions would require that

$$\left(\frac{\partial V_M^{a\mu}}{\partial S_p^{a\mu}}\right)_0 = 0 \tag{19}$$

if all the $S_p^{a\mu}$ are independent. For a molecule in the crystal however the equilibrium conditions will be different since the molecule is under the strain of the intermolecular potential. We shall discuss further this point later on.

The intermolecular potential V_I can be expressed, ignoring three and multi-body interaction, as the sum of all pairwise interactions between molecules. We have then [3,4]

$$V_I = \sum_{ab} \sum_{\mu\nu}{}' V_{b\nu}^{a\mu} \tag{20}$$

where $\sum_{\mu\nu}{}'$ means that $\mu \neq \nu$ if $a = b$. Two kinds of contributions to $V_{b\nu}^{a\mu}$ can be conveniently distinguished. First there are potential terms which can be directly expressed in terms of atomic positions. The most widely used potentials of this kind are the so-called atom-atom potentials in which the interaction between two molecules is split into the sum of all pairwise interactions between the atoms of one molecule and the atoms of the other. The part of the intermolecular potential that can be expressed in terms of atom-atom potentials is thus of the type

$$V_{b\nu}^{a\mu}(I) = \sum_{i}^{i\in a\mu} \sum_{j}^{j\in b\nu} V_{b\nu j}^{a\mu i}(R_{b\nu j}^{a\mu i}) \tag{21}$$

each $V_{b\nu j}^{a\mu i}$ being a function of the atom-atom distance $R_{b\nu j}^{a\mu i}$ between the atom i on the molecule $a\mu$ and the atom j on the molecule $b\nu$. Analytical forms of semiempirical nature of the atom-atom potentials are available for several kinds of atom pairs. Typical semiempirical atom-atom potentials are the well-known potentials of Lennard-Jones

$$V_{b\nu j}^{a\mu i} = \varepsilon[\alpha^2(R_{b\nu j}^{a\mu i})^{-12} - 2\alpha(R_{b\nu j}^{a\mu i})^{-6}] \tag{22}$$

or of Buckingham

$$V_{b\nu j}^{a\mu i} = A_{ij}\exp(-B_{ij}R_{b\nu j}^{a\mu i}) - C_{ij}(R_{b\nu j}^{a\mu i})^{-6} \tag{23}$$

widely used in the literature [3,4]. These potentials, which include an attractive and a repulsive term falling off very rapidly with the distance, furnish a satisfactory description of the short range forces in solids. Owing to the fact that the short range forces represent by far the largest contribution to the intermolecular potential, the atom-atom part is often sufficient alone for a correct calculation of the dynamical properties of molecular crystals.

The second type of contribution to the intermolecular potential arises from terms involving overall molecular properties, such as the dipole-dipole or the quadrupole-quadrupole potentials, which are not easily reduced to the sum of atomic contributions. Although in principle all such terms are at work in a crystal, their relative importance is different. For instance the quadrupole-quadrupole potential is effective roughly over the same range of distances covered by the atom-atom potentials and thus the relative contributions are strongly overlapped. Owing to the semiempirical nature of the atom-atom potentials, i.e. to the fact that the parameters A, B and C of (23) are adjusted to fit physical properties of the crystal, it is unnecessary to introduce a quadrupole-quadrupole term so long as one is interested in the heuristic power of these intermolecular potentials. In other words the use of a quadrupole-quadrupole potential, together with atom-atom potentials, amounts to a different parametrisation of these latter ones. On the contrary dipole-dipole potentials extend over a much larger range of distances and thus account for long range forces which are not included in the atom-atom potentials.

For these reasons we shall discuss here a form of the intermolecular potential which includes only atom-atom and dipole-dipole contributions [5]. Other terms of the total potential will be ignored either because they are considered as incorporated in the atom-atom potential or because their influence on the dynamical properties is too small to justify an independent evaluation. For the dipole-dipole term we shall use the general expression [6]

$$V^{a\mu}_{b\nu}(II) = -\vec{\mu}_{a\mu}\cdot\vec{T}\cdot\vec{\mu}_{b\nu} \qquad (24)$$

where $\vec{\mu}_{a\mu}$ is the total dipole on molecule $a\mu$ in a molecule-fixed reference system and $\vec{T}$ is the usual field propagation matrix.

The total potential of interaction between two molecules is thus the sum of (21) and (24)

$$V^{a\mu}_{b\nu} = \sum_{ij} V^{a\mu i}_{b\nu j}(R^{a\mu i}_{b\nu j}) - \vec{\mu}_{a\mu}\cdot\vec{T}\cdot\vec{\mu}_{b\nu} \qquad (25)$$

There is however a fundamental difference between the two terms of (25). The short range atom-atom part is a function of the interatomic distance $R^{a\mu i}_{b\nu j}$ only and is thus easily expressed in terms of the Cartesian coordinates of the atoms through the relation

$$R^{a\mu i}_{b\nu j} = [\sum_{\alpha} (X^{b\nu j}_{\alpha} - X^{a\mu i}_{\alpha})^2]^{\frac{1}{2}} \qquad (26)$$

where $X^{a\mu i}_{\alpha}$ is the Cartesian coordinate of the i-th atom in the $a\mu$-th molecule referred to a space-fixed reference system

(α = X, Y, Z). The long range dipole-dipole term is instead a function of the distance between the centres of mass of the molecules, of their orientation in the space-fixed system and of the molecular structure. The dipole-dipole term is therefore not directly expressed in terms of the Cartesian coordinates of the atoms and thus a series of coordinate transformations must be performed in order to reduce the complete potential to the same coordinate basis. These transformations will be discussed in the next section.

3. GENERAL THEORY

For the choice of a suitable coordinate basis for a molecular crystal we start from a crystal-fixed Cartesian system (X,Y,Z). We use the symbol $X_\alpha^{a\mu i}$ for the Cartesian coordinate in the α-th direction of atom i in the molecule at the site in the a-th unit cell and symbol $\bar{X}_\alpha^{a\mu i}$ for the corresponding equilibrium value. Displacement coordinates for the atoms in the crystal-fixed system are then defined by the relation:

$$U_\alpha^{a\mu i} = X_\alpha^{a\mu i} - \bar{X}_\alpha^{a\mu i} \tag{27}$$

The potential energy of the crystal, expressed in a Taylor expansion of the displacement coordinates (27) has the form:

$$V = \sum_{a\mu}\sum_{i}\sum_{\alpha}\left(\frac{\partial V}{\partial X_\alpha^{a\mu i}}\right)_o U_\alpha^{a\mu i} + \tfrac{1}{2}\sum_{a\mu}\sum_{b\nu}\sum_{ij}\sum_{\alpha\beta}\left(\frac{\partial^2 V}{\partial X_\alpha^{a\mu i}\,\partial X_\beta^{b\nu j}}\right)_o U_\alpha^{a\mu i} U_\beta^{b\nu j} + \ldots\ldots \tag{28}$$

and the kinetic energy is:

$$T = \tfrac{1}{2}\sum_{a\mu}\sum_{i}\sum_{\alpha} m_i (\dot{U}_\alpha^{a\mu i})^2 \tag{29}$$

We define now for each molecule another Cartesian basis (x,y,z) with the origin in the centre of mass and oriented along the principal inertia axes of the molecule. We denote by $x_\tau^{a\mu i}$ the Cartesian coordinate of atom i of molecule $a\mu$ in the τ direction and by $\bar{x}_\tau^{a\mu i}$ the corresponding equilibrium value. Displacement coordinates in the molecule-fixed system are given by:

$$u_\tau^{a\mu i} = x_\tau^{a\mu i} - \bar{x}_\tau^{a\mu i} \tag{30}$$

The position in space of the molecular system is specified in the crystal-fixed system by the three Cartesian coordinates $X_\alpha^{a\mu}$ of the centre of mass and by the direction cosines $\Omega_{\alpha\tau}^{a\mu}$ between the axes of the molecule-fixed and the axes of the crystal-fixed system. The equilibrium values of $X_\alpha^{a\mu}$ and $\Omega_{\alpha\tau}^{a\mu}$ are indicated by the symbols $\bar{X}_\alpha^{a\mu}$ and $\Lambda_{\alpha\tau}^{a\mu}$ respectively. The displacement coordinates

$$U_\alpha^{a\mu} = X_\alpha^{a\mu} - \bar{X}_\alpha^{a\mu} \tag{31}$$

of the centre of mass can be used to describe the three degrees of translational freedom of the molecule and the variation of the direction cosines

$$\lambda^{a\mu}_{\alpha\tau} = \Omega^{a\mu}_{\alpha\tau} - \Lambda^{a\mu}_{\alpha\tau} \tag{32}$$

to describe the three degrees of rotational freedom. Since the set of direction cosines is redundant, it is convenient to express them in terms of only three rotational coordinates defined as infinitesimal rotations $\Theta^{a\mu}_{\tau}$ around the three molecular axes at equilibrium. The relation between the $\lambda^{a\mu}_{\alpha\tau}$ and the rotations $\Theta^{a\mu}_{\tau}$ is, to the second order [7].

$$\lambda^{a\mu}_{\alpha\tau} = \sum_{\rho\sigma} \Lambda^{a\mu}_{\alpha\sigma}\, \delta_{\sigma\tau\rho}\, \Theta^{a\mu}_{\rho} - \tfrac{1}{2} \sum_{\rho\sigma} \sum_{\chi\omega} \Lambda^{a\mu}_{\alpha\sigma}\, \delta_{\sigma\omega\rho}\, \delta_{\tau\omega\chi}\, \Theta^{a\mu}_{\rho}\Theta^{a\mu}_{\chi} \tag{33}$$

where $\delta_{\sigma\omega\rho}$ is the Levi Civita symbol which assumes the values

$$\delta_{\sigma\omega\rho} = \begin{cases} 0 & \text{unless } \sigma \neq \omega \neq \rho \\ +1 & \text{if } \sigma\omega\rho \text{ are in the cyclic order xyz} \\ -1 & \text{if } \sigma\omega\rho \text{ are in the cyclic order zyx} \end{cases}$$

If the three infinitesimal rotations Θ_x, Θ_y and Θ_z are used as rotational coordinates it is convenient to use as translational coordinates the Cartesian displacements of the centre of mass along the same axes x, y and z of the molecule-fixed system. These displacements will be indicated by the symbol $U^{a\mu}_{\tau}$ (τ = x, y, z) and are related to the displacements (31) by the simple equations

$$U^{a\mu}_{\alpha} = \sum_{\tau} \Lambda^{a\mu}_{\alpha\tau} U^{a\mu}_{\tau} \qquad \begin{matrix} \tau = x,y,z \\ \alpha = X,Y,Z \end{matrix} \tag{34}$$

The Cartesian displacements $u^{a\mu i}_{\tau}$ of the atoms in the molecule-fixed system are also redundant for the description of the 3N-6 internal degrees of freedom. The six redundancies among them are the well-known Sayvetz conditions [8]

$$\sum_{i} m_i u^{a\mu i}_{\tau} = 0 \tag{35}$$

$$\sum_{i} m_i (\bar{x}^{a\mu i}_{\tau} u^{a\mu i}_{\sigma} - \bar{x}^{a\mu i}_{\sigma} u^{a\mu i}_{\tau}) = 0$$

From the Cartesian displacements in the molecule-fixed system we define now a set of internal coordinates $S^{a\mu}_{p}$ (p = 1,2 .... , 3N-6) for each molecule, through the relation [8]

$$S^{a\mu}_{p} = \sum_{i}\sum_{\tau} B^{a\mu i}_{p\tau} u^{a\mu i}_{\tau} + \tfrac{1}{2} \sum_{\tau\tau'} \sum_{ii'} B^{a\mu ii'}_{p\tau\tau'} u^{a\mu i}_{\tau} u^{a\mu i'}_{\tau'} \tag{36}$$

The calculation of the coefficients $B^{a\mu i}_{p\tau}$ and $B^{a\mu ii'}_{p\tau\tau'}$ of (36) can be found in standard texts of vibrational spectroscopy [8]. Finally we define a set of normal coordinates for each molecule through the relations

$$u_\tau^{a\mu i} = \sum_m L_{\tau m}^i q_m^{a\mu} \qquad \begin{array}{l}\tau = x,y,z \\ m = 1,2, \ldots , 3N-6\end{array} \tag{37}$$

$$U_\tau^{a\mu} = (M)^{-\frac{1}{2}} t_\tau^{a\mu} \tag{38}$$

$$\Theta_\tau^{a\mu} = (I_\tau)^{-\frac{1}{2}} r_\tau^{a\mu} \tag{39}$$

In terms of these normal coordinates the kinetic energy matrix is the unity matrix whereas the potential energy has the form

$$V = \sum_{a\mu} \sum_{\ell} \left(\frac{\partial V}{\partial Q_\ell^{a\mu}}\right)_o Q_\ell^{a\mu} + \tfrac{1}{2} \sum_{a\mu} \sum_{b\nu} \sum_{\ell h} \left(\frac{\partial^2 V}{\partial Q_\ell^{a\mu} \partial Q_h^{b\nu}}\right)_o Q_\ell^{a\mu} Q_h^{b\nu} + \ldots . \tag{40}$$

where for simplicity we have used the symbols $Q_\ell^{a\mu}$ or $Q_h^{b\nu}$ to denote all types of normal coordinates defined through (37) to (40). Thus for $\ell,h = 1,2, \ldots , 3N-6$ the coordinates $Q_\ell^{a\mu}$ and $Q_h^{b\nu}$ denote internal normal coordinates of the type (37), for $\ell,h = 3N-5$, $3N-4$ and $3N-3$, they denote the external translational normal coordinates (38) and for $\ell,h = 3N-2$, $3N-1$ and $3N$ they denote the external rotational normal coordinates (39).

By substitution of (16) in (40) we obtain then

$$V = \sum_{a\mu} \sum_{\ell} \left[\left(\frac{\partial V_M}{\partial Q_\ell^{a\mu}}\right)_o + \left(\frac{\partial V_I}{\partial Q_\ell^{a\mu}}\right)_o\right] + \tfrac{1}{2} \sum_{a\mu} \sum_{b\nu} \sum_{\ell h} \left[\left(\frac{\partial^2 V_M}{\partial Q_\ell^{a\mu} \partial Q_h^{a\mu}}\right)_o \delta_{ab} \delta_{\mu\nu} + \left(\frac{\partial^2 V_I}{\partial Q_\ell^{a\mu} \partial Q_\ell^{b\nu}}\right)_o\right] Q_\ell^{a\mu} Q_h^{b\nu} + \ldots\ldots \tag{41}$$

where we have taken into account the fact that the intramolecular potential V_M is a sum of the type (17) over all molecules and that each term is a function of the normal coordinates of one molecule only.

The equilibrium conditions are then, being the $Q_\ell^{a\mu}$ independent [4]:

$$\left(\frac{\partial V_M}{\partial Q_\ell^{a\mu}}\right)_o + \left(\frac{\partial V_I}{\partial Q_\ell^{a\mu}}\right)_o = 0 \qquad \text{for } \ell = m = 1,2,\ldots,3N-6 \tag{42}$$

$$\left(\frac{\partial V_I}{\partial Q_\ell^{a\mu}}\right)_o = 0 \qquad \text{for } \ell = 3N-5,\ldots,3N$$

and the potential reduces to

$$V = \tfrac{1}{2} \sum_{a\mu} \sum_{b\nu} \sum_{\ell h} \left[\left(\frac{\partial^2 V_M}{\partial Q_\ell^{a\mu} \partial Q_h^{a\mu}} \right)_o \delta_{ab} \delta_{\mu\nu} + \left(\frac{\partial^2 V_I}{\partial Q_\ell^{a\mu} \partial Q_h^{b\nu}} \right)_o \right] Q_\ell^{a\mu} Q_h^{b\nu} \tag{43}$$

The equilibrium condition (42) tells us that the internal stress for a molecule in the crystal is balanced by the external forces acting on the molecule and due to the external potential. In an isolated molecule the internal potential is stress-free and thus $(\partial V_M/\partial Q_\ell) = 0$. In a crystal however it is the total potential that is stress-free and therefore

$$\left(\frac{\partial V_M}{\partial Q_\ell^{a\mu}} \right)_o = - \left(\frac{\partial V_I}{\partial Q_\ell^{a\mu}} \right)_o \tag{44}$$

For actual calculations we need to reduce the general expressions for the potential derivatives in (43) to known quantities. By using as intermediate the expansion of the internal coordinates [8]

$$S_p^{a\mu} = \sum_m L_{pm} q_m^{a\mu} + \tfrac{1}{2} \sum_{mn} L_{pmn} q_m^{a\mu} q_n^{a\mu} \tag{45}$$

we have

$$\left(\frac{\partial V_M}{\partial q_m^{a\mu}} \right) = \sum_p \left(\frac{\partial V_M}{\partial S_p^{a\mu}} \right)_o L_{pm} \tag{46}$$

$$\left(\frac{\partial^2 V_M}{\partial q_m^{a\mu} \partial q_n^{a\mu}} \right)_o = \sum_{pq} \left(\frac{\partial^2 V_M}{\partial S_p^{a\mu} \partial S_q^{a\mu}} \right)_o L_{pm} L_{qn} + \sum_p \left(\frac{\partial V_M}{\partial S_p^{a\mu}} \right)_o L_{pmn}$$

i.e.

$$\left(\frac{\partial^2 V_M}{\partial q_m^{a\mu} \partial q_n^{a\mu}} \right)_o = \sum_{pq} f_{pq} L_{pm} L_{qn} + \sum_p \left(\frac{\partial V_M}{\partial S_p^{a\mu}} \right)_o L_{pmn} \tag{47}$$

where

$$f_{pq} = \left(\frac{\partial^2 V_M}{\partial S_p^{a\mu} \partial S_q^{a\mu}} \right)_o \tag{48}$$

are the usual force constants in internal coordinates and

$$L_{pm} = \left(\frac{\partial S_p^{a\mu}}{\partial q_m^{a\mu}} \right)_o \qquad L_{pmn} = \left(\frac{\partial^2 S_p^{a\mu}}{\partial q_m^{a\mu} \partial q_n^{a\mu}} \right)_o \tag{49}$$

Multiplication of (46) by L^{-1}_{qm} and summing over m yields

$$\left(\frac{\partial V_M}{\partial S_q^{a\mu}}\right)_o = \sum_m \left(\frac{\partial V_M}{\partial q_m^{a\mu}}\right)_o L^{-1}_{qm} \tag{50}$$

By substitution of (50) in (47) and by recalling [8] that

$$\sum_{pq} f_{pq} L_{pm} L_{qn} = \lambda_m \delta_{mn} \tag{51}$$

where $\lambda_m = 4\pi^2\nu_m^2$, we obtain

$$\left(\frac{\partial^2 V_M}{\partial q_m^{a\mu} \partial q_n^{a\mu}}\right) = \lambda_m \delta_{mn} + \sum_m \sum_p \left(\frac{\partial V_M}{\partial q_m^{a\mu}}\right)_o L_{pmn} L^{-1}_{pm} \tag{52}$$

We notice that the second term in (52) vanishes for isolated molecules since $(\partial V_M/\partial q_m) = 0$. By substitution of (52) in (43) and by use of (44) we have

$$V = \tfrac{1}{2} \sum_{a\mu} \sum_{b\nu} \sum_{\ell h} \left[\lambda_\ell \delta_{ab} \delta_{\mu\nu} \delta_{\ell h} + \left(\frac{\partial^2 V_I}{\partial Q_\ell^{a\mu} \partial Q_h^{b\nu}}\right)_o - \sum_m \sum_p \left(\frac{\partial V_I}{\partial Q_m^{a\mu}}\right)_o L_{p\ell h} L^{-1}_{pm} \right] Q_\ell^{a\mu} Q_h^{b\nu} \tag{53}$$

In this equation $\lambda_\ell = 0$ when $\ell = 3N-5,..,3N$ and the sum over m extends only over the internal normal coordinates.

For the solution of the dynamical problem for a piece of crystal containing L unit cells, with L large enough to fulfil the cyclic boundary conditions, we define crystal normal coordinates belonging to the irreducible representations of the translational group [9]

$$Q_\ell^{\mu}(\vec{k}) = \frac{1}{\sqrt{L}} \sum_a Q_\ell^{a\mu} e^{2\pi i \vec{k}.\vec{r}_a} \tag{54}$$

where $\vec{k}$ is the wavevector and $\vec{r}_a$ is the position of the a-th unit cell in the crystal-fixed reference system.

The inverse transformation of (54) is

$$Q_\ell^{a\mu} = \frac{1}{\sqrt{L}} \sum_k Q_\ell^{\mu}(\vec{k}) e^{-2\pi i \vec{k}.\vec{r}_a} \tag{55}$$

and thus, by substitution of (55) in (53) we obtain [9] using (20)

$$V = \tfrac{1}{2}\sum_{k}\sum_{\mu\nu}\sum_{\ell h}[\lambda_\ell\delta_{\mu\nu}\delta_{\ell h} + F^{\mu\ell}_{\nu h}(\vec{k})]Q^{\mu}_{\ell}(\vec{k})^{*}Q^{\nu}_{h}(\vec{k}) \tag{56}$$

where

$$F^{\mu\ell}_{\nu h}(\vec{k}) = \delta_{\mu\nu}\sum_{b}\sum_{\lambda}{}'\left[\left(\frac{\partial^2 V^{1\mu}_{b\lambda}}{\partial Q^{1\mu}_{\ell}\partial Q^{b\mu}_{h}}\right)_o - \sum_m\sum_p L_{p\ell h}L^{-1}_{pm}\left(\frac{\partial V^{1\mu}_{b\lambda}}{\partial Q^{1\mu}_{m}}\right)_o\right]$$
$$+ \sum_b{}'\left(\frac{\partial^2 V^{1\mu}_{b\nu}}{\partial Q^{1\mu}_{\ell}\partial Q^{b\nu}_{h}}\right)_o e^{-2\pi i\vec{k}.\vec{r}_b} \tag{57}$$

in which $\sum'$ means that $b \neq 1$ if $\nu = \mu$ or $\lambda = \mu$, and 1 defines a unit cell chosen as origin.

By insertion of (56) in the equations of motion we obtain the secular equation

$$|F^{\mu\ell}_{\nu h}(\vec{k}) - (\lambda(\vec{k}) - \lambda_\ell)\delta_{\mu\nu}\delta_{mn}| = 0 \tag{58}$$

and by solution of this equation for each value of k, we obtain the corresponding crystal frequencies $\lambda(\vec{k}) = 4\pi^2\nu^2(k)$. In order to solve the secular equation (58) we need to evaluate the elements $F^{\mu\ell}_{\nu h}(\vec{k})$ of the dynamical matrix and thus we need to calculate the derivatives occurring in (57). In the next two sections we shall discuss this problem for the atom-atom and for the dipole-dipole model potentials.

4. ATOM-ATOM POTENTIALS

The atom-atom potentials are by far the most widely used intermolecular potentials in the treatment of the lattice dynamics of molecular crystals. Calculations have been performed for aliphatic and aromatic hydrocarbons, for several benzene derivatives, for polymers and for some small molecules [9-16]. Since the atom-atom potential between two atoms of two different molecules is, according to (22) and (23), a function only of the distance between the atoms, the matrix elements occurring in (29) can be written, using (21), in the form

$$\left(\frac{\partial V^{a\mu}_{b\nu}}{\partial Q^{a\mu}_{\ell}}\right)_o = \sum_{ij}\left(\frac{\partial V^{a\mu i}_{b\nu j}}{\partial Q^{a\mu}_{\ell}}\right)_o = \sum_{ij} V^{a\mu i\,\prime}_{b\nu j}\left(\frac{\partial R^{a\mu i}_{b\nu j}}{\partial Q^{a\mu}_{\ell}}\right)_o \tag{59}$$

$$\left(\frac{\partial^2 V^{a\mu}_{b\nu}}{\partial Q^{a\mu}_{\ell}\partial Q^{b\nu}_{h}}\right)_o = \sum_{ij} V^{a\mu i\,\prime\prime}_{b\nu j}\left(\frac{\partial R^{a\mu i}_{b\nu j}}{\partial Q^{a\mu}_{\ell}}\right)_o\left(\frac{\partial R^{a\mu i}_{b\nu j}}{\partial Q^{b\nu}_{h}}\right)_o$$
$$+ \sum_{ij} V^{a\mu i\,\prime}_{b\nu j}\left(\frac{\partial^2 R^{a\mu i}_{b\nu j}}{\partial Q^{a\mu}_{\ell}\partial Q^{b\nu}_{h}}\right)_o \tag{60}$$

where

$$V_{b\nu j}^{a\mu i\,'} = \left(\frac{\partial V_{b\nu j}^{a\mu i}}{\partial R_{b\nu j}^{a\mu i}}\right)_o \qquad V_{b\nu j}^{a\mu i\,''} = \left(\frac{\partial^2 V_{b\nu j}^{a\mu i}}{\partial R_{b\nu j}^{a\mu i\,2}}\right)_o \tag{61}$$

The derivatives (61) are directly evaluated from the analytical expressions (22) and (23). The derivatives of the atom-atom distance with respect to normal coordinates are obtained using as intermediate the Cartesian coordinates of the atoms in the space-fixed system.

The atom-atom distance $R_{b\nu j}^{a\mu i}$ is given by

$$R_{b\nu j}^{a\mu i} = [\sum_{\alpha}(X_{\alpha}^{b\nu j} - X_{\alpha}^{a\mu i})^2]^{\frac{1}{2}} \tag{62}$$

and in turn the Cartesian coordinates in the space-fixed system can be related to the normal coordinates through the relation [5]

$$X_{\alpha}^{a\mu i} = \bar{X}_{\alpha}^{a\mu i} + \sum_{\tau}[\Lambda_{\alpha\tau}^{a\mu}(U_{\tau}^{a\mu} + u_{\tau}^{a\mu i}) + \lambda_{\alpha\tau}^{a\mu}\bar{x}_{\tau}^{a\mu i}] \tag{63}$$

By substitution of (37)-(39) in (63) we obtain the relation between the Cartesian coordinates in the space-fixed system and the normal coordinates, in the form

$$\begin{aligned} X_{\alpha}^{a\mu i} = {} & \bar{X}_{\alpha}^{a\mu i} + (M)^{-\frac{1}{2}}\sum_{\tau}\Lambda_{\alpha\tau}^{a\mu}\, t_{\tau}^{a\mu} + \sum_{m}[\sum_{\tau}\Lambda_{\alpha\tau}^{a\mu}\, L_{\tau m}^{i}]q_{m}^{a\mu} \\ & + \sum_{\rho}[(I_{\rho})^{-\frac{1}{2}}\sum_{\tau\sigma}\Lambda_{\alpha\sigma}^{a\mu}\,\bar{x}_{\tau}^{a\mu i}\delta_{\sigma\tau\rho}]r_{\rho}^{a\mu} \\ & - \tfrac{1}{2}\sum_{\rho\sigma}[(I_{\rho}I_{\sigma})^{-\frac{1}{2}}\sum_{\tau\omega\chi}\Lambda_{\alpha\chi}^{a\mu}\delta_{\chi\omega\rho}\delta_{\tau\omega\sigma}\bar{x}_{\tau}^{a\mu i}]r_{\rho}^{a\mu}r_{\sigma}^{a\mu} \end{aligned} \tag{64}$$

We have then

$$\left(\frac{\partial R_{b\nu j}^{a\mu i}}{\partial Q_{\ell}^{a\mu}}\right)_o = \sum_{\alpha}\left(\frac{\partial R_{b\nu j}^{a\mu i}}{\partial X_{\alpha}^{a\mu i}}\right)_o\left(\frac{\partial X_{\alpha}^{a\mu i}}{\partial Q_{\ell}^{a\mu}}\right)_o \tag{65}$$

$$\begin{aligned} \left(\frac{\partial^2 R_{b\nu j}^{a\mu i}}{\partial Q_{\ell}^{a\mu}\partial Q_{h}^{b\nu}}\right)_o = {} & \sum_{\alpha\beta}\left(\frac{\partial^2 R_{b\nu j}^{a\mu i}}{\partial X_{\alpha}^{a\mu i}\partial X_{\beta}^{b\nu j}}\right)_o\left(\frac{\partial X_{\alpha}^{a\mu i}}{\partial Q_{\ell}^{a\mu}}\right)_o\left(\frac{\partial X_{\beta}^{b\nu j}}{\partial Q_{h}^{b\nu}}\right)_o \\ & + \delta_{ab}\delta_{\mu\nu}\sum_{\alpha}\left(\frac{\partial R_{b\nu j}^{a\mu i}}{\partial X_{\alpha}^{a\mu i}}\right)_o\left(\frac{\partial^2 X_{\alpha}^{a\mu i}}{\partial Q_{\ell}^{a\mu}\partial Q_{h}^{a\mu}}\right)_o \end{aligned} \tag{66}$$

The derivatives of $R_{b\nu j}^{a\mu i}$ with respect to the Cartesian coordinates of the atoms are obtained from (62) and are

$$\left(\frac{\partial R^{a\mu i}_{b\nu j}}{\partial X^{b\nu j}_{\alpha}}\right)_o = -\left(\frac{\partial R^{a\mu i}_{b\nu j}}{\partial X^{a\mu i}_{\alpha}}\right)_o = (X^{b\nu j}_{\alpha} - X^{a\mu i}_{\alpha})/R^{a\mu i}_{b\nu j} = S^{ij}_{\alpha} \quad (67)$$

$$\left(\frac{\partial^2 R^{a\mu i}_{b\nu j}}{\partial X^{a\mu i}_{\alpha}\partial X^{b\nu j}_{\beta}}\right)_o = (\delta_{\alpha\beta} - S^{ij}_{\alpha}S^{ij}_{\beta})/R^{a\mu i}_{b\nu j} = D^{ij}_{\alpha\beta} \quad (68)$$

By means of (64)-(68) we can then construct the desired derivatives (59) and (60). The general expressions for each atom-atom potential are

$$\left(\frac{\partial V^{a\mu i}_{b\nu j}}{\partial Q^{a\mu}_{\ell}}\right)_o = -\sum_{\alpha} V^{a\mu i\prime}_{b\nu j} S^{ij}_{\alpha}\left(\frac{\partial X^{a\mu i}_{\alpha}}{\partial Q^{a\mu}_{\ell}}\right)_o \quad (69)$$

$$\left(\frac{\partial^2 V^{a\mu i}_{b\nu j}}{\partial Q^{a\mu}_{\ell}\partial Q^{b\nu}_{h}}\right)_o = -\sum_{\alpha\beta} P^{ij}_{\alpha\beta}\left(\frac{\partial X^{a\mu i}_{\alpha}}{\partial Q^{a\mu}_{\ell}}\right)_o\left(\frac{\partial X^{b\nu j}_{\beta}}{\partial Q^{b\nu}_{h}}\right)_o \quad (70)$$

$$\left(\frac{\partial^2 V^{a\mu i}_{b\nu j}}{\partial Q^{a\mu}_{\ell}\partial Q^{a\mu}_{h}}\right)_o = \sum_{\alpha\beta} P^{ij}_{\alpha\beta}\left(\frac{\partial X^{a\mu i}_{\alpha}}{\partial Q^{a\mu}_{\ell}}\right)_o\left(\frac{\partial X^{a\mu i}_{\beta}}{\partial Q^{a\mu}_{h}}\right)_o$$

$$- \sum_{\alpha} V^{a\mu i\prime}_{b\nu j} S^{ij}_{\alpha}\left(\frac{\partial^2 X^{a\mu i}_{\alpha}}{\partial Q^{a\mu}_{\ell}\partial Q^{a\mu}_{h}}\right)_o \quad (71)$$

where

$$P^{ij}_{\alpha\beta} = V^{a\mu i\prime\prime}_{b\nu j} S^{ij}_{\alpha}S^{ij}_{\beta} + V^{a\mu i\prime}_{b\nu j} D^{ij}_{\alpha\beta} \quad (72)$$

The first derivative with respect to $Q^{b\nu}_{h}$ can be obtained from (69) by changing the sign and replacing $X^{a\mu i}_{\alpha}$ with $X^{b\nu j}_{\alpha}$. The second derivative with respect to two normal coordinates on molecule $b\nu$ can be obtained from (71) by changing the sign of the second term and by replacing $X^{a\mu i}_{\alpha}$ and $X^{a\mu i}_{\beta}$ by $X^{b\nu j}_{\alpha}$ and $X^{b\nu j}_{\beta}$ respectively.

In the specific case of the translational, rotational and vibrational normal coordinates defined before, these expressions become:

$$\left(\frac{\partial V^{a\mu i}_{b\nu j}}{\partial q^{a\mu}_{m}}\right)_o = -\sum_{\tau}\sum_{\alpha} V^{a\mu i\prime}_{b\nu j} S^{ij}_{\alpha} \Lambda^{a\mu}_{\alpha\tau} L^{i}_{\tau m} \quad (73)$$

$$\left(\frac{\partial V^{a\mu i}_{b\nu j}}{\partial t^{a\mu}_{\tau}}\right)_0 = -\sum_{\alpha}(M)^{-\frac{1}{2}} V^{a\mu i'}_{b\nu j} S^{ij}_{\alpha} \Lambda^{a\mu}_{\alpha\tau} \tag{74}$$

$$\left(\frac{\partial V^{a\mu i}_{b\nu j}}{\partial r^{a\mu}_{\rho}}\right)_0 = -\sum_{\tau\sigma}\sum_{\alpha} V^{a\mu i'}_{b\nu j} S^{ij}_{\alpha} \Lambda^{a\mu}_{\alpha\tau} \bar{x}^{a\mu i}_{\sigma} \delta_{\tau\sigma\rho} (I_{\rho})^{-\frac{1}{2}} \tag{75}$$

$$\left(\frac{\partial^2 V^{a\mu i}_{b\nu j}}{\partial q^{a\mu}_{m} \partial q^{b\nu}_{n}}\right)_0 = -\sum_{\tau\alpha}\sum_{\alpha\beta} P^{ij}_{\alpha\beta} \Lambda^{a\mu}_{\alpha\tau} \Lambda^{b\nu}_{\beta\sigma} L^{i}_{\tau m} L^{j}_{\sigma n} \tag{76}$$

$$\left(\frac{\partial^2 V^{a\mu i}_{b\nu j}}{\partial t^{a\mu}_{\tau} \partial t^{b\nu}_{\sigma}}\right)_0 = -\sum_{\alpha\beta} (M)^{-1} P^{ij}_{\alpha\beta} \Lambda^{a\mu}_{\alpha\tau} \Lambda^{b\nu}_{\beta\sigma} \tag{77}$$

$$\left(\frac{\partial^2 V^{a\mu i}_{b\nu j}}{\partial r^{a\mu}_{\rho} \partial r^{b\nu}_{\rho'}}\right)_0 = -\sum_{\tau\sigma}\sum_{\tau'\sigma'}\sum_{\alpha\beta} (I_{\rho} I_{\rho'})^{-\frac{1}{2}} P^{ij}_{\alpha\beta} \Lambda^{a\mu}_{\alpha\tau} \Lambda^{b\nu}_{\beta\tau'} \delta_{\tau\sigma\rho} \delta_{\tau'\sigma'\rho'} \bar{x}^{a\mu i}_{\sigma} \bar{x}^{b\nu j}_{\sigma'} \tag{78}$$

$$\left(\frac{\partial^2 V^{a\mu i}_{b\nu j}}{\partial q^{a\mu}_{m} \partial t^{b\nu}_{\sigma}}\right)_0 = -\sum_{\tau}\sum_{\alpha\beta} (M)^{-\frac{1}{2}} P^{ij}_{\alpha\beta} \Lambda^{a\mu}_{\alpha\tau} \Lambda^{b\nu}_{\beta\sigma} L^{i}_{\tau m} \tag{79}$$

$$\left(\frac{\partial^2 V^{a\mu i}_{b\nu j}}{\partial q^{a\mu}_{m} \partial r^{b\nu}_{\rho}}\right)_0 = -\sum_{\chi}\sum_{\tau\sigma}\sum_{\alpha\beta} (I_{\rho})^{-\frac{1}{2}} P^{ij}_{\alpha\beta} \Lambda^{a\mu}_{\alpha\chi} \Lambda^{b\nu}_{\beta\tau} \bar{x}^{b\nu j}_{\sigma} \delta_{\tau\sigma\rho} L^{i}_{\chi m} \tag{80}$$

$$\left(\frac{\partial^2 V^{a\mu i}_{b\nu j}}{\partial t^{a\mu}_{\chi} \partial r^{b\nu}_{\rho}}\right)_0 = -\sum_{\tau\sigma}\sum_{\alpha\beta} (MI_{\rho})^{-\frac{1}{2}} P^{ij}_{\alpha\beta} \Lambda^{a\mu}_{\alpha\chi} \Lambda^{b\nu}_{\beta\tau} \bar{x}^{b\nu j}_{\sigma} \delta_{\tau\sigma\rho} \tag{81}$$

$$\left(\frac{\partial^2 V^{a\mu i}_{b\nu j}}{\partial r^{a\mu}_{\rho} \partial r^{a\mu}_{\rho'}}\right)_0 = \sum_{\tau\sigma}\sum_{\tau'\sigma'}\sum_{\alpha\beta} (I_{\rho} I_{\rho'})^{-\frac{1}{2}} P^{ij}_{\alpha\beta} \Lambda^{a\mu}_{\alpha\tau} \Lambda^{a\mu}_{\beta\tau'} \delta_{\tau\sigma\rho} \delta_{\tau'\sigma'\rho'} \bar{x}^{a\mu i}_{\sigma} \bar{x}^{a\mu i}_{\sigma'} + \sum_{\tau\omega\chi}\sum_{\alpha} (I_{\rho} I_{\rho'})^{-\frac{1}{2}} V^{a\mu i'}_{b\nu j} S^{ij}_{\alpha} \Lambda^{a\mu}_{\alpha\tau} \delta_{\tau\omega\rho} \delta_{\sigma\omega\rho'} \bar{x}^{a\mu i}_{\rho} \tag{82}$$

The first derivatives with respect to normal coordinates on $b\nu$ are obtained from (73)-(75) by changing the sign and by replacing $\Lambda_{\alpha}^{a\mu}$ with $\Lambda_{\alpha\tau}^{b\nu}$ and $\bar{x}_{\sigma}^{a\mu i}$ with $\bar{x}_{\sigma}^{b\nu j}$.

The second derivatives for two coordinates on the same molecule are obtained from (76), (77) and (79)-(81) by changing the sign and by replacing aμ with bν or vice versa. In the case of rotational coordinates only the second term in (71) is different from zero according to (64). For this reason the expression for the second derivative with respect to two rotational normal coordinates on aμ has been given explicitly in (82). The corresponding expression for both coordinates on bν is obtained from (82) by changing the sign of the second term and by replacing aμ with bν.

5. DIPOLE-DIPOLE POTENTIAL

The general expression for the interaction of two dipoles on two different molecules is [6]

$$V_{b\nu}^{a\mu} = -\vec{\mu}_{a\mu}\cdot\vec{T}\cdot\vec{\mu}_{b\nu} \tag{83}$$

where $\vec{\mu}_{a\mu}$ is the total dipole on molecule aμ and $\vec{T}$ is the field propagation matrix. If the vector product in (83) is done, we obtain

$$V_{b\nu}^{a\mu} = -\sum_{\alpha\beta}\sum_{\tau\sigma}\mu_{\tau}^{a\mu}\mu_{\sigma}^{b\nu}\Omega_{\alpha\tau}^{a\mu}\Omega_{\beta\sigma}^{b\nu}T_{\alpha\beta} \tag{84}$$

where $\mu_{\tau}^{a\mu}$ is the component of the molecular dipole in the direction τ of the molecule-fixed system and [5,6]

$$T_{\alpha\beta} = \frac{\partial}{\partial R_{\alpha}^{AB}}\frac{\partial}{\partial R_{\beta}^{AB}}\frac{1}{R^{AB}} \tag{85}$$

R^{AB} being the distance between the centres of mass of the two molecules and R_{α}^{AB}, R_{β}^{AB} the components in the crystal-fixed reference system. For simplicity the labels aμ and bν have been contracted to the labels A and B respectively.

A nice feature of the dipole-dipole potential is that each factor of the product in (84) depends only upon one type of normal coordinates. For instance $T_{\alpha\beta}$ is a function only of the distance between the molecules

$$R^{AB} = [\sum_{\alpha} R_{\alpha}^{AB\,2}]^{\frac{1}{2}} = [\sum_{\alpha}(X_{\alpha}^{b\nu} - X_{\alpha}^{a\mu})^{2}]^{\frac{1}{2}} \tag{86}$$

and thus will depend only on the translational normal coordinates since, according to (31), (34) and (38)

$$X_{\alpha}^{a\mu} = \bar{X}_{\alpha}^{a\mu} + (M)^{-\frac{1}{2}}\sum_{\tau}\Lambda_{\alpha\tau}^{a\mu}t_{\tau}^{a\mu} \tag{87}$$

In the same way the direction cosines $\Omega_{\alpha\tau}^{a\mu}$ and $\Omega_{\beta\sigma}^{b\nu}$ depend only on the rotational normal coordinates and the components $\mu_{\tau}^{a\mu}$ and $\mu_{\sigma}^{b\nu}$ of the molecular dipole moment depend only on the internal normal coordinates. In particular the components of the dipole moment can be expanded in terms of the internal normal coordinates in the usual form [8]

$$\mu^{a\mu} = \bar{\mu}_{\tau}^{a\mu} + \sum_{m} \left(\frac{\partial \mu_{\tau}^{a\mu}}{\partial q_{m}^{a\mu}} \right)_{0} q_{m}^{a\mu} + \ldots \tag{88}$$

Substitution of (88) in (84) yields

$$V_{b\nu}^{a\mu} = - \sum_{\alpha\beta} \bar{\mu}_{z}^{2} \Lambda_{\alpha z}^{a\mu} \Lambda_{\beta z}^{b\nu} T_{\alpha\beta} - \sum_{\alpha\beta} \sum_{\tau\sigma} \sum_{mn} \left(\frac{\partial \mu_{\tau}}{\partial q_{m}^{a\mu}} \right)_{0} \left(\frac{\partial \mu_{\sigma}}{\partial q_{n}^{b\nu}} \right)_{0} \Lambda_{\alpha\tau}^{a\mu} \Lambda_{\beta\sigma}^{b\nu} T_{\alpha\beta} q_{m}^{a\mu} q_{n}^{b\nu}$$

$$- \sum_{\alpha\beta} \sum_{\tau} \sum_{m} \bar{\mu}_{z} \left(\frac{\partial \mu_{\tau}}{\partial q_{m}} \right)_{0} T_{\alpha\beta} \left(\Lambda_{\alpha z}^{a\mu} \Lambda_{\beta\tau}^{b\nu} q_{m}^{b\nu} + \Lambda_{\alpha\tau}^{a\mu} \Lambda_{\beta z}^{b\nu} q_{m}^{a\mu} \right) \tag{89}$$

where for simplicity the permanent dipole moment of the molecule has been assumed to lie along the z axis and where the labels $a\mu$ or $b\nu$ have been omitted from quantities such as $\bar{\mu}_z$ and $(\partial \mu_\tau / \partial q_m)$ since these are molecular properties and thus independent of the position of the molecule.

The first term in (89) is the interaction between the permanent dipoles on the molecules and is a function only of the external coordinates. This term vanishes for non-polar molecules and influences only the lattice vibrations. The second term is a transition dipole-dipole interaction and depends only upon the internal normal coordinates. This term is important in the calculation of the splitting of intense infrared bands and has been utilised in many actual calculations [17-19]. The last term couples the internal infrared active vibrations with lattice modes and is often neglected in the calculation of crystal vibrations, even if its effect in some cases can be relevant. Again this term vanishes for non-polar molecules.

The derivatives of the dipole-dipole potential with respect to the normal coordinates can now be constructed from (89), using (86), (87) and (33). We have for the first derivatives with respect to normal coordinates on molecule $a\mu$

$$\left(\frac{\partial V_{b\nu}^{a\mu}}{\partial q_{m}^{a\mu}} \right)_{0} = - \sum_{\alpha\beta} \sum_{\tau} \bar{\mu}_{z} \left(\frac{\partial \mu_{\tau}}{\partial q_{m}} \right)_{0} T_{\alpha\beta} \Lambda_{\alpha\tau}^{a\mu} \Lambda_{\beta z}^{b\nu} \tag{90}$$

$$\left(\frac{\partial V_{b\nu}^{a\mu}}{\partial t_{\tau}^{a\mu}} \right)_{0} = - (M)^{-\frac{1}{2}} \sum_{\alpha\beta} \sum_{\gamma} \bar{\mu}_{z}^{2} \Lambda_{\alpha z}^{a\mu} \Lambda_{\beta z}^{b\nu} \Lambda_{\gamma\tau}^{a\mu} T_{\alpha\beta\gamma} \tag{91}$$

$$\left(\frac{\partial V_{b\nu}^{a\mu}}{\partial r_{\rho}^{a\mu}} \right)_{0} = - (I_{\rho})^{-\frac{1}{2}} \sum_{\alpha\beta} \sum_{\sigma} \bar{\mu}_{z}^{2} \Lambda_{\alpha\sigma}^{a\mu} \Lambda_{\beta z}^{b\nu} T_{\alpha\beta} \delta_{\sigma z \rho} \tag{92}$$

If aμ is replaced by bν, the sign of (91) is reversed. For the second derivatives with respect to two normal coordinates on the same molecule, we have

$$\left(\frac{\partial^2 V^{a\mu}_{b\nu}}{\partial q^{a\mu}_m \partial q^{a\mu}_n}\right)_o = 0 \tag{93}$$

$$\left(\frac{\partial^2 V^{a\mu}_{b\nu}}{\partial q^{a\mu}_m \partial t^{a\mu}_\tau}\right)_0 = (M)^{-\frac{1}{2}} \sum_{\alpha\beta} \sum_{\gamma} \sum_{\sigma} \bar{\mu}_z \left(\frac{\partial \mu_\sigma}{\partial q_m}\right) \Lambda^{a\mu}_{\alpha\sigma} \Lambda^{b\nu}_{\beta z} \Lambda^{a\mu}_{\gamma\tau} T_{\alpha\beta\gamma} \tag{94}$$

$$\left(\frac{\partial^2 V^{a\mu}_{b\nu}}{\partial q^{a\mu}_m \partial r^{a\mu}_\rho}\right)_0 = -(I_\rho)^{-\frac{1}{2}} \sum_{\alpha\beta} \sum_{\sigma\tau} \bar{\mu}_z \left(\frac{\partial \mu_\sigma}{\partial q_m}\right) \Lambda^{a\mu}_{\alpha\tau} \Lambda^{b\nu}_{\beta z} T_{\alpha\beta} \delta_{\tau\sigma\rho} \tag{95}$$

$$\left(\frac{\partial^2 V^{a\mu}_{b\nu}}{\partial t^{a\mu}_\tau \partial t^{a\mu}_\sigma}\right)_0 = -(M)^{-1} \sum_{\alpha\beta} \sum_{\gamma\delta} \bar{\mu}^2_z \Lambda^{a\mu}_{\alpha z} \Lambda^{b\nu}_{\beta z} \Lambda^{a\mu}_{\gamma\tau} \Lambda^{a\mu}_{\delta\sigma} T_{\alpha\beta\gamma\delta} \tag{96}$$

$$\left(\frac{\partial^2 V^{a\mu}_{b\nu}}{\partial t^{a\mu}_\tau \partial r^{a\mu}_\rho}\right)_0 = (MI_\rho)^{-\frac{1}{2}} \sum_{\alpha\beta\gamma} \sum_{\sigma} \bar{\mu}^2_z \Lambda^{a\mu}_{\alpha\sigma} \Lambda^{b\nu}_{\beta z} \Lambda^{a\mu}_{\gamma\tau} T_{\alpha\beta\gamma} \delta_{\sigma z\rho} \tag{97}$$

$$\left(\frac{\partial^2 V^{a\mu}_{b\nu}}{\partial r^{a\mu}_\rho \partial r^{a\mu}_{\rho'}}\right)_0 = \tfrac{1}{2}(I_\rho I_{\rho'})^{-\frac{1}{2}} \sum_{\alpha\beta} \sum_{\sigma\omega} \bar{\mu}^2_z \Lambda^{a\mu}_{\alpha\sigma} \Lambda^{b\nu}_{\beta z} T_{\alpha\beta} (\delta_{\sigma\omega\rho} \delta_{z\omega\rho'} + \delta_{\sigma\omega\rho'} \delta_{z\omega\rho}) \tag{98}$$

If the derivatives refer to molecule bν the sign of (94) and (97) is reversed.

The derivatives with respect to one coordinate on aμ and one on bν are:

$$\left(\frac{\partial^2 V^{a\mu}_{b\nu}}{\partial q^{a\mu}_m \partial q^{b\nu}_n}\right)_o = -\sum_{\alpha\beta} \sum_{\sigma\tau} \left(\frac{\partial \mu_\tau}{\partial q_m}\right)\left(\frac{\partial \mu_\sigma}{\partial q_n}\right) \Lambda^{a\mu}_{\alpha\tau} \Lambda^{b\nu}_{\beta\sigma} T_{\alpha\beta} \tag{99}$$

$$\left(\frac{\partial^2 V^{a\mu}_{b\nu}}{\partial q^{a\mu}_m \partial t^{b\nu}_\tau}\right)_0 = -(M)^{-\frac{1}{2}} \sum_{\alpha\beta\gamma} \sum_{\sigma} \left(\frac{\partial \mu_\sigma}{\partial q_m}\right) \bar{\mu}_z \Lambda^{a\mu}_{\alpha\sigma} \Lambda^{b\nu}_{\beta z} \Lambda^{b\nu}_{\gamma\tau} T_{\alpha\beta\gamma} \tag{100}$$

$$\left(\frac{\partial^2 V^{a\mu}_{b\nu}}{\partial q^{a\mu}_m \partial r^{b\nu}_\rho}\right)_0 = -(I_\rho)^{-\frac{1}{2}} \sum_{\alpha\beta} \sum_{\sigma\tau} \left(\frac{\partial \mu_\sigma}{\partial q_m}\right) \bar{\mu}_z \Lambda^{a\mu}_{\alpha\sigma} \Lambda^{b\nu}_{\beta\tau} T_{\alpha\beta} \delta_{\tau z\rho} \tag{101}$$

$$\left(\frac{\partial^2 V^{a\mu}_{b\nu}}{\partial t^{a\mu}_{\tau}\partial t^{b\nu}_{\sigma}}\right)_0 = (M)^{-1} \sum_{\alpha\beta}\sum_{\gamma\delta} \bar{\mu}^2_z \Lambda^{a\mu}_{\alpha z}\Lambda^{b\nu}_{\beta z}\Lambda^{a\mu}_{\gamma\tau}\Lambda^{b\nu}_{\delta\sigma} T_{\alpha\beta\gamma\delta} \tag{102}$$

$$\left(\frac{\partial^2 V^{a\mu}_{b\nu}}{\partial t^{a\mu}_{\tau}\partial r^{b\nu}_{\rho}}\right)_0 = (MI_\rho)^{-\frac{1}{2}} \sum_{\alpha\beta\gamma}\sum_{\sigma} \bar{\mu}^2_z \Lambda^{a\mu}_{\alpha z}\Lambda^{b\nu}_{\beta\sigma}\Lambda^{a\mu}_{\gamma\tau} T_{\alpha\beta\gamma}\delta_{\sigma z\rho} \tag{103}$$

$$\left(\frac{\partial^2 V^{a\mu}_{b\nu}}{\partial r^{a\mu}_{\rho}\partial r^{b\nu}_{\rho'}}\right)_0 = (I_\rho I_{\rho'})^{-\frac{1}{2}} \sum_{\alpha\beta}\sum_{\sigma\sigma'} \bar{\mu}^2_z \Lambda^{a\mu}_{\alpha\sigma}\Lambda^{b\nu}_{\beta\sigma'} T_{\alpha\beta}\delta_{\sigma z\rho}\delta_{\sigma' z\rho'} \tag{104}$$

If the labels are reversed (100) and (103) change sign. In the above expressions the only undefined quantities are:

$$T_{\alpha\beta\gamma} = \frac{\partial}{\partial R^{AB}_\alpha}\frac{\partial}{\partial R^{AB}_\beta}\frac{\partial}{\partial R^{AB}_\gamma}\frac{1}{R^{AB}} = 3(-5S^{AB}_\alpha S^{AB}_\beta S^{AB}_\gamma + 3\sum_3 S^{AB}_\alpha \delta_{\beta\gamma})/(R^{AB})^4 \tag{105}$$

and

$$T_{\alpha\beta\gamma\delta} = \frac{\partial}{\partial R^{AB}_\alpha}\frac{\partial}{\partial R^{AB}_\beta}\frac{\partial}{\partial R^{AB}_\gamma}\frac{\partial}{\partial R^{AB}_\delta}\frac{1}{R^{AB}} = 3(35S^{AB}_\alpha S^{AB}_\beta S^{AB}_\gamma S^{AB}_\delta - 5\sum_6 S^{AB}_\alpha S^{AB}_\beta \delta_{\gamma\delta} + \sum_3 \delta_{\alpha\beta}\delta_{\gamma\delta})/(R^{AB})^5 \tag{106}$$

where

$$S^{AB}_\alpha = (X^B_\alpha - X^A_\alpha)/R^{AB} \tag{107}$$

6. REFERENCES

1. M. Born and K. Huang, Dynamical theory of crystal lattices, Oxford University Press, 1954.
2. B. Donovan and J.F. Angress, Lattice Vibrations, Chapman and Hall, 1971.
3. J. Zak, in S. Califano (Editor), Lattice dynamics and intermolecular forces, Proceedings of the International School of Physics "E. Fermi", Academic Press, 1975.
4. N. Neto, G. Taddei, S. Califano and W.H. Walmsley, Mol. Phys., 31 (1976) 457.
5. N. Neto, R. Righini, S. Califano and S.H. Walmsley, to be published.
6. J.C. Decius, J. Chem. Phys., 49 (1968) 1387.
7. S.H. Walmsley, in S. Califano (Editor), Lattice dynamics and intermolecular forces, Proceedings of the International School of Physics "E. Fermi", Academic Press, 1975.
8. S. Califano, Vibrational States, Wiley, 1976.
9. G. Taddei, H. Bonadeo, M. Marzocchi and S. Califano, J. Chem. Phys., 58 (1973) 966.

10. G. Taddei, H. Bonadeo and S. Califano, Chem. Phys. Letters, 13 (1972) 136.
11. D.A. Dows, L. Hsu, S.S. Mitra, O. Brafman, M. Hayek, W.B. Daniels and R.K. Crawford, Chem. Phys. Letters, 22 (1973) 595.
12. M. Cangeloni and V. Schettino, Mol. Cryst. Liq. Cryst., 31 (1975) 219.
13. H. Bonadeo and E. D'Alessio, Chem. Phys. Letters, 19 (1973) 117.
14. R. Righini and S. Califano, to be published.
15. L.C. Brunel and D.A. Dows, Spectrochim. Acta, 30A (1974) 929.
16. R. Righini and S. Califano, Chem. Phys., 17 (1976) 45.
17. D.A. Dows and V. Schettino, J. Chem. Phys., 58 (1973) 5009.
18. V. Schettino and P.R. Salvi, Spectrochim. Acta, 31A (1975) 399, 411.
19. D.P. Craig and V. Schettino, Chem. Phys. Letters, 23 (1973) 315.

CHAPTER 19

VIBRATIONAL SPECTRA OF SOLIDS

I.R. Beattie

1. ISOTROPIC MEDIA

This chapter is concerned with the vibrational (infrared and Raman) spectra of crystalline materials. When a collimated beam of natural light falls on to a plane surface of an isotropic medium which is non-absorbing various phenomena are observed. Part of the light is reflected and part is refracted. For incidence normal to the surface the reflectivity R is given in terms of the refractive index n by

$$R = \frac{(n - 1)^2}{(n + 1)^2} \qquad (1)$$

As the angle of incidence is varied so the reflectivity varies and changes may occur so that the polarisation characteristics of those reflected and refracted rays differ from one another and from the incident light. Consider the plane of incidence (also the plane of the paper) as shown in Figure 1 for an

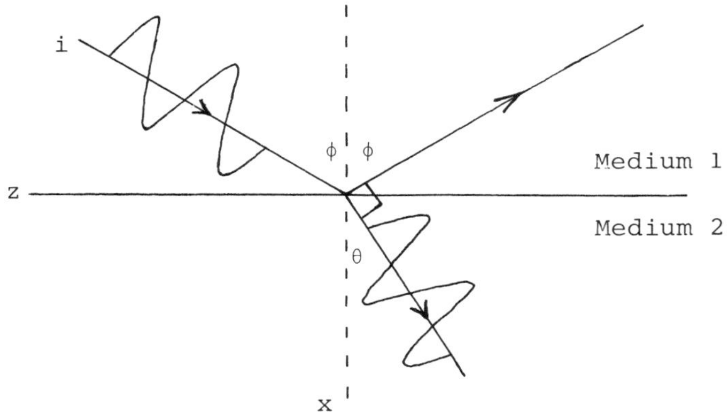

Figure 1. Reflection at the Brewster angle showing zero amplitude for the electric vector in the plane of incidence (xz plane)

incident light ray with the electric vector oscillating solely in the plane of the paper and with maximum amplitude A_{ixz}. If the refracted and reflected rays are at right angles to one another then $(\phi + \theta) = \pi/2$. Under these conditions the light is (ideally) completely refracted and the angle ϕ is termed the Brewster angle. Natural light may conveniently be thought of as elliptically polarised with the principal axes of the ellipse constantly changing in direction and magnitude. Thus when natural light falls on a surface at the Brewster angle the reflected radiation is polarised with the electric vector perpendicular to the plane of incidence.†

If A_{ixz} and A_{ixy} are the maximum amplitudes of the incident ray in the plane of incidence and perpendicular to the plane of incidence, then for the reflected ray, the corresponding maximum amplitudes are given by

$$A_{rxz} = A_{ixz} \frac{\tan(\phi - \theta)}{\tan(\phi + \theta)} \qquad (2)$$

$$A_{rxy} = -A_{ixy} \frac{\sin(\phi - \theta)}{\sin(\phi + \theta)} \qquad (3)$$

It will be noted that when $(\phi + \theta) = \pi/2$, $\tan(\phi + \theta) = \infty$ and $A_r = 0$, leading to Brewster's Law. An elementary way of looking at this is that the incident field causes vibrations of electrons in medium 2, the vibrations being in the direction of the electric vector of the transmitted wave and hence perpendicular to the propagation direction. The vibrating electrons give rise to the reflected wave, propagated in medium 1. Clearly if $(\phi + \theta) = \pi/2$ there can be no reflected wave polarised with the electric vector in the plane of incidence. At visible wavelengths the refractive index of glass is about 1.5, the Brewster angle is around 56° and the total reflectivity for natural light incident at 45° is approximately 5%.

Where light is propagated from a medium of higher refractive index to one of lower refractive index the phenomenon of total internal reflection can occur. If the angle of incidence ϕ exceeds a critical value ϕ_c given by

$$\sin \phi_c = n_2/n_1 \quad (< 1) \qquad (4)$$

no light enters the second medium. Further, there is a phase change (jump) on total internal reflection which is different for the components polarised with the electric vector in the plane of incidence (||) and with the electric vector perpendicular (⊥) to the plane of incidence. For glass of refractive index 1.51 a relative phase difference of 45° is found between the || and ⊥ components of the totally internally reflected ray for angles of incidence of 48°37' or 54°37'. Thus two successive reflections at either of these angles introduces a phase difference of 90° and converts linearly polarised light to circularly polarised light [1], when the initial polarisation direction is at 45° to the plane of incidence.

The previous discussion has relied on geometrical optics and for single crystals which are individually handled can be considered to represent reasonably accurately situations likely to be explored by the chemist. We have neglected diffraction effects, arising from interference phenomena. For example, we have ignored the effect of particle size. For

† For historical reasons this is unfortunately frequently referred to as plane polarised in the plane of incidence. (In fact the plane of incidence contains the magnetic vector).

spheres which are very much smaller than the wavelength of the incident light the total scattered light intensity is inversely proportional to the fourth power of the wavelength. The effect of radius of a sphere (r), wavelength of the incident light (λ) and refractive index of the scattering medium (n) on the total scattered light intensity is very complicated. Figure 2 shows [2] the relationship between the parameter

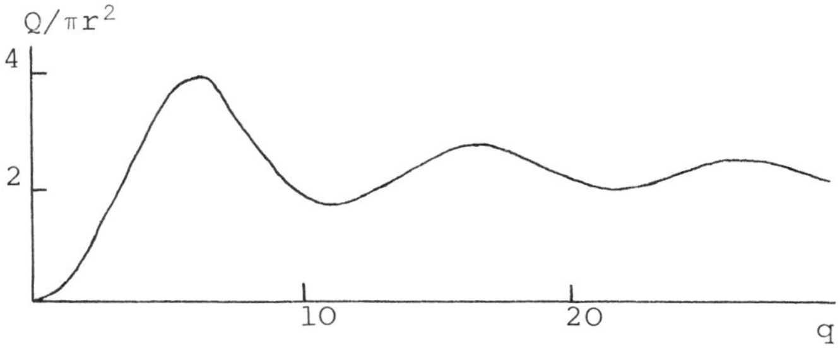

Figure 2. Relation between scattering cross section $Q/\pi r^2$ and a function of the wavelength $q = 2\pi r/\lambda$. The calculations refer to a dielectric of refractive index 1.33 and minor (but significant) minima and maxima have been smoothed (after ref. 2).

$2\pi r/\lambda$ and the scattering cross section Q defined as the ratio:

$$Q = \frac{\text{rate of dissipation of energy}}{\text{rate at which energy is incident on unit cross section}}$$

Finally the effect of absorption bands in the material under examination must be noted. In the region of an absorption band the refractive index undergoes major changes (anomalous dispersion) leading to correspondingly large changes in the reflectivity. At normal incidence the reflectivity is now given by

$$R = \frac{(n-1)^2 + k^2}{(n+1)^2 + k^2} \qquad (5)$$

where k is the absorption index.

A brief consideration of the features given above will make one reflect on the somewhat reckless behaviour of chemists in taking mull or disc spectra on finely powdered materials of random shape and size, which have absorption bands in the region under study, which are not isotropic and where the surrounding medium (mulling agent or matrix of the disc) is chosen normally without considerations of refractive index. I am sure physicists must feel sometimes that God is in league with the chemists!

2. ANISOTROPIC MEDIA

The refractive index of a crystal is not in general independent of the direction in the crystal. It is convenient to represent the variation of refractive index with direction by a triaxial ellipsoid called the indicatrix. Where the material is cubic the ellipsoid is a sphere and there is no optic axis - the material is isotropic. If the material has two directions perpendicular to one another that are equivalent in all respects (for example a tetragonal crystal) the indicatrix is an ellipsoid of revolution. Any section of the ellipsoid is also an ellipse <u>except</u> for sections perpendicular to the optic axis which are circles. Thus, viewed down the optic axis (also the four fold axis) the tetragonal crystal appears to the observer to be isotropic. For crystals of lower symmetry there are three independent axes at right angles for the triaxial ellipsoid. In such a case there are <u>two</u> sections of the indicatrix which are circular. The directions perpendicular to these two sections are the optic axes and the crystal is termed biaxial (Figure 3). Again, if the material is

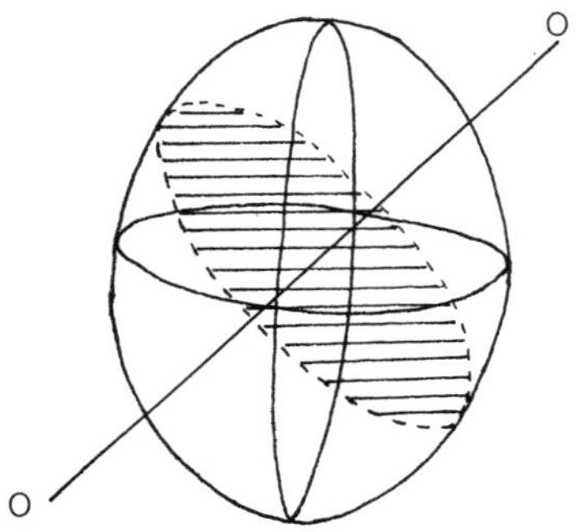

Figure 3. One of the two optic axes of a biaxial crystal (O-O). The shaded portion is the circular section of the triaxial ellipsoid perpendicular to the optic axis.

viewed along either of these axes it appears to the observer to be isotropic. For crystals of orthorhombic or higher symmetry, the indicatrix and the crystallographic axes coincide. At the other extreme for triclinic crystals there is no necessary relationship between the crystallographic axes chosen and the indicatrix axes [3].

When light falls on a crystal it is generally split into two rays of different polarisation characteristics, one or both of which will not obey the normal laws of refraction. The rays which do not obey the normal laws of refraction are termed "extraordinary". Ideally any splitting of the incident (or observed) radiation should be avoided when studying the vibrational infrared and Raman spectra of crystals. In practice this can relatively easily be overcome. Providing the electric vector of the incident light and the observed light both lie along principal axes of the appropriate elliptic section of the indicatrix, no splitting will occur. These are the "extinction" directions found by rotating the appropriately oriented crystal under the polarising microscope.

If the material has absorption bands in the region under study the refractive indices may undergo major changes. For crystals of orthorhombic or higher symmetry, the magnitude of the indicatrix axes may change but the direction will not. For crystals of lower symmetry not only will the magnitude change, but also the direction (where this is not symmetry directed). Where absorption or reflectance measurements are being made it must also be remembered that there is the additional problem of pleochroism (anisotropy in the absorption index). Only for crystals of orthorhombic or higher symmetry need all the axes of the indicatrix coincide with those of the absorption index.

Finally, we may note that because chemists frequently handle materials that are air sensitive, alignment, grinding thin sections etc. may represent serious difficulties. Beattie and Gilson [4] carried out single crystal Raman studies on randomly oriented single crystals of gallium trichloride in a vacuum ampoule, using the principal axes of the elliptic section of the indicatrix but disregarding the crystallographic axes. A complete assignment of the Raman bands to symmetry classes was possible by studying a variety of crystals in various orientations. The technique essentially was to observe which groups of bands behave similarly. In this case there was only one Ga_2Cl_6 molecule in the primitive cell [5].

3. VIBRATIONS IN SOLIDS

Chemists who carry out vibrational infrared and Raman spectra on solids frequently think of only the vibrations in solids relevant to their experiment. Physicists tend to consider an entire assembly. This leads to some difficulty in communication between the two groups.

Consider a cubic lattice along the 100, 111 or 110 directions. Figures 4 (a) and (b) represent sections through

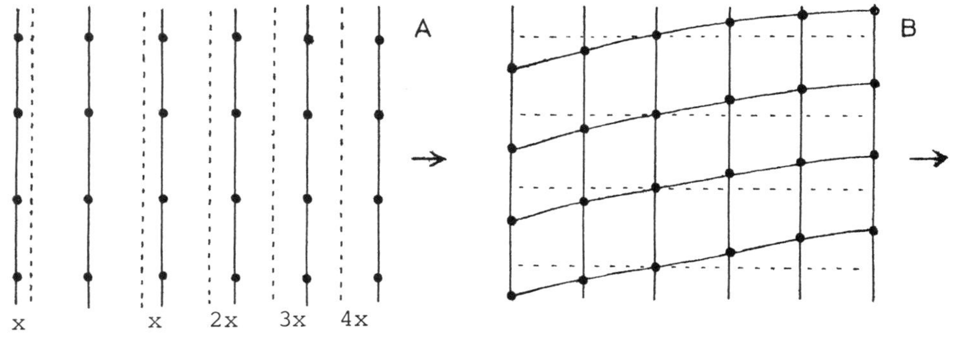

Figure 4 (a). Longitudinal wave with crystal planes undergoing displacements along the propagation direction (b) Transverse wave with crystal planes undergoing displacements perpendicular to the propagation direction. In case (a) dotted lines represent undisplaced planes; in case (b) intersection of dotted lines with planes represent undisplaced atom positions.

such a crystal, perpendicular to the direction of the planes. In Figure 4 (a) a longitudinal wave is shown, the displacements occurring along the propagation direction. In Figure 4 (b) a transverse wave is shown where the displacements now occur perpendicular to the propagation direction [6]. In a similar manner to that familiar for the molecular oscillator problem we may assume that restoring force is proportional to displacement and that only nearest neighbour interactions are important. Under these conditions it may be shown that the frequency is given by

$$2\pi\nu = \left(\frac{4k}{M}\right)^{\frac{1}{2}} |\sin(Kd/2)| = \omega \tag{6}$$

where k is the force constant between adjacent atoms
M is the mass of an atom
K is the wave vector defined by $\omega = cK/n = 2\pi n/\lambda_{vac}$
c is the velocity of light
n is the refractive index of the crystal
d is the separation between successive planes.

Note that the value of d will vary for a given lattice, depending on the direction of K. A plot of $\omega/(4k/M)^{\frac{1}{2}}$ against K is shown in Figure 5.†

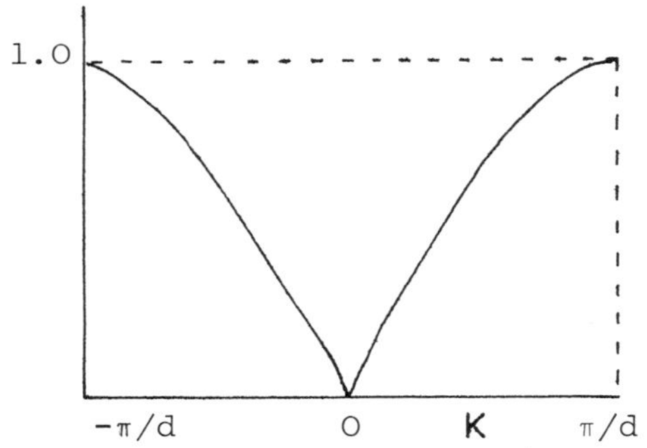

Figure 5. Plot of $\omega/(4kM)^{\frac{1}{2}}$ versus wave vector for the first Brillouin zone assuming interaction only between nearest neighbour planes (after ref. 6).

Consider a crystal such as diamond which has two atoms per unit cell. It is now found that there are so called optical and acoustical branches in the dispersion relation (shown in Figure 6) with longitudinal or transverse character. In the case shown the transverse phonons* are degenerate leading to a total of six branches (two non-degenerate plus two degenerate). Of these four, one is optical and three are acoustical. In general for a crystal of p atoms per unit cell there are 3p phonon branches in the dispersion relation of which 3p-3 are optical and 3 are acoustical. (This clearly is a reminder of the 3n-6 modes of vibration for a molecule.)

A study of the spectrum of a solid using infrared radiation

---

† It is meaningless to say atoms are out of phase by greater than $\pm\pi$. A relative phase of 1.32π is identical to -0.68π. Thus the limits of K are $-\pi/d < K < \pi/d$.

* The quantum of energy in an elastic wave is termed a phonon by analogy with the photon which is the quantum of energy in an electromagnetic wave.

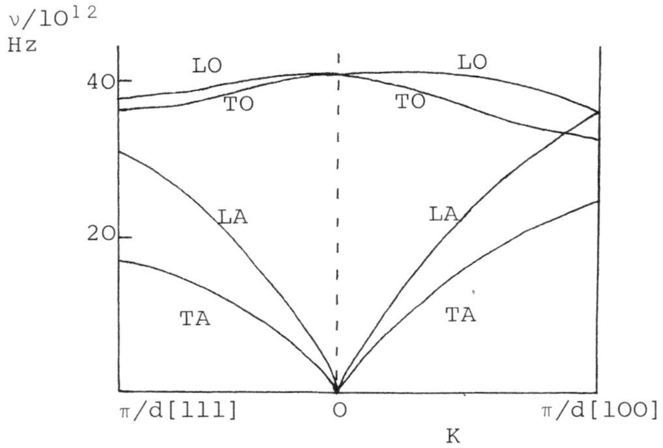

Figure 6. Plot of frequency versus wave vector for |111| and |100| planes of diamond (after J.L. Warren, R.G. Wenzel and J.L. Yarnell, Inelastic Scattering of Neutrons, IAEA, Vienna, 1965, Vol. 1, p. 361).

means use of wavelengths of the order of 10 μm (1,000 cm^{-1}). This wavelength is infinite compared to unit cell dimensions of a few hundred picometers. Thus for our purposes $K \simeq 0$ and all atoms move in phase. In particular, atoms related solely by a primitive translation move identically. Further, as can be seen from the diagram, the frequency of the acoustical modes will be close to zero at $K \simeq 0$ (sometimes referred to as the zone centre).

Figure 7 shows transverse optical and acoustical modes for a linear array of single positive and negative charges. The corresponding longitudinal modes are also shown. It is essential to realise that the electromagnetic radiation (which is transverse) interacts only with the transverse optical mode (Figure 7 (b)) in absorption experiments at normal incidence.† A complete analysis of infrared reflectance studies defines all the optical paramters of the solid and identifies both transverse optical (TO) and longitudinal optical (LO) modes.

† In a "Nujol mull" not only may band maxima and contours be in error, but in addition longitudinal optical (LO) modes may be observed.

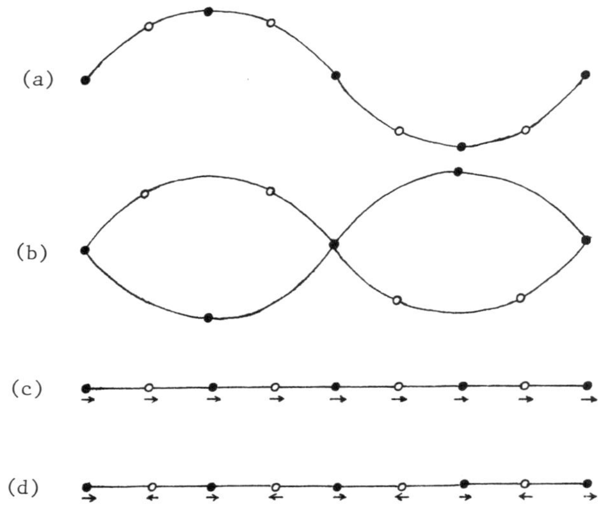

Figure 7. (a) transverse acoustical (b) transverse optical (c) longitudinal acoustical (d) longitudinal optical modes for a linear diatomic lattice considered along the lattice direction. Note that (c) and (d) may be regarded as zone centre modes (K = O). For (a) and (b) K = $\pi/4$. o and ● refer to the two different atoms.

4. POLYATOMIC GROUPS IN CRYSTALS

When a molecule is part of a crystalline array, the vibrational spectrum may be considered at a variety of levels of sophistication. In a particularly simple case we could assume all the molecules to be aligned in the same direction. If the molecular vibrational wave functions are further assumed to be unperturbed relative to the gas (which is a gross approximation) then we merely have to consider a molecular crystal as a collection of (independent) orientated gas-phase molecules. In practice the molecule is usually in an environment of lower symmetry than the symmetry of the free molecule. Thus we must identify the site symmetry of the molecule. This implies the identification of the symmetry elements of the crystal that apply to the molecule. Here the International Tables for X-ray crystallography will be found invaluable [7]. Having identified the site symmetry it is now possible to use correlation tables to correlate the effect of the lowered symmetry of the molecule (in the crystal) on the molecular vibrations. For example lowering the symmetry of the molecule from C_{4v} to C_{2v} will remove all degeneracies. It is impossible to stress too strongly here the necessity to ensure when using correlation tables that the molecular axes of the higher symmetry and lower symmetry molecule are correctly "lined up" for the tables used.

The next improvement in the approximation is to note that if there is more than one molecule in the primitive unit cell, then correlation splitting is normally allowed. Essentially, one molecule is aware of the presence of one or more other molecules. If there are two molecules then every internal mode of these molecules can couple in-phase and out-of-phase leading to 2(3n-6) "internal" modes of vibration for a non-linear molecule.

In considering the vibrational frequencies of an isolated molecule, the translational and rotational motions are removed by considering molecular fixed axes and the Born-Oppenheimer approximation. In a crystal the translational motions are the three acoustic waves (giving rise to Brillouin scattering) and the lattice modes, both arising from movement of the molecules regarded as a point. The molecular rotations become "librations" in the crystal. They (together with the translations and lattice modes) are frequently referred to as external modes. To a chemist librations may be considered as arising from the rocking of a rigid molecule in the crystal. Clearly for a linear molecule there are only two librations, which may be degenerate.

The above discussion is essentially the approach of a chemist who wishes to visualise species such as $TiCl_4$ or $Me_2SnCl_4{}^{2-}$ in the structure. A physicist is more likely to be concerned with crystals considered as extended arrays of atoms. For a centrosymmetric crystal the prediction of optically active modes can be made rigorously using group theory. The procedure is precisely analogous to that carrying out a conventional point group analysis of a molecule. The appropriate paper on the X-ray structure determination will normally give the space group in both the systematic notation and in the Schoenflies notation e.g. $P2_1/n2_1/m2_1/a$ - D_{2h}^{16} . The factor group contains the symmetry operations remaining after primitive translations have been equated to the identity operation, the isomorphic point group is D_{2h}. One then considers the effect of the relevant symmetry operations on the Cartesian coordinates and takes the product of these with the number of atoms unmoved by the symmetry operation to lead to the reducible representation. In carrying out such an analysis it is important firstly to remember that it is the primitive cell that is important in vibrational spectroscopy. Crystallographers frequently make use of the centred cell. Secondly an atom transformed by a factor group operation into another atom which is related to the first solely by one or a succession of primitive translations is regarded as invariant to that symmetry operation [8]. Thus for sodium dithionate dihydrate the space group number is 62, Pnma - D_{2h}^{16} . The cell is of course primitive as denoted by the symbol P and is centrosymmetric. The site symmetry of the dithionate ions is C_s (one plane of symmetry) whereas the symmetry of the free ion is D_{3d} .

Point group analysis of $S_2O_6{}^{2-}$ under D_{3d} symmetry leads to [9]

$$\Gamma_{ion} = 3a_{1g} + 3e_g + a_{1u} + 2a_{2u} + 3e_u$$

for the vibrational modes (3n-6 = 18). From correlation tables (noting that σ_{xz} of the free ion is also the xz plane of the crystal) we find:

free ion   ion in crystal site   four ions in crystal

D_{3d} $\rightarrow$ C_s $\rightarrow$ D_{2h}

a_{1g}

e_g a' $a_g + b_{2g} + b_{1u} + b_{3u}$

a_{1u}

a_{2u} a'' $a_u + b_{2u} + b_{1g} + b_{3g}$

e_u

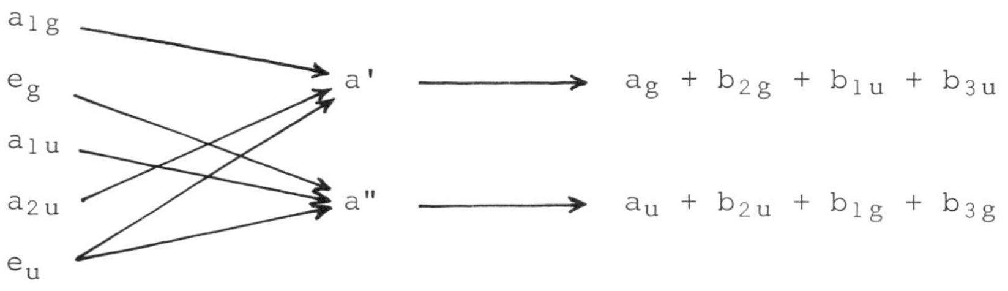

giving the internal modes as:

$$11a_g + 7b_{1g} + 11b_{2g} + 7b_{3g} + 11b_{1u} + 7b_{2u} + 11b_{3u} + 7a_u .$$

For the crystal containing four ions under D_{2h}^{16} symmetry using factor group analysis:

$$\Gamma_{cryst} = 14a_g + 10b_{1g} + 14b_{2g} + 10b_{3g} + 10a_u + 14b_{1u}$$
$$+ 10b_{2u} + 14b_{3u}$$

giving the expected 96 modes derived from 3 x 4 x 8 degrees of freedom.

The lattice modes obtained from regarding the four dithionate ions as points are given by

$$\Gamma_{lat} = 2a_g + b_{1g} + 2b_{2g} + b_{3g} + a_u + 2b_{1u} + b_{2u} + 2b_{3u}$$

giving the expected twelve symmetry species from 4 x 3 degrees of freedom. Three of these are pure translations[†] (acoustic modes) found directly from the D_{2h} character tables as

$$\Gamma_{trans} = b_{1u} + b_{2u} + b_{3u}$$

The D_{3d} free ion clearly has three possible rotations, two about x and y which are degenerate (e_g) and one about z (a_{2g}). Again both of these can be found from D_{3d} character tables. Correlating we find for the librations

free ion   ion in crystal site   four ions in crystal

D_{3d} $\rightarrow$ C_s $\rightarrow$ D_{2h}

a_{2g} $\rightarrow$ a'' $\rightarrow$ $b_{1g} + b_{3g} + a_u + b_{2u}$

e_g $\rightarrow$ a' $\rightarrow$ $a_g + b_{2g} + b_{1u} + b_{3u}$

Hence,

[†] Correctly they involve the Na^+ ions and H_2O molecules as well.

$$\Gamma_{lib} = a_g + 2b_{1g} + b_{2g} + 2b_{3g} + 2a_u + b_{1u} + 2b_{2u} + b_{3u} ,$$

the expected twelve modes.

Hence we may summarise the modes of $Na_2S_2O_6.2H_2O$ remembering we have, to simplify the analysis, neglected the sodium ions and the water molecules:

Acoustic Modes: (giving rise to Brillouin scattering)

$$b_{1u} + b_{2u} + b_{3u}$$

Optically Active Modes:

Lattice Modes apart from translations:

$$2a_g + b_{1g} + 2b_{2g} + b_{3g} + a_u + b_{1u} + b_{3u}$$

expected to occur below 100 cm^{-1}.

Librations:

$$a_g + 2b_{1g} + b_{2g} + 2b_{3g} + 2a_u + b_{1u} + 2b_{2u} + b_{3u}$$

expected to occur below 150 cm^{-1} but likely to be intense for modes deriving from the e_g "rotation" of the D_{3d} dithionate ion.

Internal Modes:

(Raman active) $11a_g + 7b_{1g} + 11b_{2g} + 7b_{3g}$

of these $6a_g + 3b_{1g} + 6b_{2g} + 3b_{3g}$ derive directly from the

$3a_{1g}$ and $3e_g$ modes of the free D_{3d} ion and are expected to be intense in the Raman effect. Further they will appear in regions close to those found for the appropriate free ion modes. The remaining Raman active bands derive from infrared allowed fundamentals of the free ion and are likely to be weakly allowed in the Raman effect.

(Infrared active) $11b_{1u} + 7b_{2u} + 11b_{3u} + 7a_u$

of these $4a_u + 4b_{2u} + 5b_{1u} + 5b_{3u}$ derive directly from the

$a_{1u} + 2a_{2u} + 3e_u$ of the free D_{3d} ion. The remaining infrared active bands derive from Raman allowed fundamentals of the free ion.

For non-centrosymmetric crystals where Raman bands may also be infrared active the situation is much more complex. A simplified description of the elegant work by Damen, Porto and Tell [10] on zinc oxide has been given in a recent review by the author [11]. For an excellent survey of the Raman spectra of solids a recent review by Wilkinson includes a bibliography on crystals [12].

5. SINGLE CRYSTAL RAMAN SPECTRA

Single crystal Raman measurements may be carried out using conventional 90° or 180° scattering geometries for example. Propagation is arranged so that the electric vector of the incident (normally) plane polarised light is along a principal axis of the appropriate elliptic section of the indicatrix. (These orientations are easily defined by examining the crystal under a polarising microscope.) Similarly analysis of the polarisation (and intensity) of the Raman radiation is carried out along either of the two elliptic axes. It should be noted that it is unwise to attempt to propagate light along an optic axis unless extreme care is taken in the measurements. The elliptic section perpendicular to the optic axis is circular. If one is slightly off axis in propagation (or if the propagating beam is not strictly parallel) interference between the ordinary and the extraordinary ray can lead to changes in polarisation characteristics of the exciting light [13].

As with a molecule, the scattering tensor of an oriented crystal may be written in the form

$$\begin{bmatrix} R_{XX} & R_{XY} & R_{XZ} \\ R_{YX} & R_{YY} & R_{YZ} \\ R_{ZX} & R_{ZY} & R_{ZZ} \end{bmatrix}$$

where the terminology R_{ij} refers to the Raman scattering element for the crystal for the appropriate vibrational mode under study. The axes X,Y and Z are both crystal and laboratory fixed axes for the purposes of this experiment. It must be appreciated that in terms of the Raman tensor, X,Y,Z refer to <u>polarisation</u> (not propagation) directions of the light. Thus for an R_{XY} term, light incident polarised along X is, after scattering, polarised along Y.

It is clear that for an ideal experiment, we can make experimental measurements that will define the relative magnitudes of R_{XX}, R_{YY}, R_{ZZ}, R_{XY}, R_{XZ}, R_{YZ}. Note that the intensity of the scattered radiation depends on $R_{ij}{}^2$. The intensity of the radiating dipole thus follows a $\cos^2\theta$ law where θ is the angle between the observation direction and the normal to the oscillating dipole. This leads in essence to the generation of a figure like a doughnut representing the scattered light intensity (see Figure 8).

The only feature of difficulty now is that the chemist is interested in the molecule (with scattering tensor components α'_{ij}) and has made measurements on the crystal. In a particularly simple case molecular (xyz) and crystal/laboratory (XYZ) axes may coincide. In general this is not so and a transformation† of axes is necessary [14].

† This transformation need by made for one molecule only in a polymolecular unit cell; the crystal symmetry then takes care of the rest automatically.

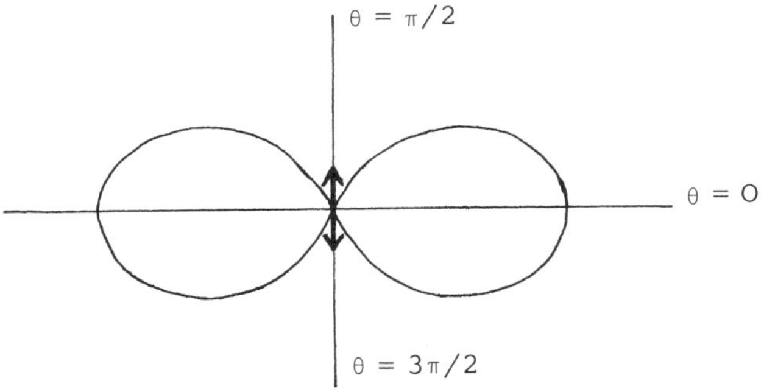

Figure 8. Radiation emission from an oscillating dipole. The "doughnut" is obtained by rotation of the figure about the dipole axis.

$$R_{XYZ} = T\alpha_{xyz}'T^{t} \tag{7}$$

where T is the appropriate transformation matrix and T^t is the transpose (and also the inverse) of T. On the basis that the molecule carries its scattering tensor over from the gas-phase into the crystal it is then formally possible to calculate relative magnitudes of α'_{ij} terms. This is analogous to the oriented gas-phase approximation introduced by Pimentel [15] for infrared spectroscopy.

6. SINGLE CRYSTAL INFRARED MEASUREMENTS

Here there are two main approaches possible: absorption spectroscopy and reflectance spectroscopy. To a chemist, the disadvantage of absorption spectroscopy is principally the need to cut thin slices of the material under examination. Anyone who has tried to make thin sections (down to a thousandth of an inch i.e. 25 μm) knows that it is not particularly easy even for air stable compounds. Many crystals of interest to chemists can only be handled under anhydrous conditions in the absence of oxygen. A thickness of 25 μm is comparable to the wavelength of infrared radiation. When a beam of infrared radiation enters a crystal it is in general split into two rays, one of which may obey the normal laws of refraction (the ordinary ray). It must be realised that when the two rays leave the crystal similar relationships apply. The extraordinary ray regains the direction of the original incident light (if both faces of the slice are parallel) and because it is in an isotropic medium, it is no longer "extraordinary". There will, however, be both a spatial difference and a phase difference generated between the two rays during passage through a crystal. This phase difference is given by 360 $(\Delta n) d/\lambda$ (in degrees) where Δn is the refractive index difference (the birefringence), d the thickness of the crystal and λ the wavelength. Clearly for small Δn or thin crystals the phase difference (and the spatial separation) may be small. However, the magnitude is difficult to predict in the region of an absorption band. Recently Adams and Trumble [16] have studied triclinic crystals where the two molecules in the unit cell were related by an inversion centre. Birefringence was neglected and molecular axes were used for the observations. This approach is similar to that adopted in early studies on oriented polymer films. It is interesting to note that for thin films at oblique incidence, both LO and TO modes may be observed in absorption [17].

The introduction of tunable infrared lasers will transform reflectance spectroscopy in crystals. Thus using laser diodes or parametric mixing in alkali metal vapours, much of the infrared spectrum of interest to chemists can be covered (down to about 280 cm^{-1} at present)†. Thus it will be possible to study the reflectance spectra of small (less than 1 mm face) crystals at selected angles over a wide frequency range using a highly collimated beam. As with many branches of spectroscopy the full impact of lasers has yet to be appreciated. It will be possible to completely determine the optical (dielectric) parameters of materials over a very wide range of frequencies. From the analysis of the reflectivity one derives the refractive index and absorption index, together with the identification of both LO and TO modes. The relevant equations are:

$$n(\nu) = \frac{1 - R(\nu)}{1 + R(\nu) - 2\sqrt{R(\nu)}\cos\theta(\nu)} \qquad (8)$$

$$k(\nu) = \frac{-2\sqrt{R(\nu)}\sin\theta(\nu)}{1 + R(\nu) - 2\sqrt{R(\nu)}\cos\theta(\nu)} \qquad (9)$$

where n, k and R are the real refractive index, absorption index and reflectivity respectively. The bracketed ν refers to the particular frequency at which the reflectivity is measured. The difficulty lies in the phase angle θ, which is unknown. However, it may be found from the reflected intensity measurements using what are known as the Kramers-Kronig inversion relationships, which are described in detail elsewhere [18].

7. CONCLUSIONS

The purpose of this article has been to describe the problems associated with studying the vibrational spectra of molecules or complex ions in solids; the difficulties of studies with Nujol mulls; the elegance of single crystal studies and, in particular, the ease of single crystal Raman measurements even on highly reactive materials.

Figure 9 shows some of the results [4,19] that may be obtained using single crystal studies in the Raman and the infrared effect. A pertinent question is,

---

† In the case of tunable laser diodes, resolution of 10^{-6} cm^{-1} is potentially possible.

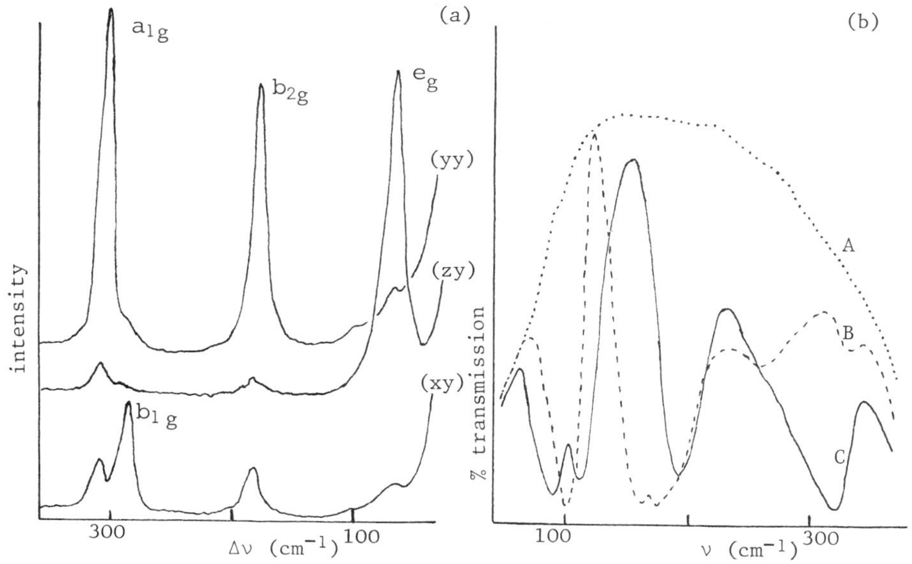

Figure 9. (a) Single crystal Raman spectra of $(NH_4)_2PtCl_4$. (ij) refer to incident and scattered polarisation, where x, y and z are crystallographic axes. a_{1g}, b_{1g}, b_{2g} are internal modes of $PtCl_4{}^{2-}$, the nomenclature b_{1g}/b_{2g} depends on choice of molecular axis system (courtesy Dr. T.R. Gilson).

(b) Far infrared spectra of a single crystal of K_2PtCl_4. A, background; B, electric vector parallel to crystal z; C, electric vector parallel to crystal x (reproduced from ref. 19 with permission).

"Why do this research?"
Perhaps the most obvious answer is,
"To determine the symmetry of modes of vibrations of molecules and complex ions."
"Why?"
"So that it is possible to determine a force-field for the molecule or ion."
"Why?"
"So that we can learn something about the bonding."
"But even for diatomics there is no obvious relationship between force constant and bond strength. Further, the X-ray crystallographic study has already defined electron density distributions which help to determine the nature of the bonding. In addition the force constants you determine are normally harmonic approximations because of the complexity of the problem."

It is very easy for chemists to do elegant single crystal studies, but frequently I am not clear why they were carried out. Similarly, it is easy to write qualitative papers on Raman and infrared spectra of crystalline materials in Nujol mulls. If you can get crystals, do an X-ray single crystal structure determination - which will be unambiguous.

8. REFERENCES

1. For background reading in this area see: M. Born and E. Wolf, Principles of Optics, Pergamon Press, Oxford, 1975; M. Garbuny, Optical Physics, Academic Press, New York, 1965.
2. B. Goldberg, J. Opt. Soc. Amer., 43 (1953) 1221.
3. See for example, N.H. Hartshorne and A. Stuart, Crystals and the Polarizing Microscope, Arnold, London, 1970.
4. I.R. Beattie and T.R. Gilson, Proc. Roy. Soc., A307 (1968) 407.
5. S.C. Wallwork and I.J. Worrall, J. Chem. Soc., (1965) 1816.
6. For general reading see C. Kittel, Introduction to Solid State Physics, Wiley, New York, 1971.
7. N.F.M. Henry and K. Londsdale (Editors), International Tables for X-ray Crystallography, Kynoch Press, Birmingham, 1952.
8. Gilson gives an excellent account of factor group analyses in T.R. Gilson and P.J. Hendra, Laser Raman Spectroscopy, Wiley-Interscience, London, 1970; other useful articles are: R.S. Halford, J. Chem. Phys., 14 (1946) 8; S.S. Mitra and P.J. Gielesse, Progress in Infrared Spectroscopy, Plenum Press, New York, 1963.
9. I.R. Beattie, M.J. Gall and G.A. Ozin , J. Chem. Soc. A, (1969) 1001; I.R. Beattie, Essays in Structural Chemistry, Macmillan, London, 1971.
10. T.C. Damen, S.P.S. Porto and B. Tell, Phys. Rev., 142 (1966) 570.
11. I.R. Beattie, Chem. Soc. Rev., 4 (1975) 107.
12. G.R. Wilkinson, in Molecular Spectroscopy, Volume 3, Chemical Society (Specialist Periodical Report), London, 1975.
13. S.P.S. Porto, J.A. Giordmaine and T.C. Damen, Phys. Rev., 147 (1966) 608.
14. See for example, I.R. Beattie and G.A. Ozin, J. Chem. Soc. A, (1969) 542.
15. G.C. Pimentel, J. Chem. Phys., 19 (1951) 1536.
16. D.M. Adams and W.R. Trumble, J. Chem. Soc. Dalton, (1974) 690; Inorg. Chim. Acta, 13 (1975) 17.
17. D.W. Berreman, Phys. Rev., 130 (1963) 2193.
18. See for example, G.R. Wilkinson, in R.G.J. Miller and B.C. Stace (Editors), Spectroscopy, 2nd edition, Heyden, London, 1972.
19. D.M. Adams and D.C. Newton, J. Chem. Soc. A, (1969) 2998.

CHAPTER 20

DETERMINATION OF BARRIERS TO INTERNAL ROTATION AROUND SINGLE BONDS

James R. Durig

1. INTRODUCTION

Torsional oscillations, or internal rotations involve the twisting about a bond of one part of the molecule (the top or internal rotor) with respect to the rest of the molecule (the frame). In Table 1 are listed the methyl torsional barriers

Table 1. Barriers (kcal mol^{-1}) to methyl rotation of some haloethanes as a function of halogen substitution

molecule	F (cm^{-1})	V_3 (gas)	V_3 (solid)	molecule	F (cm^{-1})	V_3 (gas)	V_3 (solid)
CH_3CH_3	10.705	2.93		CH_3CH_2Br	5.883	3.71	4.53
CH_3CH_2F	6.431	3.33		CH_3CHBr_2	5.244	4.33	
CH_3CHF_2	5.468	3.18	3.97	CH_3CBr_3	5.297	5.78	6.0
CH_3CF_3	5.44	3.2	4.51	CH_3CCl_2F	5.361	4.42	
CH_3CH_2Cl	6.067	3.72	4.48	CH_3CBr_2Cl	5.304	5.70	5.70
CH_3CHCl_2	5.511	4.13	4.77	CH_3CF_2Cl	5.70	4.43	4.80
CH_3CCl_3	5.299	5.43	5.49	CH_3CF_2Br	5.329	5.06	4.99
CD_3CCl_3	2.681	5.40		CH_3CH_2I	5.874	3.67	4.43

for a number of haloethanes. Such rotation was at one time thought to be potential free. However, thermodynamic measurements gradually provided convincing evidence that potential barriers had to be surmounted in turning from one conformation to another in most molecules. For example, in ethane the rotation of one methyl group with respect to the other about the C-C bond is hindered by [1,2] a barrier of 2.928 kcal mol^{-1}. Experimentally, it has been found that internal rotational barriers vary considerably in value and the magnitudes for three-fold barriers may range from a few hundred calories to values in excess of six kilocalories. Despite the fact that the potential barriers are relatively small compared to bond energies or the total energy of the molecule, there are a number of thermodynamic properties which are markedly influenced by them. For example, the heat capacity, entropy, and equilibrium constants contain an appreciable contribution from the internal rotation. Studies relating to the magnitude and origin of the barriers to internal rotation [3] were initiated in the 1930's;

however, a completely satisfactory theoretical interpretation of the forces which give rise to these potential barriers is only now beginning to unfold.

Although the theoretical rationalisation of the potential barrier is a very formidable problem, progress in this area has certainly been restricted in the past by the lack of reliable experimental data. At present the most accurate methods for determining barriers to internal rotation around single bonds appear to be infrared, Raman, and microwave techniques, since these methods provide information on the torsional energy level separations. It should be emphasised that the torsional data are obtained from these methods in quite different ways. For example, in the microwave spectrum the observed perturbations on the pure rotational transitions are correlated to the torsional barrier height by either the splitting or intensity methods [4,5]. However, the actual energy level separations are measured by the infrared and Raman techniques, since the torsional vibrations are generally found in the region below 300 cm^{-1}. Thus, the vibrational data can generally be interpreted more rapidly than the microwave data, but both methods should provide comparable barrier heights. The purpose of this chapter will be to illustrate the manner in which torsional barriers can be evaluated from vibrational data for a variety of molecules.

As previously stated, the ultimate objective in most barrier height measurements is the understanding of the forces which give rise to restricted rotation and thus, studies on the isolated molecule are preferred. Since the torsional modes occur at such low energies, Boltzmann calculations indicate that several of the excited states will also be occupied. Therefore, under favourable conditions it is possible to observe a series of transitions in the vapour state which can be assigned to the ground and excited energy levels in the torsional potential function. With such data one can determine the shape of the potential well, in addition to measuring the barrier height. When a molecule has more than a single internal rotator, however, the problem is more complex and a unique assignment of the spectrum becomes much more difficult.

Types of internal rotors fall into two categories - symmetric tops and asymmetric tops. For symmetric tops, a rotation about the top-frame bond of $2\pi/n$ (where n is an integer) will bring the top to a position symmetrically equivalent to or indistinguishable from the original configuration. It is, thus, possible to speak of the foldness of the top in terms of n. For example, a methyl (CH_3) group is a threefold symmetric top (with local C_{3v} symmetry) since a rotation about the carbon-to-frame bond of 120° will result in an orientation completely superimposable upon the initial orientation. Twofold tops include phenyl, $-NO_2$ and $-BF_2$ groups (of local C_{2v} symmetry). When a rotation of 360° (i.e. when n = 1) is the only operation that results in a symmetrically equivalent position for the top, it is known as an asymmetric rotor. Examples of asymmetric tops include amino ($-NH_2$), phosphino ($-PH_2$), alcohol (-OH) and thiol (-SH) groups, when they are bonded to an asymmetric frame. In the case of a symmetric frame, the top with the highest degree of symmetry prevails, and when two tops of different foldness are bonded directly to one another, the resultant foldness is the product of the two individual top's foldness. For instance, CH_3BF_2 would be classified as a sixfold internal rotor while ethane, CH_3CH_3, would be threefold.

The foldness of the internal rotor relates directly to the potential function governing the torsional oscillation. A threefold rotor implies that there are three symmetrically equivalent positions of minimum energy between 0 and 2π. Thus, a potential function which has three minima in this range is dictated in this case. Similarly, CH_3BF_2, being a sixfold rotor, would require a potential function with six minima.

For symmetric tops, the wells in the potential function are all equivalent and occur every $2\pi/n$. Since the wells are equivalent, energy levels within each well will be n-fold degenerate as a result of quantum mechanical tunnelling. For a typical threefold torsional potential function, the energy levels are split into an A component and a doubly degenerate E component. As the energy levels approach the top of the barrier, the splitting between A and E levels increases (see Figure 1).

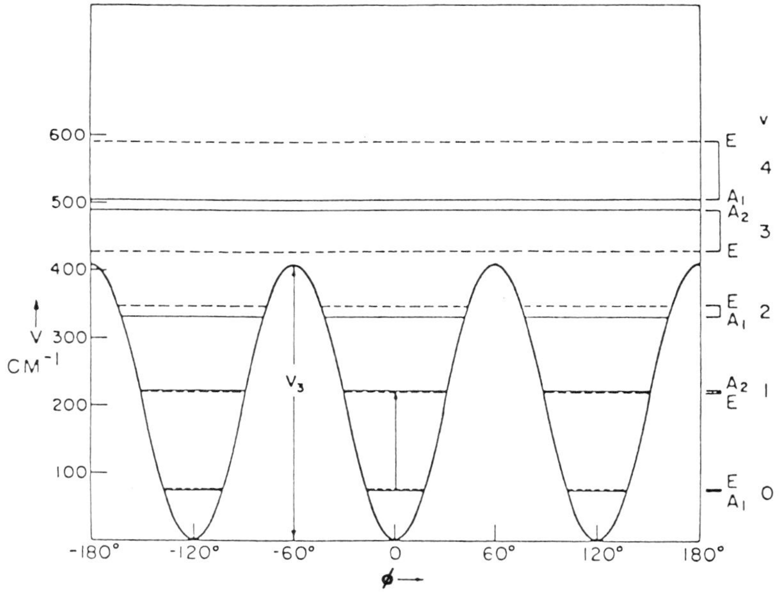

Figure 1. Threefold rotor potential function (reproduced from ref. 12 with permission).

This leads to different possible frequencies for A ← A, E ← A, and E ← E transitions. In some cases, the barrier to internal rotation is low enough so that energy levels near the top of the barrier are populated and A-E splittings may be resolved in the vibrational spectrum.

In general, for asymmetric rotors the wells are no longer symmetrically equivalent. However, for a molecule like ethylphosphine, the phosphino rotor can align itself trans to the methylene group or in either of the two equivalent gauche conformations. This equivalency of the two gauche forms makes energy levels degenerate in the gauche wells while those in the trans well are non-degenerate. Depending on the relative magnitudes of the V_n terms, the various wells in the potential function of asymmetric rotors may be of different relative energies. This leads to the possibility of one conformer being more stable than other possible conformers (see Figure 2).

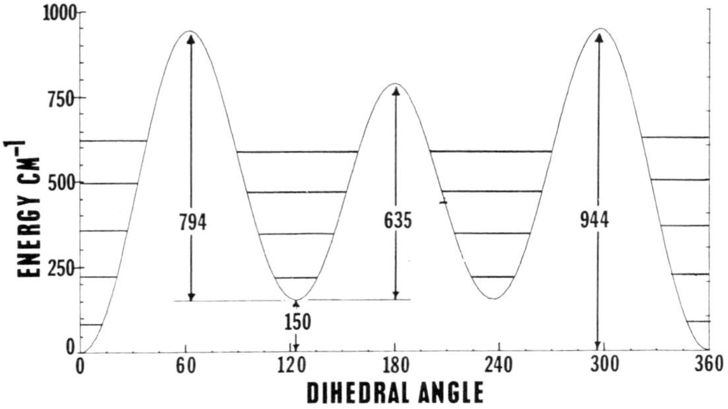

Figure 2. Calculated conformational potential function for $CH_3CH_2PD_2$ (reproduced from ref. 19 with permission).

Experimentally, the far infrared spectra of gases generally result in relatively broad and weak bands for the torsional vibrations. Therefore, the assignment of the v = 1 ← v = 0 transition is often uncertain which results in a barrier height of questionable value. The breadth of the band results from the overlap of the unresolved transitions. This is particularly true if the symmetry of the molecule is such that the torsional motion gives rise to a band with a B-type contour. The best data can usually be obtained for torsional motions which give rise to C-type band contours with pronounced Q branches for the ground and excited state transitions. Torsional fundamentals are notoriously weak in the infrared spectrum since their motion usually results in a very small dipole change. Isotopic substitution is often essential if one hopes to actually assign the torsional oscillation confidently for many molecules. This is particularly true if the molecule has other low frequency motions. Until very recently, the standard method of directly observing torsional vibrations has been infrared spectrosocpy by either measuring the torsional vibration itself or its combination with other normal modes.

Recent studies have shown that it is also possible to assign the torsional vibrations with more certainity in some cases by studying the compounds at liquid nitrogen temperatures. The low temperature effectively "depopulates" the upper energy levels and one observes only the v = 1 ← v = 0 transition. Furthermore, it is apparent that the bands recorded at lower temperatures have much greater intensities than their vapour phase counterparts, and the frequency shift upon condensation is often not appreciable [4]. In fact, the low temperature torsional studies on a variety of molecules display the same barrier height trends as observed for the molecules when studied as gases [5-8]. Thus, the ease with which the low temperature data can be obtained and the fact that reliable barrier height trends have resulted suggest that the torsional studies of molecular crystals might also be quite useful in the eventual understanding of some of the forces which contribute to internal rotation barriers.

It should be pointed out that gas phase Raman spectroscopy has recently been added [2,7] as another technique for direct observation of torsional frequencies. Gas phase torsional vibrations have not been previously studied in the Raman effect for various reasons. Firstly, gas phase Raman spectroscopy has only become a practical tool with the advent of high-powered lasers and improved detection systems. Secondly, torsional fundamentals in the gas phase are usually either disallowed in the Raman effect or have broad and extremely weak band contours from which virtually no information can be extracted. Finally, overtones in the Raman spectrum are ordinarily very weak and quite often not observed at all.

However, it has recently been shown [8] that the $\Delta v = 2$ transitions of low-frequency large amplitude vibrations, like ring-puckering, linear bending modes, or torsional motions are observable in the Raman effect due to mechanical anharmonicity and, in particular, electrical anharmonicity. Another contributing factor is that the $\Delta v = 2$ transitions are totally symmetric and give rise to sharp Q branches, whereas the fundamentals are not usually totally symmetric and do not have Q branches in the gas phase Raman spectrum. So in practice, $\Delta v = 1$ transitions of the large amplitude torsional motions are not usually observed unless the torsional mode is totally symmetric as it is for gauche $CH_3CH_2PH_2$ and other similar molecules. For these cases it is easy to identify the torsional modes for the gauche isomer by comparing the Raman spectrum with the far infrared data; the torsional modes which appear in the infrared spectrum but have no counterpart in the Raman spectrum can be assigned as arising from the trans isomer. For other molecules, the overtone region of the torsional mode is the most informative and series of peaks have been assigned to $\Delta v = 2$ transitions of the methyl torsional mode for a number of molecules [2,7,9]. The succeeding double jumps are much more intense than one would expect from the Boltzmann factors, but electrical and mechanical anharmonicity of the large-amplitude torsional motions apparently accounts for the unusual intensity of the second, third and fourth "hot bands". Therefore a combination of the infrared and Raman spectra appears to be the best way to directly study the torsional motions.

2. SINGLE-TOP SYMMETRIC ROTOR

The appropriate equations necessary for the evaluation of the torsional barriers will be reviewed and some examples will be listed. The rotation of a single group (i.e. $-CH_3$, $-CF_3$, SiH_3) relative to a reference framework such as -CHO, CH_2CN, CH_2Cl, is the simplest and most studied type of torsional motion. In ethyl chloride the stable conformation is a staggered one, but there are three equivalent such forms each separated by 120°. Clearly, to pass from one equivalent form to another entails the crossing of an energy barrier (see Figure 1). The simplest mathematical function which will reproduce such a potential variance upon rotation is a cosine function. By assuming the problem to be one-dimensional, the quantum mechanical energy solutions are readily obtainable, and they have been discussed in detail by many previous workers [10,11,12,13]. The eigenvalue problem will be outlined below and the interested reader is referred to the above cited studies for additional details.

The model employed is a rigid symmetric top (i.e. CH_3 group) attached to a rigid frame which may be completely asymmetric. There are four rotational degrees of freedom, three for overall rotation and one for the hindered internal rotation of the two groups. The axis of internal rotation coincides with the unique axis of the symmetric top. Since the top has a threefold symmetry axis, the potential energy hindering rotation may be expressed by a Fourier expansion:

$$V(\phi) = (V_3/2)(1 - \cos 3\phi) + (V_6/2)(1 - \cos 6\phi) + \ldots \quad (1)$$

where V_3 is the height of the threefold barrier, V_6 of the sixfold, and so forth. In the above expression ϕ represents the angle of internal rotation. The V_6 term does not affect the barrier height because its value is zero at both the minimum and maximum of the potential well (see Figure 3). Thus, the barrier height is simply given by V_3; the sixfold term merely governs the shape of the well. For a positive value of V_6, the maximum is broader and the well is narrower. Obviously, a negative sixfold term would have the opposite effect and these results are graphically illustrated in Figure 3. Although the procedure for numerical evaluation of the respective coefficients in Equation (1) will not be discussed, it should be noted that experimental results indicate that $0 < V_6/V_3 < 0.05$ and $V_6 \gg V_9 \gg$ the higher order terms. Thus, the potential function is adequately represented by $V(\phi) = (V_3/2)(1 - \cos 3\phi)$ for most molecules with a single internal rotational degree of freedom with threefold symmetry.

The Hamiltonian for this system may be expressed as:

$$H = H_r + F(p - P)^2 + (V_3/2)(1 - \cos 3\phi) \quad (2)$$

where the following definitions are applicable:

H_r = rigid rotor Hamiltonian

$F(cm^{-1}) = h/8\pi^2 cI_r = 2.80 \times 10^{-39}/I_r$ (g cm^2)

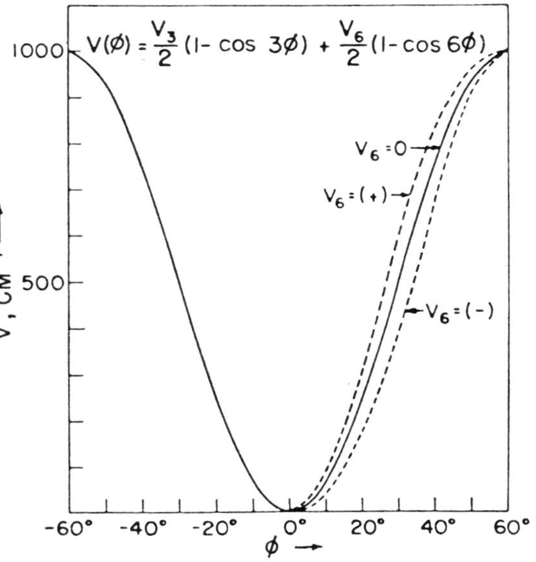

Figure 3. Effect of V_6 on the shape of the potential curve; ϕ is the angle of rotation; V_3 = 1000, V_6 = +100, 0 and -100 cm^{-1} (reproduced from ref. 13 with permission).

$I_r = I_\phi(1 - \sum_g \lambda_g^2 \frac{I_\phi}{I_g})$ the reduced moment of inertia for internal rotation

I_ϕ = moment of inertia of the internal top

I_g = g-th principal moment of inertia of the entire molecule

λ_g = cosine of the angle between the axis of the internal top and the g-th principal axis of inertia of the molecule

(p - P) = the relative angular momentum of the top and the frame

The eigenvalue problem associated with the torsional motion may be transformed into the well established Mathieu equation, if the cross term, -2FPp, is neglected.

$$\frac{d^2\phi_{v\sigma}}{dx^2} + (b_{v\sigma} - s\cos 2x)\phi_{v\sigma} = 0 \qquad (3)$$

where,

$b_{v\sigma}$ = $(4/9)E_v/F$ an eigenvalue of the Mathieu equation

s = $(4/9)V_n/F$ dimensionless parameter of the Mathieu equation

2x = $3\phi + \pi$

v = torsional quantum number

σ = sublevel index

The energy levels for restricted internal rotation are:

$$E_v = (9/4)Fb_v \tag{4}$$

For very high barriers, the energy levels would be triply degenerate because the three potential valleys are identical. For most molecules, however, the barriers are sufficiently low that the degeneracy of the levels is removed by quantum mechanical tunnelling. Removal of the degeneracy results in two levels, one singly (A_1 or A_2) and one doubly degenerate (E). The splitting for the A-E components increase as the levels approach the top of the barrier (see Figure 1). Transitions about the barrier correspond to those for a group undergoing free rotation about a fixed axis; as such, uniform spacings are expected.

The observed infrared frequency, $\bar{\nu}$, is the difference between two of the energy levels:

$$\bar{\nu}\ (\mathrm{cm}^{-1}) = E_{v'\sigma'} - E_{v\sigma} = (9/4)F(b_{v'\sigma'} - b_{v\sigma}) \tag{5}$$

If the $v = 1 \leftarrow v = 0$ torsional transition is observed in the infrared spectrum, and if there is enough structural information to determine F, $\Delta b_{v\sigma}$ can be calculated. From $\Delta b_{v\sigma}$, a dimensionless parameter s can be obtained from tables of solutions for the Mathieu equation. Table 2 lists $\Delta b_{v\sigma}$ vs s for s values ranging between 12 and 200, and this is the range in which one normally deals. More extensive tables of the Mathieu equation parameters may be found elsewhere [11]. By employing the definition of s, the barrier height may be deduced:

$$V_3\ (\mathrm{cm}^{-1}) = (9/4)Fs \tag{6}$$

The principal source of error in the evaluation of the barrier height will be in the value of F, if the vibrational frequencies have been correctly assigned.

The preceding expressions relating the torsional barrier height to the observed infrared frequency can be generalised so as to be applicable for any rotor with an n-fold axis of symmetry. For this general case, the potential expression can be written as:

$$V(\phi) = (V_n/2)(1 - \cos n\phi) \tag{7}$$

where V_n is the height of the n-fold barrier. The appropriate energy solutions then become:

$$E_v = (1/4)n^2Fb_{v\sigma} \tag{8}$$

In a similar way, the n-fold potential barrier is again related to F and s:

$$V_n\ (\mathrm{cm}^{-1}) = (1/4)n^2Fs \tag{9}$$

In addition to this Mathieu equation procedure, one can also estimate the barrier height by assuming that the torsional oscillation is harmonic. This approximation is valid, however,

TABLE 2

Mathieu Function Eigenvalues ($b_{v\sigma}$) for the First Five Energy Levels versus the Mathieu Equation Parameter, s

s	b_{0A}	b_{0E}	b_{1A}	b_{1E}	b_{2A}	b_{2E}	b_{3A}	b_{3E}	b_{4A}	b_{4E}
12	3.166	3.202	9.277	8.668	12.045	13.852	22.273	17.586	22.339	27.996
16	3.719	3.735	10.747	10.412	14.829	16.270	24.452	20.003	24.650	30.169
20	4.200	4.207	12.099	11.914	17.449	18.578	26.648	22.572	27.096	32.395
24	4.631	4.635	13.351	13.247	19.870	20.719	28.845	25.258	29.689	34.679
30	5.213	5.214	15.072	15.027	23.123	23.635	32.108	29.352	33.823	38.219
36	5.738	5.738	16.641	16.620	25.983	26.274	35.303	33.333	38.161	41.902
40	6.063	6.063	17.618	17.605	27.717	27.915	37.381	35.858	41.105	44.428
46	6.522	6.522	18.997	18.991	30.121	30.231	40.410	39.410	45.516	48.276
50	6.811	6.811	19.867	19.863	31.618	31.692	42.367	41.624	48.416	50.841
56	7.224	7.224	21.109	21.107	33.736	33.778	45.207	44.737	52.650	54.619
60	7.487	7.487	21.901	21.899	35.078	35.107	47.037	46.692	55.375	57.062
66	7.866	7.866	23.039	23.039	37.001	37.018	49.692	49.475	59.290	60.590
70	8.108	8.108	23.770	23.769	38.232	38.243	51.403	51.244	61.779	62.852
76	8.460	8.460	24.827	24.827	40.010	40.016	53.888	53.788	65.335	66.120
80	8.687	8.687	25.509	25.509	41.154	41.159	55.494	55.420	67.594	68.223
86	9.016	9.016	26.500	26.500	42.817	42.820	57.831	57.784	70.832	71.277
90	9.230	9.230	27.142	27.141	43.892	43.894	59.344	59.309	72.901	73.252
96	9.541	9.541	28.077	28.077	45.460	45.462	61.553	61.530	75.885	76.129
100	9.743	9.743	28.685	28.685	46.478	46.479	62.986	62.970	77.805	77.996
110	10.232	10.232	30.153	30.153	48.934	48.935	66.448	66.440	82.397	82.500
120	10.698	10.698	31.555	31.555	51.280	51.280	69.752	69.748	86.738	86.794
130	11.146	11.146	32.899	32.899	53.528	53.528	72.918	72.915	90.875	90.905
140	11.576	11.576	34.193	34.193	55.690	55.690	75.961	75.960	94.837	98.854
150	11.992	11.992	35.441	35.441	57.776	57.776	78.894	78.894	98.648	98.657
160	12.394	12.394	36.648	36.648	59.792	59.792	81.730	81.730	102.325	102.330
170	12.783	12.783	37.817	37.817	61.746	61.746	84.476	84.476	105.883	105.886
180	13.161	13.161	38.953	38.953	63.642	63.642	87.142	87.142	109.333	109.334
190	13.529	13.529	40.057	40.057	65.486	65.486	89.732	89.732	112.684	112.685
200	13.887	13.887	41.132	41.132	67.282	67.282	92.255	92.255	115.945	115.946

only if the first excited state is well below the top of the barrier. It is generally considered that the harmonic treatment is useful if the barrier is greater than 2.5 kcal mol^{-1} [14]. When the cosine function is written in series form, the mathematical treatment is straightforward and will be summarised below in terms applicable to any n-fold rotor:

$$1 - \cos X = (1/2!)X^2 - (1/4!)X^4 + \ldots\ldots \qquad (10)$$

where $X = n\phi$. The X^4 and higher order terms are zero in this approximation; therefore the expression for the potential energy becomes

$$V(\phi) = (1/4)n^2\phi^2V_n \qquad (11)$$

The frequency of the one-dimensional oscillator is

$$\bar{\nu}(cm^{-1}) = (1/2\pi c)(k/I_r)^{\frac{1}{2}} \qquad (12)$$

from which the force constant k is seen to be:

$$k = 4\pi^2c^2\bar{\nu}^2I_r \qquad (13)$$

By definition the force constant is the second derivative of the potential energy, so we can obtain the following relationship:

$$k = \partial^2V(\phi)/\partial\phi^2 = (1/2)n^2V_n \qquad (14)$$

The barrier height can be written in terms of experimentally accessible quantities by substituting Equation (13) into Equation (14):

$$V_n\ (cm^{-1}) = 8\pi^2cI_r\bar{\nu}^2/n^2h = \bar{\nu}^2/n^2F \qquad (15)$$

Thus, the barrier to internal rotation may be simply computed if the torsional fundamental and F are known. It is important to emphasise two assumptions that are inherent in the above treatment: (1) the torsional vibration is not mixed with the other fundamentals; and (2) the barrier is symmetrical. The first assumption appears to be reasonable, since recent studies suggest that torsional modes often shift by a factor of 1.4 upon deuteration [15-17]. It is noted that any perturbation of the torsional mode would probably result in a higher frequency, since the torsional fundamental is often the vibration of lowest energy. This, of course, would result in an apparently greater barrier.

3. SINGLE-TOP ASYMMETRIC ROTOR

If the barrier is not symmetric (i.e. asymmetric tops), the potential function can no longer be expressed by a single parameter. The torsional function of such a system is usually expressed in terms of the angle of twist ϕ as:

$$V(\phi) = \tfrac{1}{2}\sum_i V_i(1 - \cos i\phi) \qquad (16)$$

where terms up to V_6 are usually retained [18-23]. The problem then becomes that of evaluating the potential energy contributions of the V_1, V_2, V_3 and possibly higher terms. Although such a calculation requires several pieces of experimental data with at least one transition in each "stable" well, often only the $v = 1 \leftarrow v = 0$ transition in the central potential well is known. In such cases, it may be possible to estimate the different barrier heights on the basis of symmetry considerations but, in general, these estimates are of limited value.

It should be instructive to consider the usual approach to the asymmetric rotor porblem. Initially, the potential function is expressed in a Fourier series:

$$V(\phi) = (V_1/2)(1 - \cos\phi) + (V_2/2)(1 - \cos 2\phi) + (V_3/2)(1 - \cos 3\phi) \tag{17}$$

which is truncated after the threefold term since these terms will be the largest by far and should be a good first approximation. The Hamiltonian would then be of the form:

$$H = \frac{d}{d\phi} F(\phi) \frac{d}{d\phi} + \tfrac{1}{2} \sum_{i=1}^{3} (1 - \cos i\phi) \tag{18}$$

where $F = \dfrac{h}{8\pi^2 c I_r}$ and I_r is the reduced moment of inertia.

The calculation of F demands a knowledge of the molecular geometry at all angles of ϕ and experimentally one is usually limited to the known geometry of two forms (trans, cis and/or gauche). In order to make the best use of the information, F values are calculated for several values of ϕ using each structure. A weighted (proportional to ϕ) composite of the F values from each structure is then fitted to a Fourier cosine series in ϕ,

$$F(\phi) = F_o + \sum_{i=1}^{5} F_i \cos i\phi \tag{19}$$

where the series converges very rapidly and in some cases only the first three terms are retained.

A computer program similar to the one described by Lewis et al. [18] is used to calculate the energy levels and wave functions associated with the above Hamiltonian. The Hamiltonian matrix is set up in a plane rotor ($\exp il\phi$) basis and rotated to the symmetry adapted sine-cosine basis. Seventy to ninety basis functions are usually used in the calculations. The program should be written to handle potential terms up to V_6 and kinetic terms up to F_5. Preliminary values of the first three potential constants are estimated and the program calculates the torsional energy levels together with the wave functions which indicate the potential well in which each level lies. Then, by comparison with the experimental vibrational spacings, the program refines the potential constants by iteration using least squares. Repeated checks are essential to insure that the energy levels remain correctly labelled during the iterative process, and this labelling of the levels (related closely to the assignment of the experimental transitional frequencies) determines whether the final result defines a trans-cis or a trans-gauche-gauche potential curve. The error limits on the potential constants are estimated from the variance-covariance matrix. It should be stressed that if the experimental data are limited, then they may be insufficient to determine uniquely the potential function and the structure of the less stable conformer.

It is also possible to utilise the harmonic approximation in which case the potential expression becomes:

$$V(\phi) = 1/4[V_1 + 4V_2 + 9V_3]\phi^2 \qquad (20)$$

for small ϕ. This expression is generally written in the following form since one often observes only the $v = 1 \leftarrow v = 0$ transition for the most stable conformer:

$$V(\phi) = (1/4)V^*\phi^2 \quad \text{where} \quad V^* = V_1 + 4V_2 + 9V_3 \qquad (21)$$

Since the force constant for this harmonic oscillator is $(\frac{1}{2})V^*$ and the effective mass is $I_r [V^* \ (cm^{-1}) = 8\pi^2 c I_r \bar{\nu}^2/h]$, the observed $v = 1 \leftarrow v = 0$ frequency can be related to V^* by:

$$V^* \ (cm^{-1}) = \bar{\nu}^2/F \qquad (22)$$

Observation of a central maximum for the torsional fundamental can thus lead to the evaluation of only a single potential parameter V^* for molecules with non-symmetric barriers. However it should be pointed out that such a calculation has led to some confusion in the past.

It is interesting to consider the magnitudes of the various potential terms for several asymmetric rotors. These are listed in Table 3 and the most stable conformer is listed along with the values for the various potential constants. The V_3 terms for both the sulphur and selenium containing molecules have very similar values as do the values for the corresponding term for the amine and phosphine compounds. However, the potential for ethylamine contains a V_1 term, whereas the ethylphosphine potential has a V_2 term. In the expressions above, the V_1 appears only with the coefficient one and is thus the most subject to slight errors in the experimental measurement of the frequencies. Since the higher order terms are associated with much larger coefficients, these terms are expected to show much less error than the V_1 term.

Table 3. Calculated potential constants (cm^{-1}) for a number of molecules which contain asymmetric rotors

molecule	most stable form	V_1	V_2	V_3	V_4	V_5	V_6	ref.
CH_3CH_2SH	gauche	0	-170.9	483.5	0	0	-21.1	15
CH_3CH_2SD	gauche	0	-162.1	476.6	0	0	-14.4	15
CH_3CH_2SeH	gauche	0	-92.4	432	0	0	-20	16
CH_3CH_2SeD	gauche	0	-91.7	408.6	0	0	-7.8	16
$CH_3CH_2NH_2$	trans	218	0	751	0	0	-52	17
$CH_3CH_2ND_2$	trans	279	0	707	0	0	-12	17
$CH_3CH_2PH_2$	trans	0	270	830	0	0	-58	19
$CH_3CH_2PD_2$	trans	0	207	785	0	0	-25	19
$C_2H_2O_2$	trans	1182	1114	0	-56	0	0	20
$CH_2CHCHCH_2$	trans	600	2068	273	-49	0	0	21
$C_6H_5CHCH_2$	planar	0	622.8	0	27.0	0	0	22

4. MULTI-TOP ROTORS

Although there are now several groups studying the torsional spectra of multi-top molecules, the theoretical analysis of the resulting data must still be considered in the development stages because of the complexity of the problem. One can readily appreciate the difficulties that might arise with such systems, particularly after observing the theoretical development of the single-top example in the previous section. In the case of a molecule which has more than a single rotor, the potential energy restricting the internal rotation arises from two sources: (1) interaction with the framework, and (2) interaction with the other rotors. The earlier thermodynamic and low resolution spectroscopic studies generally assumed that the internal rotational degrees of freedom were independent and equivalent. Such a treatment resulted in an average one-dimensional potential barrier computed with assumption that the rotor-rotor interaction is negligible. Advances in instrumentation have now made it necessary to treat the multi-top molecules in a somewhat more quantitative fashion, since in many cases either excited state transitions were resolved or fundamentals resulting from the different torsional modes were assigned. The additional data permit certain conclusions to be drawn regarding the nature and magnitude of the forces interacting in these systems. For a more detailed discussion of the multi-top problem, the reader should see references 23 and 24.

5. REFERENCES

1. S. Weiss and G.E. Leroi, J. Chem. Phys., 48 (1968) 962.
2. J.R. Durig, W.E. Bucy, L.A. Carreira and C.J. Wurrey, J. Chem. Phys., 60 (1974) 1754.
3. J.D. Kemp and K.S. Pitzer, J. Chem. Phys., 4 (1936) 749.
4. D.A. Dows, in M.M. Labes, D. Fox and A. Weissberger (Editors), Physics and Chemistry of the Organic Solid State, Vol. 1, Interscience, New York, 1963, p. 667.
5. J.R. Durig, C.M. Player, Jr. and J. Bragin, J. Chem. Phys., 54 (1971) 460.
6. J.R. Durig, C.M. Player, Jr. and J. Bragin, J. Chem. Phys., 52 (1970) 4224.
7. J.R. Durig, W.E. Bucy and C.J. Wurrey, J. Chem. Phys., 60 (1974) 3293.
8. C.J. Wurrey, L.A. Carreira and J.R. Durig, in J.R. Durig (Editor), Vibrational Spectra and Structure, Vol. 5, Elsevier Scientific Publishing Co., Amsterdam, 1976.
9. J.R. Durig, A.D. Lopata and C.J. Wurrey, J. Raman Spectr., 3 (1975) 345.
10. J.E. Kilpatrick and K.S. Pitzer, J. Chem. Phys., 17 (1949) 1065.
11. D.R. Herschbach, J. Chem. Phys., 31 (1959) 91.
12. W.G. Fateley and F.A. Miller, Spectrochim. Acta, 17 (1961) 857.
13. W.G. Fateley and F.A. Miller, Spectrochim. Acta, 19 (1963) 611.
14. D.R. Lide, Jr. and D.E. Mann, J. Chem. Phys., 29 (1958) 914.
15. J.R. Durig, S.M. Craven, K.K. Lau and J. Bragin, J. Chem. Phys., 54 (1971) 479.
16. J.R. Durig, S.M. Craven and J. Bragin, J. Chem. Phys., 52 (1970) 5663.
17. J.R. Durig, S.M. Craven and J. Bragin, J. Chem. Phys., 52 (1970) 2046.
18. J.D. Lewis, T.B. Malloy, Jr., T.H. Chao and J. Laane, J. Mol. Struct., 12 (1972) 427.
19. J.R. Durig and A.W. Cox, Jr., J. Chem. Phys., 63 (1975) 2303.
20. J.R. Durig, W.E. Bucy and A.R.H. Cole, Can. J. Phys., 53 (1975) 1832.
21. L.A. Carreira, J. Chem. Phys., 62 (1975) 3851.
22. L.A. Carreira and T.G. Towns, J. Chem. Phys., 63 (1975) 2015.
23. J.R. Durig, S.M. Craven and W.C. Harris, in J.R. Durig (Editor), Vibrational Spectra and Structure, Vol. 1, Marcel Dekker, Inc., New York, 1972.
24. W.J. Orville-Thomas (Editor), Internal Rotation in Molecules, John Wiley and Sons, New York, 1974.

CHAPTER 21

VIBRATIONAL SPECTRA OF TRANSITION METAL COORDINATION COMPOUNDS AND THEIR ANALYSIS

A. Müller

1. INTRODUCTION

The investigation of transition metal coordination compounds has attracted much interest in the last few years [1]. It has been found that several metal complexes play an important role in biological processes [2]. Several new physical methods [3] have been applied to study metal ligand bonding, which is important for the understanding of complex formation. One classical method which is still very important is vibrational spectroscopy even though several new techniques (experimental as well as theoretical) have been developed. Vibrational spectroscopy presents the best possibility of studying complexes which are only stable under extreme conditions, such as in matrices at very low temperatures. Therefore, it should be worthwhile to summarise the corresponding facts and trends in the field of vibrational spectra of transition metal coordination compounds in a very condensed form.

2. STUDY OF COMPLEX FORMATION

The study of complex formation in solution is one of the most important classical subjects of coordination chemistry. By measuring stability constants with classical physical methods the complex formation can be well understood [4]. This type of investigation is also important for the preparation of transition metal complexes. With the increasing importance of different physical methods (especially spectroscopic) very interesting new aspects of complex formation under unusual conditions (as in matrices, in the gas phase and in melts) can be studied. The use of spectroscopic methods in these cases gives the only possibility to study the formation of many interesting compounds, which are mostly stable only under the above mentioned extreme conditions (at high temperatures, under high pressures or at very low temperatures). Vibrational spectroscopy can be used to study complex formation qualitatively (see Table 1) or quantitatively (determination of stability constants from the intensities of bands, even in aqueous solution [5]).

Table 1. (continued).

Phase	Example	Proof by	Ref.
Surface (chemisorption of gases on metallic surfaces)	CO (g) + M $\longrightarrow$ 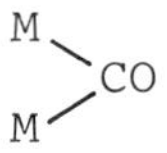	ν(CO): M = Cu (2128 cm^{-1}) Pt (2070) Pd (2053) Ni (2033)	k

a L.H. Jones and R.A. Penneman, J. Chem. Phys., 22 (1954) 965.
b See ref. 15.
c L.W. Daasch, Spectrochim. Acta, 15 (1959) 726; J. Chem. Phys., 28 (1958) 1005.
d H.A. Øye, E. Rytter, P. Klaboe, S.J. Cyvin, Acta. Chem. Scand., 25 (1971) 559.
e F.P. Emmenegger, C. Rohrbasser and C.W. Schläpfer, Inorg. Nucl. Chem. Lett., 12 (1976) 127.
f H. Huber, E.P. Kündig, M. Moskovits and G.A. Ozin, J. Amer. Chem. Soc., 95 (1973) 332.
g A.J. Rest and J.R. Sodeau, J. Chem. Soc. Chem. Comm., (1975) 696.
h this problem is going to be investigated by the present author; see also reaction produced between NiF_2 and N_2: D.A. Van Leiersburg and C.W. DeKock, J. Phys. Chem., 78 (1974) 134; J. Amer. Chem. Soc., 94 (1972) 3235.
i A. Müller and V. Flemming, unpublished.
j J.R. Ferraro, K. Nakamoto, J.T. Wand and L. Lauer, Chem. Comm., (1973) 266.
k R.P. Eischens, W.A. Pliskin and S.A. Francis, J. Chem. Phys., 22 (1954) 1786; J. Phys. Chem., 60 (1956) 194.

Table 1. Complex formation

Phase	Example	Proof by	Ref.
Solution	(a) $Ag^+ + nCN^- \xrightarrow{H_2O} [Ag(CN)_n]^{(n-1)-}$ (n = 2,3,4)	(a) analysis of ν(CN) stretching vibrations	a
	(b) $Ni^{2+}(aq) + NH_3 \longrightarrow [Ni(NH_3)_6]^{2+} + (aq)$	(b) ν(NiN)	b
	(c) molecular complex $AgClO_4$/benzene	(c) ir spectrum	c
Melt	$Al_2Cl_7^-$ acts as bidentate ligand in $MCl_2/AlCl_3$ melts	Raman spectrum of the melt	d
Gas phase	$CuCl_2(s) + 2AlCl_3(g) \rightleftarrows (CuCl_2.2AlCl_3)(g)$	Raman gas spectrum	e
Matrix (a) metal molecular reaction	$Ni(at) + nN_2 \xrightarrow{N_2/Ar} Ni(N_2)_n$ (n = 1-4)	matrix isolation spectrum	f
(b) photolysis	$[W(CO)_5py] \xrightarrow[Ar,\ 12\ K]{h\nu(320-390\ nm)} [W(CO)_5] + py$	matrix isolation ir spectrum	g
(c) molecular complexes	MnO_3Cl - HCl ?	matrix isolation ir spectrum	h
Solid	$[Ni(NH_3)_6](ReO_3S)_2 \xrightarrow{T} [Ni(NH_3)_4(ReO_3S)_2] + 2NH_3$	ir spectrum	i
Solid under pressure	$[Ni(PPh_2Bz)Br_2]_{tetr.} \xrightarrow{20000\ atm}$ planar complex	high pressure ir spectrum	j

3. MOLECULAR VIBRATIONS OF FREE AND COMPLEXED LIGANDS

Characterisation of vibrations and frequency shifts due to coordination

If ligands are coordinated to a central atom it is in general not difficult to distinguish between ligand internal vibrations and skeletal vibrations as the frequencies of the ligand vibrations are usually similar to those of the free ligands. The observed frequency shift can be caused by force constant change and kinematic coupling (between skeletal and ligand vibrations [6]). The first effect is in general larger and is of course important for the understanding of the nature of the bonding.

Vibrational modes of coordination compounds can be classified as:

(a) ligand vibrations
(b) ligand framework couplings
(c) framework vibrations.

For example, in $M(AB)_4$ complexes of T_d symmetry (e.g. $Ni(CO)_4$), the four diatomic ligands AB give rise to four ligand vibrations in the complex

$$\Gamma(\text{lig}) = A_1 + F_2$$

whereas the ligand framework couplings may be described as linear bendings

$$\Gamma(\text{cpl}) = E + F_1 + F_2$$

and the skeletal modes may be identified as the vibrational modes of an XY_4 type molecule

$$\Gamma(\text{fr}) = A_1 + E + 2F_2$$

Corresponding symmetry coordinates can be constructed.

More complicated complexes can be treated in the same way and by using a group theoretical method the overall symmetry coordinates of the whole complex can be constructed from those of the ligands having a different local symmetry. This allows one to transfer directly an <u>initial</u> symmetry force constant matrix F^0 of the free ligand in a normal coordinate analysis [7]. This method is similar to the so called LSFF (local symmetry force field) method and it allows one to study the effect of coordination on ligand vibrations due to force constant changes and kinematic effects. The initial force field would also include roughly estimated constants for the skeletal force constants and those related to ligand framework coupling vibrations (for details see also ref. 6).

Force constant changes could be estimated according to:

$$\underline{\Delta F} = (\underline{L}_o^{-1})\ \underline{\Delta\Lambda}\underline{L}_o^{-1} \qquad (1)$$

(which can be derived from first order perturbation theory [8];

all symbols have their usual meaning [9]) or more strictly if isotopic data are known, one should estimate $\underline{\Delta F}$ according to (when $(\underline{\Delta\Lambda}_0)_i \neq 0$ for all i)

$$\underline{\Delta F} = (\underline{L}_0^t)^{-1}[\underline{\Lambda}_0(\underline{\Delta\Lambda}_0)^{-1}\underline{\Delta\Lambda} - \underline{\Lambda}_0]\underline{L}_0^{-1} \quad (2)$$

given by first order perturbation theory. In the above equations the index zero corresponds to the initial solution.

It should be noted that ligand internal vibrations are often very dependent on intermolecular interactions (different anions in crystals or different hydration in solution). These effects can have a larger influence on ligand vibrations than coordination to different metals [10]. In the case of ammine complexes the $\nu(NH_3)$, $\delta(NH_3)$ and $\rho(NH_3)$ vibrations are strongly dependent on the type of anion (shifts of the order of 100 cm^{-1}). This effect has been systematically studied for $\rho(NH_3)$ [10].

Change of band intensities due to complexation

In addition to frequency shifts, usually a change in band intensities of ligand vibrations is observed on complexation. In general there are three different mechanisms responsible for infrared intensity changes:

(1) change of local symmetry and normal coordinates of the ligand (due to vibrational coupling)
(2) equilibrium electrostatic polarisation (the charge distribution in the ligand changes)
(3) vibronic effects (reorientation of electrons during a vibration).

We observe for example an enormous increase of intensity for the ν(CO) band from free to complexed CO. The intensities also vary for different central atoms. These effects can be interpreted with respect to the nature of bonding (see section 7). Intensity changes due to complexation of SCN^- with metals are also observed. These data are extremely useful to distinguish different kinds of linkage isomerism (see Table 3).

Splitting of bands due to complexation

Besides change of frequencies and intensities, bands of the free ligand may split (e.g. if the local symmetry of the ligand in the complex is lower than that of the free ligand). This effect is important for the determination of the type of coordination (section 6; see also ref. 1a).

4. VIBRATIONAL CONSTANTS

This section briefly summarises the possible methods for the characterisation of complexes with vibrational constants [11]. The knowledge of constants like frequency shifts, infrared and Raman intensities, anharmonicities, Coriolis-, centrifugal distortion- and ℓ-doubling constants as well as the inertia defect and mean amplitudes of vibration, is in general necessary for the characterisation of the structure and dynamical properties of a compound [11]. In some cases the measured data can be directly correlated to the molecular structure

Table 2. (continued).

	Isotopes	$\Delta m/m$	Example	Obs. shift	Vibration	Ref.
Ta	+ + +	-	-	-	-	
W	182/186	0.022	WF_6	(∿1.3)	$\nu_3(F_{1u})$	b
Re	185/187	0.011	ReO_4^-	(∿0.5)	$\nu_3(F_2)$	b
Os	186/192	0.032	OsO_4	(∿1.6)	$\nu_3(F_2)$	b
Ir	191/193	0.011	$[Ir(NH_3)_6]^{3+}$	(∿0.6)	$\nu_3(IrN)(F_{1u})$	b
Pt	192/198	0.031	$[Pt(NH_3)_4]^{2+}$	(∿1.3)	$\nu_{as}(PtN)(E_u)$	b
Au	only one stable	-	-	-	-	
Hg	198/204	0.030	$[Hg(NH_3)_2]^{2+}$	(∿1.1)	$\nu_{as}(HgN)$	b

a F. Königer, R. Carter and A. Müller, Spectrochim. Acta, 32A (1976) 891.
b estimated.
c K.H. Schmidt and A. Müller, J. Mol. Struct., 22 (1974) 343.
d K. Nakamoto, C. Udovich and J. Takemoto, J. Amer Chem. Soc., 92 (1970) 3973.
e A. Cormier, K. Nakamoto, P. Christophliemk and A. Müller, Spectrochim. Acta, 30A (1974) 1059.
f A. Müller, H.H. Heinsen, K. Nakamoto, A. Cormier and N. Weinstock, Spectrochim. Acta, 30A (1974) 1661.
g see ref. 14.
h F. Königer and A. Müller, J. Mol. Spectr., in press.
i see ref. 18.
j A. Müller et al., unpublished.

Table 2. Metal isotope shifts of transition metal (3d, 4d and 5d) complexes (cm^{-1})

	Isotopes	Δm/m	Example	Obs. shift	Vibration	Ref.
Sc	only one stable	-	-	-	-	
Ti	46/50	0.087	$TiCl_4$	10.8	$\nu_3(F_2)$	a
V	50/51	0.020	VCl_4	(∿2.9)	$\nu_3(F_2)$	b
Cr	50/53	0.060	$[Cr(NH_3)_6]^{3+}$	4.0	$\nu_{as}(CrN)(F_{1u})$	c
Mn	only one stable	-	-	-	-	
Fe	54/57	0.056	$Fe(acac)_3$	5.0	$\nu(FeO)$	d
Co	only one stable	-	-	-	-	
Ni	58/62	0.069	$[Ni(CS_3)_2]^{2-}$	7.2	$\nu_{as}(NiS)(B_{3u})$	e
Cu	63/65	0.032	$[Cu(NH_3)_4]^{2+}$	2	$\nu_{as}(CuN)(E_u)$	c
Zn	64/68	0.063	$[Zn(^{92}MoS_4)_2]^{2-}$	3	$\nu_{as}(ZnS)(E)$	f
Y	only one stable	-	-	-	-	
Zr	90/94(96)	0.045	-	-	-	
Nb	only one stable	-	-	-	-	
Mo	92/100	0.087	$[^{58}Ni(MoS_4)_2]^{2-}$	6.0	$\nu_{as}(MoS)(B_{3u})$	g
Tc	no stable	-	-	-	-	
Ru	96/104	0.083	RuO_4	6.6	$\nu_3(F_2)$	h
Rh	only one stable	-	-	-	-	
Pd	104/110	0.058	$[PdCl_6]^{2-}$	3.5	$\nu_3(F_{1u})$	i
Ag	107/109	0.019	$[Ag(NH_3)_3]^{2+}$	(∿1)	$\nu_{as}(AgN)(E_u)$	b
Cd	110/114	0.036	$[Cd(NH_3)_4]^{2-}$	1	$\nu_3(CdN)(F_2)$	j
Hf	174/180	0.035	$HfCl_6{}^{2-}$	(∿1.4)	$\nu_3(F_{1u})$	b

or nature of bonding (e.g. intensities; see section 7). But on the other hand, these data are necessary for the determination of force constants (Chapter 16). The force field itself fully characterises the compound in the electronic ground state, as in general the above mentioned constants can be determined from the force constants. The force field itself provides one of the best possibilities of characterising the chemical bonding in complexes.

Frequencies and isotope shifts

(a) The metal ligand vibration

The metal ligand vibration ν(ML) is very important for the characterisation of the metal-ligand interaction [1c]. ν(ML) increases with increasing oxidation number of M and decreasing coordination number and is also dependent on other substituents at M as well as on the relative position of ligands (e.g. trans effect).

(b) Metal isotope shifts

The metal isotope technique has recently been used with success for the important problem of assignment of metal ligand vibration as well as for the evaluation of force constants [12, 13]. For example the analysis of the spectra of the complexes [14,15]

$[^{58}Ni(^{92}MoS_4)_2]^{2-}$, $[^{58}Ni(^{100}MoS_4)_2]^{2-}$

$[^{62}Ni(^{92}MoS_4)_2]^{2-}$, $[^{62}Ni(^{100}MoS_4)_2]^{2-}$

could be studied in more detail with respect to the mixing between metal sulphur stretching vibrations in the metal sulphide array S_2 Mo S_2 Ni S_2 Mo S_2 by measuring the corresponding metal isotope shifts.

The measurement of the metal isotope shift is also important because of the fact that central atom substitution data are in general very effective in fixing the force constants [16,17], while isotope shifts of low lying vibrations (such as metal ligand vibrations) are in general very useful for the calculation of force constants as will be shown in the next section. Examples of isotope shifts for transition elements are given in Table 2.

(c) Isotope shifts for ligand internal vibrations

For the assignment of ligand internal vibrations and for the determination of force fields many isotope shifts have been measured in the past (H/D, $^{15}N/^{14}N$, $^{13}C/^{12}C$, $^{18}O/^{16}O$, and $^{37}Cl/^{35}Cl$) [13].

5. FORCE FIELDS

The most serious problem, which we have to take into consideration is that there are (because of the large number of atoms in most complexes) normally many more unknown force constants than experimental data. In addition, we cannot determine most of the molecular constants mentioned in the last section as many of the coordination compounds are solids, for which only isotope shifts can be measured [18].

Because of this, many papers have been published dealing with the use of mathematical approximation methods or model force fields like the Urey-Bradley Force Field (Chapter 16) both of which are very problematical with respect to an interpretation of the calculated force constants (for some basic principles of force field calculations of coordination compounds see refs. 6,18).

Exact (complete) force fields

Though the number of these calcuations which have been done up to now is limited, some calculations have been reported even for species with n (order of secular equation) = 4 [19]. In these cases many frequency shifts (e.g. for $Ni(CO)_4$ and $[Zn(CN)_4]^{2-}$: $^{64/68}Zn$, $^{12/13}C$, $^{14/15}N$ and $^{16/18}O$) and vibrational-rotational interaction constants have been used. Exact force constants are known for several carbonyls, tetracyano complexes, tetroxides, tetraoxoanions, tetrathioanions and halogeno complexes. Some calculations have even been done with harmonic frequencies. Also in some cases, the values of the interaction constants allowed interesting interpretations (e.g. $f_{NiC/CN}$ of $Ni(CO)_4$ is much higher than $f_{ZnC/CN}$ of $[Zn(CN)_4]^{2-}$ [19]). Most of the exact force field data are summarised in ref. 14.

In most transition metal coordination compounds the metal-ligand vibration has a low frequency. It can be theoretically proven that in those cases even small shifts determined with a large uncertainty (e.g. $\Delta\nu = 3 \pm 1$ cm^{-1}) are useful in fixing the force field [20]. This has important consequences for many solids where no higher accuracy is possible (see Chapter 12).

Approximate force fields

Approximation methods in connection with additional data can sometimes be used to calculate force constants of coordination compounds with success. The point mass model and the HLFS (high-low frequency separation) method together with additional data like isotope shifts have been used. Several examples have been published and discussed in the literature; the advantage of these methods is that it can be physically proven whether it is possible to apply the methods or not (e.g. validity of the Teller-Redlich rule for the truncated problem) - see Chapter 16.

Correlation of force constants with other physical properties (e.g. skeletal constants of ammine complexes)

Whereas the old values for the skeletal force constants for this class of compounds reported in the literature show large deviations [21], it is now possible to calculate a physically reasonable set of values for all compounds with the help of H/D, $^{14/15}N$ and metal substitution data, as has been shown by the author and co-workers [22,23]. The metal nitrogen stretching force constant could be correlated with other data like bond length, stability constants and ligand field stabilisation energies and bond energies.

(a) Bond lengths

If one compares the M-N stretching force constants with the mean bond length in hexacoordinated complexes of central atoms belonging to the same transition metal series it turns out that the force constants increase with decreasing M-N distances [21].

(b) Stability constants and ligand field stabilisation energies (LFSE)

For each step of formation of an ammine complex in aqueous solution by the displacement of water molecules by ammonia there is a thermodynamic equilibrium described by the stability constant K_n, the logarithm of which is directly related to the free energy change ΔG_n^o at the binding of the nth ligand [21]. Figure 1 contains a plot of log K_n, LFSE and f_{MN} versus the

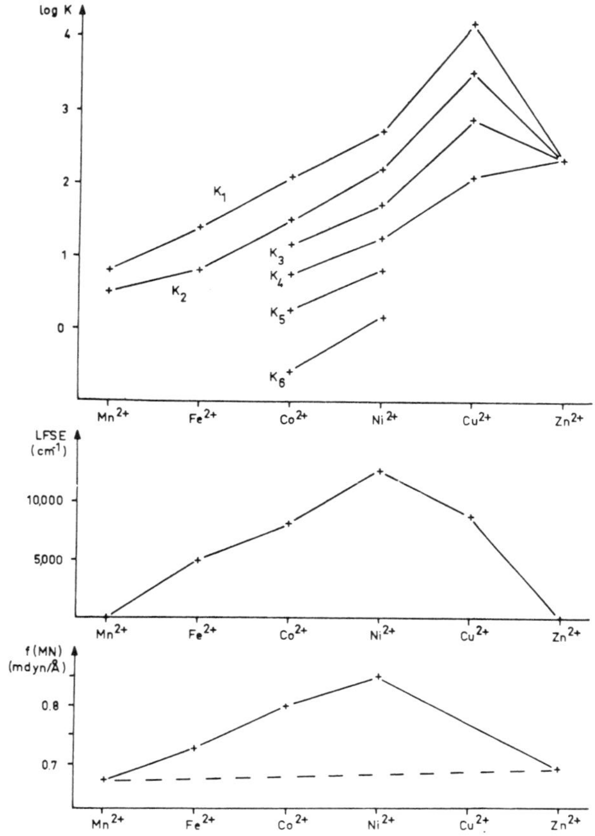

Figure 1. Plot of the logarithms of individual stability constants for ammine complexes of divalent metal ions belonging to the first transition row, of the ligand field stabilisation energies, and of the metal-nitrogen stretching force constants for the hexammines, versus atomic number(reproduced with permission from ref. 21).

atomic number of the central metal. All properties show the same trend (the inconsistency of Cu arises because $[Cu(NH_3)_6]^{2+}$ does not exist in aqueous solution) in agreement with the Irving-Williams series. It is also interesting to note that the logarithm of the mean complexity constant according to Bjerrum as defined by $(1/N)\log K_n$, which describes the affinity between ammonia and metal ions in aqueous solution, can be directly related to the force constants of all ammine complexes (see Figure 1 and ref. 21).

(c) Total bond energies

$\Delta H_f[M(NH_3)_m]^{n+}_{(g)}$, the enthalpy of formation of an ammine complex, shows a linear relation with the f_{MN} values [21].

6. STRUCTURAL DIAGNOSIS

The type of coordination of ligands

One of the problems in "Coordination Chemistry" which is frequently solved by vibrational spectroscopy is the determination of the kind of ligand coordination e.g. whether a ligand AB coordinates through A or B or as a bridged ligand [1a]. One could, for instance, distinguish whether ions like $MoOS_3^{2-}$ coordinate through O, S or as bridging ligand in novel types of transition metal complexes [24-27] (containing the above mentioned transition metal complexes themselves as ligands).

Determination of the molecular symmetry

The determination of the overall symmetry of the whole complex is difficult in general. But in several cases the symmetry can be determined from the number of bands such as in the case of carbonyls from the number of $\nu(CO)$ bands [28] or correspondingly in the case of highly symmetrical halogeno [29] or ammine complexes [21] in the usual way (complicated in the case of solids because of crystal field effects!).

Detection of conformers

Often the presence of conformational isomers can be detected (e.g. of $(h^5 - C_5H_5)Fe(CO)_2(h^1 - C_5H_5)$ where four instead of two $\nu_s(CO)$ and $\nu_{as}(CO)$ bands have been found [30]).

7. CHEMICAL BONDING

Nature of metal ligand bond

The strength of the metal-ligand bond characterises the metal-ligand interaction. Though often different effects are responsible for the metal-ligand bond strength the overall effect may be characterised by the corresponding metal-ligand force constant. The conclusions are often similar to those derived from Mössbauer , electronic (reduction in spin-orbital coupling constants and electron repulsion energies), ESR (ligand hyperfine splitting), NQR (percentage of metal ligand ionic character) and NMR (electron spin densities on ligand atoms) spectra, with which the evidence for the metal-ligand covalency can be obtained [31] (in contradiction to the simple crystal field theory). An interesting aspect is the kind of electron delocalisation present in M. Some examples should be discussed here qualitatively:

(a) Change of σ-bond strength for ML in a series ML_n with different M

For $[M(NH_3)_6]^{2+}$ complexes the σ-bond strength increases in the following order Mn < Fe < Co < Ni > Zn as inferred from the values of the f_{MN} force constants [21].

(b) Change of π-bond strength for ML in a series ML_n with different M

For $[M(III)(CN)_6]^{3-}$ complexes the M-C π-bond contribution increases with increasing number of electrons in the t_{2g} orbitals (Cr < Mn < Fe < Co) as (1) ν(MC) increases, whereas ν(CN) remains nearly constant and (2) the integrated absorption coefficient of ν(CN) increases strongly [32].

This can be explained by taking into consideration the two mesomeric valence structures for the linkage M-C-N.

(c) Influence of ligand substitutions on metal ligand π back-bonding

For $[Fe(II)(CN)_5L]^{n-}$ complexes π back-bonding decreases in the following order L = NO > NO_2 > NH_3 > H_2O (proof: analysis from vibrational [33] and Mössbauer spectra [34] where σ becomes more negative). Correspondingly decreasing M-C π back-bonding contribution was found in the series $L_3Mo(CO)_3$

L = CO > PF_3 > PCl_3 > $AsCl_3$ > $SbCl_3$ > PR_3 > NR_3 > OR_2

(π acidity series)

according to a decrease of ν(CO) [35].

The metal-metal bonds

Much interest has been shown in polynuclear complexes containing metal-metal bonds [36]. The existence of a metal-metal bond can be proved by the occurrence of a corresponding metal-metal stretching vibration (mostly with very high intensity in the Raman effect [36]). The first observation was made by Woodward in a pioneering work in 1934 [37], demonstrating that mercurous nitrate in aqueous solution gives an intense line at 169 cm^{-1} in the Raman spectrum due to an Hg-Hg bond. Force constants f_{MM} show the same trend as those for M-L bonds in general e.g. $f_{MnL} < f_{ReL}$ or $f_{RuL} < f_{OsL}$ [15,23,27] (for a simple discussion of metal-metal bonding in the light of MO theory see ref. 38). Complexes even with metal-metal quadrupole bonds (1σ, 2π and 1σ) are known [38]. In those cases the corresponding stretching vibration has a relatively high wave number [39].

8. FUTURE TRENDS

Some further typical problems which can be investigated with vibrational spectroscopy are given in Table 3. Several experimental, partly new, techniques have been used to solve problems related to coordination chemistry. These are briefly summarised in Table 4.

Table 3. Investigation of some further problems of coordination chemistry.

Subject	Example	Ref.
Structural isomerism	Distinction between $[Co(NH_3)_5NO]^{2+}$ (black monomeric) and $[Co(NH_3)_5NO]_2^{4+}$ (red)	a
Linkage isomerism	S or N coordination of SCN (distinction by ir band intensities)	b
Cis-trans isomerism		
a) effect (influence)	from ν(MCl) frequencies of $[LPdCl_3]^-$ complexes trans-effect: CO < SR_2 < PR_3	c
b) kinetics of cis-trans isomerism	$(NH_3)_2PdX_2$ (cis → trans)	d
Jahn-Teller effect		
a) distortion	$CuCl_4^{2-}$ ($T_d \rightarrow D_{2d}$)	e
b) study of Jahn-Teller active vibrations	band broadening and lowering of frequencies of $\nu_2(E_g)$ in the case of octahedral metal fluorides with electronically degenerate levels	f

a E.E. Mercer, W.A. McAllister and J.R. Durig, Inorg. Chem., 6 (1967) 1816.
b R. Larsson, Rec. Chem. Progr., 31 (1970) 171.
c R.J. Goodfellow, P.L. Goggin and D.A. Duddell, J. Chem. Soc., A (1968) 504.
d J.R. Durig and B.R. Mitchell, Appl. Spectr., 21 (1967) 221.
e B. Morosin and E.C. Lingafelter, J. Phys. Chem., 65 (1961) 50; Acta Cryst., 13 (1960) 807; D. Forster, Chem. Comm., (1967) 113; I.R. Beattie, T.R. Gilson and G.A. Ozin, J. Chem. Soc., A (1969) 534.
f B. Weinstock and G.L. Goodman, Adv. Chem. Phys., 9 (1965) 169.

Table 4. (continued).

Technique	Application	Example	Ref.
Anti-resonance Raman	identification of vibronically allowed but electronically forbidden bands	$[Co(en)_3]^{3+}$: Raman intensity of $\nu(CoN)(A_{1g})$ follows the expected PRRE frequency dependence, but is de-enhanced within the d → d absorption band at 21,500 cm^{-1} $(^1A_{1g} \rightarrow {}^1T_{1g})$	j
Infrared intensity measurements	a) bond angles from relative intensities	a) carbonyls	k
	b) determination of bond dipole moments	b) OsO_4	l
Raman intensity measurements	assignment of fundamentals by comparing measured and calculated relative Raman intensities	tetroxides and tetraoxoanions $MO_4{}^{n-}$	m
Incoherent inelastic neutron scattering	determination of phonon freqeuncies	$[Ni(NH_3)_6]I_2$	n
Vibronic spectra	a) determination of ir and Raman inactive fundamentals (from luminesence spectra)	a) $\nu_6(F_{2u})[Cr(NH_3)_6]^{3+}$	o
	b) determination of vibrational frequencies of electronic excited states (absorption spectra)	b) $\nu(CrO)$ of CrO_3X^- (X = F,Cl, Br,I)	p

Table 4. Special techniques.

Technique	Application	Example	Ref.
Single crystal Raman	assignment of lattice modes	translatory and rotatory modes of $[Pt(NH_3)_4][PtCl_4]$	a
Single crystal infrared	a) assignment of bands	a) distinction of $\pi(PtCl_4)(a_{2u})$ and $\sigma(PtCl_4)_{as}(e_u)$ in K_2PtCl_4	b
(Infrared dichroism)	b) determination of orientation of groups	b) orientation of $[Ag(CN)_2]^-$ in K_2-salt	c
Low temperature infrared	magnetic phase change with removal of rotational freedom of ligands	NH_3 in ammine complexes	d
Ir band profile method (temp. dependence of band contours)	determination of rotational barriers	rotational barrier of NH_3 in $[Ni(NH_3)_6]$-salts and $Pd(NH_3)_2Cl_2$	e
Matrix isolation	a) investigation of novel complexes	see Table 1	
	b) measurement of isotope shifts	Ru isotope shift of RuO_4	f
High pressure	a) determination of high pressure phase changes	see review	g
	b) distinction between symmetric and antisymmetric vibrations		h
High temperature	structure determination of discrete novel gaseous molecules	MoO_2Cl_2, Fe_2Cl_6	h
Resonance Raman	assignment of charge transfer transitions	$MoOS_3{}^{2-}$ (distinction between $\pi(O,S) \rightarrow d(Mo)$ and $\pi(S) \rightarrow d(Mo)$	i

Table 4. (continued).

a D.M. Adams and I.R. Hall, J. Chem. Soc. Dalton, (1973) 1450.
b D.M. Adams and D.C. Newton, J. Chem. Soc., A (1969) 2998.
c L.H. Jones, J. Chem. Phys., 26 (1957) 1578; 25 (1956) 379.
d M.B. Palma-Vitorelli, M.U. Palma, G.W.J. Drewes and C. Koerts, Physica, 126 (1960) 922; A.R. Bates and K.W.H. Stevens, J. Phys. C, 2 (1969) 1573; R.C. Leech, D.B. Powell and N. Sheppard, Spectrochim. Acta, 21 (1965) 559.
e J.M. Janik, J.A. Janik, A. Migdal and G. Pytasz, Acta Phys. Pol. Part A, 40 (1971) 741.
f F. Königer and A. Müller, J. Mol. Spectr., in press.
g D.M. Adams and S.J. Payne, Vibrational Spectroscopy of Solids at High Pressures, Chem. Soc., Annual Reports, Section A, Vol. 69, 1972.
h I.R. Beattie, K.M.S. Livingston, G.A. Ozin and D.J. Reynolds, J. Chem. Soc. A, (1970) 1210.
i A. Müller and E. Diemann, J. Chem. Phys., 61 (1974) 5469.
j P. Stein, V. Miskowski, W. H. Woodruff, J. P. Griffin, K.G. Werner, B.P. Gaber and T.G. Spiro, J. Chem. Phys., 64 (1976) 2159.
k J.G. Bullitt and F.A. Cotton, Inorg. Chim. Acta; 5 (1971) 637.
l See ref. 15.
m N. Weinstock, H. Schulze and A. Müller, J. Chem. Phys., 59 (1973) 5063.
n J.A. Janik, W. Jakob and J.M. Janik, Acta Phys. Pol., Part A, 38 (1970) 467.
o T.V. Long and D.J.B. Penrose, J. Amer. Chem. Soc., 93 (1971) 632.
p E. Königer-Ahlborn and A. Müller, to be published.

9. REFERENCES

1. Earlier published reviews on vibrational spectra of coordination compounds (but with other intentions):
   (a) K. Nakamoto, Infrared Spectra of Inorganic and Coordination Compounds, 2nd Edition, Wiley-Interscience, New York, 1970.
   (b) F.A. Cotton, The Infrared Spectra of Transition Metal Complexes, in J. Lewis and R.G. Wilkins (Editors), Modern Coordination Chemistry, Interscience Publ., New York, 1969.
   (c) D.M. Adams, Metal Ligand and Related Vibrations, Arnold, London, 1967.
2. N.N. Hughes, The Inorganic Chemistry of Biological Processes, Wiley, London, 1972.
3. H.A.D. Hill and P. Day (Editors), Physical Methods in Advanced Inorganic Chemistry, Interscience, London, 1968; P. Day (Editor), Electronic States of Inorganic Compounds: New Experimental Techniques, D. Reidel Publ. Co., London, 1975.
4. F.J.C. Rosotti, The Thermodynamics of Metal Ion Complex Formation in Solution, in J. Lewis and R.G. Wilkins (Editors), Modern Coordination Chemistry, Interscience Publ., New York, 1969.

5. e.g., Y. Tomita and K. Ueno, Bull. Chem. Soc. Japan, 36 (1963) 1069.
6. N. Mohan, S.J. Cyvin and A. Müller, Coord. Chem. Rev., 21 (1976) 221.
7. S.J. Cyvin, B.N. Cyvin, R. Andreassen and A. Müller. J. Mol. Struct., 25 (1975) 141; S.J. Cyvin, B.N. Cyvin, K.H. Schmidt, A. Müller and J. Brunvoll, J. Mol. Struct., 32 (1976) 269; B.N. Cyvin, S.J. Cyvin, K.H. Schmidt, W. Wiegeler and A. Müller, J. Mol. Struct., 30 (1976) 315.
8. See e.g., A. Müller and N. Mohan, Z. Naturforsch., in press.
9. E.B. Wilson, J.C. Decius and P.C. Cross, Molecular Vibrations, McGraw Hill, 1955.
10. A. Müller and E.J. Baran, J. Mol. Struct., 15 (1973) 283.
11. G. Herzberg, The Spectra and Structures of Simple Free Radicals - An Introduction to Molecular Spectroscopy, Cornell University Press, Ithaca, 1971.
12. K. Nakamoto, Angew. Chem. Int. Ed., 11 (1972) 666.
13. N. Mohan, A. Müller and K. Nakamoto, in R.J.H. Clark and R.E. Hester (Editors), Advances in Infrared and Raman Spectroscopy, Vol. 1, Heyden and Son, London, 1975, p. 173.
14. E. Königer-Ahlborn, A. Müller, A.D. Cormier, J.D. Brown and K. Nakamoto, Inorg. Chem., 14 (1975) 2009.
15. K.H. Schmidt and A. Müller, Coord. Chem. Rev., 14 (1974) 115.
16. D.C. McKean, Spectrochim. Acta, 22A (1966) 269; A.A. Chalmers and D.C. McKean, Spectrochim. Acta, 22A (1966) 251.
17. A. Müller, N. Mohan and K.H. Schmidt, J. Chem. Phys., 57 (1972) 1752.
18. A. Müller, N. Mohan, F. Königer and M.C. Chakravorti, Spectrochim. Acta, 31A (1975) 107.
19. L.H. Jones, R.S. McDowell and M. Goldblatt, J. Chem. Phys., 48 (1968) 2663; L.H. Jones and B.I. Swanson, J. Chem. Phys., 63 (1975) 540; K.H. Schmidt, Spectrochim. Acta, in press.
20. A. Müller, N. Mohan and F. Königer, J. Mol. Struct., 30 (1976) 297.
21. See K.H. Schmidt and A. Müller, Coord. Chem. Rev., 19 (1976) 41.
22. K.H. Schmidt, W. Hauswirth and A. Müller, J. Chem. Soc. Dalton, (1975) 343; A. Müller, K.H. Schmidt and G. Vandrish, Spectrochim. Acta, 30A (1974) 651.
23. K.H. Schmidt and A. Müller, Inorg. Chem., 14 (1975) 2183.
24. A. Müller, H.H. Heinsen and G. Vandrish, Inorg. Chem., 14 (1975) 2009.
25. E. Königer-Ahlborn and A. Müller, Angew. Chem., 86 (1974) 709.
26. E. Königer-Ahlborn and A. Müller, Angew. Chem. Int. Ed., 15 (1976) 680.
27. See also E. Diemann and A. Müller, Coord. Chem. Rev., 10 (1973) 79.
28. See e.g., P.S. Braterman, Metal Carbonyl Spectra, Academic Press, New York, 1975.
29. W. Bues, Z. Anorg. Allg. Chem., 279 (1955) 104; M.A. Bredig and E.R. van Artsdalen, J. Chem. Phys., 24 (1956) 479.
30. F.A. Cotton and I.J. Marks, J. Amer. Chem. Soc., 91 (1969) 7523.
31. F.A. Cotton and G. Wilkinson, Advances in Inorganic Chemistry, Interscience Publ., New York, 1972.

32. L.H. Jones, Inorg. Chem., 2 (1963) 777.
33. M.F. Amr, E.G. Sayed and R.K. Sheline, J. Inorg. Nucl. Chem., 6 (1958) 187; E.J. Baran and A. Müller, Chem. Ber., 102 (1969) 3915.
34. N.N. Greenwood and I.C. Gibbs, Mössbauer Spectroscopy, Chapman and Hall Ltd., London, 1971.
35. W.D. Horrocks, Jr. and R.C. Taylor, Inorg. Chem., 2 (1963) 723; F.A. Cotton, Inorg. Chem., 3 (1964) 702; W. Strohmeyer and F.J. Müller, Chem. Ber., 100 (1967) 2812.
36. T.G. Spiro, Progr. Inorg. Chem., 11 (1970) 1; J.R. Johnson, D.M. Duggan and W.M. Risen, Inorg. Chem., 14 (1975) 1053.
37. L.A. Woodward, Phil. Mag., 18 (1934) 823.
38. F.A. Cotton, Chem. Soc. Rev., 4 (1975) 27.
39. e.g., A.P. Ketteringham, C. Oldham and C.J. Peacock, J. Chem. Soc. Dalton, (1976) 1640.

CHAPTER 22

VIBRATIONAL SPECTRA OF METAL CARBONYLS

J.J. Turner

1. INTRODUCTION

Very many of the transition metals form binary carbonyls with carbon monoxide. These have either one or more transition atoms per molecule. In the latter case very large cluster compounds can be formed; however in this chapter we shall concentrate on the simple mononuclear compounds. These are listed in Table 1, together with structural data.

Table 1. Stable mononuclear transition metal carbonyls

$V(CO)_6$	$Cr(CO)_6$	$Fe(CO)_5$	$Ni(CO)_4$
	$Mo(CO)_6$	$Ru(CO)_5$	
	$W(CO)_6$	$Os(CO)_5$	
black crystals; sublime in vacuo	colourless crystals; sublime in vacuo	liquids; m.p. ∿-20°C	liquid; m.p. -25°C b.p. 43°C
O_h	O_h	D_{3h}	T_d

(All are toxic, $Ni(CO)_4$ particularly so)

These high-symmetry structures, with essentially linear M-C-O groups, are not surprising but we shall see later that other carbonyl "fragments" such as $Mo(CO)_4$ have structures of low symmetry.

The vibrational spectra of metal carbonyls show a region around 2000 cm^{-1} associated with C-O stretching; the M-C stretching, C-M-C bending and M-C-O bending vibrations are found below about 800 cm^{-1}. The infrared bands in the C-O region are extremely strong and usually, in solution, very sharp and have been much used in inorganic chemistry for diagnostic purposes. For instance $Mo(CO)_4(P\phi_3)_2$ can have the $P\phi_3$ groups trans or cis:

trans

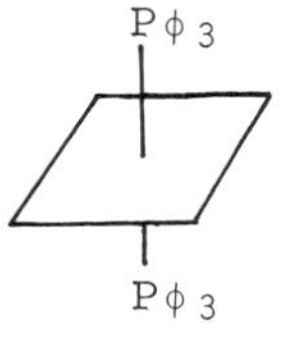

cis

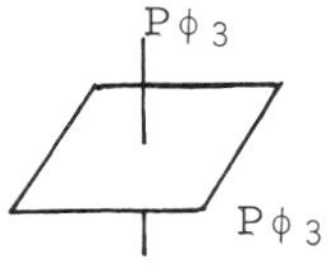

Simple group theory shows that the trans molecule of symmetry essentially D_{4h} should have only one infrared active band in the 2000 cm^{-1} region; the cis molecule of symmetry C_{2v} should show four. This is confirmed on synthesis of the two species.

More subtle problems arise when attempts are made to interpret the C-O band positions on some force field and to use the intensities of such bands to make quantitative statements about molecular structure. We first look at the force field.

2. METAL CARBONYL FORCE FIELDS

Consider a simple dicarbonyl fragment $M(CO)_2$ as shown, where we assume first that the carbonyls are not equivalent.

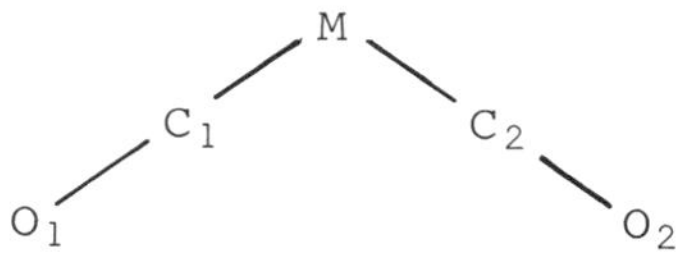

Within the usual Wilson FG matrix method the General Quadratic Valence Force Field (GQVFF) F matrix is expressed in terms of internal coordinates and contains force constants such as $2 \times f_{CO}$, $2 \times f_{MC}$, $2 \times f_{MCO}$, f_{CMC} and several interaction force constants. The G matrix is similarly unattractively complex, and there will certainly not be sufficient experimental data to solve for the whole set of force constants. If the molecule is in fact symmetric then there is some simplification but even so force field parameters exceed experimental data. Isotopic substitution of course provides more data but rarely sufficient for an absolutely complete force field. Thus it is almost inevitable that some approximations must be made. The least severe approximation will be to assume that some of the more distant interaction force constants are zero. L.H. Jones [1] has been much the most productive of very detailed GQVFF analyses. Such work is of vital importance since without it, it is impossible to assess the value of other more severe approximate methods.

Since the C-O stretching vibrations are so far removed from the other vibrations one of the most popular approximations is to assume that these vibrations are uncoupled from all other vibrations. In this approximation [2] the F matrix consists only of terms involving C-O stretching constants and interactions between them; since none of the CO groups share a common atom, the G matrix is diagonal in the reduced mass of the CO group. This has an important corollary; the bond angle (2α) in the C_{2v} molecule XY_2 can in principle be determined by the effect of isotopic substitution ($XY_2 \rightarrow \overline{XY}_2$ say) on ν_3 (strictly ω_3, the harmonic frequency):

$$\left(\frac{\bar{\omega}_3}{\omega_3}\right)^2 = \frac{m_x m_y (m_{\bar{x}} + 2m_{\bar{y}} \sin^2\alpha)}{m_{\bar{x}} m_{\bar{y}} (m_x + 2m_y \sin^2\alpha)} \qquad (1)$$

The dependence of $\bar{\omega}_3/\omega_3$ on bond angle arises since the G matrix (in this case a 1 x 1 matrix after symmetrisation) contains terms involving the bond angle. In the CO-factored force field we automatically lose this dependence so no matter how accurately the frequencies may be matched we shall be quite unable to get any structural information without going to a consideration of intensities. For a molecule such as $Cr(CO)_6$ the CO-factored force constants (see Figure 1) are:

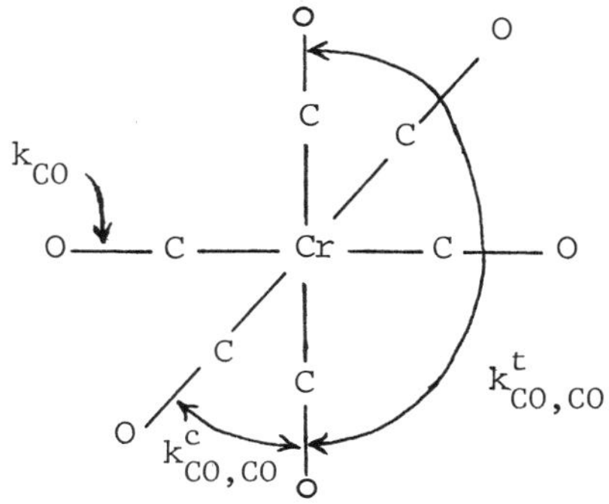

Figure 1. Structure and CO-factored force constants of $Cr(CO)_6$.

k_{CO}	-	one single C-O stretching force constant since all CO's are equivalent.
$k^t_{CO,CO}$	-	the trans CO-CO interaction constant.
$k^c_{CO,CO}$	-	the cis CO-CO interaction constant.

From group theory the C-O stretching vibrations are of symmetry a_{1g}, e_g, t_{1u}; a_{1g} and e_g are active in the Raman and t_{1u} in the IR. If all three frequencies are known then the three CO-factored force constants will be found. From the experimental data quoted in Jones' paper [1] the force constants are readily calculated by the standard FG equations and are given in Table 2 where they are compared with the corresponding values for the GQVFF constants from ref. 1.

There are some features of this data that are particularly important. For the GQVFF, the force constants can be determined using either experimental anharmonic or corrected harmonic frequencies; clearly one really needs the latter and Jones estimates these for the C-O stretching vibrations. The difference it makes is clear from the Table, particularly for the $f^t_{CO,CO}$ constant. With most carbonyl molecules there is insufficient data for the estimation of anharmonicities and approximate force fields are performed with anharmonic CO frequencies. In the case of $Cr(CO)_6$ it is possible to calculate the CO-factored force constants for both harmonic and anharmonic frequencies and these are given in Table 2. One of the most striking differences is that the ratio $k^t_{CO,CO}$: $k^c_{CO,CO}$ is $\sim$2 for the anharmonic CO-factored force field but $\sim$1 using the harmonic frequencies. It should be noted that Cotton's [2] rule of $k^t/k^c \sim 2$, which is of wide general applicability for aid in spectral analysis, is a peculiar consequence of using

Table 2. Force constants for $Cr(CO)_6$ in the gas phase[(a)]

	GQVFF			CO-factored force field	
	harmonic	anharmonic		harmonic	anharmonic
f_{CO}	17.24 ±0.07	16.74	k_{CO}	17.16	16.64
$f^c_{CO,CO}$	0.21 ±0.03	0.21	$k^c_{CO,CO}$	0.27	0.26
$f^t_{CO,CO}$	0.02 ±0.07	0.22	$k^t_{CO,CO}$	0.28	0.48
f_{MC}	2.08 ±0.08				
$f^c_{MC,MC}$	-0.019±0.003				
$f^t_{MC,MC}$	0.44 ±0.08				
$f_{MC,CO}$	0.68 ±0.07				
$f^c_{MC,CO}$	-0.05 ±0.03				
$f^t_{MC,CO}$	-0.10 ±0.07				

(a) constants in mdyn $Å^{-1}$; from reference 1.

an <u>an</u>harmonic CO-factored field. The GQVFF constants and CO-factored constants are quite different and are given different symbols in Table 2 to emphasise this; f and k respectively. We return shortly to the relationship between them.

Frequently it is possible to obtain good IR data for metal carbonyls, but Raman data is either non-existent or rather poor, partly because the carbonyls are extremely photosensitive. In such cases there may well be too few frequencies for even the CO-factored force field. Clearly for $Cr(CO)_6$ there will be only one band. In such circumstances partial isotopic (^{13}C or ^{18}O) substitution will provide more data.* Figure 2a shows the spectrum obtained by Perutz and Turner from a randomly mixed $Cr(^{13}CO)_x(^{12}CO)_{6-x}$ sample trapped in a low-temperature matrix of methane at 20 K; for comparison the spectrum of the same sample in cyclohexane solvent at room temperature (Figure 2c) shows poorer resolution and hence illustrates the value of matrix isolation for such studies - this technique of isotopic substitution combined with matrix isolation has been elegantly exploited by Müller and colleagues (see Chapter 12). Beneath the matrix spectrum is shown a theoretical spectrum (Figure 2b) calculated, following the usual iterative procedure on the anharmonic frequencies. The refinement gives k_{CO} = 16.4427 ±0.0014; $k^c_{CO,CO}$ = 0.2658 ±0.0010; $k^t_{CO,CO}$ = 0.5235 ±0.0048 mdyn $Å^{-1}$ and the fit of calculated to experimental spectra

* Note that complete substitution i.e. $Cr(^{12}C^{16}O)_6 \longrightarrow Cr(^{13}C^{16}O)_6$ will not help in this case.

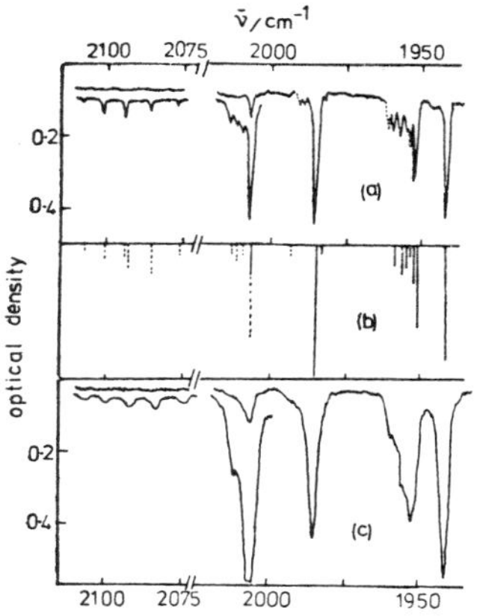

Figure 2. (a) ^{13}CO-enriched $Cr(CO)_6$ in methane matrix at 20 K; (b) theoretical spectrum (see text); (c) same sample in cyclohexane at room temperature (from ref. 7).

shows a standard deviation of 0.31 cm^{-1}. The values obtained this way are close to those obtained from Raman/IR data, the slight variation being due to environmental differences. It will also be noted that the intensity fit is very good - we return shortly to the method of calculation and significance of the intensity data.

There has been much discussion of the chemical significance of the force constants obtained in such calculations. It is clear that the only constants of value in any absolute sense (begging the question of whether any force constants in any systems have any absolute value) are those obtained by the nearest approximation to a full GQVFF - i.e. the "Jones" values [1]. In particular these are the only ones of significance when comparisons are made from a carbonyl of one structure to one of a different structure (e.g. $Cr(CO)_6$ cf. $Fe(CO)_5$). The CO-factored force field gives force constants of no absolute significance whatever and in particular, comparison of such constants between $Cr(CO)_6$ and $Fe(CO)_5$ is quite pointless. It is worth emphasising that Cotton and Kraihanzel, who really pioneered the application of the CO-factored force field, were perfectly well aware of this [3].

We can see how the k's and f's are related by considering $Cr(CO)_6$ as an example. The high symmetry of this molecule prevents any mixing of bending and stretching vibrations; for example in the A_{1g} block of the GQVFF GF matrix, the symmetry coordinates are derived from only $\Delta r(C\text{-}O)$ and $\Delta r(M\text{-}C)$. (Even with molecules of lower symmetry it can be shown that the contribution from the bending vibrations hardly affects the argument.) For the A_{1g} block we have symmetry coordinates:

$$S_1 = \frac{1}{\sqrt{6}} (\Delta r(CO)_1 + \Delta r(CO)_2 + \ldots \Delta r(CO)_6) \qquad (2)$$

$$S_2 = \frac{1}{\sqrt{6}} (\Delta r(MC)_1 + \Delta r(MC)_2 + \ldots \Delta r(MC)_6) \quad (3)$$

since $\bar{S} = \bar{L}\bar{Q}$:

$$S_1 = L_{11}Q_1 + L_{12}Q_2 \quad (4)$$

$$S_2 = L_{21}Q_1 + L_{22}Q_2 \quad (5)$$

Q_1 is the normal coordinate which is mostly described as C-O stretching and Q_2 is the corresponding M-C stretch. The L vectors derive from

$$\bar{G}\,\bar{F}\,\bar{L} = \lambda\,\bar{L} \quad (6)$$

where $\bar{G}$ and $\bar{F}$ are the symmetrical A_{1g} matrices. The frequency of the Q_1 normal mode is thus given by

$$\lambda_{CO}L_{11} = (GF)_{11}L_{11} + (GF)_{12}L_{21} \quad (7)$$

Since it can be shown that

$$\frac{L_{21}}{L_{11}} = -\frac{1}{m_C}\bigg/\left(\frac{1}{m_C} + \frac{1}{m_O}\right) \equiv -\frac{\mu_C}{\mu_{CO}} = x \quad (8)$$

$$\lambda_{CO} = (GF_{11}) + (GF)_{12}x$$

$$= \mu_{CO}|F_{11} - 2F_{12}x + F_{22}x^2| \quad (9)$$

The symmetry relationships between F's and f's allow us to rewrite this:

$$\lambda_{CO}(a_{1g}) = \mu_{CO}\,[(f_{CO} + 4f^c_{CO,CO} + f^t_{CO,CO})$$
$$- 2(f_{MC,CO} + 4f^c_{MC,CO} + f^t_{MC,CO})x$$
$$+ (f_{MC} + 4f^c_{MC,MC} + f^t_{MC,MC})x^2] \quad (10)$$

The corresponding expression using the CO-factored force field is:

$$\lambda_{CO} = \mu_{CO}\,[k_{CO} + 4k^c_{CO,CO} + k^t_{CO,CO}] \quad (11)$$

By considering the relationships for other symmetry species it is possible to deduce:

$$k_{CO} = f_{CO} + 2xf_{CO,MC} + x^2f_{MC} \quad (12)$$

and similar equations for $k^c_{CO,CO}$ and $k^t_{CO,CO}$.

Table 3 shows the calculation of k's from f's by substitution in the above expression and compares the values with those obtained directly from the CO-factored force field. More detailed analysis shows that the relationship (12) is expected to apply to vibrations of the same symmetry. The expression

Table 3. Comparison of k's and f's for $Cr(CO)_6$ harmonic frequencies in the gas phase

From Table 2		Using Eqn. (1)[(a)]	From CO-factored force field in Table 2
f_{CO}	17.24		
$f_{CO,MC}$	0.68	17.14	17.16
f_{MC}	2.08		
$f^c_{CO,CO}$	0.21		
$f^c_{CO,MC}$	-0.05	0.27	0.27
$f^c_{MC,MC}$	-0.0019		
$f^t_{CO,CO}$	0.02		
$f^t_{CO,MC}$	-0.10	0.28	0.28
$f^t_{MC,MC}$	0.44		

(a) for ^{12}C and ^{16}O, x = 0.57142

immediately shows three things:

(1) If the f's are related from one carbonyl to another there is absolutely no reason why the k's should be.
(2) The similarity of f_{CO} and k_{CO} arises because the second and third terms almost cancel. If however we were considering $Cr(^{13}C^{16}O)_6$ or $Cr(^{12}C^{18}O)_6$ the x values change and the difference f_{CO} - k_{CO} changes. This is discussed at length elsewhere [4].
(3) A corollary of (2) is that attempts to fit metal carbonyl isotopic patterns are best carried out with $^{12}C^{16}O/^{13}C^{18}O$ mixtures.

3. INTENSITIES OF METAL CARBONYL IR BANDS

If the carbonyl vibrations are completely decoupled from the rest of the molecule then the relative intensities may be used to determine structures. For example consider the fragment

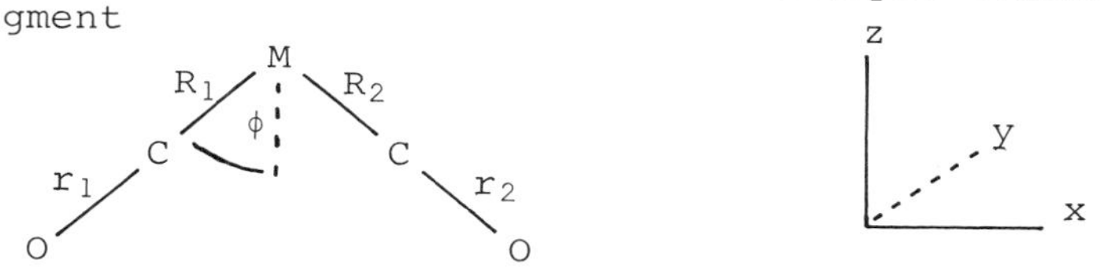

Since we are ignoring everything except the CO groups, the symmetry coordinates for the C_{2v} molecule are

$$S_1(A_1) = \frac{1}{\sqrt{2}}(\Delta r_1 + \Delta r_2) \tag{13}$$

$$S_2(B_2) = \frac{1}{\sqrt{2}}(\Delta r_1 - \Delta r_2) \tag{14}$$

Since $\bar{S} = \bar{L}\,\bar{Q}$,

$$Q_1(A_1) = L_{11}^{-1}S_1 = (1/\mu_{CO})S_1 \tag{15}$$

$$Q_2(B_2) = L_{22}^{-1}S_2 = (1/\mu_{CO})S_2 \tag{16}$$

The intensities, assuming as is almost always the case that the double harmonic approximation applies, are given by:

$$I_1(A_1) = B\left[\left(\frac{\partial\mu_x}{\partial Q_1}\right)^2 + \left(\frac{\partial\mu_y}{\partial Q_1}\right)^2 + \left(\frac{\partial\mu_z}{\partial Q_1}\right)^2\right] \tag{17}$$

$$I_2(B_2) = B\left[\left(\frac{\partial\mu_x}{\partial Q_2}\right)^2 + \left(\frac{\partial\mu_y}{\partial Q_2}\right)^2 + \left(\frac{\partial\mu_z}{\partial Q_2}\right)^2\right] \tag{18}$$

where e.g. $(\partial\mu_x/\partial Q_1)$ is the change in dipole moment along the x axis with Q_1, and B is the proportionality constant. Because of the molecular symmetry it is clear that

$$\frac{\partial\mu_x}{\partial Q_1} = \frac{\partial\mu_z}{\partial Q_2} = \frac{\partial\mu_y}{\partial Q_1} = \frac{\partial\mu_y}{\partial Q_2} = 0 \tag{19}$$

Thus

$$I_1 = B\left(\frac{\partial\mu_z}{\partial Q_1}\right)^2 \tag{20}$$

$$I_2 = B\left(\frac{\partial\mu_x}{\partial Q_2}\right)^2 \tag{21}$$

Using Equations (13)-(16), we have

$$I_1 = B\left(\frac{1}{\sqrt{2}}\frac{\partial\mu_z}{\partial\Delta r_1} + \frac{1}{\sqrt{2}}\frac{\partial\mu_z}{\partial\Delta r_2}\right)^2 \tag{22}$$

$$I_2 = B\left(\frac{1}{\sqrt{2}}\frac{\partial\mu_x}{\partial\Delta r_1} - \frac{1}{\sqrt{2}}\frac{\partial\mu_x}{\partial\Delta r_2}\right)^2 \tag{23}$$

Clearly

$$\frac{\partial\mu_z}{\partial\Delta r_1} = \frac{\partial\mu_z}{\partial\Delta r_2} \; ; \; \frac{\partial\mu_x}{\partial\Delta r_1} = -\frac{\partial\mu_x}{\partial\Delta r_2} \tag{24}$$

$$\text{so} \quad I_1 = B\left[2\left(\frac{\partial\mu_z}{\partial\Delta r_1}\right)^2\right] \tag{25}$$

$$I_2 = B\left[2\left(\frac{\partial\mu_x}{\partial\Delta r_1}\right)^2\right] \tag{26}$$

If we now assume that the change in dipole moment as the CO bond is stretched is along the CO bond then

$$\frac{\partial\mu_z}{\partial\Delta r_1} = \frac{\partial\mu}{\partial\Delta r_1}\cos\phi \; ; \quad \frac{\partial\mu_x}{\partial\Delta r_1} = \frac{\partial\mu}{\partial\Delta r_1}\sin\phi \tag{27}$$

where $\partial\mu/\partial\Delta r_1$ is the change in dipole moment along the bond. Thus

$$\frac{I_1}{I_2} = \cot^2\phi \tag{28}$$

and ϕ is determined from the two IR bands. The situation becomes somewhat more complicated if there are more than one vibration in each symmetry class. This will arise when the molecule contains inequivalent CO groups. For instance consider the planar C_{2v} fragment:

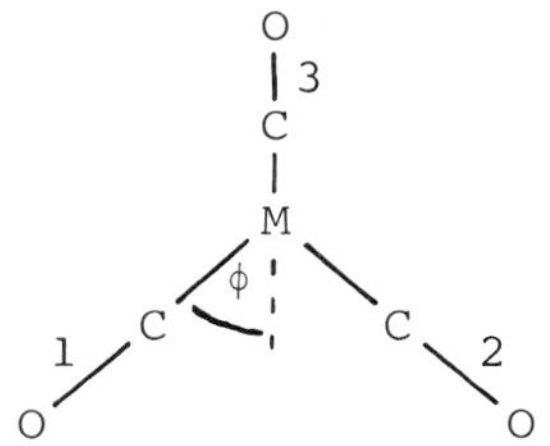

The symmetry coordinates this time are:

$$S_1(A_1) = \frac{1}{\sqrt{2}}(\Delta r_1 + \Delta r_2) \tag{29}$$

$$S_2(B_2) = \frac{1}{\sqrt{2}}(\Delta r_1 - \Delta r_2) \tag{30}$$

$$S_3(A_1) = \Delta r_3 \tag{31}$$

$$\left\{\begin{array}{l} Q_1(A_1) = L_{11}^{-1}S_1 + L_{13}^{-1}S_3 \quad (32) \\ \\ Q_3(A_1) = L_{31}^{-1}S_1 + L_{33}^{-1}S_3 \quad (33) \end{array}\right.$$

$$Q_2(B_2) = L_{22}^{-1}S_2 \tag{34}$$

where $L_{22} = \mu_{CO}$

Following essentially the same mathematics as before

$$I_1 = B\left[\left(\frac{1}{\sqrt{2}}\frac{\partial\mu_z}{\partial\Delta r_1} + \frac{1}{\sqrt{2}}\frac{\partial\mu_z}{\partial\Delta r_2}\right)L_{11} + \frac{\partial\mu_z}{\partial\Delta r_3}L_{31}\right]^2 \tag{35}$$

$$I_3 = B\left[\left(\frac{1}{\sqrt{2}}\frac{\partial\mu_z}{\partial\Delta r_1} + \frac{1}{\sqrt{2}}\frac{\partial\mu_z}{\partial\Delta r_2}\right)L_{13} + \frac{\partial\mu_z}{\partial\Delta r_3}L_{33}\right]^2 \tag{36}$$

$$I_2 = B\left[\left(\frac{1}{\sqrt{2}}\frac{\partial\mu_x}{\partial\Delta r_1} - \frac{1}{\sqrt{2}}\frac{\partial\mu_x}{\partial\Delta r_2}\right)L_{22}\right]^2 \tag{37}$$

The first thing to notice is that I_1 and I_3 are related by the force field. However since $\sum_i L_{ji}L_{j'i} = G_{jj'}$ (or $\bar{L}\tilde{L} = \bar{G}$) where the summation is carried out over L vectors of the same symmetry, $(I_1 + I_3)$ is independent of force field:

$$(I_1 + I_3) = \mu_{CO}^2\, B\left[\left(\frac{1}{\sqrt{2}}\frac{\partial\mu_z}{\partial\Delta r_1} + \frac{1}{\sqrt{2}}\frac{\partial\mu_z}{\partial\Delta r_2}\right)^2 + \left(\frac{\partial\mu_z}{\partial\Delta r_3}\right)^2\right] \tag{38}$$

We must note that there is no a priori reason why $\partial\mu/\partial\Delta r_1 = \partial\mu/\partial\Delta r_3$ since they refer to inequivalent CO groups. If we define $\partial\mu/\partial\Delta r_1 = \mu_\alpha'$ and $\partial\mu/\partial\Delta r_3 = \mu_\beta'$, and making the same assumption (Equation (27)) as before,

$$I_1 = 2B[\mu_\alpha' L_{11}\cos\phi + \mu_\beta' L_{31}]^2 \tag{39}$$

$$I_3 = 2B[\mu_\alpha' L_{13}\cos\phi + \mu_\beta' L_{33}]^2 \tag{40}$$

$$I_2 = 2B[\mu_\alpha' L_{22}\sin\phi]^2 \tag{41}$$

$$I_1 + I_3 = B[2(\mu_\alpha')^2\cos^2\phi + (\mu_\beta')^2]\mu_{CO}^2 \tag{42}$$

$$I_2 = B[2(\mu_\alpha')^2\sin^2\phi]\mu_{CO}^2 \tag{43}$$

$$\frac{I_1 + I_3}{I_2} = \frac{2(\mu_\alpha')^2\cos^2\phi + (\mu_\beta')^2}{2(\mu_\alpha')^2\sin^2\phi} \tag{44}$$

If we further assume that $\mu_\alpha' = \mu_\beta'$

$$\frac{I_1 + I_3}{I_2} = \frac{2\cos^2\phi + 1}{2\sin^2\phi} \tag{45}$$

and hence ϕ is measured. If $\mu_\alpha' \neq \mu_\beta'$ we need to solve the force field to obtain the L vectors, to measure say $I_1 + I_3/I_2$ and I_1/I_3 and we then have two observed intensity ratios and two unknowns, ϕ and the ratio μ_α'/μ_β', from which both can be obtained.

It is not infrequently found that bond angles in carbonyls determined this way are very different from the known angles determined for instance by X-ray crystallography. Clearly some of the assumption must, under some conditions, be invalid.

Assumptions

(1) We have used a CO-factored force field. It is possible to show [4] that the errors introduced by this approximation probably cancel and that therefore substantial errors are not due to this.

(2) The molecular dipole moment change as one C-O bond stretches is along the C-O bond. Almost certainly this is the real trouble. Suppose the two CO groups in $M(CO)_2$ are at 90°:

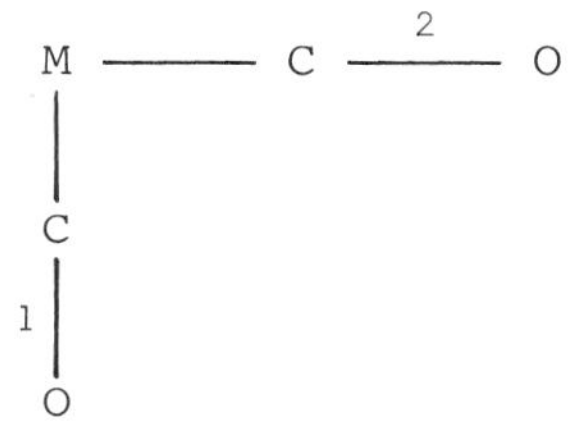

As $(CO)_1$ is stretched it is extremely probable that there is some electron flow along the perpendicular direction. Even a small contribution will make the resultant vector not co-linear with $(CO)_1$. More detailed examination of this point [5] suggests that the major errors will arise not from other CO groups but from ligands such as $P\phi_3$. Thus provided we are dealing with "pure" carbonyls the angles calculated should be reasonably reliable.

(3) The assumption $\mu_\alpha' = \mu_\beta'$ is almost always invalid so that reliable bond angle data will probably only be obtained for carbonyls with non-equivalent CO's if account is taken of this and a complete CO-factored force field determined to allow inclusion of the μ_α'/μ_β' term.

4. MATRIX-ISOLATED METAL CARBONYL FRAGMENTS

There has been considerable activity in this area; fragments have been generated by either metal atom/CO co-condensation techniques [6] or by photolytic stripping of parent carbonyls in situ [7]. Large numbers of species have been prepared by these techniques, and in all cases identification and structure determination is by spectroscopy. Hence the importance of a detailed understanding of the vibrational force field. As an example of this Figure 3 shows the IR spectrum obtained on photolysis of $Cr(CO)_6$ in CH_4 at 20 K. The theoretical spectrum shown is a very close fit and proves that $Cr(CO)_5$ has C_{4v} symmetry and that the angle between the unique CO and the

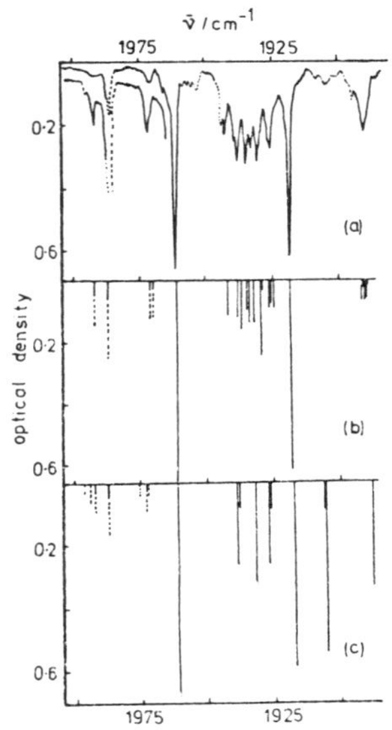

Figure 3. (a) ^{13}CO-enriched $Cr(CO)_5$ from photolysis of $Cr(CO)_6$ in methane at 20 K, (b) calculated spectrum for C_{4v} geometry and (c) for D_{3h} geometry (from ref. 7).

equatorial CO's is ∿93°. Vibrational structures determined by this photolysis method include C_{2v} $Fe(CO)_4$, C_{2v} $Mo(CO)_4$, C_{3v} $Fe(CO)_3$, C_{3v} $Mo(CO)_3$.

Acknowledgment

I thank Drs. J.K. Burdett and M. Poliakoff for helpful discussions in connection with this chapter.

5. REFERENCES

1. e.g. for $M(CO)_6$ - L.H. Jones, R.S. McDowell and M. Goldblatt, Inorg. Chem., 8 (1969) 2349.
2. F.A. Cotton and C.S. Kraihanzel, J. Amer. Chem. Soc., 84 (1962) 4432; Inorg. Chem., 2 (1963) 533.
3. e.g. F.A. Cotton, Inorg. Chem., 7 (1968) 1683.
4. J.K. Burdett, R.N. Perutz, M. Poliakoff and J.J. Turner, Inorg. Chem., 15 (1976) 1245; other papers in preparation.
5. J.K. Burdett, M. Poliakoff and J.J. Turner, in press.
6. G.A. Ozin and co-workers, many papers in Inorg. Chem., J. Amer. Chem. Soc. and Canad. J. Chem.
7. e.g. R.N. Perutz and J.J. Turner , Inorg. Chem., 14 (1975) 262; other papers in J. Amer. Chem. Soc. and J. Chem. Soc. (Dalton).

A very useful reference for a detailed examination of metal carbonyls is:

8. P.S. Braterman, Metal Carbonyl Spectra, Academic Press, London, 1975.

CHAPTER 23

INTRA- AND INTERMOLECULAR VIBRATIONS OF n-ALKANES AND POLYETHYLENE

Mitsuo Tasumi

1. INTRODUCTION

Polyethylene and its shorter-chain analogs, n-alkanes, probably form a singular class of molecules in that their vibrational spectra have been a constant subject of active studies over the past three decades. The present status of the study in this field may be understood to some extent by following the historical developments given in the bibliography. Before 1960 the theoretical framework for analysing the existing experimental data was built. During 1960-1966 measurements of high-quality spectra in the infrared and far infrared regions were made. Normal coordinate treatments for both single chains and crystals were carried out to interpret the observed band progressions, crystal-field splittings, and lattice frequencies. Since 1967, with the advent of laser Raman spectroscopy, increasing interest has been directed toward the low-frequency Raman bands. Particularly, the longitudinal acoustic mode (or the accordion vibration) has attracted the attention of polymer physicists in relation to chain folding in a single crystal. Also, the analysis of the infrared spectra of mixed crystals (normal and perdeuterated polyethylenes) has been pursued continuously. The outline of these topics will be discussed in the following sections.

2. NORMAL COORDINATE TREATMENTS OF SINGLE CHAINS

Calculations of the normal frequencies and normal modes were performed for a number of n-alkanes by Snyder and Schachtschneider using the general valence force field [15,27]. Recently a set of more refined force constants based on local symmetry coordinates was obtained by Shimanouchi et al. [74]. The latter authors utilised the data for fully deuterated n-pentane and n-hexane as well.

The normal coordinates of an infinite polyethylene chain with planar zigzag conformation may be characterised by the phase difference (δ) between adjacent methylene groups. The dynamical matrices (G and F) of infinite order may be factorised into a set of matrices of finite order (5 for the in-plane modes and 4 for the out-of-plane modes) by the use of either the complex symmetry coordinates [11] or the real ones [12]. The results of calculation for various δ values between 0 and π are shown in Figure 1, together with the assignments of nine branches (called the dispersion curves) in the figure caption. The observed frequencies of n-alkanes as well as

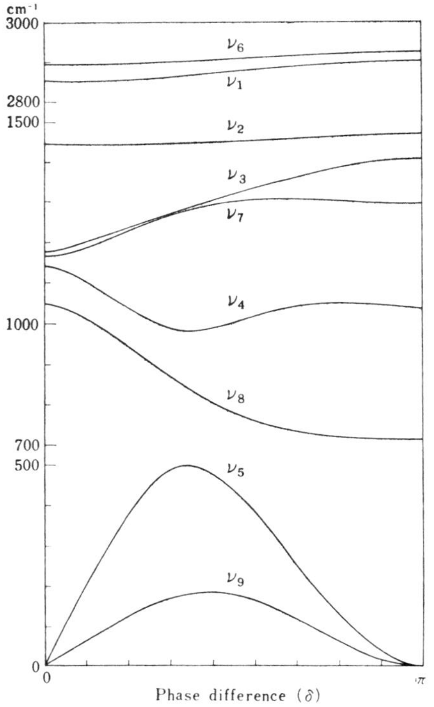

Figure 1. Dispersion curves of an infinite $(CH_2)_n$ chain. In-plane modes: ν_1, CH_2 symmetric stretching; ν_2, CH_2 scissors; ν_3, CH_2 wagging; ν_4, CC stretching - CCC bending; ν_5, CCC bending - CC stretching. Out-of-plane modes: ν_6, CH_2 antisymmetric stretching; ν_7, CH_2 rocking - CH_2 twisting; ν_8, CH_2 twisting - CH_2 rocking; ν_9, Torsion (from ref. 19).

those of polyethylene are broadly in agreement with the curves in Figure 1 [12,19].

3. NORMAL COORDINATE TREATMENTS OF CRYSTALS

First, we discuss the vibrations of the orthorhombic polyethylene crystal. On the analogy of a single chain, the crystal vibrations are generally characterised with the phase differences (δ_a, δ_b, and δ_c) along the three crystallographic axes. In more physical terms it is usual practice to define a wave vector (k) which corresponds to a set of δ_a, δ_b, and δ_c. However, we use the phase differences in order to keep consistency with the single-chain treatment. From the spectroscopic viewpoint the modes corresponding to $\delta_a = \delta_b = 0$ and $0 \leq \delta_c \leq \pi$ are most important. The normal frequencies of these modes were calculated by the present author about ten years ago using short-range (within 3 Å) intermolecular H···H repulsive force constants (Figure 2) [23,26]. Since then several groups of

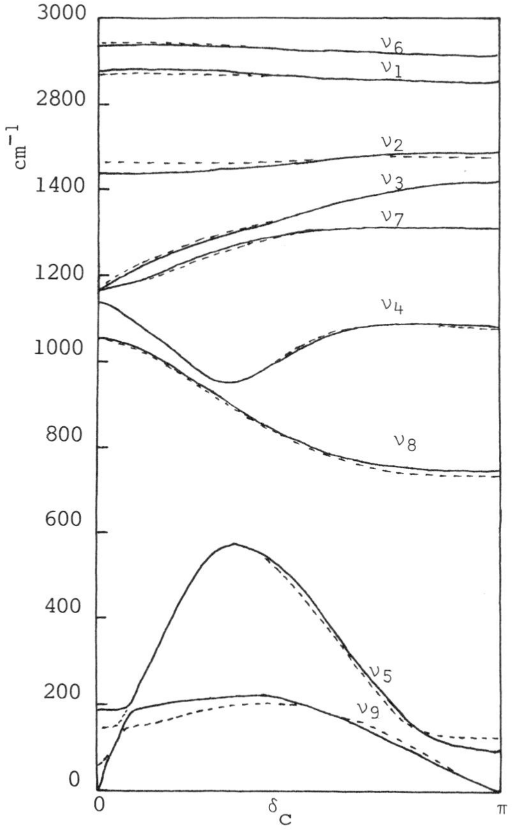

Figure 2. Dispersion curves of orthorhombic polyethylene crystal calculated with the unit-cell parameters at 100 K. ——, ν^a branch; ----, ν^b branch (from ref. 26).

authors have performed similar calculations.

In Figure 2 the following features are characteristic of the crystal vibrations:
(1) Each of the nine branches ν_1 - ν_9 splits into two sub-branches $\nu_i{}^a$ and $\nu_i{}^b$. The doublets observed for the CH_2 scissors and rocking vibration bands in the infrared spectrum of polyethylene can be accounted for by the splittings of $\nu_2(\pi)$ and $\nu_8(\pi)$.
(2) The crossing of ν_5 and ν_9 occurs near $\delta_c = 0$ and π.
(3) At $\delta_c = 0$ and π, ν_5 and ν_9 correspond to five lattice vibrations as depicted in Figure 3 and three translational modes of the whole crystal (null frequency).

Although the results of this calculation are in good agreement with the above-mentioned splittings of intramolecular vibrations and with the B_{1u} and B_{2u} translational lattice frequencies observed at 79.5 and 109 cm^{-1}, there is

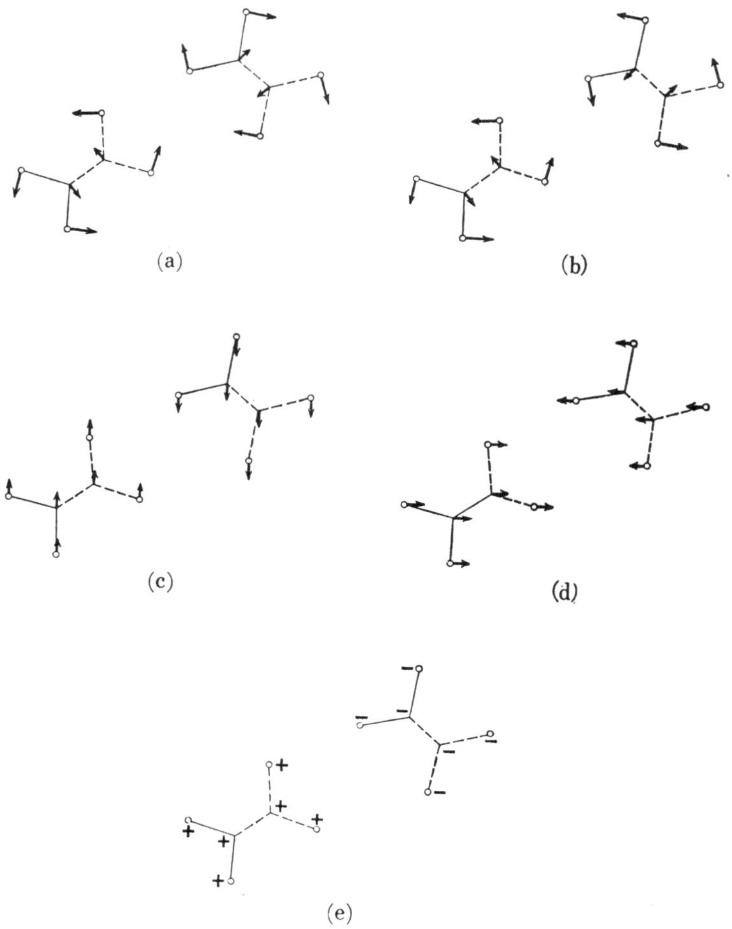

Figure 3. Lattice vibrations of orthorhombic polyethylene crystal. (a), A_g rotational mode; (b), B_{3g} rotational mode; (c), B_{1u} translational mode; (d), B_{2u} translational mode; (e) A_u translational mode (from ref. 23).

considerable discrepancy between the calculated and observed frequencies of A_g and B_{3g} rotational lattice vibrations (Table 1). Such a situation can be improved to a certain extent if we assume the Williams Set IV potential [75] for H···H, H···C, and C···C and take into account all the atom-atom interactions within 6 Å(see the column T-T in Table 1). However the discrepancy still exists for the A_g rotational mode. A similar conclusion was recently obtained by Kobayashi and Tadokoro in a more systematic approach (to be published in J. Chem. Phys.). These results are somewhat inconsistent with the calculations of Twisleton and White (cited in ref. 59) and of Wu and Nicol [55] which give smaller discrepancies. On the other hand, the

Table 1. Observed and calculated lattice frequencies of orthorhombic polyethylene

Mode	Observed		Calculated	
	Infrared[a] (2.0 K)	Raman[b] (77 K)	T-K[c] (77 K)	T-T[d] (77 K)
A_g	-	133.0	181.7	152.2
B_{3g}	-	108.2	138.8	114.9
B_{1u}	79.5	-	80.6	81.1
B_{2u}	109	-	113.3	102.3
A_u	-	-	57.5	39.5

(a), ref. 29; (b) ref. 59; (c), ref. 26;
(d), H. Takeuchi and M. Tasumi, unpublished results (1976).

observed large temperature dependences of the rotational lattice frequencies (particularly that of the A_g mode [59]) agree with the trends predicted by the calculation [31].

In triclinic n-alkanes there is only one molecule per unit cell. Accordingly, the low-frequency Raman spectra of triclinic n-alkanes are quite different from those of orthorhombic and monoclinic crystals [56]. Takeuchi et al. successfully analysed the low-frequency Raman spectra of triclinic n-alkanes of C_8, C_{10}, C_{12}, C_{14}, C_{16}, and C_{18} on the basis of straightforward normal coordinate treatments using the Williams Set IV potential. Most observed Raman bands were assigned to the three rotational lattice modes and intramolecular skeletal vibrations. Computer programs for such normal coordinate treatments were developed by Takeuchi and the input and output of these programs are outlined in Table 2. The programs are quite useful for computing the normal frequencies of molecular crystals in general including polymer crystals.

Table 2. Computer programs (CVOA & CVDD) for treating the crystal vibrations*.

Input

(a) Structure data
Unit cell dimension and atomic positions (of one molecule in the unit cell) in the unit-cell coordinate system, or in the Cartesian coordinate system

(b) Symmetry relationship among molecules in a unit cell
Matrices for rotational operations
Vectors for translational operations

(c) Intramolecular coordinates and force field
Local symmetry coordinates plus intramolecular force field, or
Normal coordinates and normal frequencies of a free molecule

Table 2. (continued)

(d) Intermolecular force field
Functional forms of atom-atom interactions, and/or
Force constants relating to intermolecular or inter-unit bonding (e.g. hydrogen bond, bond between adjacent units in a polymer chain, etc,)

Output

(a) With CVOA
Frequencies of intramolecular and lattice modes for $k = 0$
Eigenvectors and Cartesian displacements
Potential energy distributions
Jacobian

(b) With CVDD
Frequencies and eigenvectors for $k \neq 0$
Density of states

* These programs are used with a HITAC 8800/8700 system at the Computer Centre of the University of Tokyo. The same programs are available also at Imperial College, London.

4. MIXED CRYSTAL ANALYSIS

As an extension of the normal coordinate treatment of the polyethylene crystal, the present author and Krimm proposed a spectroscopic method for discerning whether the chain folding in a polyethylene single crystal is along (110) or (200) planes [31]. The method is based on the prediction that the splittings of CH_2 scissors and rocking bands in the infrared spectra of normal polyethylene diluted in deuterated polyethylene (and vice versa) depend on the direction of chain folding. After this work a number of experimental studies have been developed by Krimm and co-workers [35,43,54,65,69] and now it seems clear that the method is a useful tool for studying the chain folding.

5. LONGITUDINAL ACOUSTIC VIBRATION

In 1949 Shimanouchi and Mizushima reported the Raman spectra of solid n-alkanes where they found a series of bands (a band for each n-alkane) at frequencies (< 500 cm^{-1}) inversely proportional to the chain length. These bands were assigned to the fundamental longitudinal vibrations of rod-like molecules. In 1967 Schaufele and Shimanouchi observed the Raman bands arising from not only the fundamental but also overtones of longitudinal vibrations for a number of n-alkanes. These authors derived an experimental formula correlating the chain length with the fundamental and overtone frequencies.

The fundamental which is intense in the Raman effect is called the longitudinal acoustic (LA) mode or the accordion vibration after the similarity of its motion with that of an accordion. However, the LA mode is not a simple accordion-like motion since it should be described as the totally symmetric CC stretching rather than the totally symmetric CCC bending [48]. As shown in Figure 4 for n-octadecane, the contribution

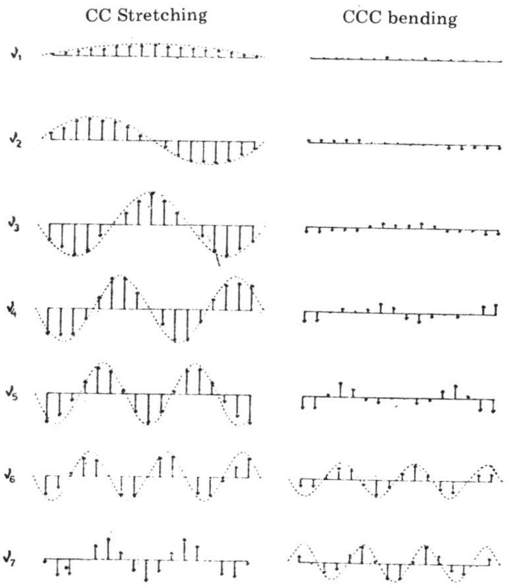

Figure 4. Schematic representation of the contributions of CC stretching and CCC bending coordinates (not the atomic displacements) to the normal coordinates of the LA mode (m=1) and its overtones (m=2-7) (from ref. 48).

of CC stretching increases relative to that of CCC bending with decreasing vibration order (m). For m=1 (the LA mode) the contribution of CC stretching is close to 60% in the potential energy distribution. In such a case the mode can be characterised by the phase difference $\delta_m = m\pi/n$ (n is the number of carbon atoms). In Figure 5 the observed Raman frequencies of longitudinal modes of various n-alkanes are plotted against δ_m and are compared with the calculated ν_5 dispersion curve [48]. Clearly, excellent agreement between the observed and calculated dispersion curves is obtained, particularly in the region of $\delta_m < 30°$. This ensures the validity of using the phase difference $m\pi/n$ for the LA mode and lower overtones.

In 1971 Peticolas et al. observed a low-frequency Raman band from polyethylene single crystals [49,50]. The frequency of this Raman band varied between 10 and 40 cm^{-1}, depending upon the sample treatment. They assigned the Raman band to the LA mode of the all-trans segment in a polyethylene crystal. Several groups of authors have examined the behaviour of this low-frequency Raman band using polyethylene samples prepared under a variety of conditions. As a result it is now established that the length of all-trans segment determined from the LA frequency agrees with the value obtained from low-angle X-ray diffraction within experimental errors [63,64,66].

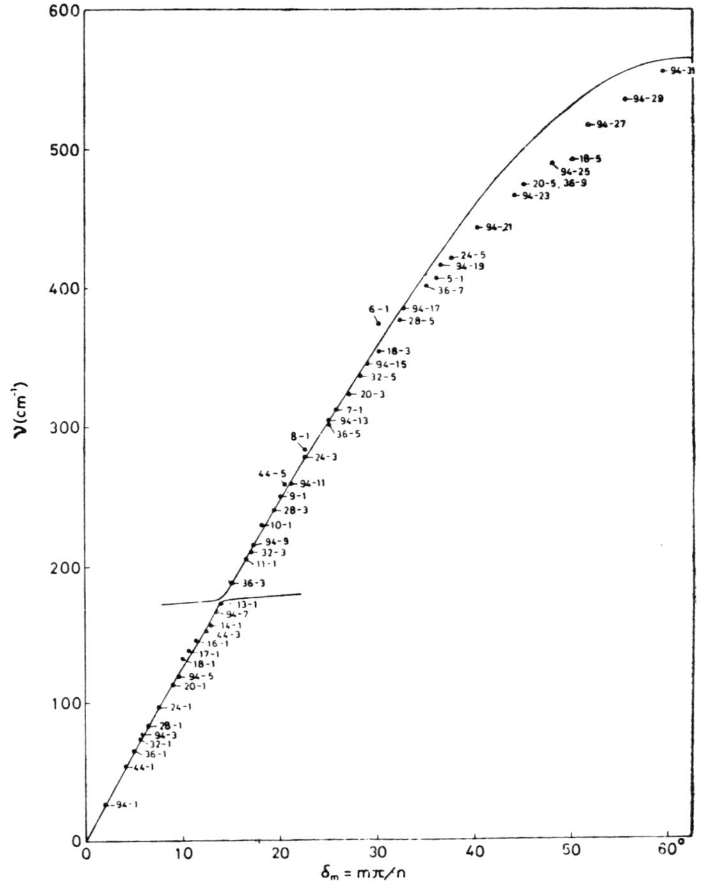

Figure 5. Observed Raman frequencies of n-alkanes vs. $\delta_m = m\pi/n$. The assignment is indicated by n-m, where n is the number of carbon atoms and m is the vibration order. The full curves indicate the calculated ν_5 dispersion relation (from ref. 48).

6. FUTURE WORK

In concluding this chapter, the author feels that <u>new approaches</u> to the following problems should be initiated:

(1) Intermolecular potentials in polyethylene and n-alkane crystals.
(2) The effect of chain folding on the LA frequency of a polyethylene crystal.
(3) The conformations of polyethylene and n-alkane molecules in the liquid state and in solutions.

7. CLASSIFIED BIBLIOGRAPHY

The abbreviations indicate the following classifications: *IR*, infrared; *FIR*, far infrared; *R*, Raman; *R-LF*, Raman spectra in the low-frequency region; *Th*, theoretical; *NCT*-1, normal coordinate treatments for single chains; *NCT*-2, normal coordinate treatments for crystals; *INT*, intensities; *MC*, mixed crystal analysis.

1949

1. *R*. *NCT*-1. Raman frequencies of n-paraffin molecules, S. Mizushima and T. Shimanouchi, J. Am. Chem. Soc., 71 (1949) 1320-1324.
2. *R*. *NCT*-1. The constant frequency Raman lines of n-paraffins, T. Shimanouchi and S. Mizushima, J. Chem. Phys., 17 (1949) 1102-1106.

1956

3. *IR*. *Th*. Infrared spectra of high polymers. II. Polyethylene, S. Krimm, C.Y. Liang and G.B.B.M. Sutherland, J. Chem. Phys., 25 (1956) 549-562.

1957

4. *IR*. *R*. Vibrational spectra of polyethylenes and related substances, J.R. Nielsen and A.H. Woollett, J. Chem. Phys., 26 (1957) 1391-1400.

1960

5. *IR*. *R*. *Th*. Infrared spectra of high polymers, S. Krimm, Fortschr. Hochpolym.-Forsch. (Adv. Polym. Sci.), 2 (1960) 51-172.
6. *IR*. The b_{1u} methylene wagging and twisting fundamentals in crystalline polyethylene, J.R. Nielsen and R.F. Holland, J. Mol. Spectrosc. 4 (1960) 488-498.
7. *IR*. Vibrational spectra of crystalline n-paraffins. Part I. Methylene rocking and wagging modes, R.G. Snyder, J. Mol. Spectrosc., 4 (1960) 411-434.

1961

8. *IR*. Dichroism and interpretation of the infrared bands of oriented crystalline polyethylene, J.R. Nielsen and R.F. Holland, J. Mol. Spectrosc., 6 (1961) 394-418.
9. *IR*. *Th*. Vibrational spectra of crystalline n-paraffins. II. Intermolecular effects, R.G. Snyder, J. Mol. Spectrosc., 7 (1961) 116-144.

1962

10. *IR*. Infrared spectra of single crystals. Part I. Orthorhombic n-$C_{24}H_{50}$, monoclinic n-$C_{36}H_{74}$, and triclinic n-$C_{18}H_{38}$ and n-$C_{20}H_{42}$, R.F. Holland and J.R. Nielsen, J. Mol. Spectrosc., 8 (1962) 383-405.
11. *NCT*-1. A method for the complete vibrational analysis of isolated chain, T.P. Lin and J.L. Koenig, J. Mol. Spectrosc., 9 (1962) 228-243.

12. *NCT*-1. Normal vibrations and force constants of polymethylene chain, M. Tasumi, T. Shimanouchi and T. Miyazawa, J. Mol. Spectrosc., 9 (1962) 261-287.

1963

13. *R*. Raman spectra of polyethylenes, R.G. Brown, J. Chem. Phys., 38 (1963) 221-225.
14. *IR*. Vibrational analysis of the n-paraffins. I. Assignments of infrared bands in the spectra of C_3H_8 through n-$C_{19}H_{40}$, R.G. Snyder and J.H. Schachtschneider, Spectrochim. Acta, 19 (1963) 85-116.
15. *NCT*-1. Vibrational analysis of the n-paraffins. II. Normal coordinate calculations, J.H. Schachtschneider and R.G. Snyder, Spectrochim. Acta, 19 (1963) 117-168.
16. *NCT*-1. A refined treatment of normal vibrations of polymethylene chain, M. Tasumi, T. Shimanouchi and T. Miyazawa, J. Mol. Spectrosc., 11 (1963) 422-432.

1964

17. *FIR*. Study of the 73 cm^{-1} band in polyethylene, A.O. Frenzel and J.P. Butler, J. Opt. Soc. Am., 54 (1964) 1059-1060.
18. *FIR*. Infrared-active interchain vibration in polyethylene, J.E. Bertie and E. Whalley, J. Chem. Phys., 41 (1964) 575-576.
19. *IR*. Infrared spectra of n-higher alcohols, M. Tasumi, T. Shimanouchi, A. Watanabe and R. Goto, Spectrochim. Acta, 20 (1964) 629-666.
20. *IR*. *NCT*-1. Stereoregulated polydideuteroethylene. II. Infrared spectra and normal vibration analysis, M. Tasumi, T. Shimanouchi, H. Tanaka and S. Ikeda, J. Polym. Sci. Part A, 2 (1964) 1607-1631.

1965

21. *INT*. Group moment interpretation of the infrared intensities of crystalline n-paraffins, R.G. Snyder, J. Chem. Phys., 42 (1965) 1744-1763.
22. *FIR*. Assignment of the 71 cm^{-1} band in polyethylene, S. Krimm and M. Bank, J. Chem. Phys., 42 (1965) 4059-4060.
23. *NCT*-2. Crystal vibrations and intermolecular forces of polymethylene crystals, M. Tasumi and T. Shimanouchi, J. Chem. Phys., 43 (1965) 1245-1258.

1966

24. *IR*. *NCT*-1. Molecular vibrations of irregular chains. I. Analysis of infrared spectra and structures of polymethylene chains consisting of CH_2, CHD, CD_2 groups, M. Tasumi, T. Shimanouchi, H. Kenjo and S. Ikeda, J. Polym. Sci. Part A-1, 4 (1966) 1011-1021.
25. *IR*. *NCT*-1. Molecular vibrations of irregular chains. II. Configurations of polydideuteroethylenes, M. Tasumi, T. Shimanouchi and S. Ikeda, J. Polym. Sci. Part A-1, 4 (1966) 1023-1029.

1967

26. *NCT-2.* Crystal vibrations of polyethylene, M. Tasumi and S. Krimm, J. Chem. Phys., 46 (1967) 755-766.
27. *IR. NCT-1.* Vibrational study of the chain conformation of the liquid n-paraffins and molten polyethylene, R.G. Snyder, J. Chem. Phys., 47 (1967) 1316-1360.
28. *R-LF.* Longitudinal acoustical vibrations of finite polymethylene chains, R.F. Schaufele and T. Shimanouchi, J. Chem. Phys., 47 (1967) 3605-3610.
29. *FIR.* Inter-molecular vibrations of crystalline polyethylene and long-chain paraffins, G. Dean and D.H. Martin, Chem. Phys. Lett., 1 (1967) 415-416.

30. *IR. R.* A revised assignment of the B_{2g} methylene wagging fundamental of the planar polyethylene chain, R.G. Snyder, J. Mol. Spectrosc., 23 (1967) 224-228.

1968

31. *NCT-2. MC.* Vibrational analysis of chain folding in polyethylene crystals, M. Tasumi and S. Krimm, J. Polym. Sci. Part A-2, 6 (1968) 995-1010.
32. *FIR.* Lattice-frequency studies of crystalline and fold structure in polyethylene, M.I. Bank and S. Krimm, J. Appl. Phys., 39 (1968) 4951-4958.
33. *NCT-1.* Vibrational analysis of random polymers, M. Tasumi and G. Zerbi, J. Chem. Phys., 48 (1968) 3813-3820.
34. *R.* Chain shortening in polymethylene liquids, R.F. Schaufele, J. Chem. Phys., 49 (1968) 4168-4175.

1969

35. *R. IR.* Raman spectrum of polyethylene and the assignment of the B_{2g} wag fundamental, R.G. Snyder, J. Mol. Spectrosc., 31 (1969) 464-465.
36. *IR.* Infrared study of lamellar linking by cilia in polyethylene single-crystal mats, M.I. Bank and S. Krimm, J. Appl. Phys., 40 (1969) 4248-4253.
37. *IR.* Mixed crystal infrared study of chain folding in crystalline polyethylene, M.I. Bank and S. Krimm, J. Polym. Sci. Part A-2, 7 (1969) 1785-1809.
38. *Th.* Group theoretical treatment of crystal vibrations; application to orthorhombic polyethylene, T. Kitagawa and T. Miyazawa, Bull. Chem. Soc. Jpn., 42 (1969) 3437-3447.

1970

39. *R.* Polarised Raman spectra of oriented polyethylene, V.B. Carter, J. Mol. Spectrosc., 34 (1970) 356-357.
40. *R.* Interpretation of the Raman spectrum of polyethylene and deuteropolyethylene, R.G. Snyder, J. Mol. Spectrosc., 36 (1970) 222-231.
41. *R.* Raman scattering in crystalline polyethylene, F.J. Boerio and J.L. Koenig, J. Chem. Phys., 52 (1970) 3425-3431.
42. *R.* Laser-vibrational scattering by polymers, R.F. Schaufele, J. Polym. Sci. D, Macromol. Rev., 4 (1970) 67-90.

43. *IR.* Mixed crystal infrared study of chain segregation in polyethylene, M.I. Bank and S. Krimm, J. Polym. Sci. Part B, 8 (1970) 143-148.
44. *Th. NCT-2.* Frequency distribution and dispersion curves of crystal vibrations of perdeuterated polyethylene, T. Kitagawa and T. Miyazawa, Polym. J., 1 (1970) 471-479.

1971

45. *R. NCT-1.* Laser Raman scattering by the $C_{34}H_{68}$ ring molecule, M. Tasumi, T. Shimanouchi and R.F. Schaufele, Polym. J., 2 (1971) 740-746.
46. *R. NCT-1.* Laser Raman scattering by $C_{34}H_{68}$ in the low-frequency region, R.F. Schaufele and M. Tasumi, Polym. J., 2 (1971) 815-816.
47. *NCT-1.* Vibrational spectrum of cyclic $C_{34}H_{68}$ and chain folding in polyethylene single crystals, S. Krimm and J. Jakes, Macromolecules, 4 (1971) 605-609.
48. *R-LF. NCT-1.* Skeletal vibrations of chain molecules, T. Shimanouchi and M. Tasumi, Indian J. Pure Appl. Phys., 9 (1971) 958-961.
49. *R-LF.* Raman scattering from longitudinal acoustical vibrations of single crystals of polyethylene, W.L. Peticolas, G.W. Hibler, J.L. Lippert, A. Peterlin and H. Olf, Appl. Phys. Lett., 18 (1971) 87-89.
50. *R-LF.* Laser Raman and X-ray study of the two-phase structure of polyethylene single crystals, A. Peterlin, H.G. Olf, W.L. Peticolas, G.W. Hibler and J.L. Lippert, Polym. Lett., 9 (1971) 583-589.

1972

51. *R. Th.* The laser Raman spectrum of polyethylene. The assignment of the spectrum to fundamental modes of vibration, M.J. Gall, P.J. Hendra, C.J. Peacock, M.E.A. Cudby and H.A. Willis, Spectrochim. Acta, 28A (1972) 1485-1496.
52. *NCT-2.* Lattice vibrations of polyethylene, D.I. Marsh and D.H. Martin, J. Phys. C: Solid State Phys., 5 (1972) 2309-2316.
53. *R.* Raman spectroscopy, rotational isomerism, and the "rotator" phase transition in n-alkanes, J.D. Barnes and B.M. Fanconi, J. Chem. Phys., 56 (1972) 5190-5192.
54. *IR.* Infrared spectra of polyethylene-poly(ethylene-d_4) mixed crystal systems, S. Krimm and J.H.C. Ching, Macromolecules, 5 (1972) 209-211.

1973

55. *R-LF.* Low frequency modes in the Raman spectra of polyethylene and paraffins. I. Lattice vibrations and their pressure dependence, C.-K. Wu and M. Nicol, J. Chem. Phys., 58 (1973) 5150-5162.
56. *R-LF.* Low frequency Raman-active lattice vibrations of n-paraffins, H.G. Olf and B. Fanconi, J. Chem. Phys., 59 (1973) 534-544.
57. *R-LF.* Effects of polymorphism on the Raman-active longitudinal acoustic mode frequencies of n-paraffins, F. Khoury, B. Fanconi, J.D. Barnes and L.H. Bolz, J. Chem. Phys., 59 (1973) 5849-5857.

58. *R-LF*. Low frequency intramolecular and lattice vibrations of n-$C_{36}H_{74}$ as studied by Raman scattering, G. Vergoten, G. Fleury, M. Tasumi and T. Shimanouchi, Chem. Phys. Lett., 19 (1973) 191-194.
59. *R-LF*. Low frequency Raman spectra of single-crystal textured polyethylene, R.T. Hartley, W. Hayes and J.F. Twisleton, J. Phys. C: Solid State Phys., 6 (1973) 167-170.
60. *IR. NCT-1*. Conformational structure of polyethylene chains from the infrared spectrum of the partially deuterated polymer, R.G. Snyder and M.W. Poore, Macromolecules, 6 (1973) 708-715.

1974

61. *R. NCT-2*. Raman spectra of n-alkane crystals: lattice vibrations of n-hexane, n-heptane and n-octane, L.-C. Brunel and D.A. Dows, Spectrochim. Acta, 30A (1974) 929-940.
62. *R-LF. NCT-2*. Low frequency Raman-active vibrations of triclinic n-paraffins, H. Takeuchi, T. Shimanouchi, M. Tasumi, G. Vergoten and G. Fleury, Chem. Phys. Lett., 28 (1974) 449-453.
63. *R-LF*. Laser Raman study of the longitudinal acoustic mode in polyethylene, H.G. Olf, A. Peterlin and W.L. Peticolas, J. Polym. Sci. Polym. Phys. Ed., 12 (1974) 359-384.
64. *R-LF*. Annealing studies of solution and bulk crystallised polyethylene using the Raman-active longitudinal acoustical vibrational mode, J.L. Koenig and D.L. Tabb, J. Macromol. Sci. - Phys., B9 (1974) 141-161.
65. *IR. MC*. Mixed-crystal infrared study of chain organisation in polyethylene crystallised under orientation and pressure, S. Krimm, J.H.C. Ching and V.L. Folt, Macromolecules, 7 (1974) 537-538.

1975

66. *R-LF*. Low frequency Raman spectroscopic study on lamellar polymer crystals, M.J. Folkes, A. Keller, J. Stejny, P.L. Goggin, G.V. Fraser and P.J. Hendra, Colloid & Polymer Sci., 253 (1975) 354-361.
67. *R-LF*. Polarised Raman scattering from the longitudinal acoustic mode in polyethylene, G.V. Fraser, A. Keller and D.P. Pope, J. Polym. Sci. Polym. Lett. Ed., 13 (1975) 341-344.
68. *R-LF*. Influence of methyl group branches on the longitudinal acoustical mode frequencies in alkanes and polyethylene, B. Fanconi and J. Crissman, J. Polym. Sci. Polym. Lett. Ed., 13 (1975) 421-426.
69. *IR. MC*. Mixed-crystal infrared studies of annealed poly (ethylene) single crystals, J.H.C. Ching and S. Krimm, Macromolecules, 8 (1975) 894-897.

1976

70. *R*. Polarised Raman spectra of single-crystal n-$C_{36}H_{74}$, M. Kobayashi, T. Uesaka and H. Tadokoro, Chem. Phys. Lett., 37 (1976) 577-581.

Neutron scattering studies

71. Neutron scattering and normal vibrations of polymers, T. Kitagawa and T. Miyazawa, Fortschr. Hochpolym.-Forsch. (Adv. Polym. Sci.) 9 (1972) 335-414.
72. Study of low frequency motions of extended chains in polyethylene by neutron inelastic scattering, H. Berghmans, G.J. Safford and P.S. Leung, J. Polym. Sci. Part A-2, 9 (1971) 1219-1234.
73. Low frequency molecular vibrations in solid n-paraffins by neutron inelastic scattering: n-pentane, n-hexane, n-heptane and n-octane, K.W. Logan, H.R. Danner, J.D. Gault and H. Kim, J. Chem. Phys., 59 (1973) 2305-2308.

Other references

74. T. Shimanouchi, H. Matsuura and I. Hirada, J. Phys. Chem. Ref. Data, submitted.
75. D.E. Williams, J. Chem. Phys., 47 (1967) 4680.

CHAPTER 24

MOLECULAR DYNAMICS AND VIBRATIONAL SPECTRA OF POLYMERS

Giuseppe Zerbi

1. INTRODUCTION

The purpose of this chapter is to give a general introduction to the problem of the interpretation of the vibrational (infrared and Raman) spectrum of polymeric materials and to describe some mathematical techniques which we think useful for the solution of the problem; a few examples will also be discussed.

Polymeric materials are generally considered to originate from the world of organic chemistry (either natural or synthetic); many inorganic materials as well as hydrogen-bonded systems however can be considered as made up by a repetition of structural sub-units. The techniques discussed here can equally be applied to these systems. The only difference is that organic systems possess a larger variety of structural features which make the problem more complex, as discussed in the pages which follow. We will first discuss the more complex case of organic polymers (where covalent forces are mainly acting); simplifications can be introduced when treating hydrogen-bonded or inorganic systems.

The ideal process of formation of a polymer chain occurs generally in the following steps: (1) The monomeric unit suitable for polymerisation under the influence of a catalyst forms a chain in which each unit is covalently bonded to the others. The type of linking (e.g. head-to-tail, head-to-head, etc.) is determined by the catalyst chosen. (2) When the monomeric unit contains an asymmetric carbon atom, the selective factors affecting the mechanism of polymerisation must supervise the stereospecificity of the reaction [1,2]. (3) After the chain has reached its final chemical structure intramolecular interactions force the molecule to take up the minimum energy conformation [3]. The geometry obtained is the balance of several forces whose origin is still not yet completely clear [4,5]. Semiempirical [6] or ab initio [7] calculations try to predict the minimum energy structure and the other possible less stable minima which may occur when the environment of the polymer chain is changed. The minimum energy geometry is generally helical and can be described by a few parameters related to bond length, bond angle and torsional angle [8]. Such a helix defines an infinite one-dimensional array of chemical units which can be generated by a screw symmetry operation along a given axis. Such an array can be considered as a "one-dimensional perfect crystal". (4) The one-dimensional crystal just described can finally pack into a three dimensional lattice whose parameters depend on the extent of intermolecular interactions whose origin is still matter of extensive studies

[6]. Generally Van der Waals type forces are operative; such interactions are at least one order of magnitude weaker than the intramolecular covalent forces. The symmetry operations of the lattice define the structure of the crystal just as in the case of any system in the solid state.

The chemical and physical processes described above would generate a perfect system where a chemically and stereochemically pure sequence of units is organised in a conformationally perfectly ordered one-dimensional array which packs with no lattice imperfections. In nature actually all these processes occur partially and the actual polymer chains contain several chemical defects (change of type of linking), stereochemical disorder (e.g. syndiotactic sequences in an isotactic polymer, stereoblocks polymers, etc.), conformational disorder (e.g. kinks, jogs, folds, etc.) and lattice defects (e.g. dislocations, etc.). A feature which is intrinsically connected to the nature of polymeric organic chains is the fact that in order to crystallise in a suitable lattice the very long macromolecular chain has to fold into itself very many times to form a single crystallite. The kinetics of crystallisation and the derived morphology is a field of active research. For what we are concerned with it becomes clear that even a single crystal contains a certain concentration of "amorphous" (more exactly of irregular) structures which become an intrinsic and unavoidable structural property of the solid state of an organic polymer system.

The detailed understanding of the vibrational spectrum, when diagnostic chemical vibrational correlations are neglected, requires first the treatment of the molecular dynamics of these systems, second the understanding of the selection rules in terms of overall symmetry or local symmetry, and third the prediction of the absorption coefficient in infrared and the scattering power in the Raman, both related to the corresponding transition moments.

In this chapter, following the above criteria, we will discuss: (1) the case of perfect one-dimensional crystals, (2) the case of perfect three-dimensional crystals, and (3) the case of disordered polymers. Cases (1) and (2) require a mathematical technique completely different from that which is necessary for case (3).

2. PERFECT ONE-DIMENSIONAL CRYSTAL: SINGLE CHAIN POLYMER

While we refer to more detailed accounts of polymer spectroscopy for a comprehensive treatment [9], we recall here the classical concepts of the normal vibrations of a one-dimensional monoatomic or diatomic chain which can be found in any book of solid state physics [10]. Let us take a linear chain made up of equal particles of mass m at a (repeat) distance d and joined by weightless springs which obey Hooke's law, with force constant f. We restrict to longitudinal motions along the chain axis taken as the x axis. If the particle m_n is displaced from its equilibrium position the equation of motion for the nth mass is

$$m\ddot{x}_n = f(x_{n+1} + x_{n-1} - 2x_n) \tag{1}$$

Solutions of Equation (1) are of the type

$$x_n = Ae^{i(\omega t + knd)} \tag{2}$$

which describe a motion periodic in time and space with circular frequency ω. k is the so-called wave vector, with $|\bar{k}| = 2\pi/\bar{\lambda}$ and in the direction of propagation of the wave; $\bar{\lambda}$ is the wavelength of the spatial wave. Differentiation of Equation (2) and substitution into (1) gives a solution if

$$\omega = \pm(4f/m)^{\frac{1}{2}}\sin(kd/2) \tag{3}$$

This relation is called a dispersion relation and the plot of ω vs. k represents the dispersion curve.

Equation (3) is periodic with period $k = 2\pi/d$. Physically Equation (2) describes a wave whose wavelength $\bar{\lambda} = 2\pi/k$ propagates through the chain. Each particle vibrates with frequency $\omega(k)$. A band of frequencies from $\omega = 0$ to $\omega_{max} = (4f/m)^{\frac{1}{2}}$, corresponding to $k = 0$ and $k = \pm\pi/d$ respectively, can be propagated through the chain. The displacements of two successive and equivalent particles are described by the ratio

$$x_j/x_{j+1} = e^{-ikd} \tag{4}$$

At $k = 0$ all particles move in phase with each other and the motion corresponds to a rigid translation of the whole chain. At $k = \pm\pi/d$ particles move out-of-phase with each other with $\bar{\lambda} = 2d$. For a general k the phase difference between the displacements of two successive (translationally equivalent) units is given by

$$\phi = kd \tag{5}$$

The concept of phase difference is very important in the case of polymer vibrations since most of the simplest cases are treated in terms of the phase difference and not in terms of k. The phase can be more easily visualised than k for an immediate understanding of the motions of the atoms. Motions described by values of k outside the range $\pm\pi/d$ simply reproduce the motions already described by values of k within the defined range. The range of k values $-\pi/d \leqslant k \leqslant +\pi/d$ is called the first Brillouin zone for the one-dimensional lattice.

We next take the one-dimensional lattice made up by two kinds of particles, with masses M and m, located at odd and even positions in the chain respectively, at a distance d, joined as before by weightless elastic springs of force constant f. Only nearest neighbour interactions are taken into account. Again only longitudinal motions along the chain axis x are considered. Two equations of motion can be written, one for each particle of the repeating unit.

$$m\ddot{x}_{2n} = f(x_{2n+1} + x_{2n-1} - 2x_{2n}) \tag{6}$$

$$M\ddot{x}_{2n+1} = f(x_{2n+2} + x_{2n} - 2x_{2n+1})$$

The possible solutions of Equation (6) are of the form

$$x_{2n} = Ae^{i(\omega t + 2nkd)} \tag{7}$$

$$x_{2n+1} = Be^{i[\omega t + (2n+1)kd]}$$

After differentiation and substitution into Equation (6) one obtains a set of simultaneous equations

$$-\omega^2 mA = 2fB \cos kd - 2fA \tag{8}$$

$$-\omega^2 MB = 2fA \cos kd - 2fB$$

whose non-trivial solutions are given by the values of ω^2 for which the determinant of the coefficients of A and B vanishes:

$$\begin{vmatrix} 2f - m\omega^2 & -2f \cos kd \\ -2f \cos kd & 2f - M\omega^2 \end{vmatrix} = 0 \tag{9}$$

The roots of Equation (9) are given by the dispersion relation

$$\omega^2 = f\left(\frac{1}{m} + \frac{1}{M}\right) \pm f\left[\left(\frac{1}{m} + \frac{1}{M}\right)^2 - \frac{4\sin^2 kd}{Mm}\right]^{\frac{1}{2}} \tag{10}$$

and the solutions $\omega_{\pm}^2$ are collected into two branches, one called the acoustical branch (- sign) and the other the optical branch (+ sign). The understanding of the type of motions along these branches is essential for the study of polymer vibrations and spectra. The ratio of the amplitudes at a given k is obtained by substituting the roots ω_-^2 and ω_+^2 of Equation (9) into (8). At k = 0, A = B for the acoustical branch and A = -(M/m)B for the optical branch. On the acoustical branch at k = 0 particles move rigidly together giving rise to a rigid translational longitudinal acoustic motion of all repeating units like when a sound wave is transmitted by the material. It is intuitive that the slope at k = 0 of the acoustical branch can be related to the elastic constant of the linear crystal [10]. For the optical branch the centre of mass of the unit cell does not move and if particles are oppositely charged in a rigid ion approximation (which is never the case for an organic polymer) interaction with an electromagnetic wave will take place.

The concepts just discussed can be extended for the treatment of real organic one-dimensional polymer chains. In doing this we consider the polymer chain as isolated from its neighbouring chains in the actual sample we are studying. In such a case we account for the intramolecular interactions and neglect the intermolecular interactions which are certainly weaker for organic polymers, as already stated. Moreover since the repeat unit is made up of many atoms it is clear that we cannot restrict our study to motions only along the chain axis but we must allow for motions in all directions even if k is only along x.

For making calculations on one-dimensional organic polymers feasible and physically more understandable the treatment in terms of Wilson's internal coordinates has been introduced. The well known GF method has been modified in order to take into account k dependent quantities [11,12].

Let us consider an ideal and isolated polymer chain whose geometry can be constructed by applying a screw-symmetry operation to the starting chemical repeat unit. Let the chemical repeat unit contain p atoms and the translational repeat unit contain n chemical units. Let $\phi(\theta,\ell)$ be the screw symmetry operation defined by a rotation of an angle θ about the axis and a translation of ℓ along the same axis. The translational repeat unit is obtained by applying a certain number of times such a symmetry operation, this number depending on the geometry of the polymer chain [8]. More precisely $\theta = 2\pi(m/n)$ where m is the number of turns of the helix within the translational repeat unit.

Let R_i^n describe the ith internal displacement coordinate belonging to the nth chemical unit. In the quadratic approximation the vibrational potential energy of the system $V = V(R_i^n)$ can be expanded in a Taylor series about the equilibrium configuration and the expansion truncated to the second order when the harmonic approximation is adopted, just as in the case of molecules:

$$V = V_o + \sum_{n,i} (F_R)_i^n R_i^n + \frac{1}{2} \sum_{\substack{n,n' \\ i,k}} (F_R)_{ik}^{nn'} R_i^n R_k^{n'} \tag{11}$$

where

$$(F_R)_i^n = \left(\frac{\partial V}{\partial R_i^n}\right)_{eq} \qquad (F_R)_{ik}^{nn'} = \left(\frac{\partial^2 V}{\partial R_i^n \, \partial R_k^{n'}}\right)_{eq}$$

At the equilibrium $(F_R)_i^n = 0$ when the coordinates are independent and V_o can be removed by suitable shifting of the axis. The potential energy can then be written as

$$2V = \sum_{\substack{n,n' \\ i,k}} (F_R)_{ik}^{nn'} R_i^n R_k^{n'} \tag{12}$$

where

$$(F_R)_{ik}^{nn'} = (F_R)_{ki}^{nn'} \tag{13}$$

The periodicity of the chain requires that

$$(F_R)_{ik}^{nn'} = (F_R)_{ik}^{s} \tag{14}$$

where $s = |n-n'|$ is the distance of interaction. Substitution of Equation (14) into (12) gives

$$2V = \sum_{\substack{n\\i,k}} (F_R)^o_{ik} R^n_i R^n_k + \sum_{\substack{n,s\\i,k}} [(F_R)^s_{ik} R^n_i R^{n+s}_k + (F_R)^s_{ki} R^n_i R^{n-s}_k] \tag{15}$$

The kinetic energy 2T of the infinite polymer in terms of the momenta p^n_i conjugated to R^n_i can be written in an analogous way in terms of the kinetic energy matrix G.

Hamilton's equations of motion can be written using Equation (15) and the corresponding one for 2T. A system of an infinite number of second order differential equations in the unknowns R^{n+s}_i is obtained. Its solution is given by a plane wave,

$$R^{n+s}_i = A_i \exp[-i(\omega t + s\phi)] \tag{16}$$

In Equation (16) A_i is independent of n, ϕ is the phase shift between two adjacent rototranslationally equivalent internal coordinates, and ω is the circular frequency. Substitution of Equation (16) into the system of second order differential equations leads to a set of 3p simultaneous linear equations in the unknown amplitudes A_i whose non-trivial solutions are given by the 3p values of ω^2 (for each ϕ) for which the determinantal equation

$$|G_R(\phi)F_R(\phi) - \omega^2(\phi)E| = 0 \tag{17}$$

vanishes. In Equation (17)

$$G_R(\phi) = (G_R)^o + \sum_s [(G_R)^s e^{is\phi} + (\tilde{G}_R)^s e^{-is\phi}] \tag{18}$$

$$F_R(\phi) = (F_R)^o + \sum_s [(F_R)^s e^{is\phi} + (\tilde{F}_R)^s e^{-is\phi}] \tag{19}$$

Let us point out a few important points related to the previous equations:

(i) Equation (17) is of 3p th degree in ω^2; there are 3p characteristic roots $\omega^2 = 4\pi^2c^2\nu^2$ for each value of the phase difference ϕ. The dispersion relation is then a multiple-valued function with 3p branches. The function is periodic with period 2π and the first Brillouin zone is comprised between $(-\pi \leqslant \phi \leqslant \pi)$.
(ii) Of the 3p branches two always reach zero for $\phi = 0$. These branches correspond to the acoustical ones previously discussed for simpler models. It may happen that because of the geometry of the molecule one or both acoustical branches reach zero again for $\phi \neq 0$. The other 3p-2 branches correspond to optical branches since, for some particular value of ϕ depending on the geometry of the chain, they give rise to spectroscopic activity in infrared and/or Raman*.
(iii) Attention has to be paid to Equations (18) and (19).

---

* The number of genuine normal modes for a one-dimensional polymer is 3p-4, only one rotation about the chain axis being allowed with zero frequency.

The phase-dependent force constant matrix $F_R(\phi)$ is made up by a term $(F_R)^o$ which represents the force constant matrix of the chemical unit isolated from (or unperturbed by) its neighbours, the second term comprises the perturbations by the neighbouring units. These perturbations are modulated by the phase factor and depend on s, i.e. the distance of intramolecular interaction. While in principle no limitation exists on s, practical experience has shown that for covalent organic systems dispersion curves can be reproduced even if only nearest neighbour interactions are considered.

The experimental evidence for the last statement is mostly based on experience derived from the optical spectra of several polymer systems. In reality phonon dispersion curves can be mapped only from neutron scattering experiments on stretch-oriented polymers and few good data are available. However because of the experimental limitations of neutron experiments only low energy branches can be studied and thus generally only longitudinal phonons travelling along the chain axis can be revealed. To our knowledge a few low energy experimental dispersion curves are available for perdeuteropolyethylene [13] and polytetrafluoro ethylene [14]. Attempts to find transverse phonons for polyethylene have been reported [15]. Another method for plotting dispersion curves is the study of model compounds of various lengths. Assuming that perturbations by the end groups are negligible (or can be accounted for) [16] the normal modes of such finite chain molecules can be related to the phase difference between the vibrations of adjacent units, hence ω vs. ϕ can be plotted. Model compounds with a structure similar to the polymer studied must be available. The most famous case is that of polyethylene, thoroughly treated first by Snyder and Schachtschneider [17] and later revised or revisited by several authors [18]. While these authors were able to map dispersion curves of polyethylene from the study of linear paraffins for all but the two lowest acoustical branches, the recent availability of laser Raman spectra in the very low frequency region has given a plot of the longitudinal acoustic and the torsional branches [19-21].

For the cases discussed there seems no need to extend the interaction in the potential energy beyond the nearest neighbour. This fact has also been verified for the case of diamond, Si and Ge [22] and hexagonal D_2O ice [23] for which experimental dispersion curves were well described by neutron experiments. If force constants are a qualitative measure of the electron re-distribution along the chemical bonds during vibrations [24] it can be concluded that in covalent organic systems where σ bonds make up the main network the interactions decay very rapidly and become practically a localised effect. This may not be the case when highly delocalised π electron systems occur such as in the case of polyenes. Recent resonance Raman studies on polyene-type pigments responsible for the mechanism of vision seem to indicate that π electron delocalisation along many bonds does affect the force constants, hence the frequency of the C=C stretching [25]. When the nature of the chemical bond departs from typical covalency, interactions at longer distances are required as in the case of crystalline tellurium [26].

Once the $\omega_i(k)$ have been calculated their appearance in the vibrational spectrum (infrared and/or Raman) can be predicted. Conservation of momentum in the interaction with electromagnetic radiation implies that only those $\omega_i(k)$ with $k = 0$ may be active in the infrared and/or Raman spectrum. Group theory (we refer to specialised books for the detailed discussion of this problem [27]) tells us which of the $\omega_i(0)$ are actually active.

The statement that only $k = 0$ motions are active means that translationally equivalent atoms must move in phase for any optically active frequency. If the translational repeat unit is made up of several identical chemical units, generated the one from the other by application of the screw rotation $\phi(\theta,\ell)$ along the polymer axis, then corresponding atoms in two adjacent chemical units have a definite phase relationship ϕ and the dispersion curves $\omega_i(k)$ can be redescribed as $\omega_j(\phi)$. Care must be taken with the fact that i runs to 3pn, j runs to 3p. Thus the description in terms of ϕ shows a lower number of dispersion curves; in this case the optically active frequencies may be found not only for $\phi = 0$ (IR and Raman) but also for $\phi = \theta$ (IR and Raman) and $\phi = 2\theta$ (Raman).

The concept of $k = 0$ motions may be restated in terms of the phase difference between rototranslationally equivalent chemical units in a polymer chain referring to the geometry (conformation) of the macromolecule. The condition of in-phase motion of translationally equivalent atoms is reached in a helical polymer by successive application of the screw rotation $\phi(\theta,\ell)$ on the displacements of the starting chemical repeat unit. The existence of other elements of symmetry in addition to the screw rotational axis introduces more restrictive selection rules for the modes with the phase differences indicated above. No confusion will arise in working out selection rules if group theory on line-groups is applied [28].

It has to be pointed out however, just as in the case of simple molecules, that while group theory predicts the number of spectroscopically allowed species it does not provide any quantitative prediction of the actual absorption coefficient in the infrared or scattering power in the Raman for each normal mode.

To account for the experimental intensity suitable numbers must be used in the evaluation of the transition moments. If compared with the enormous effort made in force constant calculations, this problem seems to have received very little attention in the past because of the lack of a suitable model for describing the dipole moment changes in infrared or the polarisability changes in the Raman. Attempts have been recently made also by us for a quantitative prediction of intensities. We refer to Chapter 14 of this book for a comprehensive discussion of the problem; in section 6 of this chapter some results on a regular polymer will be presented.

If the sample of the polymer is suitably oriented (by stretching, extrusion, etc.) measurements of the spectra in polarised light in the infrared, or under suitable geometrical conditions in the Raman, allow the symmetry species of the various modes to be recognised; $\phi = 0$ and $\phi = \theta$ modes in the infrared give rise to a transition moment parallel and perpendicular to the chain axis respectively. Symmetry species in the Raman can also be determined when proper averaging is done because of random orientation of molecules about the chain axis. Based on these principles many studies on molecular orientations in bulk samples have been carried out [29,30].

3. PERFECT POLYMER IN THREE DIMENSIONS

As previously stated, neglecting for the time being the folded region of the sample, polymers are able to crystallise in a three-dimensional network because of highly directional intermolecular forces generally represented as pairwise atom-atom interactions [6].

Lattice dynamics of these systems can be treated with a technique very similar to that discussed in section 2 of this chapter. While it is very useful and of more direct physical meaning to describe the potential energy in terms of internal coordinates the whole calculation in 3 dimensions becomes greatly simplified when Cartesian displacement coordinates are adopted [31]. A $\underline{k}$ dependent F_R matrix is constructed in a way analogous to Equation (19)

$$F_R(\underline{k}) = F^0 + \sum_s \tilde{F}^s e^{-i\underline{k}.\underline{t}(s)} + F^s e^{i\underline{k}.\underline{t}(s)} \tag{20}$$

The $\underline{k}$ dependent $F_x(\underline{k})$ matrix expressed in terms of Cartesian coordinates can be then derived as

$$F_x(\underline{k}) = \tilde{B}(\underline{k})F_R(\underline{k})B(\underline{k}) \tag{21}$$

where

$$B(\underline{k}) = \sum_{j=-m}^{+m} B_j e^{-i\underline{k}.\underline{t}(j)}$$

is the transformation matrix from Cartesian to internal phonon coordinates. The dynamical matrix in terms of mass-weighted Cartesian coordinates can be written as

$$D(\underline{k}) = M^{-\frac{1}{2}}F_x(\underline{k})M^{-\frac{1}{2}} \tag{22}$$

The roots of the corresponding eigenvalue equation provide the dispersion curves for a given crystal, once a proper set of $\underline{k}$ values is inserted:

$$[D(\underline{k}) - \omega^2(\underline{k})E]L_x(\underline{k}) = 0 \tag{23}$$

The technique described above can equally be applied to molecular or polymer lattices. A few particular points should be pointed out in the case of polymers:

(i) The quantity $\underline{k}$ appearing in all the expressions above is the wave-vector with components k_a, k_b and k_c which must be specified for a three-dimensional crystal. If two or one-dimensional lattices are to be studied only two or one

components of the wave vector are to be considered.
(ii) The complete knowledge of the molecular dynamics of a crystalline polymer becomes an almost impossible task, even for the simplest materials, since the crystallographic unit contains in general many atoms and a very large number of points in $\underline{k}$ space is needed for plotting the dispersion curves. Even if one decides to do the calculations for special symmetry points in the Brillouin zone [32] and plot dispersion curves along chosen symmetry lines, the calculation becomes cumbersome because in general the symmetry of polymer crystals is rather low. The simplest cases which have been treated are orthorhombic polyethylene [33], orthorhombic polyoxymethylene [34] and hexagonal polytetrafluoroethylene [14]. A three dimensional covalent polymer of high symmetry like diamond allows instead a thorough study [22].
(iii) Because of the existence of intermolecular forces and because of the increasing number of chains in the unit cell we expect the occurrence in the spectrum, as in the case of organic crystals, of factor group splittings, correlation field splittings and external lattice modes (in the low energy region of the spectrum) [35]. All these features have been actually observed for a very few simple polymers. It has to be pointed out, however, that because of the flexibility of the helical polymer chain the distinction between internal and external modes in the low energy region of the spectrum is sometimes very difficult because modes may be strongly coupled. The separation of external and internal modes which belong to the same symmetry species, while possible for rigid organic molecules, becomes impossible for polymers since internal modes occur sometimes at equal or even lower energy than the energies of the external modes.
(iv) The splitting of the intramolecular (line-group) modes because of intermolecular (space-group) interactions are generally very small, and are not observed for all modes even when symmetry would allow. The motions of the groups of atoms which perform comparatively large amplitude motions and which are located at the exterior of the polymer chain are generally candidates for exhibiting factor group splittings. Inter chain distances are also an important factor, the bulkier the substituent on the chain, the larger is the chain-chain distance and the smaller should be the possible factor group splitting [36].

Selection rules for three-dimensional crystals are the same as those discussed for one-dimensional crystals. If q is the number of atoms in the unit cell $3q-3$ is the number of $\underline{k} = 0$ vibrations which are potentially infrared and/or Raman active. Of these only the motions which belong to an irreducible representation of the factor group of the space group which contains the translations or the components of the polarisability tensor will actually appear in the infrared and Raman spectra. For polymers space-groups analysis correlated with the results of line-group analysis is necessary for the prediction of the spectroscopic activity of the many modes.

4. CHAIN REGULARITY AND CRYSTALLINITY

The concepts discussed so far find interesting applications in practical work when the structural nature of a polymeric material has to be studied. From what we have discussed previously a clear distinction between intra and intermolecular effects can be made.

If a one-dimensional crystal (single chain polymer) is considered, phonons propagating only along the chain axis are involved and their dispersion depends only on intramolecular coupling forces. The so-called splitting between $k = 0$ modes observed in the spectra ($\phi = 0, \theta, 2\theta$) originates from the existence of an infinite one-dimensional lattice, with a particular regular geometry (conformation) which depends on the chemical nature of the atoms involved. The rototranslational regularity of the infinite helical polymer thus generates what we have defined as "regularity bands" [37]; these bands inform us that a large portion of the polymer chain in the sample has taken up a helical structure whose parameters can be defined.

It becomes clear then that "regularity bands" do not require the sample to be solid; in amorphous, liquid and solution it may well occur that chains coil regularly to such an extent that the dynamics and selection rules of an infinite one-dimensional helical array are approached.

Information on the three-dimensional arrangements of chains packed in a crystal lattice can only be obtained if factor group splitting and correlation field splitting of line group modes and/or lattice modes are identified with certainty. Bands certainly associated with the existence of a three-dimensional order have been named by us "crystallinity bands" [37] and are due to ordered intermolecular couplings. In the case of polymers intermolecular coupling may exist as soon as a cluster of chains is formed where three-dimensional periodicity is achieved.

With the above distinction it is clear that in principle vibrational spectra are able to distinguish between regularity and crystallinity as opposite of irregularity and amorphicity. Very often confusion on the origin of the bands is found in the literature and the nature of the polymer is wrongly described.

5. DISORDERED POLYMERS

The four steps in the formation of a crystalline macromolecule which have been described in section 1 of this chapter never occur in an ideal way. Physical and chemical evidence indicates the existence of changes in the type of linking, chain branching, imperfect type of stereospecificity, generation of conformational disorder, errors in a three-dimensional packing. In reality even the cleanest polymer must be considered a system containing a sizeable portion of disordered material. Since most of the chemical and physical properties which make a polymer an industrially useful material strongly depend on the type and concentration of the disorder, efforts have been made to use also vibrational spectroscopy to characterise such materials at the molecular level.

As a matter of fact the vibrational spectra of polymers show many additional bands which cannot be accounted for in terms of a perfect ideal system. Efforts have been made to understand the so-called "amorphous" bands; in the past industrial researchers were eager to find a quick qualitative answer for analytical purposes. More recently some efforts have been made to understand the spectra from the view-point of molecular and lattice dynamics [38-42].

The mathematical techniques adopted so far for the treatment of a perfect polymer cannot be applied for this case since they were all based upon the existence of a translational symmetry either in one or three dimensions. The periodicity of the chain has allowed the introduction of the phase factor ($\underline{k}$ vector) and simplification of the equations. The lack of periodicity poses the problem of treating the vibrations of a very large number of atoms organised in space in any possible geometry.

In order to overcome this difficulty in the case of simple inorganic crystals many authors have treated the case of crystals containing a very low concentration of impurities thus reducing the problem to that of the vibrations of a perfect lattice perturbed by the vibrations of a few generally non-interacting defects. Very elegant analytical treatments based on Green's function methods have been developed and have allowed an understanding of the physics of the phenomenon which has found many experimental verifications [43]. More recently attempts have been made to apply Green's function methods also to disordered polymers and especially to polyethylene containing a few non-interacting conformational defects [44]. The very large amount of algebraic treatment required for the solution of the problem with just one isolated conformational defect, however, forced the authors to introduce drastic approximations in the model of the chain thus making the model physically unrealistic. Moreover the problem becomes intractable when one has to consider a real chemical system which contains a large number of various kinds of defects.

From the study of the dynamics of glasses and disordered ice carried out at the National Physical Laboratory by Dean and collaborators a very useful numerical technique has been proposed [45] which has opened the possibility of the study of the vibrations of real polymers in a more detailed quantitative way [38,42,46-49]. The study of disordered polymers approaches that of glasses with the difference that in glasses the network consists of the same type of bonds in three dimensions; in the case of organic polymers a covalently bonded worm-like chain wanders around in space and its interactions with neighbouring chains occur via weak van der Waals type interactions.

Moreover, the structure of the polymer is such that the whole large dynamical matrix can be constructed from a generating chemical sub-unit by suitably introducing any desired type, concentration and distribution of disorder [39,46]. This is not so easy in the case of a three-dimensional glass.

Let us define the density of vibrational states $g(\omega)$ as the fraction of normal modes with frequencies in the interval between ω and $\omega + d\omega$ in the limit $d\omega \to 0$. When the number of atoms per repeat chemical unit in a single chain polymer is large or when three-dimensional polymers are treated the determination of the density of states in the whole Brillouin zone becomes practically impossible by the so called "root sampling method" since the required number of points is enormously large. In the case of diamond 1,440,000 points have been calculated using the high symmetry of the system [22]. For the ideal model some of the peaks of the frequency spectrum corresponding to the high concentration of vibrational states in the neighbourhood of the limiting $\underline{k} = 0$ modes must coincide with the absorption bands in the infrared and/or scattering in the Raman. Other peaks do not find any coincidence in the vibrational spectrum and can only be revealed by inelastic neutron scattering experiments which are able to plot directly the experimental $g(\omega)$ [9]. The coincidence between the optical spectra and the calculated $g(\omega)$ is only thought of at present in terms of frequencies, no comparison can yet be made with peak heights since transition moments have not yet been introduced in the treatment (see section 6 of this chapter).

The numerical technique adopted by Dean [45] is the Negative Eigenvalue Theorem (NET) which allows computation of the number of eigenvalues of a given, even very large, matrix in a chosen frequency interval (ω_1, ω_2). The intervals can be restricted to any desired accuracy thus reaching that of a single isolated exact eigenvalue when necessary. Thus the histogram of $g(\omega)$ can be calculated with any desired accuracy. Approximate or exact eigenvectors can also be obtained by the so called "Inverse Iteration Method". The vibrational displacements can then be obtained. We refer to ref. 41 for the discussion of the mathematical details. Such a method allows calculation of $g(\omega)$ also for very large matrices (we have calculated $g(\omega)$ for dynamical matrices of order up to 3600 in reasonable and acceptable computing time); remember that these dynamical matrices are in general band matrices with a limited number of co-diagonals.

For the understanding of the dynamical and spectroscopic properties of a disordered polymer let us proceed by steps which are somewhat related to what is actually found in nature. Let us take first a chain molecule of finite length with r atoms which may be a section of our real polymer. The 3r-6 normal modes will occur with frequencies which lie on the dispersion curves calculated for the infinite chain when end effects are properly accounted for. To each wave-like motion propagating along the chain a certain phase shift (or a certain $\underline{k}$ vector) can be assigned [17]. Each of the modes may now be active because of the lack of translational symmetry. The intensity of each normal mode along a given dispersion branch will slowly approach the intensities of the limiting modes at $\underline{k} = 0$. If the polymer contains a distribution of chains of different length, the envelope of all $\underline{k} \neq 0$ motions gives rise to asymmetrical wings at one side of $\underline{k} = 0$ bands, their width depending on the excursion of the dispersion curve. One can then, in general, say that the whole $g(\omega)$ becomes activated because of the lack of translational periodicity.

If a defect is introduced with its natural vibrational frequency such a motion may be transmitted by the one-dimensional perfect host lattice or may be soon damped depending on the value of its frequency and on the geometrical environment of the defect oscillator. One can then speak of out-of-band or gap-modes for frequencies which are not transmitted and can speak of "resonance modes" for those modes which may be transmitted by the perfect host lattice. Extra characteristic bands will then appear in the spectrum due to gap or out-of-band modes, and the vibrational density of states within the range of frequencies transmitted by the host lattice will be modified. While gap or out-of-band modes correspond to modes highly localised in space at the defect (the defect vibrates practically alone), resonance modes require a cooperative coupling throughout the chain.

If the defect is a "molecular defect", thus containing several atoms or various types of conformations different from that of the host lattice, a more complex dynamical coupling may take place and extra frequencies may appear in the spectrum. The extreme case is that of a highly disordered molecular system where the host lattice is destroyed and only a very complex pattern of frequencies due to a disordered coupling can be calculated and eventually experimentally observed. This is the case of glasses and also of liquid or "amorpohus" polymers. If in such a disordered system a few structures reminiscent of the ordered structure can be found [50,51] weak $\underline{k} = 0$ modes will be observed thus indicating the existence of one-dimensional or three-dimensional ordered clusters.

The study of the dynamics of organic polymers has been and is one of the subjects of research in our laboratory. Polyethylene [38,39], polytetrafluoroethylene [48,49] and polyvinylchloride [46,47] have been treated and their dynamics studied and compared with the actual vibrational spectrum.

<u>Example: polyvinylchloride</u>

We report here the case of polyvinylchloride (PVC) as an example [46,47]. Ideal PVC would consist of a head-to-tail sequence of the $-CHCl-CH_2$ units, with a syndiotactic structure, coiled in a chain whose energy minima either correspond to: (1) a trans-planar conformation (sequence of ...TTTT... internal rotational angles) or (2) a sequence of the type ...TTGGTTGG... . An isotactic structure is also possible, (3), and would generate a three fold helix with a sequence of ...TGTGTG... conformations.

The structure of the irreducible representation at $\underline{k} = 0$ and the corresponding spectroscopically active modes are the following:

model (1)	C_{2v} point group	$9A_1(ir,R)+7A_2(R)+9B_1(ir,R)+7B_2(ir,R)$
model (2)	D_2 point group	$17A(R)+17B_1(ir,R)+17B_2(ir,R)+17B_3(ir,R)$
model (3)	C_3 point group	$16A(ir,R)+17E(ir,R)$

Dispersion curves and densities of states for the three ideal models are shown in Figure 1.

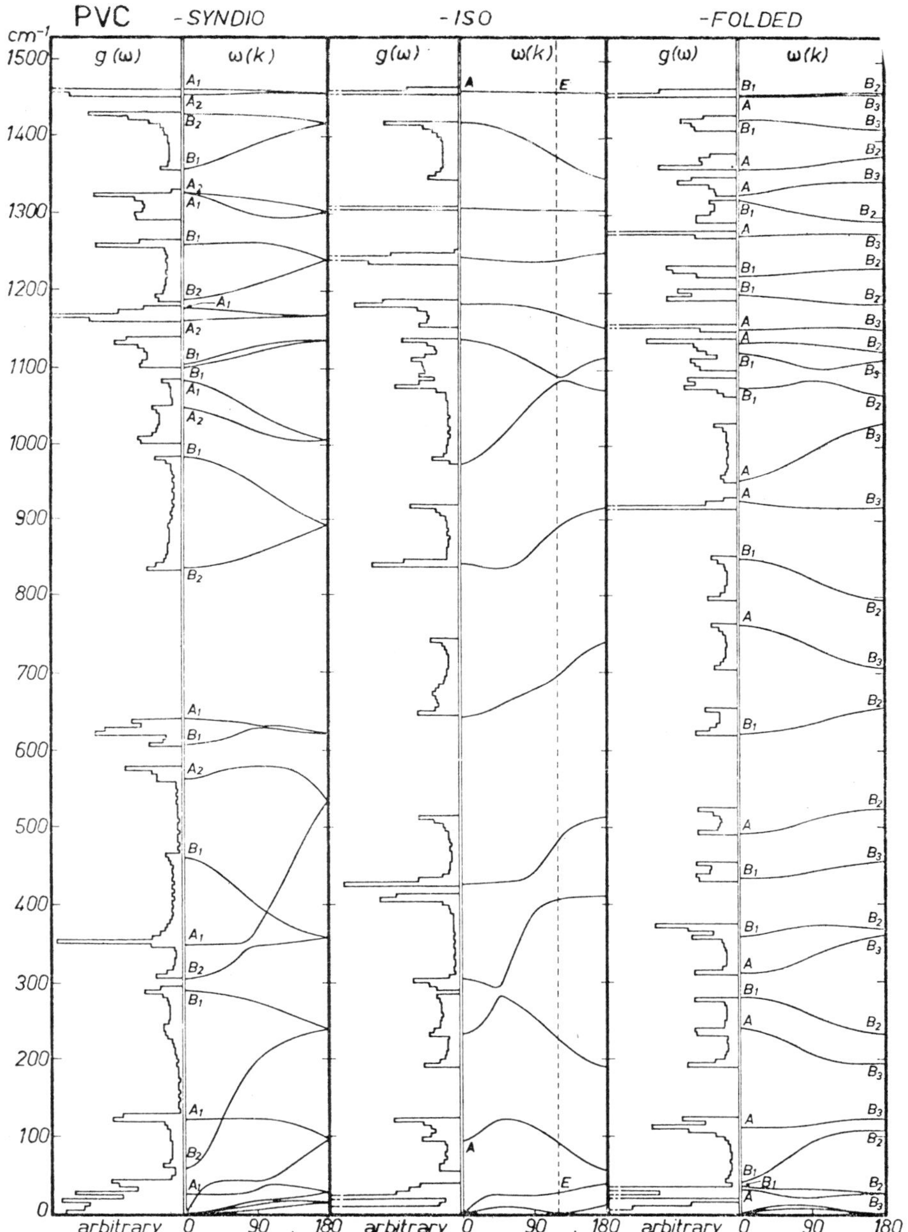

Figure 1. Dispersion curves ω(k) and density of states g(ω) from 0 to 1500 cm^{-1} of three possible models of PVC single chain (reproduced with permission from ref. 46).

It has been generally assumed that the planar zig-zag structure is the more likely structure. Let us focus our attention on the A_1 in-phase and B_1 out-of-phase C-Cl stretchings of model (1) which occur as strong bands in the infrared and have been assigned to bands at 640 and 604 cm^{-1} respectively. The calculated $g(\omega)$ for the C-Cl dispersion curve can be seen in Figure 1, and the band of frequencies which can be transmitted by the perfect lattice is very narrow comprising only about 40 cm^{-1}. No other frequencies are expected to occur for a large frequency range above (up to 835 cm^{-1}) and for about 30 cm^{-1} below (see Figure 1).

Since a real sample of PVC is shown to contain many chemical, conformational and configuration irregularities we have carried out calculations by introducing in a perfect host lattice of syndiotactic trans-planar PVC conformational, configurational (and conformational) and chemical defects with different concentrations and distribution.

From Figure 2 it is clear that the head-to-head chemical

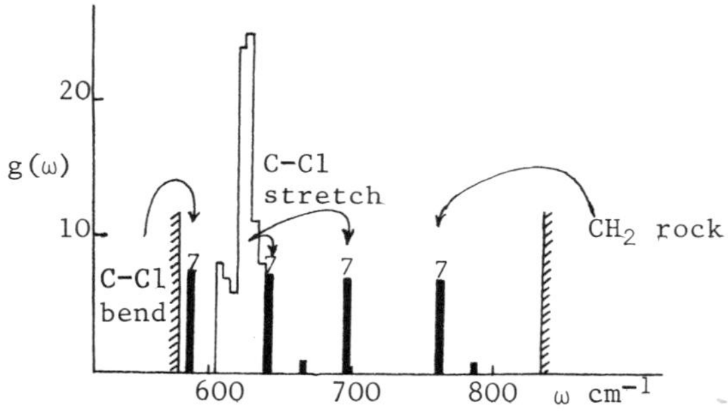

Figure 2. Density of states $g(\omega)$ of syndiotactic planar zig-zag chain of polyvinylchloride containing seven isolated head-to head defects. The chain consists of 200 carbon atoms (i.e. 100 chemical repeating units) (reproduced with permission from ref. 47).

error clearly generates gap-modes both for CH_2 rocking, C-Cl stretching and C-Cl bending modes. From Figure 3 the frequency at 763.0 cm^{-1} corresponds to the rocking of two adjacent CH_2 groups surrounded by two C-Cl groups. The frequency is very characteristic of such a CH_2-CH_2 chemical defect and is highly localised in space along the chain. The C-Cl stretching mode at 695.1 cm^{-1} moves many more chemical units along the chain but can still be considered as an isolated oscillation. The gap mode at 589.6 cm^{-1} extracted from the very large frequency band due to C-Cl bending, though occurring in a gap, is not very much damped but carries very many units during the motion.

In an analogous way gap-modes are generated above the frequency band of the C-Cl stretching modes when conformational and configurational defects are introduced in the host lattice (Figure 4). Depending on the distance between defects along the chain intramolecular coupling may occur thus producing splitting of the modes which then become characteristic of the

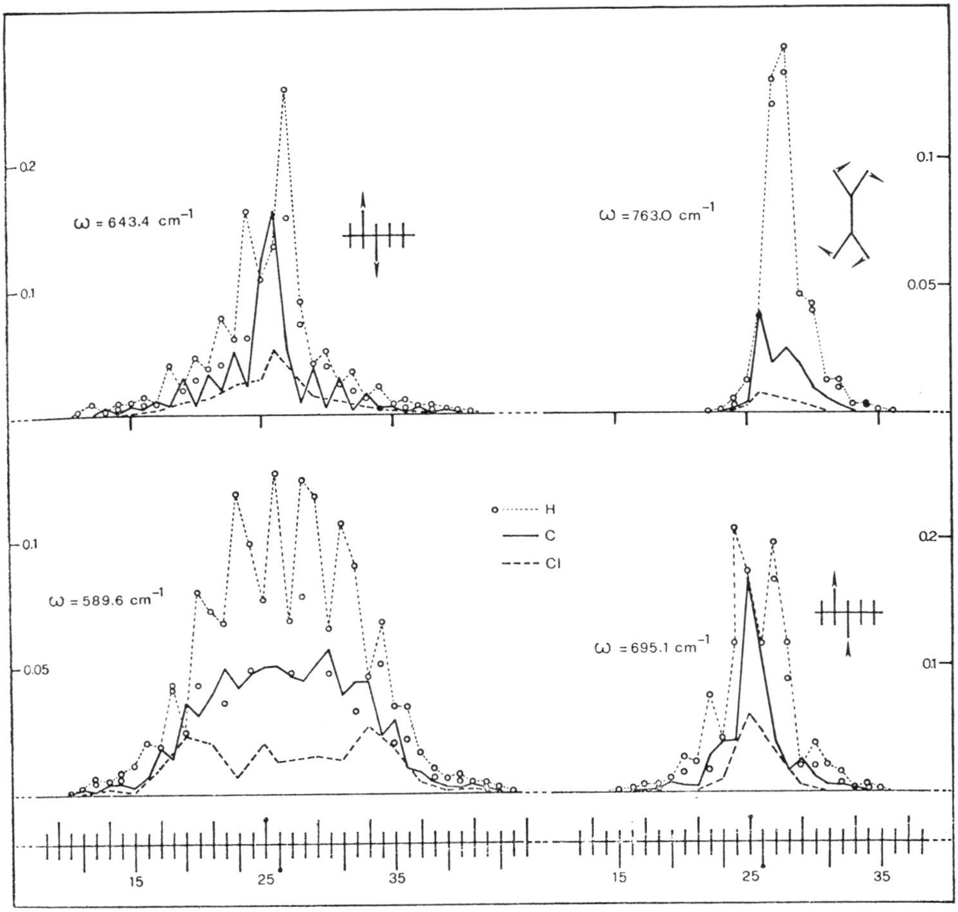

Figure 3. Eigenvectors of the gap frequencies of syndiotactic PVC chain with one head-to-head defect placed in the middle of the chain. The model contains 50 units. The total displacements of atoms from equilibrium are shown and the main type of motion sketched (reproduced with permission from ref. 47).

sequence of defects introduced and of their distance. Figure 5 gives a comparison between the calculated g(ω) for a realistic model of disordered PVC and the observed spectrum in this region.

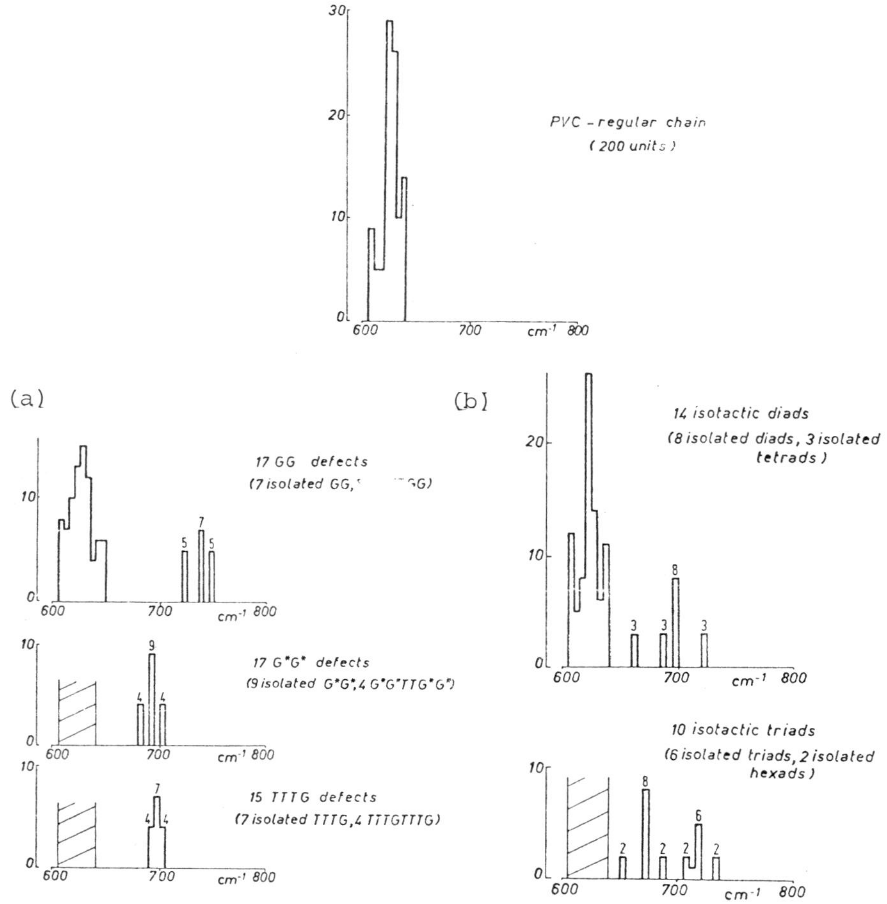

Figure 4. Comparison between the density of states in the C-Cl stretching region of a conformationally and configurationally regular syndiotactic PVC chain with that of (a) conformationally and (b) configurationally impure chains. Notice the correspondence of the population of defects and the number of gap frequencies indicated in the histograms (reproduced with permission from ref. 46).

6. INTENSITIES

Figure 5 precisely describes the limitation in the use of only $g(\omega)$ for the interpretation of the vibrational spectrum of a disordered system. Since no dipole or polarisability weighting factor for each normal mode has been introduced, as already stated, no comparison between observed and calculated peak heights can be made. It must be clear that while the experimental spectrum maps the dipole weighted density of states in the infrared and the polarisability weighted density of states in the Raman, our calculation of $g(\omega)$ assumes the dipole factor to be unity. Of the many gap-modes predicted for the

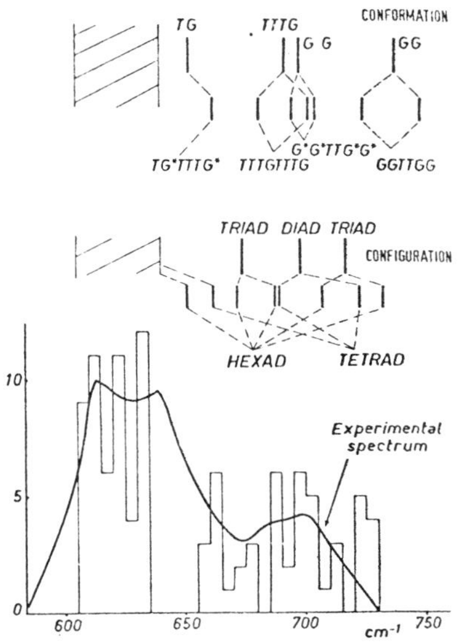

Figure 5. Density of vibrational states for a realistic model of configurationally disordered PVC (Bernoullian parameter P_m = 0.47). Comparison with the experimental spectrum. At the top of the figure a schematic representation of calculated gap frequencies is given (reproduced with permission from ref. 46).

disordered model of PVC of Figure 5, some may occur as strong bands, others may completely disappear. A treatment of the intensity problem becomes then mandatory.

The same problem has always been found in the interpretation of the vibrational spectra of molecules; frequency fitting between observed and calculated spectrum has always neglected the intensity of each band. However, while for molecules and/ or regular polymers, symmetry states very clearly which bands are active and which are inactive, in the case of disordered materials only a precise knowledge of the weighting factor may give the same information.

An example may be quoted at this point: we have studied the lattice dynamics of HCl/DCl mixed crystals [42]. The compounds crystallise forming planar zig-zag chains as shown in Figure 6. The X atoms may be hydrogens or deuteriums in Bernoullian distribution; the relative concentration is known from the composition of the isotopic mixture. The geometrical parameters are known and so is the force field. Still, the $g(\omega)$ calculated (Figure 7(b)) does not reproduce the infrared spectrum (Figure 7(a)). A very rough attempt has been made to weight $g(\omega)$ by a dipole function $M(\omega)$. The latter function has been evaluated in the approximation of rigid ions, which in this case may be not too unrealistic. The weighted $g(\omega)$ reproduces the spectrum much better, as can be seen in Figure 7(c). It is important to notice that, in cases where the kind and

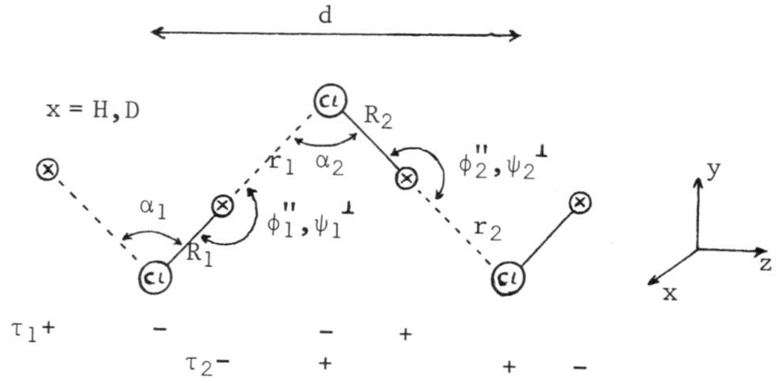

Figure 6. Geometry and internal coordinates for crystalline HCl or DCl in the orthorhombic modification (reproduced with permission from ref. 42).

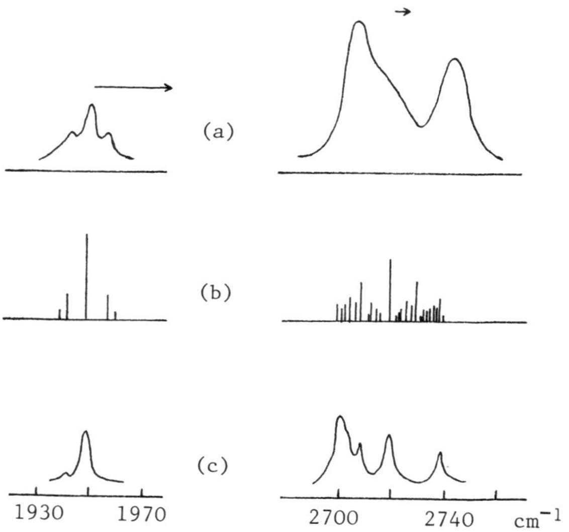

Figure 7. (a) Infrared spectrum in the stretching region, of a mixed HCl/DCl crystal containing 25% of DCl; (b) $g(\omega)$ and (c) dipole weighted $g(\omega)$ (assuming a Lorentzian band shape $\Delta\omega_{\frac{1}{2}} = 2\ cm^{-1}$) for a single mixed chain of 25% DCl in HCl (reproduced with permission from ref. 42).

concentration of defects is unknown, comparison of only $g(\omega)$ with the spectrum may lead to erroneous conclusions.

To account for infrared and Raman intensities a model for the charge distribution in the molecule and its fluctuation during vibrations should be worked out. We refer to other chapters of this book for a discussion on the problem of infrared and Raman intensities in isolated molecules. As stated in Chapter 14 our main hope in the development of a suitable model for vibrational intensities is to be able to collect a set of intensity parameters to be transferred to chemically similar molecules, hence also to polymers for the prediction of the intensities of $\underline{k} = 0$ modes as well as for the introduction of the right intensity factor in the calculated g(ω).

As a preliminary result in this direction we show in Figures 8 and 9 the calculated infrared intensities [52] of

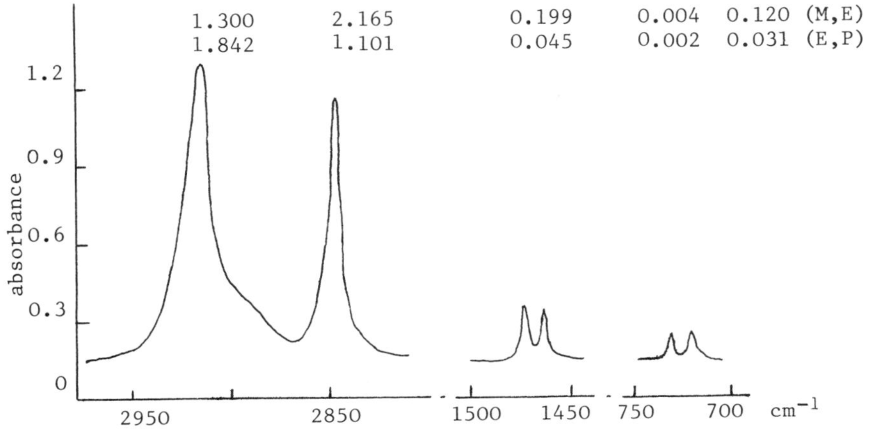

Figure 8. Infrared spectrum of polyethylene. The numbers at the top give the calculated intensities (reproduced with permission from ref. 52).

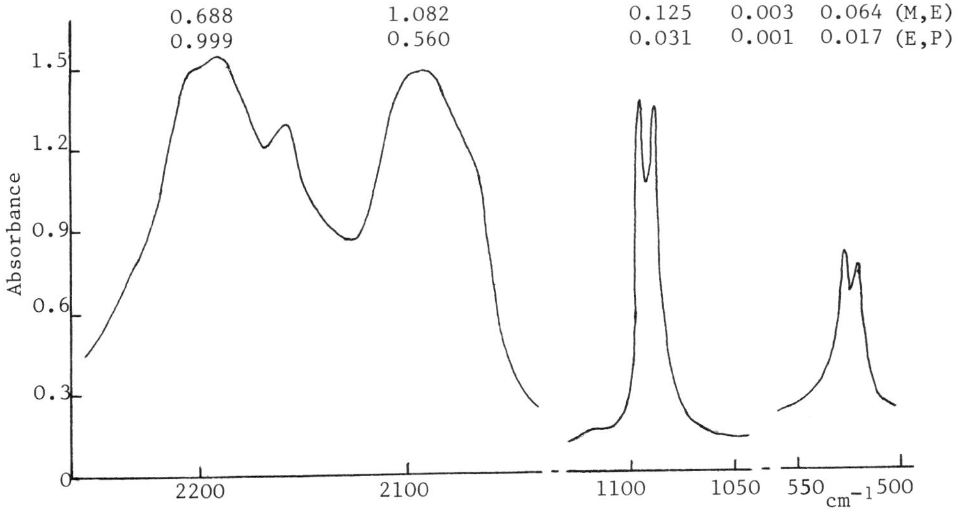

Figure 9. Infrared spectrum of perdeuteropolyethylene. The numbers at the top give the calculated intensities (reproduced with permission from ref. 52).

polyethylene and perdeuteropolyethylene; the parameters used have been refined on a number of hydrocarbons, as described in Chapter 14. The calculations have been made neglecting the three-dimensional array of chains since at the moment we do not know parameters for describing the electro-optical interactions among chains. As can be seen, the fitting is more than acceptable.

In Figures 10 and 11 the calculated Raman intensities [52]

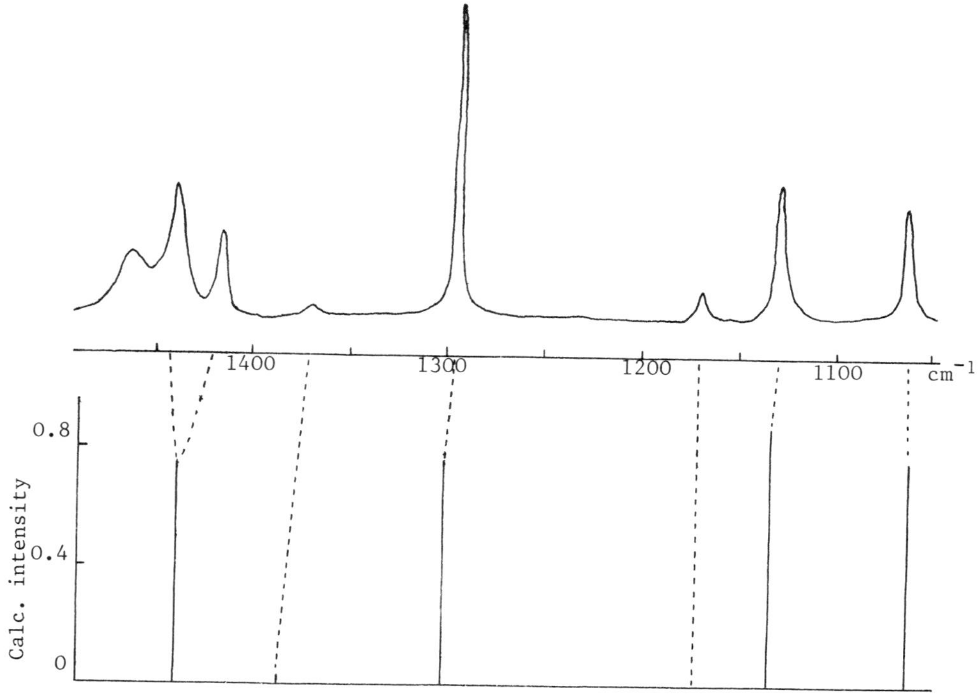

Figure 10. Top: Raman spectrum of polyethylene in the CH_2 deformation and skeletal stretching region. Bottom: graphical representation of the calculated intensities (reproduced with permission from ref. 52)

of polyethylene and perdeuteropolyethylene (bending and skeletal region) are reported; the parameters used have been refined on Raman intensities of CH_4 and deuterated derivatives, on C_6H_{12} and C_6D_{12} (only bending and skeletal regions). Here again the calculations have been made for the single chain model. The results are quite impressive and have also carried to an immediate application which we quote as example: the very low calculated intensity of the CD_2 rock suggests assigning this motion to the 998 cm^{-1} shoulder and not to the 970 cm^{-1} peak, as previously generally assumed; this agrees with the recent investigations by Koenig and Boerio [53] who assign the doublet at 990-970 cm^{-1} to the CD_2 bend split by factor group splitting.

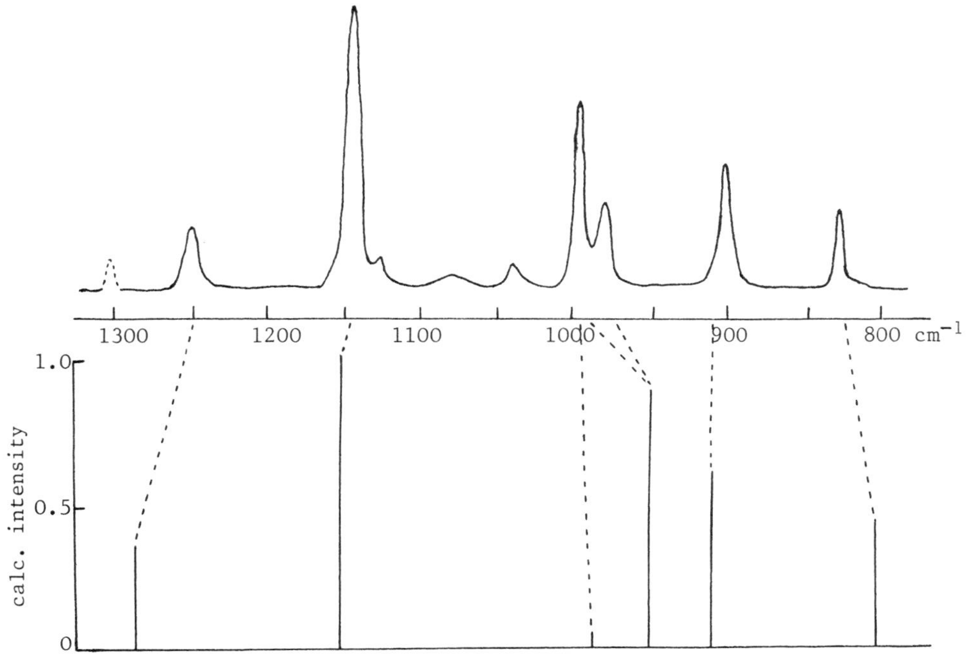

Figure 11. Top: Raman spectrum of perdeuteropolyethylene in the CD_2 deformation and skeletal stretching region. Bottom: graphical representation of the calculated intensities (reproduced with permission from ref. 52).

7. REFERENCES

1. G. Natta and F. Danusso (Editors), Stereoregular Polymers and Stereospecific Polymerisation, Vols. 1 and 2, Pergamon Press, New York, 1967.
2. A.D. Ketley (Editor), The Stereochemistry of Macromolecules, Vol. 1, Marcel Dekker, New York, 1967.
3. P. Flory, Statistical Mechanics of Chain Molecules, Interscience, 1969.
4. C.A. Coulson, Ind. Chim. Bel., 2 (1963) 149.
5. See e.g., H.C. Longuett-Higgins, Disc. Faraday Soc., 40 (1965) 7.
6. F.A. Momany, R.F. McHuire, A.W. Burgess and H.A. Sheraga, J. Phys. Chem., 79 (1975) 2361.
7. E. Clementi, to be published.
8. P. De Santis, E. Giglio, A.M. Liquori, A. Ripamonti, J. Polymer Sci., A1 (1963) 1383.
9. S. Krimm, Fortschr. Hochpolymer. Forsch., 2 (1960) 51; G. Zerbi, Molecular Vibrations of High Polymers, in

A.D. Brame (Editor), Applied Spectroscopy Reviews, Vol. 2, Marcel Dekker, New York, 1969, p. 193.

10. See e.g., L. Brillouin, Wave Propagation in Periodic Structures, 2nd ed., Dover, New York, 1953; C. Kittel, Introduction to Solid State Physics, 2nd ed., Wiley, New York, 1956.
11. P.W. Higgs, Proc. Roy. Soc. London, A220 (1953) 472.
12. L. Piseri and G. Zerbi, J. Chem. Phys., 48 (1968) 3561.
13. L.A. Feldkamp, G. Venkataramans and J.S. King, Neutron Inelastic Scattering, Proc. Symp. Copenhagen, 1968, I.A.E.A. Vienna, Vol. 2, 1968, p. 165.
14. L. Piseri, B.M. Powell and G. Dolling, J. Chem. Phys., 57 (1973) 158.
15. J.W. White, in K.J. Ivin (Editor), Structural Studies of Macromolecules by Spectroscopic Methods, J. Wiley, 1976.
16. J. Jakes, Coll. Czech. Chem. Comm., 30 (1965) 1523.
17. R.G. Snyder and J.H. Schachtschneider, Spectrochim. Acta, 21 (1965) 169.
18. B. Fanconi, to be published.
19. R.F. Schaufele and T. Shimanouchi, J. Chem. Phys., 47 (1967) 3605.
20. H.G. Olf and B. Fanconi, J. Chem. Phys., 59 (1973) 536.
21. H. Takeuchi, T. Shimanouchi, M. Tasumi, G. Vergoten and G. Fleury, Chem. Phys. Lett., 28 (1974) 449.
22. R. Tubino, L. Piseri and G. Zerbi, J. Chem. Phys., 56 (1972) 1022.
23. P. Bosi, R. Tubino and G. Zerbi, J. Chem. Phys., 59 (1973) 4578.
24. P. Bosi, G. Zerbi and E. Clementi, J. Chem. Phys., in press.
25. A. Lewis and J. Spoonhower, in S.H. Chen and S. Yip (Editors), Spectroscopy in Biology and Chemistry, Academic Press, New York, 1974, Ch. 11.
26. B. Orel, R. Tubino and G. Zerbi, Mol. Phys., 30 (1975) 37.
27. See e.g., R. Zbinden, Infrared Spectroscopy of High Polymers, Academic Press, New York, 1964.
28. M.C. Tobin, J. Chem. Phys., 23 (1955) 891.
29. For the infrared see e.g., A. Elliot, J. Polymer Sci., C7 (1964) 37; S. Krimm, in M. Davies (Editor), Infrared Spectroscopy and Molecular Structure, Elsevier, Amsterdam, 1963.
30. For the Raman see e.g., M.J. Gall, P.J. Hendra, D.S. Watson and C.J. Peacock, Appl. Spectry, 25 (1971) 423.
31. L. Piseri and G. Zerbi, J. Mol. Spectry, 26 (1968) 259.
32. V. Heine, in I.N. Sneddon and S. Ulam (Editors), Group Theory in Quantum Mechanics, Pergamon Press, New York, 1960.
33. T. Miyazawa and T. Kitajawa, J. Polymer Sci., B2 (1964) 395; T. Kitajawa and T. Miyazawa, J. Chem. Phys., 47 (1967) 337; M. Tasumi and S. Krimm, J. Chem. Phys., 46 (1967) 755.
34. T. Kitajawa and T. Miyazawa, International Symposium on Macromolecular Chemistry, Tokyo-Kyoto, 1966, Paper 2.5.04.
35. D. Dows, in D. Fox, M.M. Labes and A. Weissberger (Editors), Physics and Chemistry of the Organic Solid State, Vol. 1, Wiley-Interscience, New York, 1963.
36. F. Ciampelli, M. Cambini and M.P. Lachi, J. Polymer Sci., C7 (1965) 213.
37. G. Zerbi, F. Ciampelli and V. Zamboni, J. Polymer Sci., C7 (1965) 141.

38. M. Tasumi and G. Zerbi, J. Chem. Phys., 48 (1968) 3813.
39. G. Zerbi, L. Piseri and F. Cabassi, Mol. Phys., 22 (1971) 241.
40. G. Zerbi, Pure Appl. Chem., 26 (1971) 495; 36 (1973) 35.
41. G. Zerbi, in S. Califano (Editor), E. Fermi Summer School on Lattice Dynamics and Intermolecular Forces, Varenna, Italy, 1972, Pergamon Press, 1975.
42. M. Gussoni and G. Zerbi, J. Chem. Phys., 60 (1974) 4862.
43. R. J. Elliott, J.A. Krumhansl and P.L. Leath, Rev. Mod. Phys., 46 (1974) 465.
44. C. Schmid and J. Hölzl, J. Phys. C, Solid State Phys., 5 (1972) L185; K. Hölzl and C. Schmid, J. Chem. Phys., 58 (1973) 5173.
45. P. Dean, Rev. Mod. Phys., 44 (1972) 1881.
46. A. Rubcic and G. Zerbi, Macromolecules, 7 (1974) 754, 759.
47. A. Rubcic and G. Zerbi, Chem. Phys. Lett., 34 (1975) 343.
48. G. Zerbi and M. Sacchi, Macromolecules, 6 (1973) 692.
49. G. Masetti, F. Cabassi, G. Morelli and G. Zerbi, Macromolecules, 6 (1973) 700.
50. G. Zerbi, M. Gussoni and F. Ciampelli, Spectrochim. Acta, A, 23 (1967) 301.
51. R.G. Snyder and M.W. Poore, Macromolecules, 6 (1973) 708.
52. S. Abbate, M. Gussoni, G. Masetti and G. Zerbi, J. Chem. Phys., in press.
53. F.J. Boerio and J.L. Koenig, J. Chem. Phys., 52 (1970) 3425.

CHAPTER 25

RAMAN SPECTROSCOPY OF NUCLEIC ACIDS AND PROTEINS

Masamichi Tsuboi

The purpose of this chapter is to show how Raman spectroscopy is used at present for obtaining information on the conformation of a nucleic acid or a protein and on the inter-molecular interactions of such molecules. It aims also to give a prospect of how Raman spectroscopy will be even more useful in future for obtaining unique and detailed pieces of information on the nucleic acid or protein conformation.

In these days one can obtain a good Raman spectrum of a purified nucleic acid or a purified protein with a tolerably small amount of the sample. For example, 0.2 ml of 5% aqueous solution is usually more than sufficient, and the minimum amount required would be about 1/10 of that. For resonance Raman spectroscopic measurement, a much more dilute solution (with nearly the same concentration as that used for observation of the absorption spectrum) is suitable. The Raman spectrum can be observed, not only of an aqueous solution, but also of a gel or an opaque pellet (of rat liver nuclei, for example). In principle, there is no limitation in the molecular weight of the macromolecule, and no limitation in the number of kinds of macromolecules involved in the sample system for a Raman spectroscopic measurement. The following sections will show what type of information can be obtained by such Raman spectroscopic measurements.

1. CONFORMATION-SENSITIVE RAMAN BANDS OF NUCLEIC ACIDS

Every purified nucleic acid gives some thirty to forty Raman bands in the 300-1700 cm^{-1} spectral region. They are mostly caused by vibrations of the base residues, but two strong ones at 1100 and 810 cm^{-1} are ascribed to the phosphate -ribose mainchain. The 1100 cm^{-1} Raman band is assignable to a localised O $\cdots$ P $\cdots$ O^- symmetric stretching vibration [1-3]. Not only its frequency but also its intensity is practically independent of the mainchain conformation [4]. Therefore, this is a useful internal standard for intensity measurements [5]. It is evident that the Raman band at 810 cm^{-1} is caused by the "A-type conformation" [4,6] . When the polynucleotide chain takes any other conformation than the A-form, this Raman band shifts out of the 807-814 cm^{-1} range and at the same time becomes weak [7-13]. The RNA double-helix is known to have the A-form mainchain, but the "A-conformation" does not necessarily occur only in the double-helical portion of an RNA structure. The percentage of RNA nucleotides in the "A-conformation" can be determined by intensity measurement of the 810 cm^{-1} Raman band, with the 1100 cm^{-1} band as the internal standard [10,14]. For each of the completely double-helical polyribonucleotides (poly (rA) . poly (rU), poly (rG) . poly (rC), etc.), the ratio of the Raman intensities I(810)/I(1100) is 1.64 ±0.04, while this is 1.53 for E. coli Q13 16S ribosomal RNA, 1.38 for E. coli arginine transfer RNA, and 1.40 for yeast phenylalanine transfer RNA. Therefore, A-form contents of these RNAs are given as 93, 84, and 85%, respectively [14-16].

In the carbonyl stretching frequency region (1750-1650 cm^{-1}), the uracil residue gives two strong Raman bands corresponding to its 2-C=O and 4-C=O. Cytosine and guanine residues, however, give only very weak Raman scatterings here. On forming the Watson-Crick type adenine-uracil base-pair, a marked change in the Raman spectrum in this region takes place:

	Random coils cm^{-1}	Double helix cm^{-1}
Poly(rA-rU) . Poly(rA-rU)	1700 (m) 1662 (s)	1677 (s) 1650 (w)
Poly rA . Poly rU	1698 (m) 1660 (s)	1680 (s) 1631 (w)

It is interesting that the carbonyl Raman frequencies depend, not only upon whether they are involved in the "horizontal" hydrogen-bonding, but also upon the "vertical" base sequence [4,12].

In a double-helical conformation of DNA and RNA, the base residues are arranged with their planes nearly perpendicular to the helix axis and parallel to one another, so that the distance between the adjacent base-planes is about 3.4 Å. Such a "stacking" of the base-planes causes a lowering of the ultraviolet absorption intensity at 260 nm (hypochromism). The stacking causes also lowerings of the intensities of some of the Raman bands of the base-residues (Raman hypochromism) [4,12,18-20]. Some examples of such Raman bands are given below with the percentage hypochromism in each band in parentheses [4,12,19,20]:

<u>Poly rA . Poly rU</u>

Uracil: 784 (30), 1231 (52), 1395 (36).
Adenine: 728 (41), 1309 (40), 1339 (10), 1380 (30), 1485 (-25), 1512 (15), 1583 (-18).

<u>Poly (rA-rU) . Poly (rA-rU)</u>

Uracil: 783 (35), 1237 (49), 1397 (35).
Adenine: 726 (46), 1302 (28), 1378 (35), 1576 (-14).

<u>Poly rA</u> 725 (35), 1303 (39), 1508 (67).

<u>Poly rC</u> 1296 (20), 1533 (10).

2. CONFORMATION-SENSITIVE RAMAN BANDS OF PROTEINS

Every purified protein gives also about thirty strong Raman bands in the 300-1700 cm^{-1} spectral region. Among these,

Amide I	1645-1680 cm^{-1}
Amide III	1235-1300 cm^{-1}
C-C stretching	900- 960 cm^{-1}

are caused by the polypeptide mainchain. These three bands are useful in estimating the amounts of the α-helical, β-form, and random coil (or non-helical) portions involved in a given protein molecule [21-39]. The α-helical conformation gives a strong amide I Raman band in the 1650-1655 cm^{-1} region, a weak amide III Raman band in the 1260-1295 cm^{-1} region, and a medium strong C-C stretching Raman band in the 900-945 cm^{-1} region. The antiparallel β-sheet conformation gives a strong amide I Raman band in the 1665-1680 cm^{-1} region, strong amide III Raman band in the 1230-1240 cm^{-1} region, and almost no Raman scattering in the 900-960 cm^{-1} region. Finally, the disordered conformation gives the amide I band near 1665 cm^{-1} , amide III band near 1245 cm^{-1} , and the C-C stretching band mostly in the 950-960 cm^{-1} region [21-24].

Among the sidechain Raman bands, those of phenylalanine, tryptophan, and tyrosine are predominant. Phenylalanine gives a very strong Raman band at 1004 cm^{-1}, and weaker ones at 1203, 1032 and 624 cm^{-1}. The tryptophan residue gives prominent Raman bands at 1620 (weak), 1553 (strong), 1433 (weak), 1358 (strong), 1016 (strong), 880 (medium), and 760 cm^{-1} (strong). Some of them, including the 1358 cm^{-1} band, seem to be sensitive to the intra-molecular environment of this residue [26, 34,38]. The tyrosine residue gives Raman bands at 1621 (weak), 1210 (strong), 1180 (medium), 850 (strong), 830 (strong), and 650 (medium) cm^{-1}. The doublet at 850 and 830 cm^{-1} is caused by a Fermi resonance of the ring breathing fundamental and the overtone of an out-of-plane vibration [40]. The intensity ratio I(850)/I(830) depends upon the intra-molecular environment. If a tyrosyl residue is on the surface of a protein molecule in aqueous solution, the phenolic hydroxyl group is considered to be a simultaneous acceptor and donor of weak hydrogen-bonds. In such a state, the doublet intensity ratio is about 10/8. On the other hand, when a tyrosyl residue is so-called "abnormal", the phenolic hydroxyl group in question is considered to be acting as a proton-donor in a strong hydrogen-bond. In such a state, the doublet intensity ratio is about 5/10 [40].

For the cystine bridge, an interesting correlation is known between the frequency of the Raman band of the S-S stretching vibration and the bridge conformation [41]. The frequency is sensitive to the internal-rotation angles around the two S-C single bonds in the CC-SS-CC chain [41,42]. For the gauche-gauche conformation it is 510 cm^{-1}, for gauche-trans or trans-gauche it is 525 cm^{-1}, and for trans-trans it is 540 cm^{-1} [41].

3. USE OF THE CONFORMATION-SENSITIVE RAMAN BANDS

The empirical rules so far described are very useful in a Raman conformational analysis of various nucleic acids and proteins, and especially in interpreting a change in the Raman spectrum of a nucleic acid or a protein caused by a certain procedure. Some examples of the biological materials subjected to such a study are given below:

Nucleic acids: calf thymus DNA [11], E. coli tRNAs [9,14,15], yeast tRNA [16], E. coli ribosomal RNAs [14,16], R17 (virus) RNA [14,43].

Proteins: hen egg white lysozyme [25-27], bovine pancreatic ribonuclease [28-30], chymotrypsinogen [31], insulin [32,33], α-lactalbumin [34,42], snake venoms [35-37], polypeptide elongation factor Tu from E. coli [38], muscle proteins [38,39].

Protein-nucleic acid complexes: R17 virus [43], Pf1 and fd viruses [44], turnip yellow mosaic virus [45], mouse myeloma chromatin [46].

4. PRERESONANCE RAMAN EFFECTS OF BASE RESIDUES AND AROMATIC AMINO ACID RESIDUES

In order to obtain further information on the nucleic acid or protein structure, a more detailed characterisation of each Raman band is desirable. It is of course effective, on the one hand, to investigate the normal mode of vibration with each Raman band by examining, for example, isotope substitution effects as well as the results of computer calculations [3,47, 48]. A unique type of characterisation, on the other hand, can be made by examining the manner of increase of the intensities of the Raman bands of the residue in question, on bringing the exciting laser frequency towards the frequency of a characteristic electronic absorption band $\tilde{A} \leftarrow \tilde{X}$ of the residue [48-50]. If a Raman band becomes outstandingly stronger, this may be taken as indicating that the equilibrium conformation of the molecule is distorted, on going from the ground ($\tilde{X}$) to the excited ($\tilde{A}$) state, along the normal coordinate for the Raman band in question. Along this line, a study of the excited state geometry has been made of uracil, cytosine, adenine, guanine, and tryptophan residues [20,50,51].

It may be pointed out here that if a Raman band is to be associated in this way with an electronic band $\tilde{A} \leftarrow \tilde{X}$, there must be an intimate relation between its intensity and the intensity of the absorption band $\tilde{A} \leftarrow \tilde{X}$. As has already been pointed out, the base-stacking in a helical structure of a nucleic acid causes on the one hand a hypochromism of the ultraviolet absorptions and on the other hand a Raman hypochromism. The details of the Raman hypochromism, however, are appreciably different in different helical structures (see Section 1). The characterisation of the Raman bands with reference to the electronic excited state may lead to an interpretation of such a difference in the Raman hypochromism.

5. RESONANCE RAMAN EFFECTS OF PROPER CHROMOPHORE GROUPS INVOLVED IN NUCLEIC ACIDS AND PROTEINS [52-62]

Many of the amino-acid transfer RNAs of Escherichia coli have 4-thiouracil residue at position 8. This residue has a strong absorption band at 330 nm, whereas the common bases (uracil, cytosine, adenine, and guanine) have absorption bands at 260 nm. Using the correct ultraviolet laser, therefore, one would expect to observe a resonance Raman effect of this minor nucleoside base. Such an effect would be valuable, because it would provide information on one particular base residue of the 70-80 bases in tRNA, and because this could be done with much smaller amount samples of tRNA than usual. In our examination with the 351.1 or 363.8 nm beam of an Ar^+ laser of tRNAs from E. coli and Thermus thermophilus HB8, what is expected above has been found to be actually the case. The resonance Raman spectrum of the intact 4-thiouracil residue is observed, however, only when a rotating cell is used. As soon as the rotation is stopped, another resonance Raman spectrum is observed, which is attributable to the 4-thiouracil residue (at position 8) subjected to a photochemical modification (intra-molecular cross-linking with the cytosine residue at position 13 of the tRNA molecule). Tyrosine tRNA of E. coli is an exception. This has guanine at position 13, instead of cytosine, so that no intra-molecular cross-linking occurs under the 351.1 or 363.8 nm irradiation; this tyrosine tRNA gives resonance Raman bands of intact 4-thiouracil residue even when a rotating cell is not used [60].

Resonance Raman spectra of metalloporphyrins in heme-proteins [61] (see also Chapter 26), retinals in rhodopsins [52-56], and copper complexes in blue copper proteins [57-59] have been extensively examined. The prosthetic groups of these proteins have strong absorption bands in the visible spectral region, and the resonance Raman effects are examined by the use of laser beams of visible light. The problem to be solved seems to have two aspects. On the one hand, we investigate the scattering mechanism of each Raman band, by examining a plot of the Raman scattering intensity versus the exciting laser frequency. In this aspect of the problem, a general theory of vibronic coupling is more or less involved. On the other hand, we inquire whether we can extract a biologically significant piece of information from a resonance Raman spectrum, which is usually obtainable with a relatively small amount of a sample. Among the Raman bands of retinal, for example, those in the 1610-1650 cm^{-1} region are used for judging whether the Schiff base linkage of retinal to opsin is protonated or not [53-55]. It has also been suggested that the Raman bands in the 990-1020 cm^{-1} and 1420-1460 cm^{-1} regions may be used for monitoring photo-isomerisation of 11-cis to all-trans [55].

It may be pointed out here that for both aspects of the problem, it is important to extend the exciting laser frequency into the ultraviolet region [62]; all of these prosthetic chromophores have absorption bands in the near-ultraviolet region, as well as in the visible light region.

6. RESONANCE RAMAN EFFECTS OF LABEL MOLECULES

Actinomycin D is an antibiotic which binds to DNA in a specific manner. It has a strong absorption band at 442 nm, and therefore resonance Raman effects are expected to take place with laser beams of visible light. In fact, a good Raman spectrum of actinomycin D is observed at a concentration of 3 x 10^{-4} mol dm^{-3} (in aqueous solution) with the 514.5 nm beam of an Ar^+ laser, and 5 x 10^{-5} mol dm^{-3} with the 457.9 nm beam. When 2 x 10^{-3} mol dm^{-3} of DNA or 5'-deoxyguanosine monophosphate (5'dGMP) is added to the solution an appreciable increase of the 1480 cm^{-1} band of actinomycin D takes place. Such a low concentration of DNA or 5'dGMP cannot cause any appreciable Raman scattering, therefore the spectral change must solely be attributed to a change in the actinomycin D structure induced by the binding to the guanine residue [20,63].

Resonance Raman spectroscopy provides a similar approach for studying small molecule-protein interactions. This has been illustrated in a study of the interaction of methyl orange with bovine serum albumin [64] and in a study of 2,4-dinitrophenyl hapten with rabbit antibodies [65]. Coenzymes such as flavin mononucleotide (FMN) and nicotinamide-adenine dinucleotide (NAD) are also good subjects of such a resonance Raman study, because they are bound to various proteins and many such bindings have biological significance. A resonance Raman scattering of FMN is observed at 1588 cm^{-1} without fluorescence disturbance, when the 363.8 or 351.1 nm beam of an Ar^+ laser is used for excitation. NADH (the hydrogen is on the position 4 of nicotinamide) has an absorption band at 340 nm, and its resonance Raman scattering at 1686 cm^{-1} is observed again with the 363.8 nm beam of an Ar^+ laser.

7. REFERENCES

1. M. Tsuboi, J. Am. Chem. Soc., 79 (1957) 1351.
2. G.B.B.M. Sutherland and M. Tsuboi, Proc. Roy. Soc. London, A239 (1957) 446.
3. T. Shimanouchi, M. Tsuboi and Y. Kyogoku, in J. Duchesne (Editor), Advances in Chemical Physics, Vol. 7, Interscience, London, 1964, Chapter 12.
4. L. Lafleur, J. Rice and G.J. Thomas, Jr., Biopolymers, 11 (1972) 2423.
5. M. Tsuboi, S. Takahashi, S. Muraishi, T. Kajiura and S. Nishimura, Science, 174 (1971) 1142.
6. S.C. Erfurth, E.J. Kiser and W.L. Peticolas, Proc. Nat. Acad. Sci. U.S.A., 69 (1972) 938.
7. G.J. Thomas, Jr., Biochim. Biophys. Acta, 213 (1970) 417.
8. G.C. Medeiros and G.J. Thomas, Jr., Biochim. Biophys. Acta, 247 (1971) 449.
9. G.J. Thomas, Jr., G.C. Medeiros and K.A. Hartman, Biochim. Biophys. Acta, 277 (1972) 71.
10. K.G. Brown, E.J. Kiser and W.L. Peticolas, Biopolymers, 11 (1972) 1855.
11. S.C. Erfurth, P.J. Bond and W.L. Peticolas, Biopolymers, 14 (1975) 247, 1259.

12. K. Morikawa, M. Tsuboi, S. Takahashi, Y. Kyogoku, Y. Mitsui, Y. Iitaka and G.J. Thomas, Jr., Biopolymers, 12 (1973) 790.
13. Y. Nishimura, K. Morikawa and M. Tsuboi, Bull. Chem. Soc. Japan, 47 (1974) 1043.
14. G.J. Thomas, Jr., and K.A. Hartman, Biochim. Biophys. Acta, 312 (1973) 311.
15. G.J. Thomas, Jr., M.C. Chen and K.A. Hartman, Biochim. Biophys. Acta, 324 (1973) 37.
16. M.C. Chen and G.J. Thomas, Jr., Biopolymers, 13 (1974) 615.
17. B.L. Tomlinson and W.L. Peticolas, J. Chem. Phys., 52 (1970) 2154.
18. B. Prescott, R. Gamache, J. Livramento and G.J. Thomas, Jr., Biopolymers, 13 (1974) 1821.
19. E.W. Small and W.L. Peticolas, Biopolymers, 10 (1971) 69; 10 (1971) 1377.
20. Y. Nishimura, Doctor Thesis, Univ. of Tokyo, 1976.
21. J.L. Koenig and B. Frushour, Biopolymers, 11 (1972) 1871.
22. T.-J.Yu, J.L. Lippert and W.L. Peticolas, Biopolymers, 12 (1973) 2161.
23. M.C. Chen and R.C. Lord, J. Am. Chem. Soc., 96 (1974) 4750.
24. B.G. Frushour and J.L. Koenig, in R.J.H. Clark and R.E. Hester (Editors), Advances in Infrared and Raman Spectroscopy, Vol. 1, Heyden, London, 1975, Chapter 2.
25. R.C. Lord and N.-T. Yu, J. Mol. Biol., 50 (1970) 509.
26. R.C. Lord and R. Mendelsohn, J. Am. Chem. Soc., 94 (1972) 2133; M.C. Chen, R.C. Lord and R. Mendelsohn, Biochim. Biophys. Acta, 328 (1973) 252.
27. N.-T. Yu and B.H.Jo, Arch. Biochem. Biophys., 156 (1973) 469.
28. R.C. Lord and N.-T. Yu, J. Mol. Biol., 51 (1970) 203.
29. N.-T. Yu, B.H. Jo and C.S. Liu, J. Am. Chem. Soc., 94 (1972) 7572.
30. N.-T. Yu and B.H. Jo, J. Am. Chem. Soc., 95 (1973) 5033.
31. J.L. Koenig and B.G. Frushour, Biopolymers, 11 (1972) 2505.
32. N.-T. Yu, C.S. Culver and D.C. O'Shea, Biochim. Biophys. Acta, 263 (1972) 1.
33. N.-T. Yu, B.H. Jo, R.C.C. Chang and J.D. Huber, Arch. Biochem. Biophys., 160 (1974) 614.
34. N.-T. Yu, J. Am. Chem. Soc., 96 (1974) 4664.
35. N.-T. Yu, B.H. Jo and D.C. O'Shea, Arch. Biochem. Biophys., 156 (1973) 71.
36. A.T. Tu, B. Prescott, C.H. Chou and G.J. Thomas, Jr., Biochem. Biophys. Res. Comm., 68 (1976) 1139.
37. I. Harada, T. Takamatsu, T. Shimanouchi, T. Miyazawa and N. Tamiya, J. Phys. Chem., 80 (1976) 1153.
38. M. Nakanishi, S. Ohta, T. Yamada, H. Shimizu and M. Tsuboi, to be published.
39. B.G. Frushour and J.L. Koenig, Biopolymers, 13 (1974) 1809.
40. M.N. Siamwiza, R.C. Lord, M.C. Chen, T. Takamatsu, I. Harada, H. Matsuura and T. Shimanouchi, Biochemistry, 14 (1975) 4870.
41. H. Sugeta, A. Go and T. Miyazawa, Chem. Lett., 83 (1972); Bull. Chem. Soc. Japan, 46 (1973) 2752.
42. M. Nakanishi, H. Takesada and M. Tsuboi, J. Mol. Biol., 89 (1974) 241.

43. K.A. Hartman, N. Clayton and G.J. Thomas, Jr., Biochem. Biophys. Res. Comm., 50 (1973) 942.
44. G.J. Thomas, Jr. and P. Murphy, Science, 188 (1975) 1205.
45. T.A. Turano, K.A. Hartman and G.J. Thomas, Jr., J. Phys. Chem., 80 (1976) 1157.
46. S. Mansy, S.K. Engtrom and W.L. Peticolas, Biochem. Biophys. Res. Comm., 68 (1976) 1242.
47. M. Tsuboi, Pure and Applied Chemistry, Suppl. Vol. 7 (1971) 145.
48. M. Tsuboi, S. Takahashi and I. Harada, in J. Duchesne (Editor), Physico-Chemical Properties of Nucleic Acids, Vol. 2, Academic Press, London, 1973, Chapter 11.
49. M. Tsuboi, S. Takahashi, S. Muraishi and T. Kajiura, Bull. Chem. Soc., Japan, 44 (1971) 2921.
50. M. Tsuboi, A.Y. Hirakawa, Y. Nishimura and I. Harada, J. Raman Spectroscopy, 2 (1974) 609.
51. A.Y. Hirakawa, M. Nakanishi, T. Matsumoto and M. Tsuboi, to be published.
52. L. Rimai, D. Gill and J.L. Parsons, J. Am. Chem. Soc., 93 (1971) 1353; M.E. Heyde, D. Gill, R.G. Kilponen and L. Rimai, J. Am. Chem. Soc., 93 (1971) 6776.
53. A. Lewis, R.S. Fager and E.W. Abrahamson, J. Raman Spectroscopy, 1 (1973) 465.
54. A. Lewis, J. Spoonhower, R.A. Bogomolni, R.H. Lozier and W. Stoeckenius, Proc. Nat. Acad. Sci. U.S.A., 71 (1974) 4462.
55. R. Mendelsohn, Nature, 243 (1973) 22; R. Mendelsohn, A.L. Verma and H.J. Bernstein, Canad. J. Biochem., 52 (1974) 774.
56. A.R. Oseroff and R.H. Callender, Biochemistry, 13 (1974) 4243.
57. O. Siiman, N.M. Young and P.R. Carey, J. Am. Chem. Soc., 96 (1974) 5583.
58. V. Miskowski, S.-P.W. Tang, T.G. Spiro, E. Shapiro and T.H. Moss, Biochemistry, 14 (1975) 1244.
59. O. Siiman, N.M. Young and P.R. Carey, J. Am. Chem. Soc., 98 (1976) 744.
60. Y. Nishimura, A.Y. Hirakawa, M. Tsuboi and S. Nishimura, Nature, 260 (1976) 173.
61. T.G. Spiro and T.M. Loehr, in R.J.H. Clark and R.E. Hester (Editors), Advances in Infrared and Raman Spectroscopy, Vol. 1, Heyden, London, 1975, Chapter 3.
62. M. Nakanishi, K. Yoshida, Y. Kagawa and M. Tsuboi, to be published.
63. L. Chinsky, P.Y. Turpin, M. Duquesne and J. Brahms, Biochem. Biophys. Res. Comm., 65 (1975) 1440.
64. P.R. Carey, H. Schneider and H.J. Bernstein, Biochem. Biophys. Res. Comm., 47 (1972) 588.
65. P.R. Carey, A. Froese and H. Schneider, Biochemistry, 12 (1973) 2198.

CHAPTER 26

RESONANCE RAMAN SPECTRA AND NORMAL COORDINATE ANALYSIS OF SOME MODEL COMPOUNDS OF HEME PROTEINS

H.J. Bernstein and S. Sunder

1. RESONANCE RAMAN SPECTROSCOPY

Vibrational modes can be classified according to their content of α^2, $\gamma_s{}^2$ and $\gamma_a{}^2$ (see Chapter 11). Since we shall be primarily concerned with planar metalloporphins [1], we can illustrate this classification for molecular point group symmetry D_{4h}, where for the various symmetry species the polarisability can be written as:

$$A_{1g}: \begin{pmatrix} S_1 & 0 & 0 \\ 0 & S_1 & 0 \\ 0 & 0 & S_2 \end{pmatrix} \quad \text{or} \quad \alpha_{xx} + \alpha_{yy},\ \alpha_{zz}$$

$$A_{2g}: \begin{pmatrix} 0 & S_3 & 0 \\ -S_3 & 0 & 0 \\ 0 & 0 & 0 \end{pmatrix} \quad \text{or} \quad \alpha_{xy} - \alpha_{yx}$$

$$B_{1g}: \begin{pmatrix} S_4 & 0 & 0 \\ 0 & -S_4 & 0 \\ 0 & 0 & 0 \end{pmatrix} \quad \text{or} \quad \alpha_{xx} - \alpha_{yy}$$

$$B_{2g}: \begin{pmatrix} 0 & S_5 & 0 \\ S_5 & 0 & 0 \\ 0 & 0 & 0 \end{pmatrix} \quad \text{or} \quad \alpha_{xy} + \alpha_{yx}$$

$$E_g: \begin{pmatrix} 0 & 0 & S_6 \\ 0 & 0 & -iS_6 \\ S_7 & -iS_7 & O \end{pmatrix} \quad \text{and} \quad \begin{pmatrix} 0 & 0 & S_6^* \\ O & 0 & iS_6^* \\ S_7^* & iS_7^* & O \end{pmatrix} \quad \text{or} \quad \alpha_{yz} \pm \alpha_{zy},\ \alpha_{xz} \pm \alpha_{zx}$$

Substituting the S values of the polarisabiliy tensor in the equations for the depolarisation ratio one obtains:-

$$\rho_{A_{1g}} = \frac{3\gamma_s^2}{45\alpha_2^2 + 4\gamma_s^2} = \frac{(3/2).2(S_1 - S_2)^2}{5(2S_1^2 + S_2^2) + (4/2).2(S_1 - S_2)^2}$$

$$= 0 \rightarrow 3/4$$

$$\rho_{A_{2g}} = 5\gamma_a^2/0 = \infty$$

$$\rho_{B_{1g}} = 3\gamma_s^2/4\gamma_s^2 = 3/4$$

$$\rho_{B_{2g}} = 3\gamma_s^2/4\gamma_s^2 = 3/4$$

$$\rho_{E_g} = \frac{3\gamma_s^2 + 5\gamma_a^2}{4\gamma_s^2} = 3/4 + 5/4 \left(\frac{S_7 - S_6}{S_7 + S_6}\right)^2 = 3/4 \rightarrow \infty$$

The depolarisation ratio can be plotted as a function of $K' = S_1/S_2$ for totally symmetric modes and of $K'' = S_6/S_7$ for non-totally symmetrical modes, as in Figures 1 and 2.

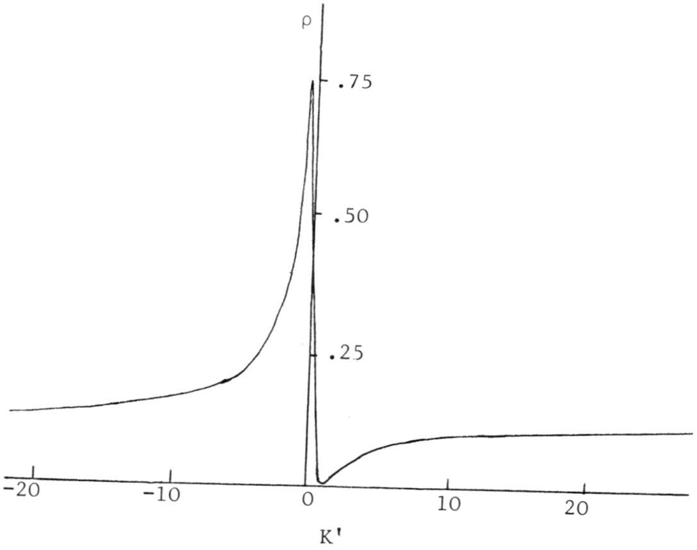

Figure 1. Depolarisation ratios for totally symmetric modes as a function of K', the ratio of S_1/S_2 in the polarisability tensor.

There are three special cases of interest - rods, spheres and discs, and we see from Table 1 the values for these limiting cases. The curves not only show how ρ changes as the relative values of S_1/S_2 and S_6/S_7 change but they also indicate the range of values one could observe for ρ as the exciting line frequency is changed.

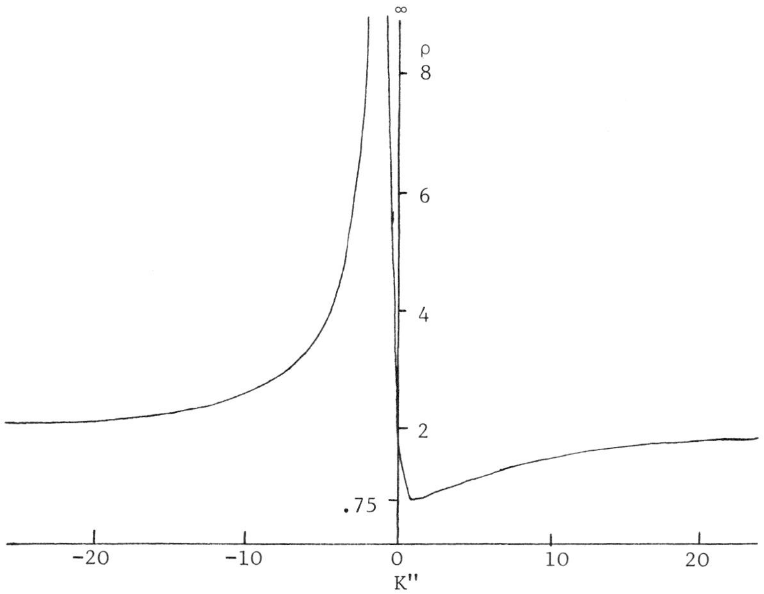

Figure 2. Depolarisation ratios for non-totally symmetric modes as a function of K", the ratio of S_6/S_7 in the polarisability tensor.

Table 1. Values of depolarisation ratios for some "special" values of K.

K	Totally symmetric	Non-totally symmetric
0	1/3	2
1	0	3/4
-1	4/7	∞
∞	1/8	2

K = 0, 1, ∞ correspond to "rod", "sphere" and "disc" shaped molecules respectively.

In order to understand the resonance Raman spectra of polyatomic molecules we have to recognise that non-totally symmetric modes can become strong and even dominate the Raman spectrum. This is possible if the two excited states can couple vibronically through a non-totally symmetric mode for which the symmetry species is contained in the direct product of allowed electronic absorption transitions [2]. For example for D_{4h}:

	A_{2u}	E_u
A_{2u}	A_{1g}	E_g
E_u	E_g	A_{1g} + [A_{2g}] + B_{1g} + B_{2g}

Since the E_g modes are out of plane vibrations, we expect the modes of the other species to dominate the spectrum. One can now look at the results given by group theory. The symmetry species, number of modes, components of polarisability tensor, depolarisation ratio, source of vibronic activity and depolarisation ratio in resonance Raman are shown in Table 2.

Table 2. "Extended" character table for the vibrational analysis of the metalloporphins of D_{4h} point group.

Sym.	No.	T	R	α_{ij}	ρ		Vibronic	RRρ
A_{1g}	9			XX+YY,ZZ	0-3/4	ip	$E_u \times E_u$	0-3/4
A_{2g}	8		R_z	XY-YX		ip	$E_u \times E_u$	∞
B_{1g}	9			XX-YY	3/4	ip	$E_u \times E_u$	3/4
B_{2g}	9			XY+YX	3/4	ip	$E_u \times E_u$	3/4
E_g	8		R_x,R_y	YZ±ZY,ZX±XZ	3/4	op	$E_u \times A_{2u}$	3/4-∞
A_{1u}	[3]							
A_{2u}	6	Z				op		
B_{1u}	[4]							
B_{2u}	[5]							
E_u	18	X,Y				ip		

[] not active in absorption nor Raman
ip in plane, op out-of-plane

It is clear that the $8A_{2g}$ modes, forbidden in the ordinary spectrum since $\alpha_{xy} - \alpha_{yx} = 0$, become allowed in the resonance Raman spectrum since $\alpha_{xy} = -\alpha_{yx}$. The vibronic activity is contained in the direct product table above, and this activity has been designated for the species in Table 2.

2. SPECTRA OF COPPER PORPHINS

We are now in a position to discuss the spectra of some copper porphins of chosen molecular point group symmetry [2], listed in Table 3. In Figure 3 we see a resonance Raman spectrum which is by no means typical. As is apparent from the dilution it has a smaller signal to noise ratio than could be obtained with our usual concentrations of around 10^{-4} mol dm^{-3}. The bands at 1583 and 1306 cm^{-1} clearly show anomalous polarisation, while bands at 1598, 1506 and 343 cm^{-1} are polarised.

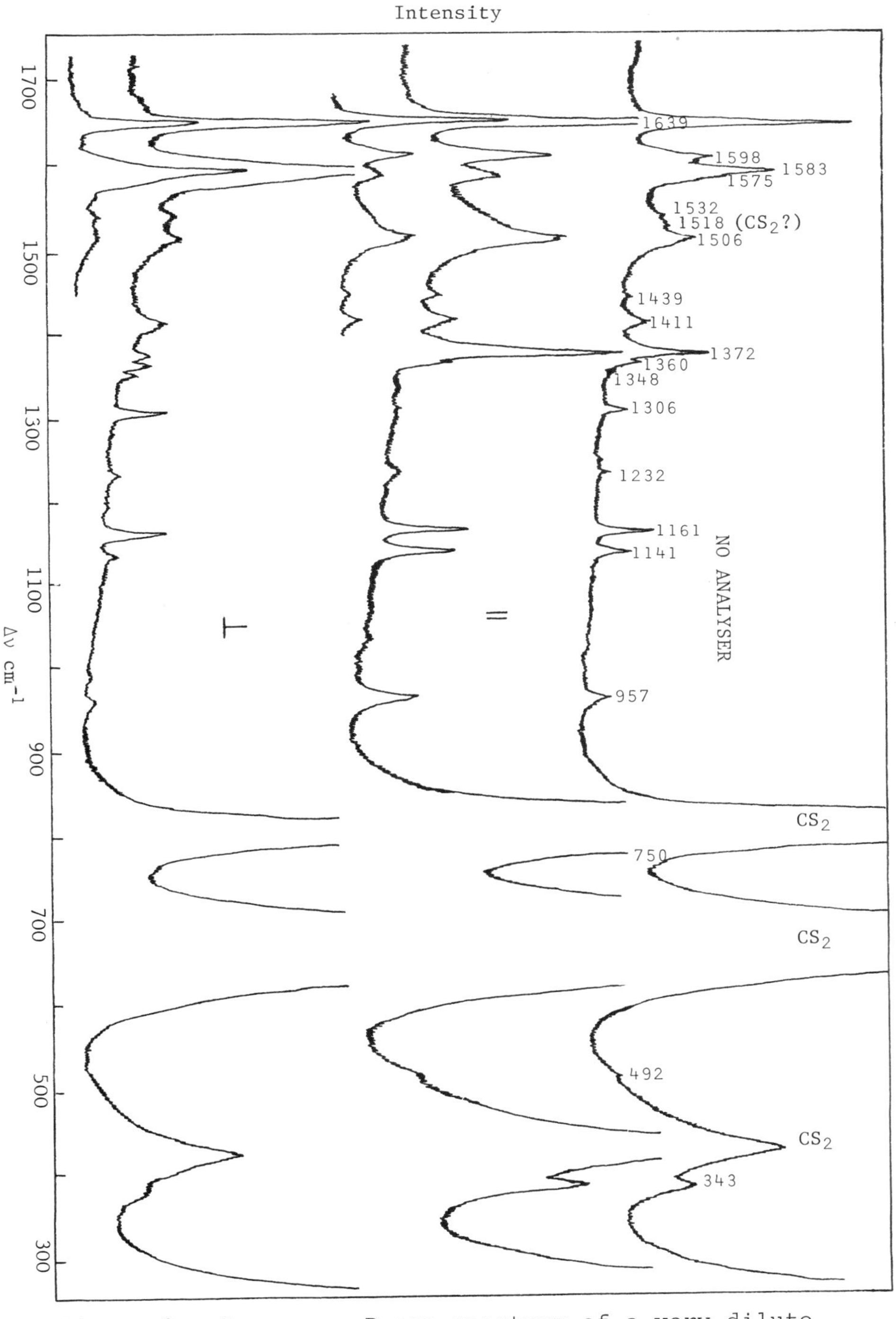

Figure 3. Resonance Raman spectrum of a very dilute solution of Cu-octamethylporphin (reproduced with permission from reference 1c).

Table 3. Symmetry and structures of various Cu-porphin model compounds.

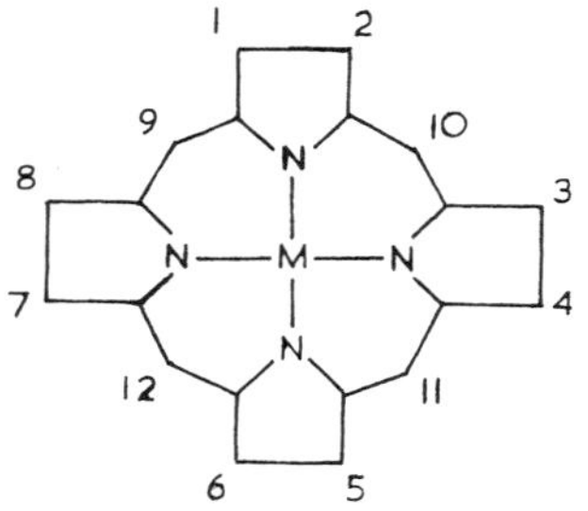

Sym.		1	2	3	4	5	6	7	8	9	10	11	12	ref.
D_{4h}	Cu porphin	H	H	H	H	H	H	H	H	H	H	H	H	2
D_{4h}	Cu porphin-d_4	H	H	H	H	H	H	H	H	D	D	D	D	2
D_{4h}	Cu octamethyl	M	M	M	M	M	M	M	M	H	H	H	H	2
D_{4h}	Cu octaMe-d_4	M	M	M	M	M	M	M	M	D	D	D	D	2
C_{4h}	Cu tetramethyl	M	H	M	H	M	H	M	H	H	H	H	H	2
C_{4h}	Cu Etioporphyrin I	M	E	M	E	M	E	M	E	H	H	H	H	9
C_{2v}	Cu Etioporphyrin IV	M	E	E	M	E	M	M	E	H	H	H	H	9
D_{2h}	Cu Etioporphyrin II	M	E	E	M	M	E	E	M	H	H	H	H	9
C_s	Cu Etioporphyrin III	M	E	M	E	M	E	E	M	H	H	H	H	9
C_s	Cu meso II	M	E	E	M	P	H	P	M	H	H	H	H	8

Carbon atoms at 2 and 10 are called b (beta) and m (meso), while the quaternary unmarked carbon is called a (alpha).

Before embarking on a normal coordinate treatment it is clear that there are not enough data to determine the force constants of a general quadratic force field. Some drastic assumptions are required to make the calculation viable. We have proceeded along the lines that the observed bond distances include effects from long range interactions represented by the molecular orbitals used to calculate them.

A one to one correspondence of bond stretching force constant with its internuclear distance would be expected to take into account the delocalisation effects (Figure 4). Thus the CC bond stretching constants used here were obtained from Figure 4 by assuming they were proportional to the bond distance [2]. The proportionality constant was determined from the least squares minimisation. Also interaction constants beyond nearest neighbours were neglected.

Lennard-Jones [3] and Coulson [4] from molecular orbital

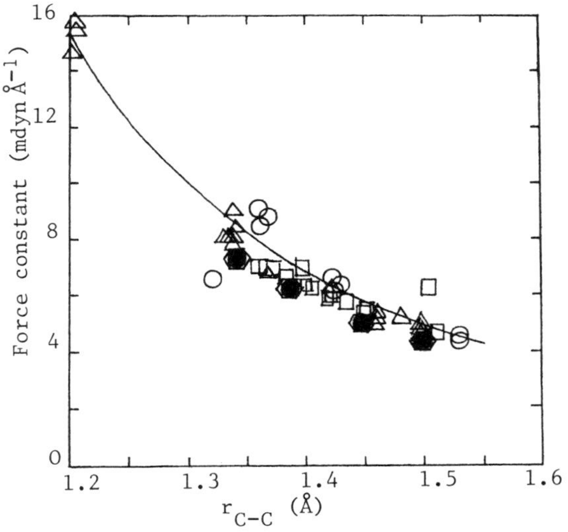

Figure 4. Plot of CC stretching force constants versus CC bond lengths. O, heterocyclic compounds with aromatic character; □, aromatic compounds; Δ, unsaturated acyclic compounds; ●, Cu-porphins and Cu-porphyrins (reproduced with permission from reference 6).

theory obtained a relation between bond distance and bond order for these delocalised bonds in conjugated hydrocarbons. Also there is a relation between force constant K, and bond order. Elimination of the bond order from these equations gives 1/K = constant + constant × bond distance. This predicted line is shown in Figure 5, for the data in the previous figure. For

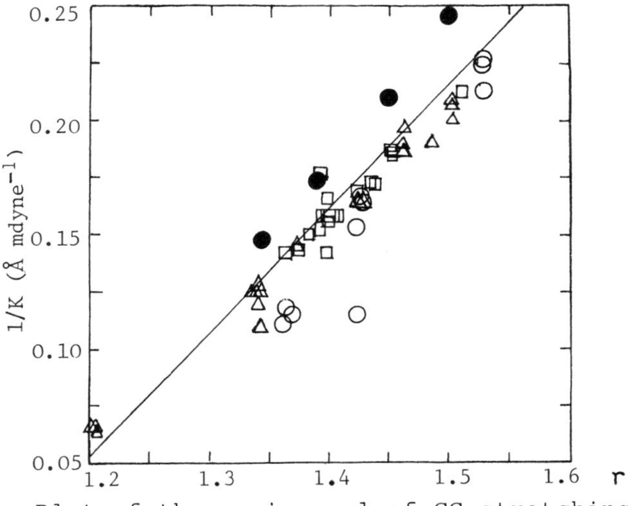

Figure 5. Plot of the reciprocal of CC stretching force constant versus CC bond length (symbols as Figure 4) (reproduced with permission from reference 6).

these molecules the skeletal angle deformation constants have also been plotted in Figure 6 against the product of the bond

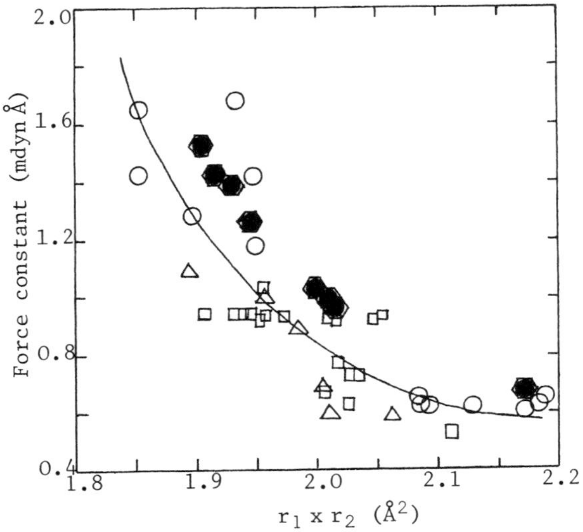

Figure 6. Plot of the skeleton deformation force constant versus the product of the two bond lengths (symbols as Figure 4)(reproduced with permission from reference 6).

distances of the bonds forming the angle. Although there is no simple theoretical justification for this as there was for the CC bonds one does see a rough correlation and the solid points are the values we have obtained in our present calculations. The calculations were made by the method of Schachtschneider [5] and the results were obtained with the parameters shown in Table 4. Some 140 frequencies were calculated for deuterium

Table 4. Force constants for copper-porphins

1.	CuN stretch[a]	0.95 mdyn $Å^{-1}$
2.	CN stretch	6.91 mdyn $Å^{-1}$
3.	CH stretch	5.12 mdyn $Å^{-1}$
4.	CC stretch[b]	4.78 mdyn $Å^{-1}$
5.	NCuN deformation[a]	0.14 mdyn $Å^{-1}$
6.	CuNC deformation[a]	0.50 mdyn $Å^{-1}$
7.	CCH deformation	0.426 mdyn $Å^{-1}$
8.	Skeleton deformation[c]	1.02 mdyn $Å^{-1}$
9.	CuN, CuN interaction	0.25 mdyn $Å^{-1}$
10.	CC, CC interaction[d]	0.42 mdyn $Å^{-1}$
11.	CC, CCH interaction[e]	0.18 mdyn
12.	CC, CCC interaction[f]	0.21 mdyn

Table 4. (continued)

a These values were kept constant in the final calculation.
b This value is for the bond ab (see footnote to Table 3). Other CC stretch force constants were obtained using Figure 4 and their respective values are bb = 6.87, am = 5.82 and the C_bC_{methyl} stretch in Cu-OMP and Cu-TMP = 4.08 mdyn $Å^{-1}$.
c This value is for $H_{(baN)}$ (see Table 3). Other skeleton deformation constants were obtained using Figure 6 and their respective values are man = 1.43, aNa = 1.51, abb = 1.24, ama = 1.37, mab = 0.98; and in Cu-OMP and Cu-TMP $(C_bC_bC_{methyl})$ = 0.96, and $(C_aC_bC_{methyl})$ = 0.68 mdyn $Å^{-1}$.
d Same value was used for $F_{CC,CN}$ and $F_{CN,CN}$. This interaction constant was used only for the stretches with a common atom.
e This interaction constant was used only for the coordinates with the common CC bond.
f Same value was used for $f_{CC,CN}$, $f_{CN,CCN}$, $f_{CN,CNC}$ and it was used only when the stretching coordinate was one of the bonds in the deformation coordinate.

---

and methyl substituted Cu-porphins and the agreement between the observed [1] and calculated [6] results is about 2.5%. In Table 5 the results calculated for Cu-octamethyl and Ni-octaethyl porphin are given and for comparison those observed for ferrocytochrome-c, and the average persistent frequencies of some heme proteins. It is clear that the calculated spectra show a good one to one correspondence with the values found in the heme proteins. Note that the inversely polarised feature at ca. 1310 cm^{-1}, arising from the α carbons in the pyrolle ring, is indeed a persistent mode. The potential energy distribution shows that the modes are very mixed in character, so that even approximately labelling the character of the modes, as arising from a particular bond, is done just as a convenience and has very little physical meaning. One can find modes with considerable CN content which should be metal sensitive and indeed this is found to be the case [6]. For example the highest B_{2g} mode around 1640 cm^{-1} depends considerably on the nature of the metal [7].

Finally one may construct a figure of merit which will help compare quickly various other calculations. This is explained in Table 6.

Acknowledgment

The work described in this chapter was done by (in alphabetical order): Drs. M. Asselin, R. Mendelsohn, S. Sunder and A.L. Verma. The first calculations were by M. Asselin and K.G. Kidd, and their final form by S. Sunder.

Table 5. Comparison of observed and calculated spectra of metalloporphins and heme proteins.

		Cu octamethyl porphin		Ni octaethyl porphin		Ferro cyto- chrome c		heme proteins
		calc[a]	obs[b]	calc[c]	obs[c]	obs[d]	obs[e]	obs[f]
A_{1g}	bs	1076	1060	1039	1025			1088
	mH	3075						
	aN	1624	1598	1605	1602	1595	1592	1605,1622
	bb	1532	1506	1526	1519	1500	1497	1480,1510
	am	1386	1372	1380	1383	1362	1362	1360,1373
	cbs	264	257	238	226		270	
	ab	878		805	806		871	915
	ring	661	684	658	674		692	675
	MN	357	343	362	364		350	353
A_{2g}	bS	1318	1306	1299	1309	1310	1315	1310,1342
	aN,am	1600	1583	1602	1603	1585	1583	1585,1553
	ab	1488	1411	1407		1400	1403	1412
	cbs	295		309				
	amH	1085	1127	1103			1130	1130
	bs,aN,am	986	995?	-				
	ring	759		783	795		752	
	ring	561		528	∿600			572?
B_{1g}	bs	1077	955	-			970	
	am	1578	1575	1659	1576			1560
	aN,am	1547	1555	1561		1547	1547	1547,1552
	bb	1458	1467	1408		1404		1399
	amH	1177	1161	1237	1159		1175	1175
	cbs	322		350	344			
	ab	718	755	748	784			755
	ring	692	716	660	751			
	MN	185	172	221	226			
B_{2g}	mH	3072						
	bs	1190	1232	1182	1220	1228	1230	1225
	aN	1619	1639	1483	1655	1626	1612	1640,1610
	ab	1441	1439	1402		1476	1387	1430
	cbs	519	492	529	496			
	am	1094	1140	1041			1086	1088
	ring	820	803	859	850		801	801
	ring	226	251	218				
	MNa	174	172	167				

a This work.
b This work.
c Kitagawa et al., to be published.
d T.G. Spiro and T.C. Strekas, J. Amer. Chem. Soc., 96 (1974) 338.
e J.M. Friedman and R.M. Hochstrasser, Chem. Phys., 1 (1973) 457.
f Persistent frequencies in heme proteins.
\- Not calc. in this approximation.

Table 6. Comparison of various normal-coordinate calculations for metalloporphins.

	obs. - calc. cm^{-1}				
	metalloporphins			metallo octaethyl porphyrins	
	ref. (a)	ref. (b)	this work	ref. (c)	our work
A_{1g}	220	58	164	31	44
A_{2g}	336	137	198	56	18
B_{1g}	437	96	165	138	74
B_{2g}	366	111	217		
Mean error	68	16	28	23	15
No. of param.	18	22	12		12
Fig. of merit*	.27	.94	1.00		

ref. (a) Nakamoto et al. (ref. 2)
ref. (b) M. Asselin, K.G. Kidd and H.J. Bernstein, unpublished work.
ref. (c) Kitagawa et al. (to be published) for NiOEt porphyrin frequencies below 900 cm^{-1}.
* Fig. of merit is assumed inversely proportional to the product of the mean deviation and the number of parameters in the potential function referred to this work at 1.00.

3. REFERENCES

1. (a) A.L. Verma and H.J. Bernstein, J. Chem. Phys., 61 (1974) 2560, [Cu-porphin]; (b) A.L. Verma, M. Asselin, S. Sunder and H.J. Bernstein, J. Raman Spectr., 4 (1976) 295, [Cu-porphin meso-d_4 and Ni-porphin]; (c) S. Sunder, R. Mendelsohn and H.J. Bernstein, J. Chem. Phys., 63 (1975) 573 [Cu-octameporphin and Cu-tetrameporphin].
2. H. Ogoshi, Y. Saito and K. Nakomoto, J. Chem. Phys., 57 (1972) 4194.
3. J.E. Lennard-Jones, Proc. Roy. Soc., A138 (1937) 280.
4. C.A. Coulson and H.C. Longuett-Higgins, Proc. Roy. Soc., A191 (1947) 39.
5. J.H. Schachtschneider, Technical reports 57-65 and 231-64, Shell Development Co., Emeryville, California, USA, (1964).
6. S. Sunder and H.J. Bernstein, J. Raman Spectr., in press.
7. A.L. Verma, R. Mendelsohn and H.J. Bernstein, J. Chem. Phys., 61 (1974) 383.
8. R. Mendelsohn, S. Sunder, A.L. Verma and H.J. Bernstein, J. Chem. Phys., 62 (1975) 37.

AUTHOR INDEX

SUBJECT INDEX